# The Mammalian Testis

**Reproductive Biology Handbooks**

Editor: C. A. Finn
Professor of Veterinary Physiology
University of Liverpool


Also published:

The Uterus

C. A. Finn and D. G. Porter

# The Mammalian Testis

## B. P. Setchell

Senior Principal Scientific Officer
Agricultural Research Council Institute
of Animal Physiology, Cambridge

Cornell University Press
Ithaca, New York

First published 1978 by Cornell University Press

International Standard Book Number 0-8014-1140-8
Library of Congress Catalog Card Number 77-79704


Text set in 1 ʰ/12 pt Photon Times, printed by photolithography,
and bound in Great Britain at The Pitman Press, Bath

# Contents

# Preface

This book aims to set out one person's understanding of how the testis works. It is directed mainly towards honours and postgraduate students, but I hope that scientists working on particular aspects of testicular function and in other fields, and seeking information, will find the book useful. In the references at the end of each chapter (in Chapter 11 at the end of each of the five main sections), my intention has been to include all the important contributions in the field, even if not all of them are referred to in the text. As I am primarily a physiologist, a certain bias in this direction is inevitable, but I have tried to stress the differences found between the testes of different species of mammals. Inevitably, most of what we know about the testis is based on experiments with rats, sheep, bulls, boars, dogs, and monkeys and on observations on humans, but I am glad to say that more species are now being studied.

The book began as a collaborative effort with my good friend and colleague Geoffrey Waites, but his other commitments unfortunately prevented him from continuing. This was a great pity as I am sure that the book would have been much better if he had been able to do so. Chapter 5 and parts of Chapter 4 are based on his early drafts and I would like to express my gratitude to him for his help over the last seventeen years. Many of the ideas set out here arose from our discussions during that time.

I am also extremely grateful to my scientific friends and colleagues too numerous to name here who have allowed me to use their published and especially their unpublished illustrations, and to those who helped me by discussion and explanation to understand points in fields where they knew more than I did. A special mention must be made of Stuart Laurie, whose assistance in drawing or redrawing many of the original illustrations has been invaluable.

The other scientists working with me, Richard Davies, Barry Hinton, Stephen Main and Lynn Pilsworth were all very helpful in discussing points of obscurity, and in carrying on the work in our laboratory while I was writing. I am also grateful to Mrs Catharine Knights for typing most of the manuscript and to the staff at Elek Books for their patience and help. Lastly, I would like to acknowledge the invaluable help given to me by my family. My wife, Marcia, typed large parts of the early drafts, and helped in many other ways. My daughter Joanna, and Keith Wright, read the whole final manuscript and made many useful suggestions. My two sons Angus and Nicholas helped me by sorting pages of typescript and reference cards. Most important, they all encouraged me when I was discouraged and, by doing my share of the household chores for me, enabled me to find the time to write.

# 1 Introduction

'Testis' is a Latin word which usually meant 'witness' or 'spectator'; from this meaning English words such as 'testify' and 'testament' were derived. It has been suggested that the use 'testes' for the male gonads reflected an ancient belief that these organs were witnesses to an act in which they took no part. It seems more likely that the Romans considered the testes to be 'witnesses' or evidence of a man's virility, and the word probably came into use as a euphemism when an earlier Latin word, which is not recorded, became too vulgar. The Romans in the time of Cicero and Celsus also used the word 'colei' for the testes. This probably also originated as a euphemism (the usual meaning of 'coleus' was 'bag', but it was also applied to the scrotum), but 'colei' was considered to be less respectable than 'testes' or the diminutive 'testiculi' which was also in common use. The Greek word '$o\rho\chi\epsilon\iota\zeta$', from which we derive the modern word 'orchitis', was also considered to be too vulgar for use in some medical texts, e.g. Herophilus but not Hippocrates, and was replaced by the word '$\delta\iota\delta\upsilon\mu o\iota$' which usually meant 'twins'. The word 'epididymis' (i.e. 'twin's extra') is derived from this.

## 1.1    Functions of the testis

The testes of mammals fulfil two functions, the production of the male gametes (spermatozoa) and of the male sex hormones (androgens). These two functions are very closely related because an adequate level of androgen production is necessary for the production of spermatozoa and because their successful delivery requires normal sexual behaviour and the development of the secondary sexual characteristics which are also under the control of androgens. The secondary sexual characteristics include the internal accessory organs (e.g. the prostate, seminal vesicles, Cowper's glands and ampullae) which contribute to the bulk of the semen; the external genitalia (e.g. the penis) which are involved mechanically in the transport of the semen to the site of insemination in the female; as well as other characteristics (e.g. muscle development, horn and antler growth, special coloration and gland secretions) which are concerned with securing female partners or repelling other competing males.

The first observations on the functions of the testes were probably incidental to accidental castration, but since before the beginning of history, it has been known that removal of the testes affected both the anatomy and behaviour of men and other animals. As a result, castration has long been used to provide eunuchs to guard women in harems and to make male domestic livestock easier to handle and less prone to yield tough or tainted meat if kept beyond puberty.

Aristotle was fully aware that the testes were essential for the virility and fertility

1

of male animals, but because he could not recognize the testes of serpents or fishes, he decided that the testes 'are merely attached [to the ducts] just like the stone weights which women hang on their looms when they are weaving. When the testes are removed, the passages are drawn up within; this is why castrated animals cannot generate.' Aristotle was also impressed by the observation that a recently castrated bull could still sire a calf and was under the mistaken impression that the testes 'are no integral part of the passages'. Galen did not subscribe to the weight theory, pointing out that if the testes were chilled or crushed, they still acted as weights but the animal was sterile. Furthermore, he pointed out that in species in which the testes were abdominal, such as birds, the testes could not act as weights.

Even as late as 1668, de Graaf thought it necessary to discuss the possibility that the testes contributed nothing to the semen although he finally rejected this theory and concluded that: 'From the observations described above, it is clear that semen is confected in the testicles.' When de Graaf wrote this, spermatozoa had not been observed, and even when they were described by Ham and van Leuwenhoeck soon after de Graaf's death (Leuwenhoeck, 1678), their origin in the testes and their function was disputed until the early years of the nineteenth century. Many authors even considered that the spermatozoa were parasitic animals unrelated to their 'host', and this opinion is enshrined in the name 'spermatozoa', which was coined by von Baer in 1827. Finally Prevost and Dumas in the 1830s showed that the spermatozoa were the part of the semen essential for its fertility and von Kölliker in 1841 demonstrated that they arose as the end product of cell division and transformation of the lining of the seminiferous tubules.

The hormone-producing function of the testis was demonstrated by Berthold (1849) in what is usually considered to be the first experiment in endocrinology. He showed that the regression of the secondary sexual characteristics which followed castration of a rooster could be prevented by re-implantation of part of the testis.

## 1.2    Location and size of the testes

Mammals are the only animals in which the testes descend from their point of origin into a scrotum. However, testicular descent occurs to a very variable extent in the various orders of mammals, ranging from virtually no change of position (in elephants and hyraxes); through migration to the caudal end of the abdominal cavity (in armadillos, whales and dolphins); migration just through the abdominal wall (in hedgehogs, moles and some seals); formation of a subanal swelling (in pigs, rodents and carnivores); to the development of the pronounced scrota (primates, ruminants and marsupials) (Figure 1.1; Carrick and Setchell, 1977).

As the testis migrates it takes its arterial and venous blood supply and nerve supply with it, so that in the adult animal with scrotal testes, the blood vessels run across the abdominal cavity, through the inguinal canal and into the scrotum. No one has yet explained satisfactorily why testicular migration takes place. There seems to be little foundation in the suggestion by Wislocki (1933) that animals with abdominal testes have a lower body temperature than animals with scrotal testes (see

Chapter 5), but presumably migration of the testis confers some evolutionary advantage on the species. Very little quantitative data has been accumulated concerning the function of the testis in those species with abdominal or inguinal testes, and until such information is available, it is difficult to make quantitative comparisons of the testes of these animals with the testes of other species in which there is a scrotum.

There is also considerable variation in the size of the testes as a proportion of body weight (each weighing between 0·02 and 0·5 per cent of body weight when fully developed) and this bears very little relationship to the position of the testis.

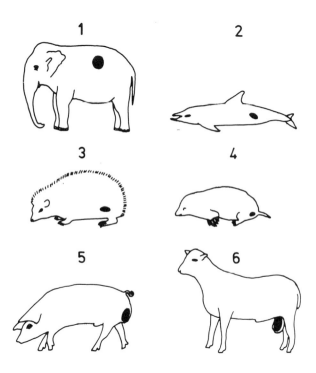

Figure 1.1 Diagram showing the approximate position of the testes in various animals. *Type 1*    Testes remain just caudal to the kidneys, as in Probiscoidea, Hyracoidaea, Sirenia, Macroscelidae, Chrysochloridae, Tenrecinae and Monotremata. *Type 2*    Testes migrate to lie near the bladder, but remain close to the dorsal abdominal wall, as in Edentata, Cetaceae, Oryzorictinae and Solenodontinae. *Type 3*    Testes migrate to or just through the ventral abdominal wall as in Erinaceus, Manis, Orycteropus, Phocidae, Tapiridae, Rhinocerotidae and Microchiroptera. *Type 4*    Testes pass caudally into a cremasteric sac near the base of the tail, without producing any external swelling, as in Soricidae and Talpidae. *Type 5*    Testes descend into a non-pendulous scrotum with no definite neck as in Rodentia, Lagomorpha, Carnivora, Equidae, Suiformes, Otariidae, Megachiroptera and Tupaia. *Type 6*    Testes descend into an obvious, pendulous scrotum with a distinct neck as in Ruminantia, Primates and most Marsupialia (Weber, 1928; Eisenberg and Gould, 1970; Harrison, 1969; Racey, 1974 and Martin, 1969.). (Reproduced from Carrick and Setchell, 1977)

## 1.3        General structure of the testis

Despite the variation in final position, the gross anatomy of the testes of various mammalian species is essentially similar. Until the time of de Graaf, it was believed that the substance of the testis was 'glandulous, pultaceous or porridge-like', but de Graaf showed convincingly that the testis consists of a series of elongated convoluted tubules, among which lie the blood vessels and the hormone-producing Leydig cells, the whole being enclosed in a tough capsule.

### 1.3.1        Seminiferous tubules and the rete testis

The seminiferous tubules are basically two-ended loops with the two ends opening into the rete testis (Figure 1.2). Each tubule is extensively convoluted and an ap-

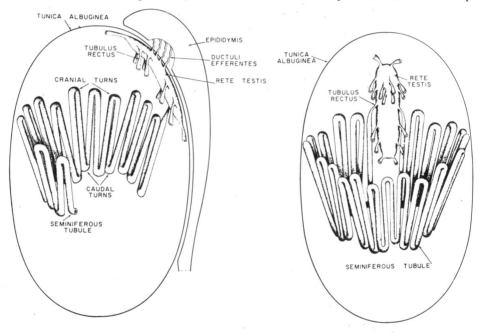

Figure 1.2 Diagram showing the arrangement of one of the seminiferous tubules and the rete testis in the testis of a rat. (Reproduced from Clermont and Huckins, 1961)

preciable proportion are branched so that they have three openings into the rete testis; a much smaller number are blind-ended with a single opening. The number of tubules varies; there are less than five in some of the dasyurid marsupials (Woolley, 1975), about thirty in the rat (Clermont and Huckins, 1961), and many more in humans and in rams. In humans, the tubules are arranged in about 300 lobules, each lobule containing 1 to 4 tubules, but this lobulation is much less marked in some other species. Each tubule in the rat is about 1 m long; in most species, the diameter is between 200 and 250 $\mu$m, but more in some marsupials.

The tubules contain the various germinal cells (see Chapter 7) and the somatic Sertoli cells. They are surrounded by a well-defined boundary tissue. This is composed of four layers. On the inner surface there is an inner non-cellular layer which in some species, e.g. the ram, consists of a number of layers. Next there is an inner cellular layer which consists of smooth muscle-like or myoid cells, which are probably responsible for the peristaltic movements of the seminiferous tubules (Roosen-Runge, 1951; Clermont, 1958; Niemi and Kormano, 1965; Wojcik, 1967; Suvanto and Kormano, 1970). Surrounding these cells is an outer non-cellular layer consisting mainly of collagen-like fibres and on the outside is an outer cellular layer which in rodents consists of cells identical with the endothelial cells lining the

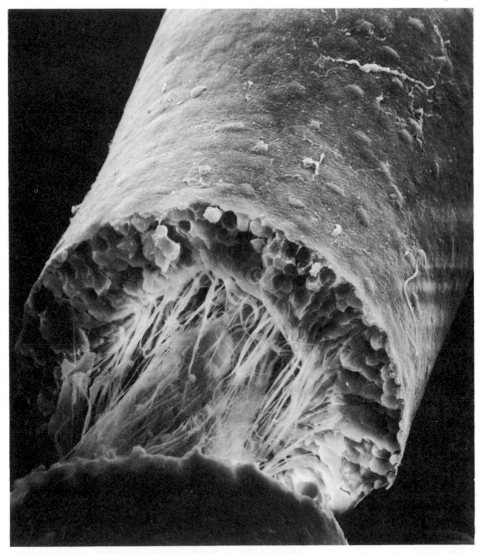

Figure 1.3 Scanning electron micrograph of an isolated seminiferous tubule of a rat, fractured to show the lumen and the tails of the spermatozoa. Note the lymphatic endothelial cells, with their nuclei, on the outside surface of tubules. (Kindly supplied by Dr A. Kent Christensen)

lymphatic sinusoids (Figure 1.3; Clermont, 1958; Brokelmann, 1960; Lacy and Rotblat, 1960; Leeson and Leeson, 1963; Baillie, 1964; Gardner and Holyoke, 1964; Ross and Long, 1966; Ross, 1967; McCord, 1970; Setchell, 1970; Bressler and Ross, 1972; Hovatta, 1972; Kormano and Hovatta, 1972; Bustos-Obregon and Holstein, 1973; de Kretser *et al.*, 1975).

The Sertoli cells, named after the man who described them in 1865, apparently do not divide after puberty (Steinberger and Steinberger, 1971; Nagy, 1972), before which time they lie immediately inside the boundary tissue of the tubules and surround the undeveloped germinal cells (see Chapter 2). In the adult, the cytoplasm of the Sertoli cells extends from the boundary tissue to the lumen of the tubule. The Sertoli cells share the surface of the boundary tissue with the spermatogonia. Pairs of Sertoli cells sandwich the spermatocytes and early spermatids and they embed the late spermatids in their luminal surface (Figure 1.4; Fawcett, 1975). Elftman (1950, 1963) likened the overall arrangement of the Sertoli cells on the boundary tissue to a 'patterned array not unlike trees in an orchard' and obviously the function of the Sertoli cells is closely entwined with the development of the germinal cells. Exactly how the Sertoli cells fulfil this nutritive or sustentacular role is not yet clear, but a number of specific functions have now been ascribed to the Sertoli cells (Fawcett, 1975), including secretion of fluid, phagocytosis, the maturation and release of spermatozoa and the synthesis of the intratubular androgen-binding protein. The Sertoli cells are the primary site of action of follicle-stimulating hormone and are involved in the transformation of steroids.

Sertoli cells have a complex nucleus with infolding or lobulation, and there is a prominent nucleolar complex, consisting of a central mass and two lateral associated bodies—also called perinucleolar spheres, satellite karyosomes or heteropyknotic bodies. The mitochondria are numerous, and have an orthodox internal appearance, although they tend to be longer and thinner than those of the germ cells. There are multiple separate Golgi elements, and numerous membrane-limited dense bodies of varying size. Both rough and smooth endoplasmic reticulum (ER) are present. The rough ER is mainly in the basal part of the cell, in the form of tubules or stacks of cisternae, while some of the smooth ER is arranged in dense masses around the developing acrosome of the spermatid (Figure 1.5). In human cells, there are two types of crystals, Charcot-Böttcher and Spangaro crystals, and in many species there are abundant lipid droplets near the bases of the cells. In some species, these lipid droplets show cyclic changes which coincide with the spermatogenic cycle (Section 7.4), and the amount of lipid is increased after X-irradiation or heat (Section 11.1). Microtubules are also abundant in the Sertoli cells at certain stages of the spermatogenic cycle. Pairs of adjacent Sertoli cells show specialized junctions which probably form the main component of the blood–testis barrier (Section 8.7).

The lumen of each tubule is filled with fluid, which carries the liberated spermatozoa off on their long journey to what de Graaf referred to as 'the place ordained by Almighty God for the reception of semen'. The fluid is moved partially by *vis a tergo* pressure from continuing secretion of fluid and partially by peristaltic movements of the seminiferous tubules themselves (see p. 5).

The seminiferous tubules open into short tubuli recti or straight tubules (Section

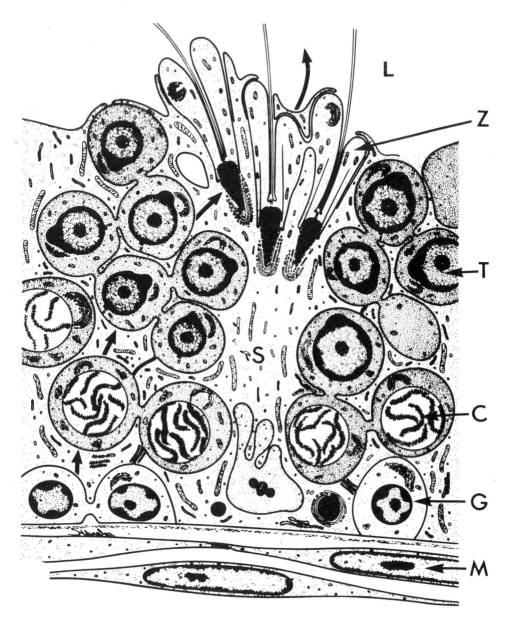

Figure 1.4  Diagram of the cells of a seminiferous tubule, with the myoid cells (M) on the outside; Sertoli cells (S) and spermatogonia (G) immediately inside the myoid cells; spermatocytes (C) and early spermatids (T) sandwiched between pairs of Sertoli cells; and the later spermatids (Z) embedded in the luminal surface of the Sertoli cells with their tails protruding into lumen of the tubules (L). The arrows show how the germinal cells migrate towards the lumen as they divide and mature. (Reproduced from Fawcett, 1974)

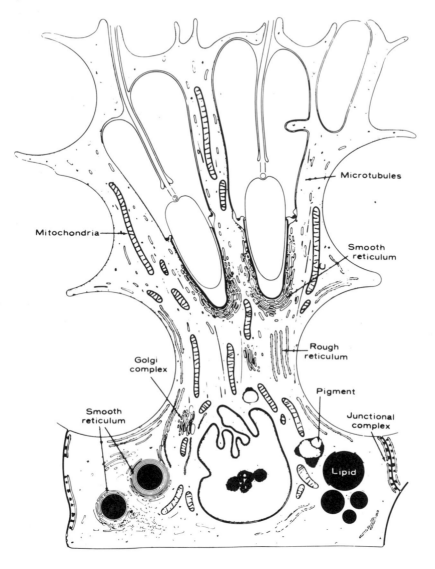

Figure 1.5 Diagram showing some details of the structure of a Sertoli cell. Note the lobulated nucleus with its prominent nucleolus, the junctional complexes between this cell and the adjacent Sertoli cells, and the prominent smooth endoplasmic reticulum associated principally with lipid droplets and the acrosome region of the later spermatids. (Reproduced from Fawcett, 1975)

8.1) which in turn open into the rete testis. The rete testis is the beginning of the excurrent duct system, and in some species is embedded in a fibrous mediastinum. The size and position of the rete is very variable between species, and so are the position and length of the efferent ducts (ductuli efferentes) which lead from the testis into the epididymis (Section 8.1).

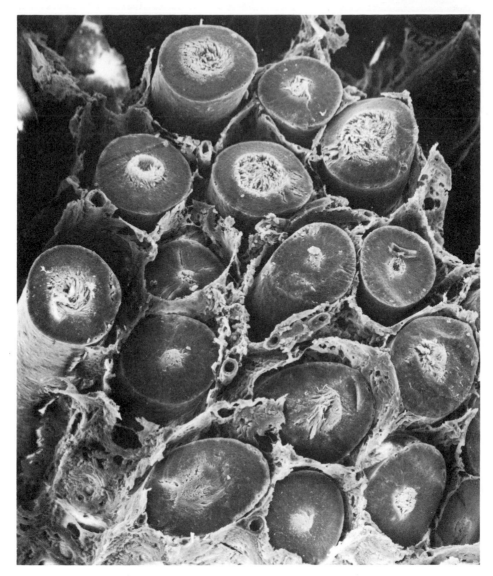

Figure 1.6 Scanning electron micrograph of a cut surface of a rat testis, showing the individual seminiferous tubules embedded in the interstitial tissue which includes the Leydig cells, blood and lymph vessels and nerves. (Kindly supplied by Dr A. Kent Christensen)

## 1.3.2    The interstitial tissue

The seminiferous tubules are basically cylindrical, and, when a number of cylinders are stacked together, this creates a series of 3-sided spaces (Figure 1.6). These spaces, known as the interstitial tissue, contain the blood vessels, lymph vessels and nerves (none of which penetrate the tubules), as well as the interstitial cells or Leydig cells. These cells were first described in man by von Kölliker (1841) and in a variety of mammals by Leydig (1850), who considered them to be modified connective tissue. Their function of producing the male sex hormone was suggested by Bouin and Ancel (1903) from a series of indirect experiments. It is now generally agreed that these are the cells which produce the majority of the steroid hormones formed *de novo* in the testis from cholesterol (Section 6.1.3.1).

The Leydig cells have an ultrastructure which is consonant with their function of steroid synthesis. They contain large amounts of smooth endoplasmic reticulum, plentiful mitochondria, a prominent Golgi complex, centrioles and a number of lipid droplets (Figure 1.7; Christensen, 1975). In humans, there are also specialized crystals, Reinke crystals, inside the Leydig cells and lipofuschin pigment granules can also be found there.

The relationship between the walls of the tubules, the Leydig cells, and the blood and lymph vessels varies among the different species. In rodents, the interstitial tissue is very sparse with large lymphatic sinusoids and the Leydig cells are clustered around the blood capillaries. In the guinea pig (Figure 1.8a) and chinchilla, the lymphatic endothelium surrounds the seminiferous tubules and the clusters of Leydig cells and blood vessels. In the rat (Figure 1.8b) and mouse, the lymphatic endothelium surrounding the Leydig cells is discontinuous, exposing the Leydig cells directly to the lymph. In man, as well as monkey, ram (Figure 1.8c), bull and elephant, clusters of Leydig cells are scattered at random in an abundant loose connective tissue which appears to be rich in interstitial fluid, and which contains one or two discrete lymphatic vessels, and blood capillaries. The Leydig cells do not seem always to be closely associated with the blood vessels. In the third type of animal, exemplified by the boar (Figure 1.8d), warthog, zebra, naked mole rat and some marsupials (e.g. the American opossum and the Australian bandicoots) the interstitial space is almost filled with closely packed Leydig cells, with occasional blood capillaries and lymphatic vessels and very little obvious connective tissue (Fawcett *et al.*, 1973; Setchell, 1977).

### 1.3.3    The capsule of the testis

The testis is encased in a tough fibrous capsule, usually referred to as the tunica albuginea. There are really three tunics: the tunica albuginea proper in the middle; a visceral tunica vaginalis on the outside; and a tunica vasculosa nearest to the parenchyma (Figure 1.9). The tunica vaginalis is the peritoneal lining which surrounds the testis and consists of a single layer of flattened mesothelial cells. The tunica vasculosa can be considered as a subtunical extension of the interstitial tissue and consists of plexiform networks of minute blood vessels held together by delicate areolar tissue.

The main tunica albuginea consists of fibroblasts and bundles of collagenous fibres running in all directions with, in some species, an appreciable number of smooth muscle cells, and nerve endings (Davis *et al.*, 1970). These muscle cells are presumably the basis for the contractions which can be elicited from the isolated tunic and which produce changes in interstitial and arterial blood pressure *in vivo*. The isolated capsule contracts in response to acetylcholine, noradrenaline,

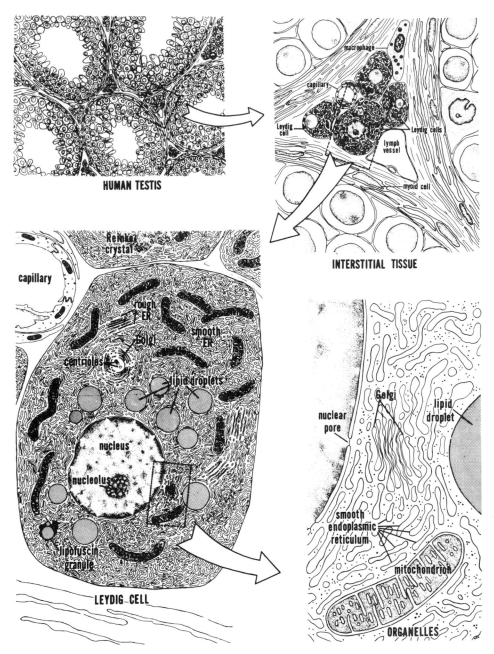

Figure 1.7 Diagram of the anatomy of the interstitial tissue and a Leydig cell from a human testis. Note the general arrangement of the tubules and the interstitial tissue; the intimate association of the Leydig cells, the blood and lymph vessels, and the walls of the tubules; the very abundant smooth endoplasmic reticulum; and the lipid droplets and the numerous mitochondria. (Reproduced from Christensen, 1975)

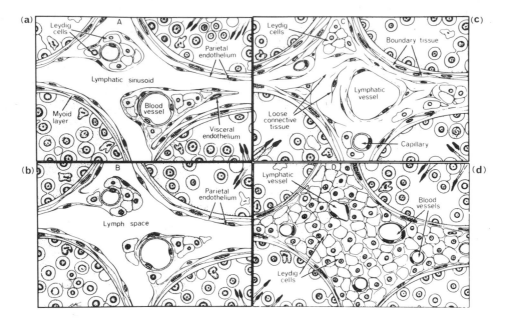

Figure 1.8 Diagram showing the variation in the anatomy of the interstitial tissue in several species. (a) Guinea pig, showing the Leydig cells clustered around blood vessels, with the whole groups of cells completely surrounded by endothelial cells and 'floating' in a lymphatic sinusoid, the contents of which also bathe the walls of the seminiferous tubules. (b) Rat, similar to the guinea pig except that the groups of Leydig cells are surrounded by an incomplete layer of endothelial cells. (c) Ram, showing the Leydig cells either in groups near a capillary or in separate clusters embedded in a loose connective tissue which also contains lymph vessels and other blood vessels. (d) Pig, showing the interstitial space crammed with closely packed Leydig cells with a few small blood and lymph vessels. (Reproduced from Fawcett, 1973)

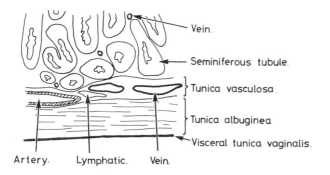
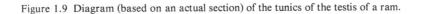

Figure 1.9 Diagram (based on an actual section) of the tunics of the testis of a ram.

adrenaline, prostaglandin F-1-$\alpha$ and sympathetic nerve stimulation and relaxes under the influence of isopropylnoradrenaline (Davis and Langford, 1969a, b, 1971). The suggestion that contractions of the capsule can assist in moving spermatozoa out of the testis has not been substantiated by any experimental observations and it seems more likely that the capsule is involved in maintaining the correct pressure inside the testis to regulate the movement of fluid out of and back into the capillaries.

## 1.4    Seasonal variations in testis function

In some mammals, e.g. human and laboratory rat, there is little consistent variation in the androgenic or spermatogenic function of the testis at different seasons of the year. However, in many species of mammals, and particularly in birds, there is a well defined breeding season, as was observed by Aristotle: 'Those birds which copulate at one season only of the year have such tiny testes when this period is over that they are almost indistinguishable, whereas during the breeding season they are very big.' Outside the breeding season, the functions of the testis may show a reduction or virtually complete cessation (Figure 1.10).

Many of the studies on this topic involve indirect measurements of androgen production, such as the size of the accessory glands, and use the size or histological appearance of the testis as an index of spermatogenic function. While these methods are not ideal, they have enabled a reasonable picture of the degree of variation involved to be obtained.

In some species, such as bats, there is an apparent dissociation of maximum androgenic activity from maximum spermatogenic activity. The testes enlarge first and spermatogenesis begins with little sign of androgen secretion, and no sexual activity. Then the tail of the epididymis becomes filled with spermatozoa ready to be ejaculated, the accessory glands begin to enlarge and sexual activity begins, while the testes, particularly the seminiferous tubules, actually regress (Figure 1.11). In fact the testis may become smaller than the tail of the epididymis with its store of spermatozoa (Courrier, 1927; Racey, 1974b; Racey and Tam, 1974).

The seasonal variation in reproductive activity is presumably achieved by variations in the rate of secretion of the pituitary gonadotrophins, but the evidence for this is very incomplete, because assay systems for plasma gonadotrophins in animals other than man and rat have only recently become available.

Experimentally, it can be shown in some species that variation in the amount of light and darkness can control the function of the testis (Hammond, 1954; Ortavant et al., 1964; Thibault et al., 1966; Clarke and Kennedy, 1967; Gaston and Menaker, 1967) while temperature seems to have little effect, unless testicular damage is produced (see Chapter 11). The situation is not simple, as the effect of a short period of illumination depends on the relation of this period to a circadian rhythm of photosensitivity (Elliott, et al., 1972; Grocock and Clarke, 1974; Stetson, et al., 1975).

Some of the effects of light are mediated via the pineal gland. For example, if hamsters are subjected to very short amounts of light (1 to 2 hours per day) or sur-

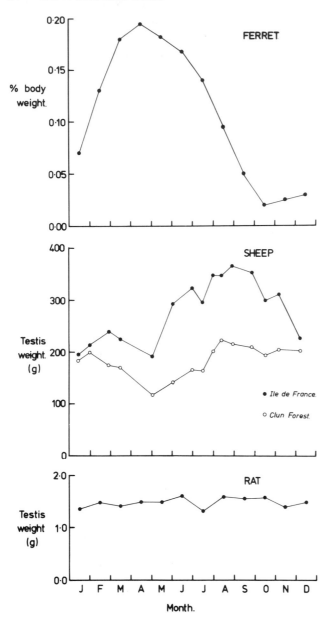

Figure 1.10 Graph of the seasonal variation in testis weight, showing: ferret, which is typical of those species with a very marked breeding season and almost complete regression of the testis for the rest of the year; ram, which is typical of a species showing a moderate seasonal variation; and the laboratory rat which shows virtually no seasonal variation even when exposed to normal variation in daylight length. Ferret data from Allanson (1932); sheep and rat data collected by R. V. Davies, S. J. Main and B. P. Setchell in Cambridge, England. The values for the sheep are the means of four animals, for each breed, the testis weight being calculated from measurements of the longitudinal and transverse diameters of the testis. Each value for the rat is the mean of between 34 and 48 adult animals (200–450g bodyweight) killed in various experiments between 1969 and 1976.

gically blinded, their testes regress; this regression can be prevented by pinealectomy, or by superior cervical gangliectomy, which denervates the pineal gland. Animals whose pineals are removed during light deprivation show regeneration of the testes

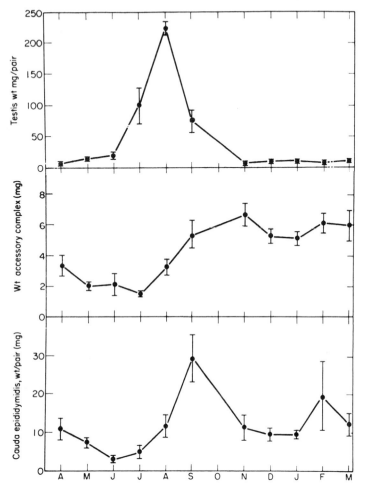

Figure 1.11 Seasonal variation in weight of the testis, the accessory glands and the tail of the epididymis of the pipistrelle bat. Note that the testis enlarges before either the epididymis or the accessory complex, and the latter remains enlarged long after the testis has regressed. In this species, spermatogenesis does not continue for long, but the spermatozoa are stored in the epididymis until ejaculation. (Reproduced from Racey and Tam, 1974)

(Figure 1.12; Hoffmann and Reiter, 1965; Reiter, 1972, 1973; Reiter and Hester, 1966). The regression of the testes of voles that are changed from 16 to 8 hours light can also be prevented by pinealectomy or by superior cervical gangliectomy (Charlton et al., 1976). In rats, blinding alone has no effect on the testes, but atrophy results when the animals are blinded and their olfactory lobes removed (anosmia) (Reiter et al., 1971); this regression is also prevented by pinealectomy. (See also Sections 6.1.5.4, 7.6.1.2 and 10.5.)

PINEAL AND SEASONAL RHYTHMS

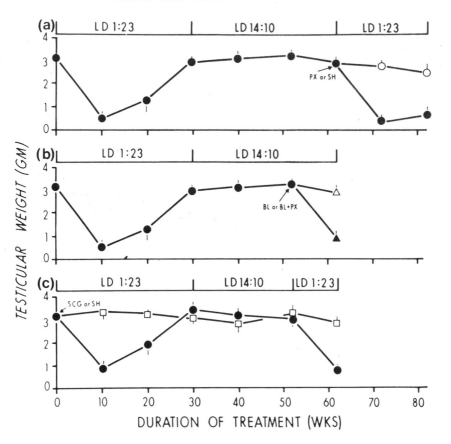

Figure 1.12 The effect of different light regimes and the pineal gland on the testis of the hamster. (a) Note the temporary decrease in testis weight during the period with 1h light and 23h darkness per day (LD 1:23), the maintenance of a large testis with 14h light and 10h dark and the abolition by pinealectomy (PX, ○—○) of the decrease in testis weight observed when sham-operated (SH, ●—●) animals were transferred again to LD 1:23. (b) Note that blinding (BL, ●—▲) and transferring to LD1:23 causes a similar degree of testicular atrophy and that the effect of blinding can be blocked by pinealectomy at the time of blinding (BL + PX, ●—△). (c) Note that superior cervical gangliectomy (SCG, □—□) also blocks the testicular atrophy seen when sham-operated (SH, ●—●) animals are transferred to LD 1:23. This ganglion is thought to carry the main sympathetic ennervation to the pineal, so that when it is removed, the pineal is effectively denervated. (Reproduced from Reiter, 1972)

## References to Chapter 1

Aldous, C. M. (1937) Notes on the life history of the snowshoe hare. *J. Mammal.* **18**, 46–57.

Allanson, M. (1932) The reproductive processes of certain mammals. III. The reproductive cycle of the male ferret. *Proc. Roy. Soc.* (London) B. **110**, 295–312.

Allanson, M. (1933) The reproductive processes of certain mammals V. Changes in the reproductive organs of the male grey squirrel (*Sciurus carolinensis*). *Phil. Trans. Roy. Soc.* (London) B. **222**, 79–96.

Allanson, M. (1934) The reproductive processes of certain mammals. VII. Seasonal variations in the reproductive organs of the male hedgehog. *Phil. Trans. Roy. Soc.* (London) B. **223**, 277–303.

Anthony, A. (1953) Seasonal reproductive cycle in the normal and experimentally treated male prairie dog, *Cynomys ludovicianus*. *J. Morphol.* **93**, 331–70.

Anthony, A. and Foreman, D. (1951) Observations on the reproductive cycle of the black-tailed prairie dog (*Cynomys ludovicianus*). *Physiol. Zool.* **24**, 242–8.

Aristotle *Historia animalium*, Book III. *Generation of Animals*, Books I, IV and V. Translated by A. L. Peck, Loeb Classical Library.

Audy, M. C. and Bonnin-Laffargue, M. (1975) L'activité endocrine du testicule chez le Blaireau européen (*Meles meles*). *Arch. Biol.* (Bruxelles) **86**, 223–32.

Audy-Relexans, M. C. (1972) Le cycle sexuel du Blaireau mâle (*Meles meles* L.). *Annals. Biol. Anim. Biochim. Biophys.* **12**, 355–66.

Baer, K. E. von (1827) *De ovi mammalium et hominis genesi*. Leopold Voss, Leipzig.

Baillie, A. H. (1964) Ultrastructural differentiation of the basement membrane of the mouse seminiferous tubule. *Quart. J. Microscop. Sci.* **105**, 203–7.

Baker, J. R. and Baker, Z. (1936) The seasons in a tropical rain forest (New Hebrides). Part 3. Fruit bats (*Pteropidae*). *J. Linn. Soc. Zool.* **40**, 123–41.

Baker, J. R. and Bird, T. F. (1936) The seasons in a tropical rain forest (New Hebrides). Part 4. Insectiverous bats (Vespertilionidae and Rhinolophidae). *J. Linn. Soc. Zool.* **40**, 143–61.

Bascom, K. F. and Osterud, H. L. (1925) Quantitative studies of the testicle. II. Pattern and total tubule length in the testis of certain common mammals. *Anat. Rec.* **31**, 159–69.

Bawa, S. R. (1963) The fine structure of the Sertoli cell of the human testis. *J. Ultrastruct. Res.* **9**, 459–74.

Beer, J. R. and Meyer, R. K. (1951) Seasonal changes in the endocrine organs and behaviour patterns of the muskrat. *J. Mammal.* **32**, 173–91.

Belt, W. D. and Cavazos, L. F. (1967) Fine structure of the interstitial cells of Leydig in the boar. *Anat. Rec.* **158**, 333–49.

Belt, W. D. and Cavazos, L. F. (1971) Fine structure of the interstitial cells of Leydig in the squirrel monkey during seasonal regression. *Anat. Rec.* **169**, 115–28.

Benda, C. (1968) Untersuchungen uber den Bau des funktionierenden Samenkanalchens einiger Saugetiere und Folgerungen fur die Spermatogenese dieser Wirbeltierklasse. *Arch. Microscop. Anat.* **30**, 49–110.

Berndtson, W. E. and Desjardins, C. (1974) Circulating LH and FSH levels and testicular function in hamsters during light deprivation and subsequent photoperiodic stimulation. *Endocrinol.* **95**, 195–205.

Berthold, A. A. (1849) Transplantation der Hoden. *Arch. Anat. Physiol. Wiss. Med.* **16**, 42–44.

Borozdin, E. K. (1966) Age and seasonal changes in testes of reindeer. *Nauchn.-Issled. Inst. Sel-kolz. Krain Sev.* **13**, 157–65. Translated in *Animal Breeding Abstr.* (1968) **36**, 128.

Bostrom, R. E., Aulerich, R. J., Ringer, P. K. and Schaible, P. J. (1968) Seasonal changes in the testes and epididymides of the ranch mink (*Mustela vision*). *Mich. State Univ. Agr. Expt.*

*Sta. Quart. Bull.* **50**, 538–558.

Bouin, P. and Ancel, P. (1903) Recherches sur les cellules interstitielles du testicule des mammifères. *Arch. Zool. Exptl. Gen.* **1**, 437–523.

Bouin, P. and Ancel, P. (1905) Le glande interstitielle du testicule chez le cheval. *Arch. Zool. Exptl. Gen.* **3**, 391–433.

Bremer, J. L. (1911) Morphology of the tubules of the human testis and epididymis. *Am. J. Anat.* **11**, 393–417.

Brokelmann, J. (1961) Surface modifications of Sertoli cells at various stages of spermatogenesis in the rat. *Anat. Rec.* **139**, 211.

Brokelmann, J. (1963) Fine structure of germ cells and Sertoli cells during the cycle of the seminiferous epithelium in the rat. *Z. Zellforsch. Mikroscop. Anat.* **59**, 820–850.

Bressler, R. S. and Ross, M. H. (1972) Differentiation of peritubular myoid cells of the testis: effects of intratesticular implantation of newborn mouse testes into normal and hypophysectomized adults. *Biol. Reprod.* **6**, 148–59.

Buhrley, L. E. and Ellis, L. C. (1975) Contractility of rat testicular seminiferous tubules *in vitro*: prostaglandin $F_{1a}$ and indomethacin. *Prostaglandins* **10**, 151–62.

Bustos-Obregon, E. (1970) On Sertoli cell number and distribution in rat testis. *Arch. Biol.* (Paris) **81**, 101–10.

Bustos-Obregon, E. (1976) Ultrastructure and function of the lamina propria of mammalian seminiferous tubules. *Andrologia* **8**, 179–85.

Bustos-Obregon, E. and Holstein, A. F. (1973) On structural patterns of the lamina propria of human seminiferous tubules. *Z. Zellforsch.* **141**, 413–25.

Canivenc, R., Bonnin-Laffargue, M. and Relexans, M. C. (1968) Cycles génitaux de quelques mustellides européens. In *Cycles génitaux saisonniers de Mammifères sauvages,* 1er Entretiens de Chize. MASSON and Cie Edt. 86–110.

Carr, I., Clegg, E. J. and Meek, G. A. (1968) Sertoli cells as phagocytes: an electron microscopic study. *J. Anat.* **102**, 501–9.

Carrick, F. N. and Setchell, B. P. (1977) The evolution of the scrotum. In *Reproduction and Evolution,* eds. J. N. Calaby and C. H. Tyndale-Biscoe, 165–70. Australian Academy of Science, Canberra.

Chaplin, R. E. and White, R. W. G. (1972) The influence of age and season on the activity of the testes and epididymides of the fallow deer, *Dama dama. J. Reprod. Fertil.* **30**, 361–9.

Chapman, D. I. and Chapman, N. G. (1970) Preliminary observations on the reproductive cycle of male fallow deer (*Dama dama* L.). *J. Reprod. Fertil.* **21**, 1–8.

Charlton, H. M., Grocock, C. A. and Ostberg, A. (1976) The effects of pinealectomy and superior cervical ganglionectomy on the testis of the vole, *Microtus agrestis. J. Reprod. Fertil.* **48**, 377–9.

Christensen, A. K. (1965) The fine structure of testicular interstitial cells in guinea pig. *J. Cell Biol.* **26**, 911–35.

Christensen, A. K. (1975) Leydig cells. In *Handbook of Physiology*, section 7, vol. V. Male Reproductive System, eds. D. W. Hamilton and R. O. Greep, 57–94. American Physiological Society, Washington.

Christensen, A. K. and Fawcett, D. W. (1961) The normal fine structure of opossum testicular interstitial cells. *J. Biophys. Biochem. Cytol.* **9**, 653–70.

Christensen, A. K. and Fawcett, D. W. (1966) The fine structure of testicular interstitial cells in mice. *Am. J. Anat.* **118**, 551–72.

Clarke, J. R. and Forsyth, I. A. (1964) Seasonal changes in the gonads and accessory reproductive organs of the vole (*Microtus agrestis*). *Gen. Comp. Endocrinol.* **4**, 233–42.

Clarke, J. R. and Kennedy, J. P. (1967) Effect of light and temperature upon gonad activity in the vole, *Microtus agrestis*. *Gen. Comp. Endocrinol.* **8**, 474–88.

Clegg, E. J. and Macmillan, E. W. (1965) The uptake of vital dyes and particulate matter by the Sertoli cells of the rat testis. *J. Anat.* (London) **99**, 219–29.

Clermont, Y. (1958) Contractile elements in the limiting membrane of the seminiferous tubules of the rat. *Exptl. Cell Res.* **15**, 438–40.

Clermont, Y. and Huckins, C. (1961) Microscopic anatomy of the sex cords and seminiferous tubules in growing and adult male albino rats. *Am. J. Anat.* **108**, 79–97.

Clewe, T. H. (1969) Observations on reproduction of squirrel monkeys in captivity. *J. Reprod. Fertil.* Suppl. **6**, 151–6.

Conaway, C. H. (1959) The reproductive cycle of the Eastern mole. *J. Mammal.* **40**, 180–94.

Connell, C. J. and Christensen, A. K. (1975) The ultrastructure of the canine testicular interstitial tissue. *Biol. Reprod.* **12**, 368–82.

Courrier, R. (1923) Cycle annuel de la glande interstitielle du testicule chez les cheiroptères. Coexistence du repos seminal et de l'activité genitale. *Compt. Rend. Soc. Biol.* **88**, 1163–6.

Courrier, R. (1927) Etude sur le déterminisme des caractères sexuels secondaires chez quelques mammifères a activité testiculaire periodique. *Arch. Biol.* (Liège) **37**, 173–334.

Crabo, B. (1963) Fine structure of the interstitial cells of the rabbit testes. *Z. Zellforsch. Mikrosckop. Anat.* **61**, 587–604.

Curtis, G. M. (1918) The morphology of the mammalian seminiferous tubule. *Am. J. Anat.* **24**, 339–94.

Davis, J. R. and Langford, G. A. (1969a) Response of the testicular capsule to acetylcholine and noradrenaline. *Nature* **222**, 386–7.

Davis, J. R. and Langford, G. A. (1969b) Response of the isolated testicular capsule of the rat to autonomic drugs. *J. Reprod. Fertil.* **19**, 595–8.

Davis, J. R. and Langford, G. A. (1971) Comparative responses of the isolated testicular capsule and parenchyma to autonomic drugs. *J. Reprod. Fertil.* **26**, 241–5.

Davis, J. R., Langford, G. A. and Kirby, P. J. (1970) The testicular capsule. In *The Testis*, ed., A. D. Johnson, W. R. Gomes and N. L. Vandemark. New York, Academic Press, vol. I, 281–337.

Delost, P. (1966) Reproduction et cycles endocriniens de l'écureuil. *Arch. Sci. Physiol.* **20**, 425–57.

Delost, P. (1968) Etude comparative de la reproduction chez quelques rongeurs sauvages non hibernants dans différentes regions de France In *Cycles génitaux saisonniers des Mammifè*res *sauvages*. Ier Entretiens de Chize. MASSON and Cie Edt. 23–50.

Duesberg, J. (1918) On the interstitial cells of the testicle in *Didelphis*. *Biol. Bull.* **35**, 175–99.

Dym, M. and Romrell, L. J. (1975) Intraepithelial lymphocytes in the male reproductive tract of rats and rhesus monkeys. *J. Reprod. Fertil.* **42**, 1–7.

Eisenberg, J. F. and Gould, E. (1970) The tenrecs: a study in mammalian behaviour and evolution. *Smithson. Cont. Zool.* **27**, 1–138.

Elder, W. H. and Finerty, J. C. (1943) Gonadotropic activity of the pituitary gland in relation to the seasonal sexual cycle of the cottontail rabbit *Sylvilagus floridanus mearnsi. Anat. Rec.* **85**, 1–16.

Elftman, H. (1950) The Sertoli cell cycle in the mouse. *Anat. Rec.* **106**, 381–93.

Elftman, H. (1963) Sertoli cells and testis structure. *Am. J. Anat.* **113**, 25–33.

Elliott, J. A., Stetson, M. H. and Menaker, M. (1972) Regulation of testis function in golden hamsters: a circadian clock measures photoperiod time. *Science* (New York) **178**, 771–3.

Evans, F. C. and Holdenried, R. (1943) A population study of the Beechey ground squirrel in central California. *J. Mammal.* **24**, 231–60.

Ewing, L. L., Green, P. M. and Stebler, A. M. (1965) Metabolic and biochemical changes in testis of cotton rat (*Sigmodon hispidus*) during breeding cycle. *Proc. Soc. Exptl. Biol. Med.* **118**, 911–13.

Fawcett, D. W. (1973) Observations on the organization of the interstitial tissue of the testis and on the occluding cell junctions in the seminiferous epithelium. *Adv. Biosci.* **10**, 83–99.

Fawcett, D. W. (1974) Interactions between Sertoli cells and germ cells. In *Male Fertility and Sterility*, eds. R. E. Mancini and L. Martini, 13–36. London, Academic Press.

Fawcett, D. W. (1975) Ultrastructure and function of the Sertoli cell. In *Handbook of Physiology*, section 7, vol. V, Male Reproductive System, eds. D. W. Hamilton and R. O. Greep, 21–55. American Physiological Society, Washington.

Fawcett, D. W. and Burgos, M. H. (1960) Studies on the fine structure of the mammalian testis. II. The human interstitial tissue. *Am. J. Anat.* **107**, 245–69.

Fawcett, D. W., Neaves, W. B. and Flores, M. N. (1973) Comparative observations on intertubular lymphatics and the organization of the interstitial tissue of the mammalian testis. *Biol. Reprod.* **9**, 500–32.

Fitch, H. S. (1948) Habits and economic relationship of the Tulare kangaroo rat. *J. Mammal.* **29**, 5–35.

Flickinger, C. J. (1967) The postnatal development of the Sertoli cells of the mouse. *Zeitschrift für Zellforschung* **78**, 92–113.

Flickinger, C. and Fawcett, D. W. (1967) Junctional specializations of the Sertoli cells in the seminiferous epithelium. *Anat. Rec.* **158**, 207–22.

Fouquet, J. P. (1968) Etude infrastructurale du cycle du glycogene dans les cellules de Sertoli du hamster. *Compt. Rend. Acad. Sci.* **267**, 545–8.

Gabe, M., Agide, R., Martoja, A., Saint-Girons, M. C. and Saint-Girons, M. (1963) Données histophysiologiques et biochimiques sur l'hibernation et le cycle annuel chez *Elyomis quercinus. Arch. Biol.* (Liège) 1–87.

Galen, *De Semine* 1, 15–16.

Gardner, P. J. and Holyoke, E. A. (1964) Fine structure of the seminiferous tubule of the Swiss mouse. I. The limiting membrane, Sertoli cell, spermatogonia and spermatocytes. *Anat. Rec.* **150**, 391–404.

Gaston, S. and Menaker, M. (1967) Photoperiodic control of hamster testis. *Science* (New York) **158**, 925–8.

George, U. (1937) Beobachtungen an Sertoli-Zellen im Hoden des Menschen und der Ratte. *Z. Mikroskop. Anat. Forsch.* **42**, 479–98.

Gilmore, D. P. (1969) Seasonal reproductive periodicity in the male Australian brush-tailed possum (*Trichosurus vulpecula*). *J. Zool.* (London) **157**, 75–98.

Gilula, N. B., Fawcett, D. W. and Aoki, A. (1976) The Sertoli cell occluding junctions and gap junctions in mature and developing mammalian testis. *Develop. Biol.* **50**, 142–68.

Goerttler, K. (1934) Die Konstruktion der Wand des menschlichen Samenleiters und ihre funktionelle Bedeutung. *Morph. Jb.* **74**, 550–80.

Gondos, B., Renston, R. H. and Conner, L. A. (1973) Ultrastructure of germ cells and Sertoli cells in the postnatal rabbit testis. *Am. J. Anat.* **136**, 427–40.

Gopalakrishna, A. (1948) Studies on the embryology of Microchiroptera. Part II. Reproduction in the male vespertilionid bat, *Scotophilus wroughtonii* (Thomas). *Proc. Indian Acad. Sci.* **27B**, 137–50.

Graaf, R. de (1668) Tractatus de virorum organis generationi inservientibus. Translated by H. D. Jocelyn and B. P. Setchell in 'Regnier de Graaf on the Human Reproductive Organs'. *J. Reprod. Fertil.* Suppl. **17**, 1–76.

Griffiths, J. (1895) Three lectures upon the testes. *Lancet* i, 791–9 and 916–20.

Grocock, C. A. and Clarke, J. R. (1974) Photoperiodic control of testis activity in the vole, *Microtus agrestis*. *J. Reprod. Fertil.* **39**, 337–47.

Groome, J. R. (1940) The seasonal modification of the interstitial tissue of the testis of the fruit bat (*Pteropus*). *Proc. Zool. Soc.* (London) A. **110**, 37–42.

Gulamhusein, A. P. and Tam, W. H. (1974) Reproduction in the male stoat (*Mustela erminea*). *J. Reprod. Fertil.* **41**, 303–12.

Gunther, W. C. (1956) Studies on the male reproductive system of the Californian pocket gopher (*Thonomys bottae navus Merrian*). *Amer. Midland Naturalist* **55**, 1–40.

Hamilton, W. J. (1939) Observations on the life history of the red squirrel in New York. *Am. Midland Naturalist* **22**, 732–45.

Hamilton, W. J. (1940) The biology of the smoky shrew (*Sorex fumeus fumeus*, Miller). *Zoologica* **25**, 473–92.

Hammond, J. (1954) Light regulation of hormone secretion *Vit. Horm.* **12**, 157–206.

Hanes, F. M. (1911) The relations of the interstitial cells of Leydig to the production of an internal secretion by the mammalian testis. *J. Exptl. Med.* **13**, 338–54.

Hanks, J. (1972) Reproduction of elephant, *Loxodonta africana*, in the Luangwa Valley, Zambia. *J. Reprod. Fertil.* **30**, 13–26.

Hansen, R. M. (1960) Age and reproductive characteristics of mountain pocket gophers in Colorado. *J. Mammal.* **41**, 323–35.

Hargrove, J. L., Johnson, J. M. and Ellis, L. C. (1971) Prostaglandin $E_1$ induced inhibition of rabbit testicular contractions *in vitro*. *Proc. Soc. Exp. Biol. Med.* **136**, 958–61.

Harrison, R. J. (1969) Reproduction and reproductive organs. In *The Biology of Marine Mammals*, ed. H. T. Anderson, 253–348. New York, Academic Press.

Harrison, R. J., Harrison Matthews, L. and Roberts, J. M. (1952) Reproduction in some pinnepedia. *Trans. Zool. Soc.* (London) **27**, 437–541.

Harrison Matthews, L. (1942) Reproduction in the Scottish wildcat *Felis silvestris grampia* Miller. *Proc. Zool. Soc.* **111B**, 59–77.

Hemmingsen, B. (1967) Postnatal development and cyclic changes in the testes of mink. *Nord. Vet. Med.* **19**, 71–80.

Hermo, L., Lalli, M. and Clermont, Y. (1977) Arrangement of connective tissue components in the walls of seminiferous tubules of man and monkey. *Am. J. Anat.* **148**, 433–46.

Hill, J. E. (1937) Morphology of the pocket gopher, mammalian genus *Thonomys. Univ. Calif. Publ. Zool.* **42**, 82–171.

Hirota, S. (1952) The morphology of the seminiferous tubules. I. The seminiferous tubules of the mouse. *Kyushu Memoirs of Medical Sciences* **3**, 121–36.

Hirota, S. (1955) The morphology of the seminiferous tubules. III. The seminiferous tubules of man. *Kyushu Journal of Medical Science* **6**, 180–7.

Hoffman, R. A. and Reiter, R. J. (1965) Pineal gland: Influence on gonads of male hamsters. *Science* **148**, 1609–11.

Holstein, A. F. (1967) Die glatte Muskulatur in der Tunica Albuginea des Hodens und ihr Einfluss auf den Spermatozoantransport in den Nebenhoden. *Anat. Anz. Ergeb. Bd.* **121**, 103–8.

Holstein, A. F. and Weiss, C. (1967) Uber die Wirkung der glatten Muskulatur in der Tunica Albuginea in Hoden des Kaninchens, Messungen des interstitiellen Druckes. *Z. Ges. Expt. Med.* **142**, 334–7.

Hooker, C. W. (1944) The postnatal history and function of the interstitial cells of the testis of the bull. *Am. J. Anat.* **74**, 1–37.

Hooker, C. W. (1948) The life history of the interstitial cells of the testis of the mouse. *Anat. Rec.* **100**, 676–7.

Hovatta, O. (1972) Contractility and structure of adult rat seminiferous tubules in organ culture. *Z. Zellforsch.* **130**, 171–9.

Huber, G. C. and Curtis, G. M. (1913) The morphology of the seminiferous tubules of mammalia. *Anat. Rec.* **7**, 207–19.

Ichihara, I. (1970) The fine structure of testicular interstitial cells in mice during postnatal development. *Z. Zellforsch. Mikroskop. Anat.* **108**, 475–86.

Illige, D. (1951) An analysis of the reproductive pattern of whitetail deer in south Texas. *J. Mammal.* **32**, 411–21.

Itoh, S., Takahashi, H. and Sakai, T. (1962) Atrophy of reproductive organs in male rats housed in continuous darkness. *J. Reprod. Fertil.* **4**, 233–4.

Joffre, M. (1976) Débit sanguin capillaire du testicule chez le renard roux (*Vulpes vulpes* L.); relation avec l'activité testiculaire pendant la période prépubère et au cours dy cycle saisonnier. Thèse de docteur és-sciences naturelles. University of Poitiers.

Johnson, E. P. (1934) Dissections of human seminiferous tubules. *Anat. Rec.* **59**, 187–99.

Johnson, J. M., Hargrove, J. L. and Ellis, L. C. (1971) Prostaglandin $F_{1\alpha}$ induced stimulation of rabbit testicular contractions *in vitro. Proc. Soc. Exp. Biol.* (N.Y.) **138**, 378–81.

Kinson, G. A. (1976) Pineal factors in the control of testicular function. In *Cellular mechanisms modulating gonadal hormone action,* eds. R. L. Singhal and J. A. Thomas, 87–139, HM & M Aylesbury.

Kirkpatrick, C. M. (1955) The testis of the fox squirrel in relation to age and seasons. *Am. J. Anat.* **97**, 229–55.

Kölliker, R. A. von (1841) *Beitrage zur Kenntnis der Geschlechtsverhaltnisse und der Samenflussigkeit wirbelloser Tiere*. Berlin.

Kormano, M. and Hovatta, O. (1972) Contractility and histochemistry of the myoid cell layer of the rat seminiferous tubules during postnatal development. *Z. Anat. Entwickl-Gesch.* **137**, 239–48.

Kostron, K. and Kukla, F. (1970) Morphological studies on non-mating male mink. *Acta Uni. Fac. Agron. Brno.* **18**, 725–32. In ABA 1972 (40 3487).

Kostron, K. and Kukla, F. (1971) Seasonal changes of testicular volume in mink. *Acta Uni. Fac. Agr. Brno.* **19**, 171–8. In ABA 1972 (40 3488).

Kretser, D. M. de (1967) The fine structure of the testicular interstitial cells in men of normal androgenic status. *Z. Zellforsch. Mikroskop. Anat.* **80**, 594–609.

Kretser, D. M. de (1968) The fine structure of the immature human testis in hypogonadotrophic hypogonadism. *Virchows Arch. Abt. B. Zellpath* **1**, 283–96.

Kretser, D. M. de, Kerr, J. B. and Paulsen, C. A. (1975) The peritubular tissue in the normal and pathological human testis. An ultrastructural study. *Biol. Reprod.* **12**, 317–24.

Lacy, D. and Rotblat, J. (1960) Study of normal and irradiated boundary tissue of the seminiferous tubules of the rat. *Exptl. Cell Res.* **21**, 49–70.

Langford, G. A. and Heller, C. G. (1973) Fine structure of muscle cells of the human testicular capsule: Basis of testicular contractions. *Science* **179**, 573–5.

Lanz, T. von and Neuhäuser, G. (1963) Metrische Untersuchungen an den Tubuli contorti des menschlichen Hodens. *Z. Anat. Entwicklungsgeschichte* **123**, 462–89.

LaVallette, St. George von (1876) Uber die Genese der Samenkorper. *Arch. Mikroscop. Anat.* **12**, 797–822.

Lechleiter, R. R. (1959) Sex ratio, age classes and reproduction of the black-tailed jack rabbit. *J. Mammal.* **40**, 63–81.

Leeson, C. R. (1963) Observations on the fine structure of rat interstitial tissue. *Acta Anat.* **52**, 34–48.

Leeson, T. S. and Adamson, L. (1962) The mammalian tunica vaginalis testis: its fine structure and function. *Acta Anat.* **51**, 226–40.

Leeson, T. S. and Cookson, F. B. (1974) The mammalian testicular capsule and its muscle elements. *J. Morphol.* **144**, 237–54.

Leeson, C. R. and Leeson, T. S. (1963) The postnatal development and differentiation of the boundary tissue of the seminiferous tubule of the rat. *Anat. Rec.* **147**, 243–59.

Leuwenhoeck, A. van (1678) Observationes de natis e semine genitali animalculis. *Phil. Trans. Roy. Soc.* (London) **12**, 1040–3.

Leydig, F. (1850) Zur Anatomie der männlichen Geschlechtsorgane und Analdrüsen der Säugethiere. *Z. Wiss. Zool.* **2**, 1–57.

Liang, D. S. (1966) Anatomical structure of the testicular tubules. *Invest. Urol.* **4**, 285–7.

Lincoln, G. A. (1971a) The seasonal reproductive changes in the red deer stag (*Cervus elaphus*). *J. Zool.* (London) **163**, 105–23.

Lincoln, G. A. (1971b) Puberty in a seasonally breeding male, the red deer stag (*Cervus elaphus* L.). *J. Reprod. Fertil.* **25**, 41–54.

Lincoln, G. A. (1974) Reproduction and 'March madness' in the brown hare (*Lepus europaeus*). *J. Zool.* (London) **174**, 1–14.

Lofts, B. (1960) Cyclical changes in the distribution of the testis lipids of a seasonal mammal (*Talpa europaea*). *Quart. J. Microscop. Sci.* **101**, 199–205.

Lubarsch, O. (1896) Uber das Vorkommen Krystallinmischer und Krystalloider Bildungen in den Zellen des menschlichen Hodens. *Arch. Pathol. Anat. Physiol.* **145**, 316–38.

Macfarlane, W. V. (1970) Seasonality of conception in human populations. *Biometeorology* **4**, 167–82.

McCord, R. G. (1970) Fine structural observations of the peritibular cell layer in the hamster testis. *Protoplasma* **69**, 283–9.

McManus, I. C. (1976) Scrotal asymmetry in man and in ancient sculpture. *Nature* **259**, 426.

Marshall, A. J. (1947) The breeding cycle of an equatorial bat (*Pteropus giganteus* of Ceylon). *Proc. Linn. Soc.* (London) **159**, 103–11.

Marshall, A. J. and Wilkinson, O. (1956) Reproduction in the Orkney vole (*Microtus orcadensis*), under a six-hour day-length and other conditions. *Proc. Zool. Soc.* (London) **126**, 391–5.

Marshall, F. H. A. (1911) The male generative cycle in the hedgehog; with experiments on the functional correlation between the essential and accessory sexual organs. *J. Physiol.* (London) **43**, 247–60.

Martin, R. D. (1969) The evolution of reproductive mechanisms in primates. *J. Reprod. Fertil.* Suppl. **6**, 49–66.

Martinet, L. (1963) Etablissement de la spermatogenèse chez le campagnol des champs *Microtus arvalis* en fonction de la durée quotidienne d'éclairement. *Ann. Biol. Animale, Biochim. Biophys.* **3**, 343–52.

Martinet, L. (1966) Modification de la spermatogenèse chez le campagnol des champs *Microtus arvalis* en fonction de la durée quotidienne d'éclairement. *Ann. Biol. Animale, Biochim. Biophys.* **6**, 301–13.

Martinet, L. (1968) Cycle saisonnier de reproduction du campagnol des champs (*Microtus arvalis*). In *Cycles génitaux saisonniers de mammifères sauvages*, Iers Entretiens de Chize. Masson and Cie Edt. 67–83.

Meschaks, P. and Nordkvist, M. (1962) On the sexual cycle in the reindeer male. *Acta Vet. Scand.* **3**, 151–62.

Michael, R. P. and Keverne, E. B. (1971) An annual rhythm in the sexual activity of the male rhesus monkey, *Macaca mulatta*, in the laboratory. *J. Reprod. Fertil.* **25**, 95–8.

Millar, R. P. and Glover, T. D. (1970) Seasonal changes in the reproductive tract of the male rock hyrax, *Procavia capensis*. *J. Reprod. Fertil.* **23**, 497–9.

Miller, R. E. (1939) The reproductive cycle in male bats of the species *Myotis lucifugus lucifugus* and *Myotis grisescens*. *J. Morphol.* **64**, 267–95.

Moore, C. R., Simmons, G. F., Wells, L. J., Zalesky, M. and Nelson, W. O. (1934) On the control of reproductive activity in an annual-breeding mammal (*Citellus tridecemlineatus*). *Anat. Rec.* **60**, 279–89.

Mossman, A. S. (1955) Reproduction of the brush rabbit in California. *J. Wildlife Management* **19**, 177–84.

Mossman, H. W., Lawlah, J. W. and Bradley, J. A. (1931) The male reproductive tract of the Sciuridae. *Am. J. Anat.* **51**, 89–155.

Mossman, H. W., Hoffman, R. A. and Kirkpatrick, C. M. (1955) The accessory genital glands of male gray and fox squirrels correlated with age and reproductive cycles. *Am. J. Anat.* **97**, 257–301.

Murakami, M. (1966) Elektronenmikroskopische Untersuchungen am interstitiellen Gewebe des Rattenhodens, unter besonderer Berucksichtigung der Leydigschen Zwichenzellen. *Z. Zellforsch. Mikroskop. Anat.* **72**, 139–56.

Nagano, T. (1966) Some observations on the fine structure of the Sertoli cell in the human testis. *Z. Zellforsch. Mikroskop. Anat.* **73**, 89–106.

Nagy, F. (1972) Cell division kinetics and DNA synthesis in the immature Sertoli cells of the rat testis. *J. Reprod. Fertil.* **38**, 389–95.

Neal, E. G. and Harrison, R. J. (1958) Reproduction in the European badger (*Meles meles* L.). *Trans. Zool. Soc.* (London) **29**, 67–130.

Neaves, W. B. (1973) Changes in testicular Leydig cells and in plasma testosterone levels among seasonally breeding rock hyrax. *Biol. Reprod.* **8**, 451–66.

Neville, A. M. and Grigor, K. M. (1976) Structure, function and development of the human testis. In *Pathology of the Testis*, ed. R. C. B. Pugh, 1–37. Oxford: Blackwell.

Nicander, L. (1963) Some ultrastructural features of mammalian Sertoli cells. *J. Ultrastruct. Res.* **8**, 190–1.

Nicander, L. (1967) An electron microscopical study of cell contacts in the seminiferous tubules of some mammals. *Z. Zellforsch. Mikroskop. Anat.* **83**, 375–97.

Niemi, M. and Kormano, M. (1965) Contractility of the seminiferous tubule of the postnatal rat testis and its response to oxytocin. *Ann. Med. Exp. Fenn.* **43**, 40–2.

Onstad, O. (1967) Studies on postnatal testicular changes, semen quality, and anomalies of the reproductive organs in the mink. *Acta Endocrinol. (Kbh.) Suppl.* **117**, 1–117.

Ortavant, R., Mauleon, P. and Thibault, C. (1964) Photoperiodic control of gonadal and hypophysial activity in domestic mammals. *Ann. N.Y. Acad. Sci.* **117**, 157–68.

Oslund, R. M. (1928) Seasonal modifications in testes of vertebrates. *Quart. Rev. Biol.* **3**, 254–70.

Pearson, O. P. (1944) Reproduction in the shrew (*Blarina brevicauda* Say). *Am. J. Anat.* **75**, 39–93.

Pfeiffer, E. W. (1956) The male reproductive tract of a primitive rodent *Aplondontia rufa. Anat. Rec.* **124**, 629–37.

Plato, J. (1897) Die interstitiellen Zellen des Hodens und ihre physiologische Bedeutung. *Arch. Mikroskop. Anat. Entwicklungsmech.* **48**, 280–304.

Prevost, J. L. and Dumas, J. B. A. (1824) Nouvelle théorie de la génération. *Ann. Sci. Nat.* (Paris) **1**, 1–29, 167–87 and 274–92.

Price, M. (1953) The reproductive cycle of the watershrew *Neomys rodiens bicolor* Shaw. *Proc. Zool. Soc.* (London) **123**, 599–621.

Racey, P. A. (1974a) Ageing and assessment of reproductive status of Pipistrelle bats, *Pipistrellus pipistrellus. J. Zool.* (London) **173**, 264–71.

Racey, P. A. (1974b) The reproductive cycle in male noctule bats, *Nyctalus noctula. J. Reprod. Fertil.* **41**, 169–82.

Racey, P. A. and Tam, W. H. (1974) Reproduction in male *Pipistrellus pipistrellus* (Mammalia: Chiroptera). *J. Zool.* (London) **172**, 101–22.

Rasmussen, A. T. (1917) Seasonal changes in the interstitial cells of the testis in the woodchuck (*Marmota monax*). *Am. J. Anat.* **22**, 475–515.

Raynaud, A. (1950) Variations saisonnières des organes génitaux des mulots (*Apodemus sylvaticus* L.) de sexe mâle. Données pondérales et histologiques. *C. Rend. Seance Soc. Biol.* **144**, 941–5.

Reddi, A. H. and Prasad, M. R. N. (1968) The reproductive cycle of the male Indian palm squirrel, *Funambulus pennanti* Wroughton. *J. Reprod. Fertil.* **17**, 235–45.

Regaud, C. (1901) Etude sur la structure des tubes seminiferes et sur la spermatogenèse chez les mammifères. *Arch. Anat. Microscop. Morphol. Exptl.* **4**, 101–55 and 231–380.

Regaud, C. (1904) Etat des cellules interstitielles du testicule chez la taupe pendant la periode de spermatogenèse et pendant l'état de repos des canalicules seminaux. *Compt. Rend. Assoc. Anat.* **56**, 54.

Reinke, F. (1896) Beitrage zur Histologic des Menschen. I. Uber Krystalloidbildungen in den interstitiellen Zellen des menschlichen Hodens. *Arch. Mikroskop. Anat. Entwicklungsmech.* **47**, 34–44.

Reiter, R. J. (1967) Effect of pineal grafts, pinealectomy, and denervation of the pineal gland on the reproductive organs of male hamsters. *Neuroendocr.* **2**, 138–46.

Reiter, R. J. (1968) The pineal gland and gonadal development in male rats and hamsters. *Fertil. Steril.* **19**, 1009–17.

Reiter, R. J. (1972) Evidence for refractoriness of the pituitary–gonadal axis to the pineal gland in golden hamsters and its possible implications in annual reproductive rhythms. *Anat. Rec.* **173**, 365–71.

Reiter, R. J. (1973) Pineal control of a seasonal reproductive rhythm in male golden hamsters exposed to natural daylight and temperature. *Endocrinology* **92**, 423–30.

Reiter, R. J. and Hester, R. J. (1966) Interrelationships of the pineal gland, the superior cervical ganglia and the photoperiod in the regulation of the endocrine systems of hamsters. *Endocrinology* **79**, 1168–70.

Reiter, R. J., Sorrentino, S., Ralph, C. L., Lynch, H. J., Mull, D. and Jarrow, E. (1971) Some endocrine effects of blinding and anosmia in adult male rats with observations on pineal melatonin. *Endocrinology* **88**, 895–900.

Relexans, M. C. and Carnivenc, R. (1967) Evolution pondérale du testicule du Blaireau européen (*Meles meles* L.) au cours du cycle génital annuel. *C. Rend. Soc. Biol.* **161**, 600–3.

Reynolds, H. G. (1960) Life history of Merriam's kangaroo rat in southern Arizona. *J. Mammal.* **41**, 48–58.

Rogers Brambell, F. W. (1935) Reproduction in the common shrew (*Sorex areneus* L.). II Seasonal changes in the reproductive organs of the male. *Phil. Trans. Roy. Soc.* (London) B. **225**, 51–62.

Rolshoven, E. (1937) Die funktionellen Strukturen des Hoden-bindegewebes. *Morph. Jahrb.* **79**, 235–74.

Rolshoven, E. (1945) Spermatogenese und Sertoli-syncytium. *Z. Zellforsch. Mikroskop. Anat. Abt. Histochem.* **33**, 439–60.

Roosen-Runge, E. C. (1951) Motions of the seminiferous tubules of rat and dog. *Anat. Rec.*

**109**, 413.

Ross, M. H. (1963) The fine structure of testicular interstitial cells in the rat. *Anat. Rec.* **145**, 277–8.

Ross, M. H. (1967) The fine structure and development of the peritubular contractile cell component in the seminiferous tubules of the mouse. *Am. J. Anat.* **121**, 523–58.

Ross, M. H. and Long, I. R. (1966) Contractile cells in human seminiferous tubules. *Science* **153**, 1271–3.

Rowlands, I. W. (1936) Reproduction of the bank vole (*Evotomys glareolus* Schreber). II. Seasonal changes in the reproductive organs of the male. *Phil. Trans. Roy. Soc.* (London) B. **226**, 99–120.

Rowlands, I. W. (1938) Preliminary note on the reproductive cycle of the red squirrel (*Sciurus vulgaris*). *Proc. Zool. Soc.* (London) A. **108**, 441–3.

Russell, L. (1977) Desmosome-like junctions between Sertoli and germ cells in the rat testis. *Am. J. Anat.* **148**, 301–12.

Russo, J. (1971) Fine structure of the Leydig cell during postnatal differentiation of the mouse testis. *Anat. Rec.* **170**, 343–56.

Saboureau, M. (1973) *Essai d'analyse quantitative et experimentale de l'activité endocrine génitale du Herisson mâle.* (Erinaceus europeus). Vemes Entretiens de Chize. MASSON and Cie Edts.

Sadleir, R. M. F. S. (1965) Reproduction in two species of kangaroo (*Macropus robustus* and *Megaleia rufa*) in the arid Pilbara region of Western Australia. *Proc. Zool. Soc.* (London) **145**, 239–61.

Sapsford, C. S. (1963) The development of the Sertoli cell of the rat and mouse, its existence as a mononucleate unit. *J. Anat.* **97**, 225–38.

Schmidt, F. C. (1964) Licht- und elektronenmikroskopische Untersuchungen am menschlichen Hoden und Nebennoden. *Z. Zelforsch. Mikroskop. Anat.* **63**, 707–27.

Schulze, C., Holstein, A. F., Schirren, C. and Körner, F. (1976) On the morphology of the human Sertoli cells under normal conditions and in patients with impaired fertility. *Andrologia* **8**, 167–78.

Sertoli, E. (1865) Dell'esistenzia di particolari cellule ramificate nei canalicoli seminiferi del testiculo humano. *Il Morgagni* **7**, 31–9.

Setchell, B. P. (1970) Testicular blood supply, lymphatic drainage and secretion of fluid. In *The Testis*, eds., A. D. Johnson, W. R. Gomes and N. L. Vandemark, I, 101–239. New York: Academic Press.

Setchell, B. P. (1977) Reproduction in male marsupials. In *The Biology of Marsupials*, eds., D. P. Gilmore and B. Stonehouse, 411–57. London: Macmillan.

Short, R. V. and Mann, T. (1966) The sexual cycle of a seasonally breeding mammal, the roebuck (*Capreolus capreolus*). *J. Reprod. Fertil.* **12**, 337–51.

Skinner, J. D. (1971) The effect of season on spermatogenesis in some ungulates. *J. Reprod. Fertil.* Suppl. **13**, 29–37.

Steinberger, A. and Steinberger, E. (1971) Replication pattern of Sertoli cells in maturing rat testis *in vivo* and in organ culture. *Biol. Reprod.* **4**, 84–7.

Stetson, M. H., Elliott, J. A. and Menaker, M. (1975) Photoperiodic regulation of hamster testis: circadian sensitivity to the effects of light. *Biol. Reprod.* **13**, 329–39.

Stieda, L. (1877) Uber den Bau des Menschen-Hoden. *Arch. Mikroskop. Anat.* **14**, 17–50.

Stieve, H. (1930) Mannliche Genitalorgane. In *Handbuch der mikroskopischen Anatomie des Menschen*, ed., W. von Mollendorff **7**, 2. Berlin: Springer.

Stoch, Z. G. (1954) The male genital system and reproductive cycle of *Elephantulus myurus jamesoni* (Chubb). *Phil. Trans. Roy. Soc.* (London) B. **238**, 99–126.

Suvanto, O. and Kormano, M. (1970) The relation between *in vitro* contractions of the rat seminiferous tubules and the cyclic stage of the seminiferous epithelium. *J. Reprod. Fertil.* **21**, 227–32.

Suzuki, F. and Racey, P. A. (1976) Fine structural changes in the epididymal epithelium of moles (*Talpa europaea*) throughout the year. *J. Reprod. Fertil.* **47**, 47–54.

Tandler, J. and Grosz, S. (1912) Ueber den Saisondimorphismus des Maulwurf hodens. *Arch. Entwicklungsmech. Organ.* **33**, 297–302.

Thibault, C., Courot, M., Martinet, L., Mauleon, P., du Mesnil du Buisson, F., Ortavant, R., Pelletier, J. and Signoret, J-P. (1966) Regulation of breeding season and estrous cycles by light and external stimuli in some mammals. *J. Anim. Sci.* **25**, *Suppl.* 119–39.

Turek, F. W., Elliott, J. A., Alvis, J. D. and Menaker, M. (1975) Effect of prolonged exposure to nonstimulatory photoperiods on the activity of the neuroendocrine–testicular axis of golden hamsters. *Biol. Reprod.* **13**, 475–81.

Turek, F. W., Alvis, J. D., Elliott, J. H. and Menaker, M. (1976a) Temporal distribution of serum levels of LH and FSH in adult male golden hamsters exposed to long or short days. *Biol. Reprod.* **14**, 630–1.

Turek, F. W., Desjardins, C. and Menaker, M. (1976b) Differential effects of melatonin on the testes of photoperiodic and nonphotoperiodic rodents. *Biol. Reprod.* **15**, 94–7.

Tsui, H. W., Tam, W. H., Lofts, B. and Phillips, J. G. (1974) The annual testicular cycle and androgen production *in vitro* in the masked civet cat (*Paguma l. larvata*). *J. Reprod. Fertil.* **36**, 283–93.

Vamburkar, S. A. (1958) The male genital tract of the Indian megachiropteran bat, *Cynopterus sphinx gangeticus*, And. *Proc. Zool. Soc.* (London) **130**, 57–77.

Vaughan, T. A. (1962) Reproduction in the plains pocket gopher in Colorado. *J. Mammal.* **43**, 1–13.

Vendrely, E., Guerillot, C., Basseville, C. and Da Lange, C. (1971) Poids testiculaire et spermatogenèse du hamster doré au cours du cycle saisonnier. *C. R. Soc. Biol.* **165**, 1562–5.

Venge, O. (1959) A short review of reproduction in the fox. *A.B.A.* **27**, 129–45.

Vilar, O., Perez del Cerro, M. I. and Mancini, R. E. (1962) The Sertoli cells as a 'bridge cell' between the basal membrane and the germinal cells. Histochemical and electron microscopic observations. *Exptl. Cell Res.* **27**, 158–61.

Weber, M. (1928) *Die Saugetiere*. Jena: G. Fischer.

Wells, L. J. (1935a) Seasonal sexual rhythm and its experimental modification in the male of the thirteen-lined ground squirrel (*Citellus tridecemlineatus*). *Anat. Rec.* **62**, 409–47.

Wells, L. J. (1935b) Prolongation of breeding capacity in males of an annual breeding wild rodent (*Citellus tridecimlineatus*) by constant low temperature. *Anat. Rec.* **64**, suppl. 1, 138.

Wells, L. J. and Zalesky, M. (1940) Effects of low environmental temperature on the reproductive organs of male mammals with annual aspermia. *Am. J. Anat.* **66**, 429–47.

Wislocki, G. B. (1933) Location of the testes and body temperatures in mammals. *Quart. Rev. Biol.* **8**, 385–96.

Wislocki, G. B. (1949) Seasonal changes in the testes, epididymis and seminal vesicles of deer investigated by histochemical methods. *Endocrinol.* **44**, 167–89.

Wojcik, K. (1967) Mechanism of spermatozoa movement in sex organs of the male. Contractibility of seminiferous tubules. *Bull. Acad. Polon. Sci., Ser. Sci. Biol.* **15**, 483–6.

Woolley, P. (1975) The seminiferous tubules in dasyurid marsupials. *J. Reprod. Fertil.* **45**, 255–61.

Wright, P. L. (1969) The reproductive cycle of the male American badger, *Taxidea taxus. J. Reprod. Fertil.* Suppl. **6**, 435–45.

Yeates, N. T. M. (1949) The breeding season of the sheep with particular reference to its modification by artificial means using light. *J. Agric. Sci.* **39**, 1–43.

Yerger, R. W. (1955) Life history notes on the eastern chipmunk *Tamias striatus lysteri* (Richardson) in central New York. *Am. Midland Naturalist,* **53**, 312–23.

# 2 Development of the testis

## 2.1      Origin of different cell types in the testis

The development of the testis is intimately associated with the growth and regression of the two fetal kidney-like organs, the pronephros and the mesonephros, and the development of the true kidney or metanephros. These three organs develop successively in a band of tissue, the nephrogenic cord, which arises in the mesoderm just ventral to the notochord. The pronephros develops first, at the cranial end of the nephrogenic cord and its duct runs back towards the cloaca. In most mammals, the pronephros soon begins to degenerate; the mesonephros develops just caudal to it and takes over its duct. When fully developed in the fetus the mesonephros is a relatively large organ extending along almost the entire length of the coelomic cavity.

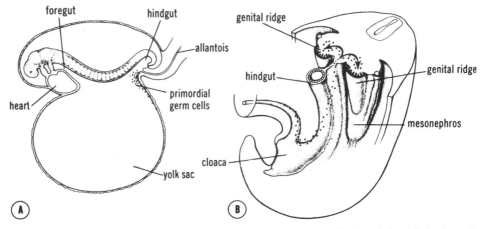

Figure 2.1 Diagram illustrating the migration of the primordial germ cells from their origin in the wall of the yolk sac along the hindgut into the genital ridge. (Reproduced from Langman, 1975, after a drawing by Witschi, 1956)

The gonad first develops as a thickening of the coelomic epithelium on the medio-ventral aspect of the mesonephros. These thickenings are known as genital ridges and are invaded by the primordial germ cells which develop in the yolk sac, lateral to the base of the vitelline artery and close to the allantoic diverticulum (Figure 2.1). These cells are identified by their large size and round shape, their high content of 'glycogen' and alkaline phosphatase. They migrate by active amoeboid movement through the hindgut wall and thence along the mesentery to the genital ridge (Witschi, 1948; Chiquoine, 1954; Mintz, 1957; Chretien, 1966). According to Gier and Marion (1969, 1970), these cells settle into the underlying coelomic epithelium and become indistinguishable from the original epithelial cells. This germinal epithelium is a low columnar epithelium, contrasting with the cuboidal cells of the surrounding coelomic epithelium. Before the germinal epithelium is fully formed at its posterior limit, some indentations begin to form near the anterior end. Gradually these indentations can be seen further along the ridge and they change progressively from hollow invaginations into solid cores of cells which grow towards the mesonephros. These cords are known as primary epithelial cords or medullary cords. The cords elongate as the gonad becomes more compact. Individual cords become more difficult to identify when they lose their connection with the germinal epithelium because of the interposition of a dense fibrous layer, which becomes the tunica albuginea. Other authors, e.g. Jost *et al.* (1972), (1973) and Jirasek (1967), found that the primordial germ cells penetrate into the gonadal mesenchyme and the cords can first be seen in the depths of the testes near the mesonephros; they therefore doubt whether the cords grow in from the epithelium.

Up to this stage, the development of the testis is indistinguishable from that of the ovary. In normal males, a gene on the Y-chromosome leads to the formation of H-Y antigen, which is thought to act as a 'cell recognition signal' that is essential for virilization of the indifferent embryonic gonad (Wachtel *et al.*, 1975; Wachtel, 1976; Ohno *et al.*, 1976).

In the ovary no tunica albuginea develops, the primary epithelial cords regress and a second set of epithelial cords give rise to the germinal cortex of the ovary. In the male, the rete testis also begins to develop at about this time; its intratesticular portion is derived from the gonadal blastema as modified portions of the sex cords and its extratesticular portion arises from 'rete blastema', which is probably of mesonephric origin (Figure 2.2; Roosen-Runge, 1961a). In females at about this

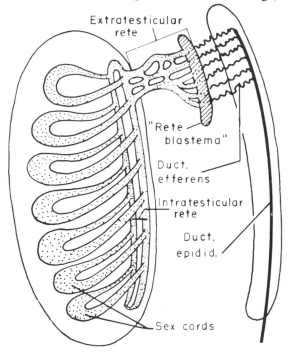

Figure 2.2 A diagram of the rat fetal testis illustrating the development of the sex cords and the dual origin of the rete testis, partly from an intratesticular source and partly from rete blastema. (Reproduced from Roosen-Runge, 1961a)

stage, the rete ovarii probably acts as a trigger for the initiation of meiosis (Byskov and Lintern-Moore, 1973; Byskov, 1974, 1976a, b; O and Baker, 1976) but the function of the rete in the male is not clear.

The next stage in development of the testis involves the inclusion of the primordial germ cells into the epithelial cords (which from now on are usually known as sex cords, seminiferous cords or seminiferous tubules). The germ cells take up a central position surrounded by the sustentacular cells and a prominent basement membrane, and they remain thus until spermatogenesis begins at the beginning of puberty (Figure 2.3; Courot, 1971).

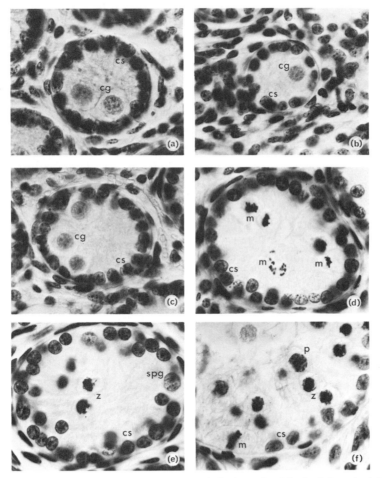

Figure 2.3  The development of the sex cords in the testis of the fetal sheep in: (a) 45-day old fetus; (b) 3-day old lamb; (c) 90-day old lamb—showing the primordial germ cells (cg) at the centre of the tubule surrounded by the Sertoli cells (cs). In (d), mitoses (m) have begun in the germinal cells, which are still in the centre of the sex cord. In (e), the germinal cells have migrated to the tubular wall and become spermatogonia (spg). Some of their daughter cells have entered meiosis and reached zygotene stage (z). In (f), some of the meiotic cells have reached pachytene (p) and spermatogonial mitoses (m) are continuing. Tubules like those in (d), (e) and (f) can be found in Ile-de-France lambs between 80 and 120 days old, i.e. between 25 and 30 kg body weight. (Reproduced from Courot, 1971)

The Leydig cells do not become obvious until some time after the formation of the seminiferous cords; then they appear as specializations of the mesenchymal cells of the stroma between the seminiferous cords, after these have separated from the germinal epithelium (Roosen-Runge and Anderson, 1959). The numbers and area of the Leydig cells increase up to parturition (or thereabouts depending on the species); thereafter they regress until they enlarge again at puberty.

## 2.2    Testicular migration

At the time of the development of the sex cords, the testis is still located in what will become the thoracic cavity. The true kidney or metanephros is differentiating at the posterior end of the abdominal cavity where cells from the posterior end of the nephrogenic cord unite with the ureteric bud, an outgrowth from the caudal end of the mesonephric duct. In succeeding stages the kidney moves forward and the testis moves back. The mesonephros degenerates and the testis appropriates its duct, the Wolffian duct, which it transforms into the efferent ducts, epididymis, ductus deferens, ampullae and seminal vesicles.

The degree of movement of the testis varies according to species, but most of the embryological observations have been made on animals with scrotal testes; little is known about what happens in those animals in which the testis migrates from its original position but remains within the abdomen. In all mammals, the testis and kidney move relative to one another so that in the adult the testis always lies caudal to the kidney, but this does not happen in birds.

In animals with scrotal testes, testicular descent has been divided into three stages: nephric displacement, transabdominal passage and inguinal passage (see Gier and Marion, 1969, 1970). In the first stage, the testis is held close to the caudal end of the abdominal cavity, by its firm posterior attachment, the posterior gonadal ligament, the mesonephric duct and the gubernaculum, while the rest of the animal grows. Meanwhile, the metanephros migrates anteriorly, dorsal to the mesonephros in eutherian mammals but ventral in marsupials. The forces causing this movement of the kidney are very ill-defined, and it is interesting to note that the kidney changes its blood supply as it migrates, whereas the testis retains its original arterial supply, which becomes elongated as the testis descends.

Transabdominal passage is a more positive stage in testicular descent and is usually said to be achieved by the pull of the gubernaculum. The gubernaculum is a fibrous band which develops retroperitoneally from the remains of the nephrogenic cord. It runs from the posterior tip of the mesonephros beyond the end of the peritoneal cavity and is connected to the testis along the mesonephros by the posterior gonadal ligament, another fibrous derivative of the nephrogenic cord. The gubernaculum contains *no* muscle fibres so the sources of its 'pull' are obscure. What probably happens is that the rest of the body continues to grow but the gubernaculum does not, so the testis changes its relative position (Figure 2.4).

The third and final stage of descent, inguinal passage, involves the formation of the processus vaginalis around the posterior attachment of the gubernaculum.

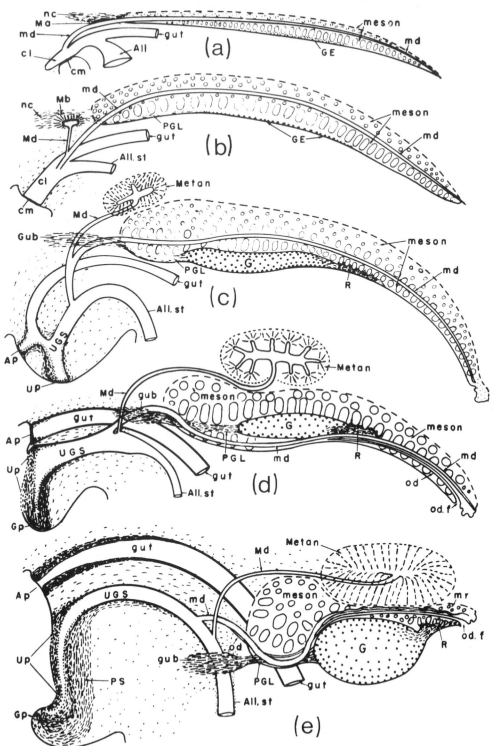

Figure 2.4

The processus develops as an evagination of the coelomic lining within a few days of the constriction of the umbilical ring and the virtual separation of the intra-embryonic and extra-embryonic coelom. Again, nothing is known of the mechanism by which the evagination is produced. Some authors talk of herniation because of an increase in fluid pressure inside the abdomen, but there is no experimental evidence for this theory. Whatever the cause, the processus enlarges to form the inguinal canal; the testis is drawn by the gubernaculum into the canal and thence into the scrotum with its blood supply and the ductus deferens following (Figure 2.5). The cremaster muscle develops outside the processus vaginalis.

Recently, Wensing (1968) has proposed an interesting alternative mechanism. He suggests that the end of the gubernaculum enlarges outside the inguinal canal by imbibition of fluid and, like a balloon being blow up with a part projecting through a narrow tube, it thereby draws itself and the attached testis through the canal (Figure 2.6).

In the human, the lower portion of the processus vaginalis persists as the tunica vaginalis of the testis but the portion near the inguinal canal becomes obliterated by apposition and fusion of its walls (Figure 2.7). In some other mammals, e.g. the domestic animals and macropod marsupials, the canal remains patent but too small for the testis to return to the abdomen, whereas in rodents and lagomorphs, it is possible to return the testis to the abdomen by gentle pressure on the scrotum, or the animal can do so by contracting its cremaster muscle.

---

Figure 2.4 Diagrams illustrating the development and the early stages in the descent of the testis in the dog. (a) 21 days—the primordial germinal cells have reached the germinal ridge on the edge of the mesonephros. (b) 23 days—the metanephros or true kidney begins to develop at the caudal end of the nephrogenic cord, joined to the cloaca by the metanephric duct. (c) 25 days—the testis is now a defined swelling on the edge of the mesonephros which has reached its maximal size. The gubernaculum is beginning to develop and the metanephros has begun to move cranially. The rete testis has appeared. (d) 28 days—the testis and metanephros continue to develop, and the mesonephros has begun to degenerate. The testis and metanephros are almost level, because of continued cranial movement of the latter. The gubernaculum is changing its point of insertion towards the inguen, and the Müllerian duct has developed. (e) 33 days—degeneration of the mesonephros and development of the metanephros and testis continue. The insertion of the gubernaculum has reached the inguen.

| Ap: | anal plate | Ma: | metanephric anlage | od.f. | fimbria of |
| All: | allantois | Mb: | metanephric bud | | Müllerian duct |
| All. st.: | allantoic stalk | Md: | metanephric duct | PGL: | posterior gonadal |
| cl: | cloaca | md: | mesonephric duct | | ligament |
| cm: | cloacal membrane | meson: | mesonephros | PS: | shaft of penis |
| G: | testis | Metan: | metanephros | R: | rete testis |
| GE: | primordial germ cells | mr: | mesonephric remnant | UGS: | urogenital sinus |
| Gub: | gubernaculum | nc: | nephrogenic cord | Up.: | urethral plate |
| Gp: | genital peduncle | od: | müllerian duct | | |

(Reproduced from Gier and Marion, 1969)

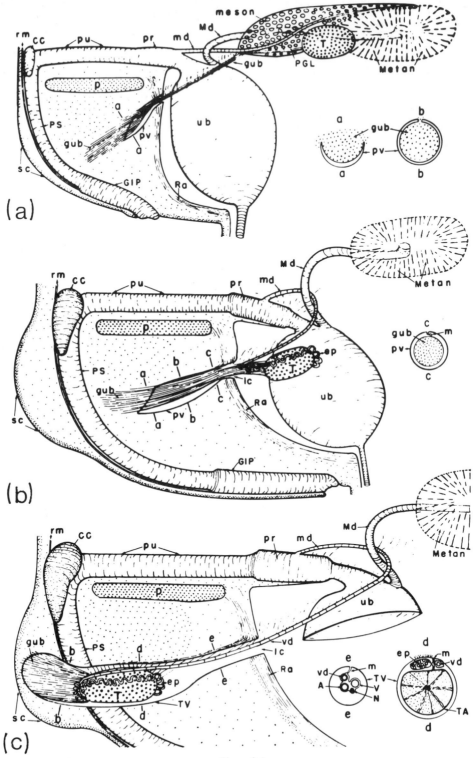

Figure 2.5

## 2.3    Function of the testis in the fetus and infant

The spermatogenic function of the testis does not begin until puberty but the fetal and prepubertal testes have several important endocrinological functions. In general, a mammal develops as a female unless directed otherwise, whereas in birds the situation is reversed and femaleness has to be imposed on a basically masculine trend (see Jost, 1953; Jost *et al.*, 1972).

### 2.3.1    Development of the Wolffian duct

As the mesonephros degenerates, its duct becomes attached to the rete testis at the urogenital union. The testis then exerts a local influence on the duct, causing it to differentiate into efferent ducts, epididymis and ductus deferens (Figure 2.8). The effect appears to be local because if a rabbit, rat or mouse fetus is unilaterally castrated before sexual differentiation has occurred, the Wolffian duct develops only on the unoperated side, and if a testis is grafted into a female fetus, near one ovary, the Wolffian duct develops only on that side. The external genitalia also remain female in the absence of the testes, but in this case the effect is not unilateral and one testis is sufficient to produce complete masculinization (see Jost, 1953; Jost *et al.*, 1973). Marsupials are born before the sexes are obviously different, but castration early in pouch life does not affect growth of the Wolffian duct. However, sexual differentiation has probably begun in these animals and more experiments are needed in younger animals (see Setchell, 1977).

The effects of the testis on the Wolffian duct seem to be due to androgen production by the testis. The fetal testis in many species is larger than the testis after birth (Figure 2.8; and see, for example, Cole *et al.*, 1935; Amoroso *et al.*, 1951; Harrison

---

Figure 2.5 Diagrams illustrating the later stages of testicular descent in the dog. (a) 36 days—the inguinal canal has formed with the gubernaculum inserted at its end, the mesonephros has degenerated further and the testis has begun to move caudally. (b) At birth—the inguinal canal has developed, still with the gubernaculum inserted at its end and the testis has moved well away from the metanephros to the mouth of the inguinal canal. (c) 26 days after birth—the testis has passed through the inguinal canal and has almost reached its final position. a, b, c, d, e Cross sections at the levels shown in (a), (b) and (c). Lettering as for Figure 2.4 with the following additions:

| | | | | | | | |
|---|---|---|---|---|---|---|---|
| A: | testicular artery | p: | pelvis | T: | testis |
| cc: | Cowper's glands | pr: | prostate gland | TA: | tunica albuginea |
| ep: | epididymis | pu: | proximal urethra | TV: | tunica vaginalis |
| G1P: | glans penis | pv: | processus vaginalis | ub: | urinary bladder |
| Ic: | inguinal canal | Ra: | rectus abdominis muscle | V: | testicular vein |
| m: | mesorchium | rm: | retractor muscle | vd: | ductus deferens |
| N: | testicular nerve | sc: | scrotum | | |

(Reproduced from Gier and Marion, 1969)

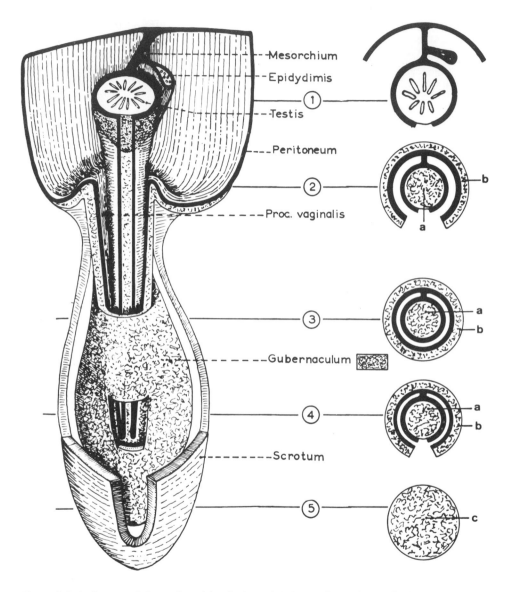

Figure 2.6 A diagram of the testis and inguinal canal during testicular descent in the pig. a, Gubernaculum proper b, Extra-vaginal part of the gubernaculum c, Infra-vaginal part of the gubernaculum. (Reproduced from Wensing, 1968)

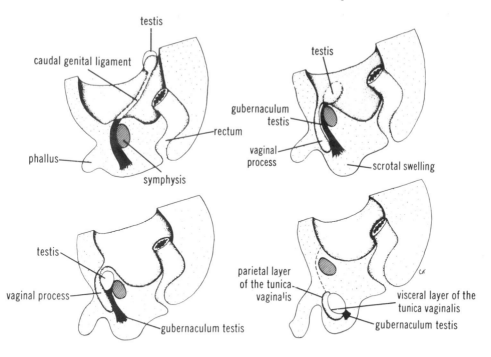

Figure 2.7 Diagrams illustrating descent of the testis in man showing the development of the vaginal process through which the testis descends. Once the testis is in the scrotum in man, the walls of the process fuse and the testis is in a small peritoneal sac isolated from the main peritoneal cavity. (Reproduced from Langman, 1975)

*et al.*, 1952) and, furthermore, higher concentrations of androgens have been demonstrated in the testes and blood of fetuses than in prepubertal animals of several species (rat, sheep, cattle: see Sections 6.1.4.2 and 6.1.5.1). Implants of testosterone or injections of testosterone into the mother can stimulate development of the Wolffian ducts in female or castrated fetuses (Jost, 1965; Short, 1974) and therefore testosterone or some closely related steroid is probably the natural active principle. Furthermore, the normal development of the Wolffian duct in rabbits can be blocked by injection of the anti-androgen cyproterone acetate (Figure 2.9)—although curiously this substance is not completely effective in rats (see Section 6.1.8). In testicular feminization, the male tract does not develop, presumably because of the inherent insensitivity of the tissues in these animals to androgens (see Section 11.5.7).

### 2.3.2 Regression of the Müllerian duct

The para-mesonephric or Müllerian duct, which gives rise to the female reproductive tract, arises as an invagination of the coelomic epithelium which grows towards and finally opens into the urogenital sinus, running roughly parallel to the mesonephric or

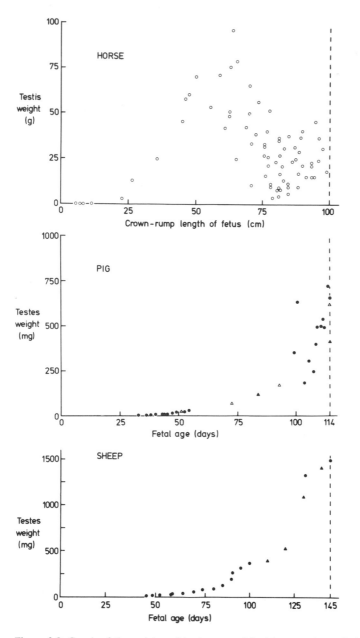

Figure 2.8 Graph of the weights of both testes of fetal horses, pigs and sheep at different fetal ages. Note the peak in the size of the horse testis before birth, whereas in the other two species shown, the testis increases before birth but does not decrease again. Data for horses from Cole *et al.* (1933); pigs from Segal and Raeside (1975); Raeside and Sigman (1975),●; Ullrey *et al.* (1965)△; and Booth (1975)▲; sheep from Attal (1969); Attal *et al.* (1972)●; and Courot (1971)▲.

Wolffian duct. In the female or castrated male, the Müllerian ducts develop into the fallopian tubes, uterus and part of the vagina. In the male, or in a female into which a testis has been grafted near the ovary, the Müllerian duct regresses and in the mature male remains only at the caudal end as part of the utriculus prostaticus and at the gonadal end as the appendix of the testis—a blind-ended duct lying on the surface of the testis near the efferent ducts (Jost, 1965).

Clearly the testis produces something which causes the Müllerian duct to regress by a local direct action. This substance is not testosterone because while development of the Wolffian duct can be produced in castrated fetuses by injections of testosterone, this treatment leaves the Müllerian ducts unaffected and the two duct systems develop side by side. Conversely, in rabbit fetuses treated with cyproterone acetate and in androgen insensitive animals (testicular feminization) the Wolffian duct does not develop because of the lack of androgen effect, but the Müllerian duct regresses normally so that these animals have testes but no other sexual organs (Figure 2.9). Research is still in progress to try and isolate the Müllerian duct inhibitor, but it seems to be a fairly large molecule which cannot pass through a dialysis membrane and it is probably a polypeptide (Josso, 1972a, 1973; Josso et al., 1975, 1977; Picard and Josso, 1976). The inhibitor substance is produced by the rat testis until the second postnatal week (Picon, 1969; Jost et al., 1972) when the first meiotic changes are beginning and by calf testes from day 110 of gestation until 6 weeks after birth (Donahoe et al., 1977). Activity could also be detected in human fetal testes less than 28 weeks old; there was variable activity in older fetal testes and none detectable in post-natal tissue (Josso, 1972b). The hormone was produced by bovine Sertoli cells cultured from fetal testis (Blanchard and Josso, 1974) and the activity in human fetal testis was unaffected by X-irradiation in vitro which reduced the numbers of gonocytes, but did not affect the Sertoli cells (Josso, 1974).

### 2.3.3    Establishment of 'hypothalamic sex' and 'behavioural sex'

We have just seen that the genital tract in mammals develops into a female tract unless influenced to become male. Similarly, the hypothalamus shows cyclical patterns of activity characteristic of the female unless it is exposed to certain steroids during a critical period. In the rat, this period is during the first few days of postnatal life, which is very convenient for the experimenter. If males are castrated at birth, ovaries grafted into them later show cyclical changes, whereas ovaries grafted into normal males become permanently luteinized. Similarly, if females or castrated males are injected with testosterone during the first days after birth, the hypothalamus does not develop its cyclical pattern of activity in later life. This effect is not produced by 5a-dihydrotestosterone, which is probably the active androgen in organs such as the prostate, where it is formed from serum testosterone. The effect of neonatally administered testosterone can be mimicked by injections of oestrogens (note that oestradiol is formed in the hypothalamus from testosterone, but not from 5a-dihydrotestosterone) and prevented with anti-oestrogens (Pfeiffer, 1936; Gorski, 1963; Harris, 1964; Barraclough, 1967; McDonald and Doughty, 1972, 1974).

In rats the adult patterns of male sexual behaviour are also set in neonatal or fetal

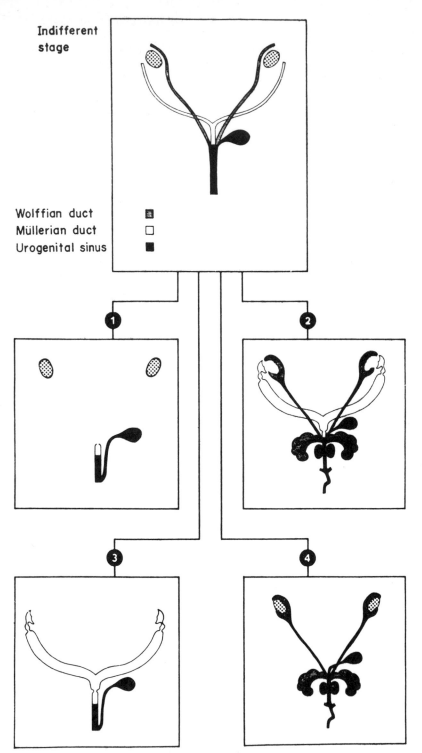

Figure 2.9

life by testicular hormones (Beach, 1975) and female sexual behaviour is reduced in female rats treated with androgens or oestrogens during the neonatal period (Doughty *et al.*, 1975; van der Schoot *et al.*, 1976).

Genetically female fetal sheep, masculinized by implanting testosterone into their mothers between 30 and 70 days of pregnancy, also show masculine sexual behaviour when mature (Clarke, 1977).

## References to Chapter 2

Allen, B. M. (1904) The embryonic development of the ovary and testis of the mammals. *Am. J. Anat.* **3**, 89–153.

Amoroso, E. C., Harrison, R. J., Harrison Matthews, L. and Rowlands, I. W. (1951) Reproductive organs of near-term and newborn seals. *Nature* (London) **168**, 771–2.

Ancel, P. and Bouin, P. (1926) Recherches experimentales sur l'origine des gonocytes dans le testicule des mammifères. *Compt. Rend. Assoc. Anat.* **21**, 1–11.

Aron, M. (1921) Sur la glande interstitielle du testicule embryonaire chez les mammifères. *Compt. Rend. Soc. Biol.* **35**, 107–10.

Attal, J. (1969) Levels of testosterone, androstenedione estrone and estradiol-17$\beta$ in the testes of fetal sheep. *Endocrinology* **85**, 280–9.

Attal, J. and Courot, M. (1963) Développement testiculaire et établissement de la spermatogenèse chez le taureau. *Ann. Biol. Animale, Biochim., Biophys.* **3**, 219–41.

Attal, J., Andre, D. and Engels, J. A. (1972) Testicular androgen and oestrogen levels during the postnatal period in rams. *J. Reprod. Fertil.* **28**, 207–12.

Baillie, A. H. (1964) Further observations on the growth and histochemistry of the Leydig tissues in the postnatal prepubertal mouse testis. **98**, 403–19.

Baillie, A. H. (1960) The interstitial cell in the testis of the foetal sheep. *Quart. J. Microscop. Sci.* **101**, 475–80.

Barraclough, C. A. (1967) Modifications in reproductive function after exposure to hormones during the prenatal and early postnatal period. In *Neuroendocrinology*, eds. L. Martini and

W. F. Ganong, vol. II, 61–100, New York: Academic Press.

Figure 2.9 Diagram illustrating the effects of testicular hormones on the development of the genital tract from the indifferent stage in the normal male (4) in which the Müllerian duct regresses and the Wolffian duct develops into the male system; in the castrate male (3) in which the Müllerian duct persists and Wolffian duct does not develop; in the castrate male treated with testosterone (2) in which both duct systems develop; and in the androgen insensitive pseudo-hermaphrodite or in an animal treated with an anti-androgen (1), in which the male structures do not develop because of the failure of androgen action, but the Müllerian duct regresses because the regression is not dependent on androgens but on other testicular hormones. (Adapted from Neumann *et al.*, 1969)

Bascom, K. F. (1923) The interstitial cells of the gonads of cattle, with especial reference to their embryonic development and significance. *Am. J. Anat.* **31**, 223–59.

Bascom, K. F. and Osterud, H. L. (1927) Quantitative studies of the testis. III: A numerical treatment of the development of the pig testis. *Anat. Rec.* **37**, 63–82.

Beach, F. A. (1975) Hormonal modification of sexually dimorphic behaviour. *Psychoendocrinology* **1**, 3–23.

Beaumont, H. M. and Mandl, A. M. (1963) A quantitative study of primordial germ cells in the male rat. *J. Embryol. Exptl. Morphol.* **11**, 715–40.

Black, V. H. and Christensen, A. K. (1969) Differentiation of interstitial cells and Sertoli cells in fetal guinea pig testes. *Am. J. Anat.* **124**, 211–38.

Blanchard, M. G. and Josso, N. (1974) Source of the anti-Müllerian hormone synthesized by the fetal testis: Müllerian-inhibiting activity of fetal bovine Sertoli cells in tissue culture. *Pediat. Res.* **8**, 968–71.

Booth, W. D. (1975) Changes with age in the occurrence of $C_{19}$ steroids in the testis and submaxiliary gland of the boar. *J. Reprod. Fertil.* **42**, 459–72.

Bouin, P. and Ancel, P. (1903) Sur la signification de la glande interstitielle du testicule embryonnaire. *Compt. Rend. Soc. Biol.* **55**, 1682–4.

Brown-Grant, K., Fink, G., Greig, F. and Murray, M. A. F. (1975) Altered sexual development in male rats after oestrogen administration during the neonatal period. *J. Reprod. Fertil.* **44**, 25–42.

Burlet, H. M. de (1921) Zur Entwicklung und Morphologie des Säugerhodens. II. Marsupialer. *Z. Anat. Entwicklungsgeschichte.* **61**, 19–31.

Burlet, H. M. de and deRuiter, H. J. (1920–21) Zur Entwicklung und Morphologie des Saugerhodens. I Der Hoden von *Mus musculus*. *Anat. Heft.* **59**, 321–83.

Burns, R. K., Jr. (1941) The origin of the rete apparatus in the opossum. *Science* **94**, 142–4.

Byskov, A. G. S. (1974) Does the rete ovarii act as a trigger for the onset of meiosis? *Nature* (London) **252**, 396–7.

Byskov, A. G. S. (1976a) Induction of meiosis in fetal mouse testis *in vitro*. *Devel. Biol.* **52**, 193–200.

Byskov, A. G. S. (1976b) The role of the rete ovarii in meiosis and follicle formation in the cat, mink and ferret. *J. Reprod. Fertil.* **45**, 201–9.

Byskov, A. G. S. and Lintern-Moore, S. (1973) Follicle formation in the immature mouse ovary: the role of the rete ovarii. *J. Anat.* **116**, 207–17.

Challis, J. R. G., Kim, C. K., Naftolin, F., Judd, H. L., Yen, S. S. C. and Benirschke, K. (1974) The concentrations of androgens, oestrogens, progesterone and luteinizing hormone in the serum of fetal calves throughout the course of gestation. *J. Endocrinol.* **60**, 107–15.

Chretien, F. C. (1966) Etude de l'origine, de la migration et de la multiplication des cellules germinales chez l'embryon de lapin. *J. Embryol. Exptl. Morphol.* **16**, 591–607.

Chiquoine, A. D. (1954) The identification, origin and migration of the primordial germ cells in the mouse embryo. *Anat. Rec.* **118**, 135–46.

Clarke, I. J. (1977) The sexual behaviour of prenatally androgenized ewes observed in the field. *J. Reprod. Fertil.* **49**, 311–5.

Clermont, Y. and Perey, B. (1957) Quantitative study of the cell population in the seminiferous tubules in immature rats. *Am. J. Anat.* **100**, 241–67.

Cole, H. H., Hart, G. H., Lyons, W. R. and Catchpole, H. R. (1933) The development and hormonal content of fetal horse gonads. *Anat. Rec.* **45**, 275–93.

Courot, M. (1971) Etablissement de la spermatogénèse chez l'agneau (*Ovis aries*): étude expérimentale de son contrôle gonadotrope; importance des cellules de la lignée Sertolienne. Thèse de doctorat d'état ès-sciences naturelles, Université Paris VI.

Davies, J., Dempsey, E. W. and Wislocki, G. B. (1957) Histochemical observations on the fetal ovary and testis of the horse. *J. Histochem. Cytochem.* **5**, 584–90.

Donahoe, P. K., Ito, Y., Price, J. M. and Hendren, W. H. (1977) Müllerian-inhibiting activity in bovine fetal newborn and prepubertal testes. *Biol. Reprod.* **16**, 238–43.

Doughty, C., Booth, J. E., McDonald, P. G. and Parrott, R. F. (1975) Effects of oestradiol-17$\beta$, oestradiol benzoate and the synthetic oestrogen RU 2858 on sexual differentiation in the neonatal female rat. *J. Endocrinol.* **67**, 419–24.

Felix, W. (1906) Die Entwickelung des Harnapparatus. In *Handbuch der vergleichenden und experimentellen Entwickelungslehre der Wirbeltiere*, ed. O. Hertwig, 3, I, 81–442. Jena: Fischer.

Felix, W. (1912) The development of the urinogenital organs. In *Manual of Human Embryology*, eds. F. Keibel and F. P. Mall, 2, 752, Philadelphia: Lippincott.

Franchi, L. L. and Mandl, A. M. (1964) The ultrastructure of germ cells in foetal and neonatal male rats. *J. Embryol. Exptl. Morphol.* **12**, 289–308.

Gier, H. T. and Marion, G. B. (1969) Development of mammalian testes and genital ducts. *Biol. Reprod.* Suppl. **1**, 1–23.

Gier, H. T. and Marion, G. B. (1970) Development of the mammalian testis. In *The Testis*, eds. A. D. Johnson, W. R. Gomes and N. L. Vandemark, New York, Academic Press, vol. I, 1–45.

Gillman, J. (1948) The development of the gonads in man, with a consideration of the role of endocrines and the histogenesis of ovarian tumors. *Contrib. Embryol.* (Carnegie Inst. Wash. Publ.) **32**, 81–131.

Gorski, R. A. (1963) Modification of ovulatory mechanisms by postnatal administration of estrogen to the rat. *Am. J. Physiol.* **205**, 842–4.

Gruenwald, P. (1934) Ueber Form und Verlauf der Keimstrange bei Embryonen der Säugetiere und des Menschen. I. Die Keimstrange des Hodens. *Z. Anat. Entwicklungsgeschichte* **103**, 1–19.

Gruenwald, P. (1942) The development of the sex cords in the gonads of man and mammals. *Am. J. Anat.* **70**, 359–97.

Hamerton, J. L., Dickson, J. M., Pollard, C. E., Grieves, S. A. and Short, R. V. (1969) Genetic intersexuality in goats. *J. Reprod. Fertil.* **7**, 25–51.

Hargitt, G. T. (1926) The formation of the sex glands and germ cells of mammals. II. The history of the male germ cells in the albino rat. *J. Morphol.* **42**, 253–303.

Harris, G. W. (1964) Sex hormones, brain development and brain function. *Endocrinology* **75**, 627–48.

Harrison, R. J., Harrison Matthews, L. and Roberts, J. M. (1952) Reproduction in some pinnepedia. *Trans. Zool. Soc.* (London) **27**, 447–531.

Holstein, A. F., Wartenberg, H. and Vossmeyer, J. (1971) Zur Cytologie der pränatalen Gonadenentwicklung beim Menschen. III. Die Entwicklung der Leydigzellen im Hoden von

Embryonen und Feten. *Z. Anat. Entwicklungsgeschichte.* **135**, 43–66.

Hoven, H. (1914) Histogenese du testicule des mammifères. Anat. Anz. **47**, 90–109.

Huckins, C. (1963) Changes in gonocytes at the time of initiation of spermatogenesis in the rat. *Anat. Rec.* **145**, 243.

Jirasek, J. E. (1967) The relationship between the structure of the testis and differentiation of the external genitalia and phenotype in man. *Ciba Colloq. Endocrinol.* **16**, 3–27.

Josso, N. (1970) Effect of cyanoketone, an inhibitor of $\Delta_5$-3$\beta$ hydroxysteroid dehydrogenase, on the reproductive tracts of male fetal rats, in organ culture. *Biol. Reprod.* **2**, 85–90.

Josso, N. (1972a) Permeability of membranes to the Müllerian-inhibiting substance synthesized by the human fetal testis *in vitro*: a clue to its biochemical nature. *J. Clin. Endocrinol. Metab.* **34**, 265–70.

Josso, N. (1972b) Evolution of the Müllerian-inhibiting activity of the human testis. Effect of fetal, perinatal and postnatal human testicular tissue on the Müllerian duct of the fetal rat in organ culture. *Biol. Neonat.* **20**, 368–79.

Josso, N. (1973) *In vitro* synthesis of Müllerian-inhibiting hormone by seminiferous tubules isolated from the calf fetal testis. *Endocrinology* **93**, 829–34.

Josso, N. (1974) Müllerian-inhibiting activity of human fetal testicular tissue deprived germ cells by *in vitro* irradiation. *Pediat. Res.* **8**, 755–8.

Josso, N., Forest, M. G. and Picard, J-Y. (1975) Müllerian-inhibiting activity of calf fetal testes: Relationship to testosterone and protein synthesis. *Biol. Reprod.* **13**, 163–7.

Josso, N., Picard, J.-Y. and Tran, D. (1977) The anti-Müllerian hormone. *Rec. Prog. Hor. Res.* **33**, 117–63.

Jost, A. (1953) Problems of fetal endocrinology: The gonadal and hypophyseal hormones. *Rec. Prog. Horm. Res.* **8**, 379–413.

Jost, A. (1965) Gonadal hormones in the sex differentiation of the mammalian fetus. In *Organogenesis*, eds. R. L. DeHaan and H. Ursprung, 611–28. New York: Holt, Rinehart & Winston.

Jost, A. (1970) Hormonal factors in the sex differentiation of the mammalian fetus. *Phil. Trans. Roy. Soc.* (London) B. **259**, 119–30.

Jost, A., Vigier, B. and Prepin, J. (1972) Freemartins in cattle: the first steps of sexual organogenesis. *J. Reprod. Fertil.* **29**, 349–79.

Lahm, W. (1922) Zur Entwicklung der interstitiellen Druse im Hoden und Ovarium. *Monatsschr. Geburtshilfe Gynaekol.* **58**, 128–40.

Langman, J. (1975) *Medical Embryology.* Baltimore: Williams and Wilkins, 3rd ed.

Leeson, C. R. and Leeson, T. S. (1963) The postnatal development and differentiation of the boundary tissue of the seminiferous tubule of the rat. *Anat. Rec.* **147**, 243–59.

McDonald, P. G. and Doughty, G. (1972) Inhibition of androgen-sterilization in the female rat by administration of an anti-oestrogen. *J. Endocrinol.* **55**, 455–6.

McDonald, P. G. and Doughty, C. (1974) Effect of neonatal administration of different androgens in the female rat: correlation between aromatization and the induction of sterilization. *J. Endocrinol.* **61**, 95–103.

MacIntyre, M. N., Hunter, J. E. and Morgan, A. H. (1960) Spatial limits of activity of fetal gonadal inductors in the rat. *Anat. Rec.* **138**, 137–47.

Mancini, R. E., Narbaitz, R. and Lavieri, J. C. (1960) Origin and development of the germinative epithelium and Sertoli cells in the human testis: cytological, cytochemical and quantitative study. *Anat. Rec.* **136**, 477–90.

Mancini, R. E., Vilar, O., Lavieri, J. C., Andrada, J. A. and Heinrich, J. J. (1963) Development of Leydig cells in the normal human testis. *Am. J. Anat.* **112**, 203–14.

Martins, T. (1943) Mechanism of the descent of the testicle under the action of sex hormones. In *Essays in Biology*, ed., T. Cowles, 387–97, University of California Press, Berkeley, California.

Matschke, G. H. and Erickson, B. H. (1969) Development and radioresponse of the prenatal bovine testis. *Biol. Reprod.* **1**, 207–14.

Merchant-Larios, H. (1975) The onset of testicular differentiation in the rat: an ultrastructural study. *Am. J. Anat.* **145**, 319–30.

Mintz, B. (1957) Germ cell origin and history in the mouse: genetic and histochemical evidence. *Anat. Rec.* **127**, 335–6.

Mintz, B. (1959) Continuity of the female germ cell line from embryo to adult. *Arch. Anat. Microscopy Morphol. Exptl.* **48**, Suppl. 155–72.

Mintz, B. (1960) Embryological phases of mammalian gametogenesis. *J. Cell. Comp. Physiol.* Suppl. 1, 31–47.

Neumann, F., Elger, W. and Steinbeck, H. (1969) Drug-induced intersexuality in mammals. *J. Reprod. Fertil.* Suppl. 7, 9–24.

Nicander, L., Abdel-Raouf, M. and Carbo, B. (1961) On the ultrastructure of the seminiferous tubules in bull calves. *Acta Morphol. Neerl. Scand.* **4**, 127–35.

Novi, A. M. and Saba, P. (1968) An electron microscopic study of the development of rat testis in the first 10 postnatal days. *Z. Zellforsch. Mikroskop. Anat.* **86**, 313–26.

O, W-S. and Baker, T. G. (1976) Initiation and control of meiosis in hamster gonads *in vitro*. *J. Reprod. Fertil.* **48**, 399–401.

Ohno, S. (1971) Simplicity of mammalian regulatory systems inferred by single gene determination of sex phenotypes. *Nature* **234**, 134–7.

Ohno, S. Christian, L. C., Watchtel, S. S. and Koo, G. C. (1976) Hormone-like role of H-Y antigen in bovine freemartin gonads. *Nature* (London) **261**, 597–9.

Pfeiffer, C. A. (1936) Sexual differences of the hypophysis and their determination by the gonads. *Am. J. Anat.* **58**, 195–225.

Peters, H. (1970) Migration of gonocytes into the mammalian gonad and their differentiation. *Phil. Trans. Roy. Soc.* (London) B. **259**, 91–101.

Picard, J. Y. and Josso, N. (1976) Anti-Müllerian hormone: estimation of molecular weight by gel filtration. *Biomedicine Express.* **25**, 147–50.

Picon, R. (1969) Action du testicule foetal sur le developpement in vitro des cannaux de Müller chez le rat. *Arch. Anat. Microscop. Morphol. Exptl.* **58**, 1–19.

Picon, R. (1976) Testicular inhibition of fetal Mullerian ducts in vitro: effect of dibutyryl cyclic AMP. *Mol. cell. Endocr.* **4**, 35–42.

Price, D. and Pannabecker, R. (1956) Organ culture studies of foetal rat reproductive tracts. *Ciba Found. Colloq. Aging* **2**, 3–13.

Price, D. and Ortiz, E. (1965) The role of fetal androgen in sex differentiation in mammals. In *Organogenesis*, eds. R. L. De Haan and H. Ursprung, 629–52, New York: Holt, Rinehart & Winston.

Raeside, J. I. and Sigman, D. M. (1975) Testosterone levels in early fetal testes of domestic pigs. *Biol. Reprod.* **13**, 318–21.

Roosen-Runge, E. C. (1961a) Rudimental 'genital canals' of the gonad in rat embryos. *Acta Anat.* **44**, 1–11.

Roosen-Runge, E. C. (1961b) The rete testis in the albino rat: its structure, development and morphological significance. *Acta Anat.* **45**, 1–30.

Roosen-Runge, E. C. and Anderson, D. (1959) The development of the interstitial cells in the testis of the albino rat. *Acta Anat.* **37**, 125–37.

Roosen-Runge, E. C. and Leik, J. (1968) Gonocyte degeneration in the postnatal male rat. *Am. J. Anat.* **122**, 275–300.

Rowlands, I. W. and Rogers Brambell, F. W. (1933) The development and morphology of the gonads of the mouse. IV. The postnatal growth of the testis. *Proc. R. Soc.* (London) B. **112**, 200–14.

Rubaschkin, W. (1912) Zur Lehre von der Keimbahn bei Säugetieren. Uber die Entwicklung der Keimdrusen. *Anat. Hefte.* **46**, 343–411.

Russo, J. and Rosas, J. de. (1971) Differentiation of the Leydig cell of the mouse testis during the fetal period—an ultra-structural study. *Am. J. Anat.* **130**, 461–80.

Sapsford, C. S. (1957) The development of the Sertoli cells. *J. Endocrinol.* **15**, lv–lvi.

Sapsford, C. S. (1962a) The development of the testis of the merino ram, with special reference to the origin of the adult stem cells. *Aust. J. Agri. Res.* **13**, 487–502.

Sapsford, C. S. (1962b) Changes in cells of the sex cords and seminiferous tubules during the development of the testis of the rat and mouse. *Aust. J. Zool.* **10**, 178–92.

Sapsford, C. S. (1963) The development of the Sertoli cell of the rat and mouse: Its existence as a mononucleate unit. *J. Anat.* **97**, 225–38.

Sapsford, C. S. (1964) Changes in the size of germ cell nuclei during the development of the testis of the ram and rat. *Aust. J. Zool.* **12**, 127–49.

Segal, D. H. and Raeside, J. I. (1975) Androgens in testes and adrenal glands of the fetal pig. *J. ster. Biochem.* **6**, 1439–44.

Schoot, P. van der, Vaart, P. D. M. van der and Vreeburg, J. T. M. (1976) Masculinization in male rats is inhibited by neonatal injections of dihydrotestosterone. *J. Reprod. Fertil.* **48**, 385–7.

Setchell, B. P. (1977) Reproduction in male marsupials. In *The Biology of Marsupials*, eds. D. P. Gilmore and B. Stonehouse, 411–57. London: Macmillan.

Short, R. V. (1969) An introduction to some of the problems of intersexuality. *J. Reprod. Fertil.* **7**, 1–8.

Short, R. V. (1970) The bovine freemartin: a new look at an old problem. *Phil. Trans. Roy. Soc.* (London) B. **259**, 141–7.

Short, R. V. (1974) Sexual differentiation of the brain of the sheep. Les Colloques de l'Institut National de la Sante et de la Recherche Medicale; Colloque International sur l'endocrinologie sexuelle de la periode perinatale. Inserm. **32**, 121–42.

Short, R. V., Smith, J., Mann, T., Evans, E. P., Hallett, J., Fryer, A. and Hamerton, J. L. (1969) Cytogenetic and endocrine studies of a freemartin heifer and its bull co-twin. *Cytogenetics* **8**, 369–88.

Steinbeck, H., Neumann, F. and Elger, W. (1970) Effect of an anti-androgen on the differentiation of the internal genital organs in dogs. *J. Reprod. Fertil.* **23**, 223–7.

Steinberger, E., Steinberger, A. and Ficher, M. (1970) Study of spermatogenesis and steroid metabolism in cultures of mammalian testes. *Rec. Prog. Horm. Res.* **26**, 547–88.

Stieve, H. (1927) Die Entwicklung der Keimzellen und der Zwischenzellen in der Hodenanlage des Menschen. Ein Beitrag zur Keimbahnfrage. *Z. Mikroskop. Anat. Forsch.* **10**, 225–85.

Torrey, T. W. (1945) The development of the urinogenital system of the albino rat. II. The gonads. *Am. J. Anat.* **76**, 375–97.

Torrey, T. W. (1947) The development of the urinogenital system of the albino rat. III. The urinogenital union. *Am. J. Anat.* **81**, 139–53.

Tyndale-Biscoe, H. (1973) *Life of Marsupials*, London: Edward Arnold.

Ullrey, D. E., Sprague, J. I., Becker, D. E. and Miller, E. R. (1965) Growth of the swine fetus. *J. Anim. Sci.* **24**, 711–17.

Wachtel, S. S. (1976) H-Y antigen: genetics and serology. *Immunol. Rev.* **33**, 33–61.

Wachtel, S. S., Ohno, S., Koo, G. C. and Boyse, E. A. (1975) Possible role for H-Y antigen in the primary determination of sex. *Nature* (London) **257**, 235–6.

Wagenen, G. van and Simpson, M. E. (1954) Testicular development in the rhesus monkey. *Anat. Rec.* **118**, 231–51.

Wells, L. J. (1943) Descent of testis: anatomic and hormonal considerations. *Surgery* **14**, 436–72.

Wensing, C. J. G. (1968) Testicular descent in some domestic mammals. I. Anatomical aspect of testicular descent. *Proc. Koninkl. Nederl. Akademie van Wetenschappen* (Amsterdam) **71**, 423–34.

Wensing, C. J. G. (1973) Testicular descent in some domestic mammals. II. The nature of the gubernacular change during the process of testicular descent in the pig. *Proc. Koninkl. Nederl. Akademie van Wetenschappen* (Amsterdam) **76**, 196–202.

Whitehead, R. H. (1904) The embryonic development of the interstitial cells of Leydig. *Am. J. Anat.* **3**, 167–82.

Whitehead, R. H. (1905) Studies of the interstitial cells of Leydig. No. 2. Their postembryonic development in the pig. *Am. J. Anat.* **4**, 193–7.

Widmaier, R. (1963) Uber die postnatale Hodenentwicklung und Keimzellreifung bei der Maus. *Z. Mikroskop. Anat. Forsch.* **70**, 215–41.

Witschi, E. (1948) Migration of the germ cells of human embryos from the yolk sac to the primitive gonadal fold. *Contrib. Embryol.* (Carnegie Inst. Wash.) **32**, 67–80.

Witschi, E. (1956) *Development of Vertebrates*, Philadelphia: Saunders.

Wolff, E. (1952) Sur la differenciation sexuelle des gonades de souris explantées in vitro. *Compt. Rend. Acad. Sci.* **234**, 1712–4.

Wolff, E. and Haffen, K. (1965) Germ cells and gonads. In *Cells and Tissues in Culture*, ed., E. N. Willmer, vol. 2, 697, New York: Academic Press.

Wyndham, N. R. (1943) A morphologic study of testicular descent. *J. Anat.* **77**, 179–88.

Yao, T. S. and Eaton, D. N. (1954) Postnatal growth and histologic development of reproductive organs in male goats. *Am. J. Anat.* **95**, 401–31.

# 3 Vascular supply and drainage of the testis

As the testis migrates from its embryological starting point, it draws its vascular supply with it so that in all adult mammals the testicular artery arises from the aorta near the renal arteries. Consequently, the artery is very elongated in those mammals with prominent scrota.

## 3.1     Anatomy of the testicular (or internal spermatic) artery and vein

### 3.1.1     Mammals with abdominal testes
The anatomy of the testicular blood vessels is much simpler in mammals with abdominal testes, although there is a certain amount of variation among the different orders.

In the monotremes, the platypus and echidna, the testis remains near the kidney but the artery is unusual in that it runs to the caudal pole of the testis at the end opposite to the efferent ducts. In the platypus, the artery and the vein are somewhat convoluted in their parallel course, though single throughout; in the echidna, the main venous drainage is from the end of the testis near the efferent ducts, and it is not parallel with the artery (Setchell, 1970). The only marsupial with abdominal testes is the marsupial mole (*Notoryctes typhlops*) and it lacks the spermatic rete characteristic of other marsupials (Barnett and Brazenor, 1958).

In the elephants and hyraxes, the testis remains just caudal to the kidney and the artery is short, single and unconvoluted (Figure 3.1). The venous drainage is also simple (Short *et al.*, 1967; Glover, 1973). In the other two orders of eutherian mammals with abdominal testes, the edentates and the cetaceae, arterial retia are found in many other parts of the body and therefore it is perhaps not surprising that there is some complexity of the testicular vasculature (Setchell, 1970). In some cetaceae, Kate Parry (personal communication) has found that there is a venous plexus in the abdominal wall next to the testis; the function of this is not known.

### 3.1.2     Animals with inguinal testes
In mammals with inguinal testes, such as hedgehogs, moles and some seals, there is usually some coiling of the artery and a pampiniform plexus surrounding it (Figure 3.2; Harrison, 1949; Setchell, 1970). In the seals, S. M. McGinnis (personal communication) has noticed that the venous drainage from the hind flippers forms a secondary venous plexus overlying the testis.

### 3.1.3     Mammals with scrotal testes
In all marsupials, except the marsupial mole, there is a pronounced spermatic arterial rete which is formed as the artery leaves the inguinal canal. The arteries reunite near

Figure 3.1 An X-ray of the arterial supply (filled with radio-opaque medium) of the testis of an African elephant (*Loxodonta africana*). Note the unconvoluted main artery and the branches running directly into the parenchyma of the testis, which is located near the kidney. (Reproduced from Glover, 1973)

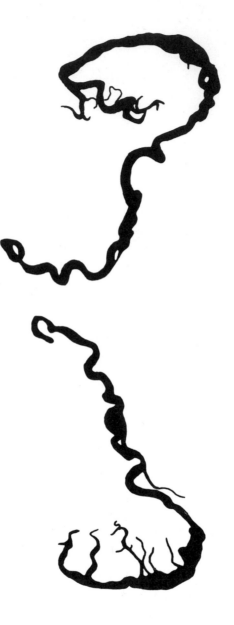

Figure 3.2 X-ray of the arterial supply (filled with radio-opaque medium) of a hedgehog (*Erinaceus europaeus*). Note the appreciable twisting of the artery and its extended course on the surface of the testis, which is located in the inguinal region. (Adapted from Harrison, 1949)

the testis into between one and four arteries which then supply the testis. The number of branches in the rete varies from about five in some of the smaller dasyurids up to almost two hundred in some of the larger macropods (Figure 3.3). These are in-

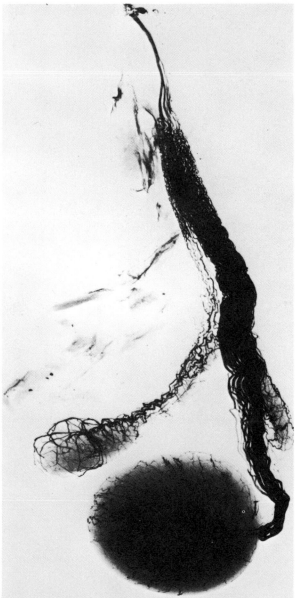

Figure 3.3 An X-ray of the arterial supply (filled with radio-opaque medium) of the testis of a wallaby (*Macropus rufogriseus*). Note how the artery branches just after it passes through the inguinal canal, to form an arterial rete of up to 200 parallel branches. These lie intermingled with a similar number of venous branches, but reunite near the testis to form 3 or 4 vessels which then branch to supply the testicular parenchyma. (Reproduced from Harrison, 1951)

terspersed among a similar number of veins, which arise by redivision of the two to four veins leaving the testis (Barnett and Brazenor, 1958; Harrison, 1949; Setchell, 1970, 1977a).

In eutherian mammals with scrotal testes, the artery, accompanied by the vein,

(a)

Figure 3.4   A cast of the testicular artery of a ram: (a) in its normal form, and (b, opposite) extended to show the extent of the coiling. The marks at the side indicate centimetres. (a) Reproduced from Waites and Moule, 1960, and (b) kindly supplied by Professor G. M. H. Waites)

(b)

Figure 3.4(b)

runs obliquely across the abdominal wall and through the inguinal canal. It then begins to coil so that, for example, in the ram, up to 7m of artery can be compressed into about 10cm of spermatic cord (Figure 3.4). The artery remains single throughout, except for branches to the head of the epididymis (Figure 3.5). The degree of coiling varies widely between species; in the human it is minimal (Figure 3.6), as it is in most carnivores, whereas in most ungulates it is quite pronounced.

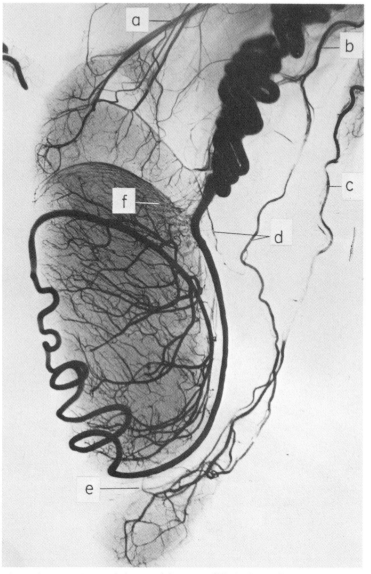

Figure 3.5 An X-ray of the arterial system (filled with radio-opaque medium) of a rat: a, superior epididymal branch and b, inferior epididymal branch of the internal spermatic artery; c, vasal (or deferential) artery; d, anastomoses between the inferior epididymal and vasal arteries; e, small arterial branch between the epididymal and testicular arteries; f, intra-albugineal venous plexus. (Reproduced from Kormano, 1967)

The veins leave the testis near the efferent ducts and then divide again to produce the many intercommunicating vessels of the pampiniform plexus. This surrounds the coils of the spermatic artery to form the vascular cone, which comprises a large part of the spermatic cord (Setchell, 1970).

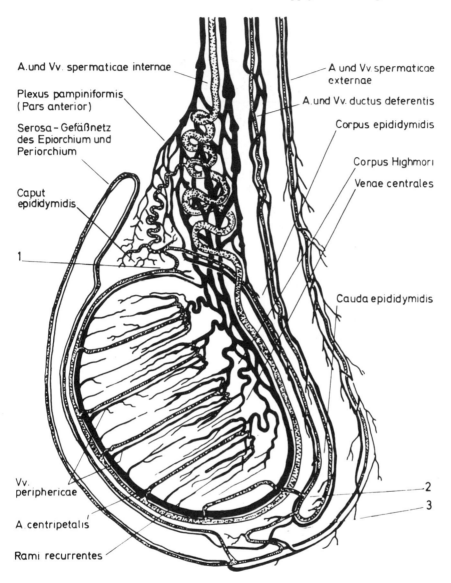

A.und Vv. spermaticae internae

Plexus pampiniformis
(Pars anterior)

Serosa – Gefäßnetz
des Epiorchium und
Periorchium

Caput
epididymidis

1

Vv.
periphericae

A. centripetalis

Rami recurrentes

A und Vv. spermaticae
externae

A.und Vv. ductus deferentis

Corpus epididymidis

Corpus Highmori

Venae centrales

Cauda epididymidis

2

3

Figure 3.6 A diagram of the arterial and venous supply of the human testis. Note the slight coiling of the artery in the spermatic cord, its course around the testis, the centripetal arteries running in to the mediastinum where they turn and branch to give rami recurrentes which supply the parenchyma. The venous drainage runs both to the surface and also to the mediastinum, and then branches again to form the pampiniform plexus surrounding the artery. The head of the epididymis is supplied from epididymal branches of the internal spermatic artery and the tail from the deferential vessels.

The external spermatic vessels supply the scrotal skin, but have some anastomoses (3) with the deferential vessels. There are also communications (2) between the vessels supplying the two lower parts of the epididymis, and these also have some connections (1) with the testicular vessels. (Reproduced from Hundeiker, 1971b)

## 3.2      Spermatic cord

### 3.2.1      Structure

As well as the artery and veins just described, the spermatic cord (Figure 3.7) contains a number of lymphatic ducts, which lie on the outside. At the back of the cord is the ductus deferens which carries the spermatozoa from their site of storage in the tail of the epididymis to the urethra. The ductus deferens is accompanied by its own

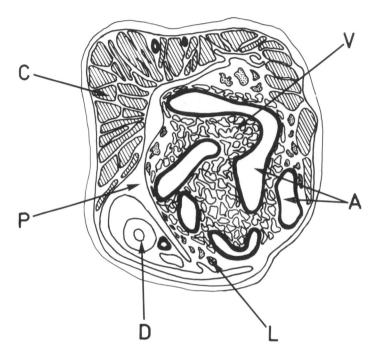

Figure 3.7 A diagram of a cross-section of one spermatic cord of a ram. Note how the artery (A) appears several times in the one cross-section, among the multiple venous branches (V) of the pampiniform plexus. Around the outside are the lymphatic vessels (L) and the cremaster muscle (C), with the ductus deferens and the associated vessels (D). The cremaster muscle is largely separated from the pampiniform plexus by the processus vaginalis (P), which persists in adults of this species (see Section 2.2).

artery and vein, the deferential vessels, which arise from the hypogastric vessels in the pelvis and supply the caudal half of the epididymis as well as the ductus deferens itself. These vessels anastomose in the middle of the epididymis with the branches of the internal spermatic artery which supply the head of the epididymis, and rarely there is a small branch running from the tail of the epididymis to the testis. The spermatic cord is surrounded by the cremaster muscle, which is a voluntary muscle. It is therefore not capable of sustained contraction, but can retract the testis against the abdominal wall during fright or sexual excitement. The sustained retraction seen when the animal is exposed to cold is due to the dartos muscle, which is a smooth muscle (see Chapter 5).

### 3.2.2    Function of the spermatic cord

Two functions have been demonstrated for the spermatic cord and another suggested. First, the pulse pressure (i.e. systolic pressure minus diastolic pressure) of the arterial blood is markedly reduced as the arterial blood traverses the cord, with only a slight reduction in mean blood pressure (Figure 3.8; Waites and Moule, 1960). Second, the arterial blood is pre-cooled as it passes through the cord by counter-current heat exchange with the cooler venous blood, which in turn is warmed up between the time it leaves the testis and reaches the inguinal canal. This arrangement ensures that the body does not lose heat through the scrotum and that the whole testis is kept at the same temperature as the scrotal skin; heat loss from the scrotal skin is essential for cooling the

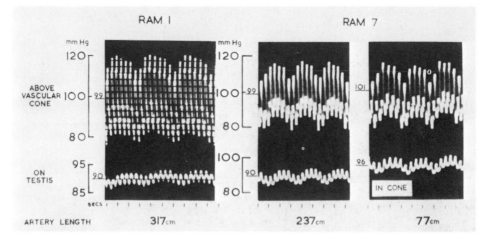

Figure 3.8 A record of the blood pressure in the testicular artery above and below the spermatic cord. Note the extreme reduction in the pulse pressure with only a small reduction in mean blood pressure. (Reproduced from Waites and Moule, 1960)

testis and the countercurrent system merely ensures that this cooling is rapidly and evenly spread through the testis (see Section 5.2.1 and Figure 5.4).

Cross and Silver (1962) suggested that exchange of substances might take place between the vein and artery in the cord. However, this has been proved to be quantitatively insignificant in the ram, boar and wallaby for the substances so far examined (Figure 3.9; Setchell and Hinks, 1969; Jacks and Setchell, 1973) but it is possibly more important in the rat (Free and Tillson, 1975; Free, 1977). There is no evidence that it is anything more complicated than simple diffusion down a concentration gradient. The only substances known to be present in widely different concentrations in testicular venous and arterial blood are oxygen and testosterone. Diffusion of the former is significant only in animals breathing oxygen (Setchell and Hinks, 1969). The small increases in the concentration of testosterone in the testicular arterial blood produced by vein to artery diffusion are trivial by comparison with the differences between the concentrations of testosterone in testicular lymph or venous and arterial blood (Lindner, 1963, 1967). However, they may be important for other structures, such as the head of the epididymis, which do not produce testosterone themselves but which

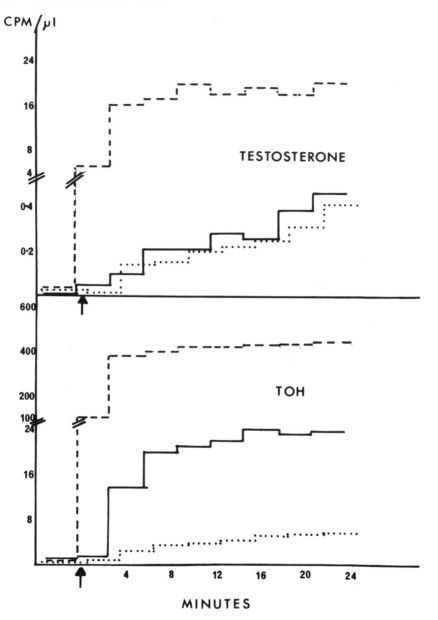

Figure 3.9 Graph of the concentrations of radioactivity in blood from the aorta ( ....... ), from the testicular artery on the surface of the testis (_____) and from the internal spermatic vein above the spermatic cord (– – –) during infusion of a mixture of ³H-water (TOH) and ¹⁴C-testosterone into a testicular vein on the testis. Note that there is appreciably more radioactivity in blood from the testicular artery than from the aorta during the infusion of ³H-water because of transfer from the vein to the artery in the spermatic cord. However this amounts to only about 20 per cent of the difference in concentration between vein and artery, and is very much less with testosterone.

receive arterial blood that has been through the spermatic cord (Setchell, 1977b). Exchange from vein to artery or vice versa of diffusible indicator is also a potential source of error in measurement of blood flow (Setchell and Linzell, 1974).

## 3.3    Blood vessels within the testis

### 3.3.1    Larger vessels

In all mammals except the elephants, hyraxes and monotremes, when the testicular artery reaches the testis, it runs along the surface of testis for quite a long way before entering the parenchyma. In many species it also courses along the anterior face of the testis before dividing, and in the rabbit the main artery and its principal branches encircle the testis three times before they enter its substance. Then, in most species, the arterial branches dive straight in and run without branching to the mediastinum near the rete testis where they form coils. Then, and only then, do the arteries ramify throughout the tissue (Figure 3.10; see also Figure 8.2).

The venous drainage does not follow the arteries. Many veins rise straight to the surface, others run to the mediastinum and form a main central vein; the proportion of the blood following these two routes depends on the species. All the veins then run directly towards the cranial pole of the testis and leave it near the efferent ducts by branching again to form the pampiniform plexus (Setchell, 1970).

In some marsupials, e.g. macropods and *Trichosurus*, the arteries run along the epididymal face of the testis over one arm of the rete, while the veins run on the opposite surface, over the other arm of the rete. In *Didelphis*, the artery enters the testis near the efferent ducts without running along its surface for any distance (Setchell, 1977a).

### 3.3.2    Lymph vessels

The distribution of lymph vessels within the testis was the subject of some dispute, but this has now been settled by the use of the electron microscope and intravascular fixation (see Section 1.3.2). The interstitial lymph vessels are either sinusoids (in rodents) or discrete lymphatic capillaries (in man, ram, boar, etc.) in the centre of the interstitial space (Fawcett *et al.*, 1969; Fawcett *et al.*, 1973). These join lymphatic trunks in the septulae in those species with lobulated testes, and then join an extensive network of lymphatic vessels in the tunica albuginea (Regaud, 1897a, b). The lymphatic vessels resorb peritoneal fluid from the scrotal cavity; this is particularly important in man in which the inguinal canal is closed after descent of the testes, but is probably also important in those species in which the scrotum is dependent and its cavity communicates with the peritoneal cavity. No quantitative data is available on the contribution of parenchymal and tunical lymphatics. However, in the boar, removing the testis from the scrotum and wrapping it in dry cotton wool has little effect on the flow of lymph from the testis (see Figure 3.14). This suggests that in this species, the tunical vessels are relatively unimportant.

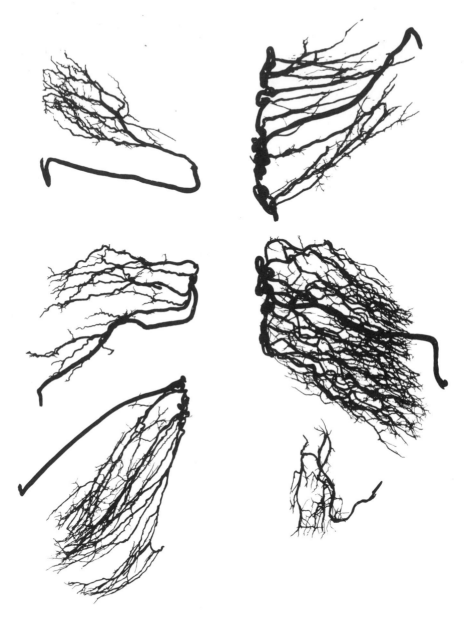

Figure 3.10 Casts of arteries from the testis of a bull. Note how in all but one example, the artery runs unbranched directly from the surface of the testis to form a coil or inflexion near the mediastinum. From there branches run back to supply the parenchyma. Only a small proportion of the parenchyma is supplied by arteries which branch as soon as they leave the surface of the testis (e.g. lower right). (Photographed from casts kindly supplied by Dr H. P. Godinho)

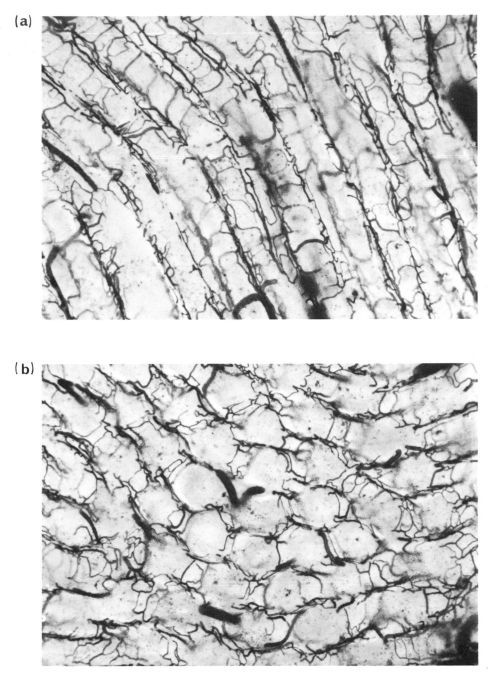

Figure 3.11 Photographs of thick sections (300 μm) of a rat testis, (a) cut approximately parallel with the seminiferous tubules, (b) cut approximately at right angles to the tubules. Blood vessels were ligated in the spermatic cord after the venous pressure had been raised for 1 minute by a cuff around the cord in an animal anaesthetized with pentobarbitone sodium. The testis was then fixed whole, sectioned and the red cells stained with benzidine and nitroprusside. Note that some capillaries lie parallel to the tubules in the interstitial spaces; others run around the walls at right angles to the length of the tubule.

### 3.3.3    Testicular capillaries

The capillaries do not enter the seminiferous tubules but are confined to the spaces in between. In the rat, the capillaries are of two types, Zwickelcapillaren which run parallel to the tubules and Quercapillaren which run around the tubules from vessels in one interstitial space to those in the next (Figure 3.11; Müller, 1957; Kormano, 1967). This arrangement develops only at puberty (Figure 3.12) and its significance is not known. However it may be that fluid is filtered by the Zwickelcapillaren into the extracellular space where it acquires a high concentration of testosterone, flows round the wall of the tubule and is reabsorbed by the Quercapillaren. For this to occur, the blood in the Zwickelcapillaren would have to be at a greater pressure than that in the Quercapillaren and no direct measurements have yet been made of capillary pressure in the testis.

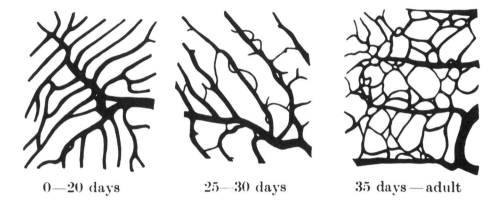

0—20 days            25—30 days            35 days — adult

Figure 3.12 The vascular anatomy around the seminiferous tubules of rats of different ages as seen from the surface of the testis. Note how the capillaries lying at right angles to the tubules are not present in the young rats but develop only at puberty. (Reproduced from Kormano, 1967)

The testicular capillaries do appear to filter a fluid with a high concentration of protein. Testicular lymph has more than two-thirds as much protein as blood plasma (Lindner, 1963; Wallace and Lascelles, 1964), much more than lymph elsewhere in the body except in the liver, and furthermore the protein concentration is not decreased by increasing lymph flow by venous congestion or by heating the testis (Section 3.5.1). Therefore the Starling hypothesis for fluid passage out of the capillaries into the tissue space and back again may not apply in the testis; according to Starling's scheme, the difference in colloid osmotic pressure between blood plasma and tissue fluid produces resorption of fluid at the venous end of the capillary, but if the fluid filtered by the testicular capillaries has a similar colloid osmotic pressure to blood plasma, other processes may be involved in resorption of fluid in the testis.

Nevertheless, the apparently high permeability of the testicular capillaries is not sufficient to produce significant displacement of the indicator dilution curves for labelled albumin and red blood cells (Setchell and Jacks, unpublished observations).

Furthermore, the testicular capillaries unlike those of other endocrine tissues, are un-fenestrated; they are classified as A-1-α, that is, there are no intercellular spaces between the endothelial cells and there is a definite basement membrane, but an incomplete covering of pericytes (Wolff and Merker, 1966). It has not been made clear whether these observations were made on Zwickelcapillaren or Quercapillaren but some later observations by Hundeiker (1971a) suggest that the two sorts of capillaries do have a similar structure.

## 3.4    Blood flow in the testis

As in several other tissues, particularly muscle, capillary or nutrient blood flow through the testis is appreciably less than total blood flow (Godinho and Setchell, 1975). No anatomical basis for this difference has been found, but it is important to keep in mind the distinction between the two types of flow. Capillary blood flow is probably more interesting in relation to testicular physiology, but total blood flow is the value which should be used in calculations of production or uptake of substances by the testis (see Table 6.1).

One factor peculiar to the testis is that the artery is so long in relation to its diameter that this in itself may pose an upper limit to blood flow. Calculations made for the ram using Poiseuille's equation, suggest that about 16ml/min would be the maximum blood flow possible in a tube like the artery with the head of pressure available (Setchell, 1970). However, this calculation is open to considerable error because it involves the radius of the artery raised to the fourth power, and it is difficult to get accurate measurements of the radius of the artery under normal conditions in conscious animals. Nevertheless, it does suggest that the control of total blood flow may involve the main artery as well as the arterioles, which are the principle site of control in other tissues. In support of this idea is evidence that the sensitivity of the artery to catecholamines is high in the cord but decreases nearer the testis (Waites et al., 1975). It also provides a mechanism for the testis to control or affect its own flow by liberating something into the vein which can cross to the artery in the spermatic cord. Although pharmacological doses of prostaglandins and 5-hydroxytryptamine affect testicular blood flow by this mechanism when infused into a testicular vein (Free and Jaffe, 1972a, b; Free and Nguyen Duc Kien, 1973), there is as yet no evidence that any naturally released substance has such an effect.

Testicular blood flow is comparatively low in relation to the metabolic requirements of the tissue. This means that the venous blood leaving the testis contains less oxygen than the blood in most veins; about 50 per cent of the haemoglobin is oxygenated (Setchell and Waites, 1964), compared with about 65 per cent for mixed venous blood. There are no capillaries in the tubules and the tubular cells have an appreciable consumption of oxygen. Diffusion of oxygen is certainly not instantaneous and it can be calculated that the oxygen tension inside the tubules is probably about 7mm Hg lower than that in the interstitium (Setchell, 1970). The oxygen tension in the interstitium of the ram testis has been calculated to be about

23mm Hg from measurements on spermatic venous blood (Free and VanDemark, 1968) but to be about 11mm Hg from direct measurements (Cross and Silver, 1962) which may have produced some artefacts. In rats, Free (1977) has recently reported values of 15mm Hg for the interstitium of the testis and 12mm Hg inside the tubules.

This means that the testis is poised on the brink of hypoxia if metabolism is increased without a corresponding increase in blood flow. This is exactly what happens if the temperature of the scrotal testes of eutherian mammals is raised to body temperature (Figure 3.13). The testis in this regard is unusual as most organs do increase their blood flow if their temperature is raised. However, if the temperature of the testis is raised above body temperature, testicular blood flow will also increase (Waites and Setchell, 1964; Setchell et al., 1966; Waites et al., 1973); the testes of marsupials increase their blood flow when heated to body temperature (Setchell and Thorburn, 1969).

The testis shows reactive hyperaemia (an increased flow after a period of occlusion) (Blombery, 1967; Free and Jaffe, 1972a). Testicular blood flow is also increased in anaesthetized rams if they are placed on their backs, presumably because of the reduced testicular venous pressure (Setchell and Waites, 1964); it can also be increased by the administration of isopropylnoradrenaline, although the increases are not as great as in other organs. Acetylcholine which is a potent vasodilator in other tissues has little effect on testicular blood flow (Setchell et al., 1966; Free and Jaffe, 1972a).

In contrast to this reluctance to increase its blood flow, the testis can reduce its blood flow very readily, often to very low values. Adrenaline and noradrenaline both cause marked vasoconstriction (Setchell et al., 1966; Free and Jaffe, 1972a), as does sympathetic nerve stimulation (Linzell and Setchell, 1969), prostaglandin $F_{2\alpha}$, $E_1$ or $E_2$ (Free and Jaffe, 1972b) and serotonin (Free and Nguyen Duc Kien, 1973). There appears to be comparatively little resting sympathetic tone in the testicular blood vessels, as blocking the constrictor alpha-receptors with phenoxybenzamine has little effect on flow (Setchell et al., 1966). On the other hand, blood flow through the isolated perfused testis is much higher than in vivo, suggesting that the blood vessels are normally partially constricted (Linzell and Setchell, 1969).

Gonadotrophins appear to have little effect on testicular blood flow, but hypophysectomy is followed by a decrease, possibly a reflection of the generally debilitated condition of these animals (Setchell, 1970). Cadmium salts have a most dramatic effect on testicular blood flow, reducing it to less than 5 per cent of control values within 12h. This decrease is probably secondary to a marked increase in vascular permeability (see Section 11.3.1.1).

## 3.5   Testicular lymph

### 3.5.1   Flow of lymph from the testis

Lymph flow from the testis is high per unit weight of tissue when compared with other organs and seems especially high when it is remembered that lymphatic vessels

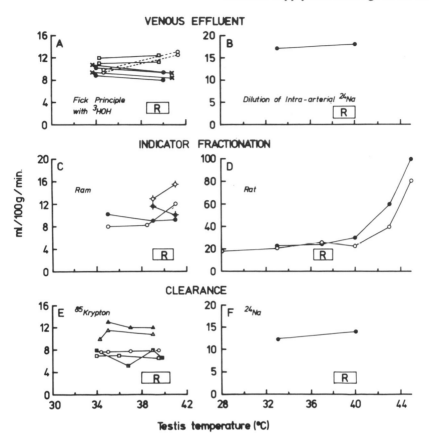

Figure 3.13 Graphs illustrating the lack of effect on testicular blood flow of testicular temperature except when it is raised above body temperature (R).

A and B are values for total venous blood flow, measured in rams by two different techniques. C to F are values for capillary of nutrient blood flow, measured in rams (C, E and F) or rats (D) by indicator fractionation or clearance of $^{85}$Kr or $^{24}$Na.

A: different symbols for four conscious rams, measurements on the same testis linked by lines (Waites and Setchell, 1964).

B: each point the mean of four measurements at 33°C and 4 measurements at 40°C on the same anaesthetized rams (Godinho and Setchell, 1975).

C: each point the mean of four values from conscious individual rams of a strain selected for plain skin (O◇) and rams of a strain selected for wrinkly skin (●,◆), after 90 min (O,●) or 150 min (◇,◆) exposure to the temperatures indicated. Rams of the wrinkly-skinned strain are more sensitive to heat infertility than those of the plain-skinned strain, which do show a slight increase in testicular blood (Fowler and Setchell, 1971).

D: each point is the mean of three values for individual anaesthetized rats, the two sets of symbols being for two different experiments. (Waites et al., 1973).

E: different symbols for five conscious rams, with repeated measurements made at various temperatures (Setchell et al., 1966).

F: as for B.

are present only in the tunica albuginea and interstitial tissue which comprise between 10 and 30 per cent of the testis, depending on the species.

The flow can be increased by raising venous pressure (Setchell and Jacks, unpublished data) and by heating the testis (Figure 3.14; Cowie *et al.*, 1964); in both cases flow increases without any decrease in protein concentration.

Exercise produced an initial rise, followed by very low flows which persisted for some time after the exercise has ended. When the testes were handled or were drawn up by the cremaster muscles during ejaculation, there were immediate and often very large (6- to 10-fold) increases in lymph flow (Cowie *et al.*, 1964).

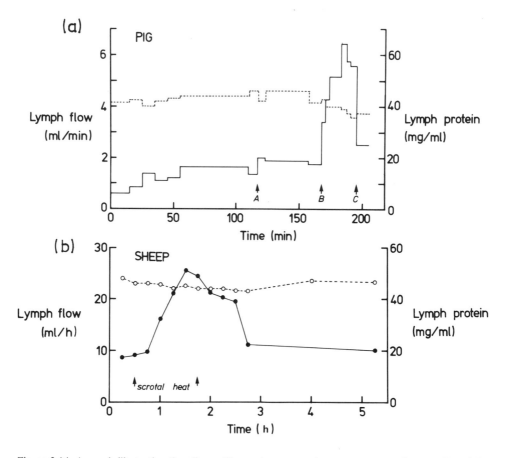

Figure 3.14 A graph illustrating the effects of increasing spermatic venous pressure (between B and C in (a)) or heating the scrotum (between arrows in (b)) on the flow ( —————— ) and protein content ( _ _ _ _ ) of testicular lymph. Note that marked increases in flow are associated with only trivial changes in protein concentration. (Data in (b) from Cowie *et al.*, 1964.) At A in (a), the testis was removed from the scrotum and wrapped in dry cotton wool.

## References to Chapter 3

Amann, R. P. and Ganjam, V. K. (1976) Steroid production by the bovine testis and steroid transfer across the pampiniform plexus. *Biol. Reprod.* **15**, 695–703.

Andres, J. (1927) Die Arterien der Keimdrüsen bei männlichen und weiblichen Versuchssaugetieren. *Z. Anat. Entwicklungsgeschichte* **84**, 445–475.

Arata, A. A. (1964) A comment on the utility of the testicular blood supply in taxonomic studies. *J. Mammal.* **45**, 493–4.

Arrou, J. (1893) Circulation artérielle du testicule: Anatomie comparée. Thèse pour le Docteur en Médecine, University of Paris.

Barnett, C. H. and Brazenor, C. W. (1958) The testicular rete mirabile of marsupials. *Aust. J. Zool.* **6**, 27–32.

Bayard, F., Boulard, P. Y., Huc, A. and Pontonnier, F. (1975) Arterio-venous transfer of testosterone in the spermatic cord of man. *J. Clin. Endocrinol. Metab.* **40**, 345–6.

Beutler, O. (1922) Das Verhalten der Arteria spermatica interna im Hoden der Haussaugetiere (Rind, Schaf, Pferd, Schwein, Hund und Katze). Inaugural-Dissertation, Tierarztliche Hochschule Hannover.

Bimar, M. (1888a) Recherches sur la distribution des vaisseaux spermatiques chez divers mammifères. *Compt. Rend. Acad. Sci.* **106**, 80–3.

Bimar, M. (1888b) Recherches sur la distribution des vaisseaux spermatiques chez les mammifères et chez l'homme. *J. Anat.* (Paris) **24**, 265–73.

Blombery, P. A. (1967) An investigation of certain aspects of the testicular vasculature. B.Sc. (med) thesis, University of Sydney, Australia.

Brie, S. J. Le (1970) Structure and function of the lymphatic system with emphasis on its role in hormone transport. *J. Reprod. Fertil.*, Suppl. 10, 123–38.

Capurro, M. A. (1902) Sulla circolazione sanguigna normale e di compenso del testicolo. *Anat. Anz.* **20**, 577–98.

Chatterjee, A. and Paul, B. S. (1968) Testicular atrophy in rats following epinephrine administration. *Endokrinologie* **52**, 406–7.

Colle, J. F. E. (1902) Arteres du testicule. Inaug.-Diss., Lille.

Collin, B. (1973) La vascularisation artérielle du testicule chez le cheval. *Zbl. Veterimärmed.* **2**, 46–53.

Courot, M. and Joffre, M. (1977) Testicular capillary blood flow in the impubertal lamb and the ram during the breeding and non-breeding seasons. *Andrologia.* **9**, 332–6.

Cowie, A. T., Lascelles, A. K. and Wallace, J. C. (1964) Flow and protein content of testicular lymph in conscious rams. *J. Physiol.* (London) **171**, 176–87.

Cross, B. A. and Silver, I. A. (1962) Some factors affecting oxygen tension in the brain and other organs. *Proc. Roy. Soc.* (London) B. **156**, 483–99.

Dahl, E. V. and Herrick, J. F. (1959) A vascular mechanism for maintaining testicular temperature by counter-current exchange. *Surg. Gynecol. Obstet.* **108**, 697–705.

Daniel, P. M., Gale, M. M. and Pratt, O. E. (1963) Hormones and related substances in the lymph leaving four endocrine glands—the testis, ovary, adrenal, and thyroid. *Lancet* I, 1232–4.

Dierschke, D. J., Walsh, S. W., Mapletoft, R. J., Robinson, J. A. and Ginther, O. J. (1975)

Functional anatomy of the testicular vascular pedicle in the rhesus monkey: Evidence for a local testosterone concentrating mechanism. *Proc. Soc. Exp. Biol.* (New York) **148**, 236–42.

Einer-Jensen, N. (1974a) Local recirculation of injected [³H] testosterone from the testis to the epididymal fat pad and the corpus epididymidis in the rat. *J. Reprod. Fertil.* **37**, 145–8.

Einer-Jensen, N. (1974b) Local recirculation of ¹³³xenon and ⁸⁵krypton to the testes and the caput epididymidis in rats. *J. Reprod. Fertil.* **37**, 55–60.

Einer-Jensen, N. and Soofi, G. (1974) Decreased blood flow through rat testis after intratesticular injection of PGF$_{2\alpha}$. *Prostaglandins* **7**, 377–82.

Einer-Jensen, N. and Waites, G. M. H. (1976) Testicular blood flow and a study of the testicular venous to arterial transfer of radioactive krypton and testosterone in the rhesus monkey. *J. Physiol.* (London) **267**, 1–15.

Everett, N. B. and Simmons, B. (1958) Measurement and radioautographic localization of albumin in rat tissues after intravenous administration. *Circulation Res.* **6**, 307–13.

Fawcett, D. W. (1963) Comparative observations on the fine structure of blood capillaries. In *The peripheral blood vessels*, eds, J. L. Orbison and D. E. Smith, 17–44, Baltimore: Williams and Wilkins Co.

Fawcett, D. W., Heidger, P. M. and Leak, L. V. (1969) Lymph vascular system of the interstitial tissue of the testis as revealed by electron microscopy. *J. Reprod. Fertil.* **19**, 109–19.

Fawcett, D. W., Neaves, W. B. and Flores, M. N. (1973) Comparative observations on intertubular lymphatics and the organization of the interstitial tissue of the mammalian testis. *Biol. Reprod.* **9**, 500–32.

Fazzari, I. (1933) La vascolarizzazione del testicolo. *Giorn. Sci. Nat. Econ.* (Palermo) **37**, 1–198.

Ferner, H. (1957) Die dissemination der Hodenzwischenzellen und zur Langerhansschen Inseln als funktionelles Prinzip fur die Samenkanalchen und das exokrine Pankreas. *Z. Mikroskop Anat. Forsch.* **63**, 35–52.

Fowler, D. G. and Setchell, B. P. (1971) Selecting Merino rams for ability to withstand infertility caused by heat. 2. The effect of heat on scrotal and testicular blood flow. *Aust. J. Exptl. Agr. Animal Husb.* **11**, 143–7.

Free, M. J. (1977) Blood supply to the testis and its role in local exchange and transport of hormones in the male reproductive tract. In *The Testis*, eds. A. D. Johnson and W. R. Gomes, IV, 39–90. New York: Academic Press.

Free, M. J. and Jaffe, R. A. (1972a) Dynamics of circulation in the testis of the conscious rat. *Am. J. Physiol.* **223**, 241–8.

Free, M. J. and Jaffe, R. A. (1972b) Effect of prostaglandins on blood flow and pressure in the testis of the conscious rat. *Prostaglandins* **1**, 483–98.

Free, M. J. and Jaffe, R. A. (1975) Dynamics of venous-arterial testosterone transfer in the pampiniform plexus of the rat. *Endocrinology* **97**, 169–77.

Free, M. J. and Nguyen Duc Kien (1973) Venous arterial interactions involving serotonin in the pampiniform plexus of the rat. *Proc. Soc. Exp. Biol.* (New York) **143**, 284–8.

Free, M. J. and Tillson, S. A. (1975) Local increase in concentration of steroids by venous-arterial transfer in the pampiniform plexus. In *Hormonal Regulation of Spermatogenesis*, eds. F. S. French, V. Hansson, E. M. Ritzen and S. N. Nayfeh, 181–94, New York: Plenum.

Free, M. J. and VanDemark, N. L. (1968) Gas tensions in spermatic and peripheral blood of rams with normal and heat-treated testes. *Am. J. Physiol.* **214**, 863–5.

Free, M. J., Jaffe, R. A., Jain, S. K. and Gomes, W. R. (1973) Testosterone concentrating mechanism in the reproductive organs of the male rate. *Nature New Biology* **244**, 24–6.

Fregly, M. J. (1962) Relationship between blood pressure and organ weight in the rat. *Am. J. Physiol.* **202**, 967–70.

Frey, H. (1863) Zur Kenntniss der lymphatischen Bahnen im Hoden. *Arch. Pathol. Anat. Physiol.* **28**, 563–9.

Gabbiani, G. and Majno, G. (1969) Endothelial microvilli in the vessels of the rat Gasserian ganglion and testis. *Z. Zellforsch.* **97**, 11–117.

Ginther, O. J., Mapletoft, R. J., Zimmermann, N., Meckley, P. E. and Nuti, L. (1974) Local increase in testosterone concentration in the testicular artery in rams. *J. Anim. Sci.* **38**, 835–7.

Glover, T. D. (1973) Aspects of sperm production in some East African mammals. *J. Reprod. Fertil.* **35**, 45–53.

Godinho, H. P. and Setchell, B. P. (1975) Total and capillary blood flow through the testes of anaesthetized rams. *J. Physiol.* (London) **251**, 19P–20P.

Godinho, H. P., Cardoso, F. M. and Nogueira, J. C. (1973) Patterns of parenchymal ramification of the testicular artery in some ruminants. *Anat. Anz.* **133**, 118–24.

Grant, R. T. and Wright, H. P. (1971) The peculiar vasculature of the external spermatic fascia in the rat; possibly subserving thermoregulation. *J. Anat.* **109**, 293–305.

Gritzuliac, B. V. (1969) On the problem of rat testicle vascularization normally and with ischemia of the testicular artery. *Arch. Histol. Anat., Embryol.* (Leningrad) **56**, 72–6.

Gutzschebauch, A. (1936) Der Hoden der Haussäugetiere und seine Hüllen in biologischer und artdiagnostischer Hinsicht. *Zeitschr. Anat. Entwicklungsgeschiche* **105**, 433–58.

Haberer, H. (1898) Uber die Venen des menschlichen Hodens. *Archs. Anat. Physiol.* 413–43.

Halley, J. B. W. (1960) Relation of Leydig cells in the human testicle to the tubules and testicular function. *Nature* **185**, 865–6.

Harrison, R. G. (1948) Vascular patterns in the testis, with particular reference to *Macropus*. *Nature* **161**, 399–400.

Harrison, R. G. (1949) The comparative anatomy of the blood supply of the mammalian testis. *Proc. Zool. Soc.* (London) **19**, 325–43.

Harrison, R. G. (1951) Applications of microradiography: the testis. In *Micro-arteriography*, ed. A. E. Barclay, 89–90, Oxford: Blackwell.

Harrison, R. G. and Barclay, A. E. (1948) The distribution of the testicular artery (internal spermatic artery) to the human testis. *Brit. J. Urol.* **20**, 57–66.

Harrison, R. G. and Weiner, J. S. (1949) Vascular patterns of the mammalian testis and their functional significance. *J. Exptl. Biol.* **26**, 304–16.

Hasumi, S. (1929a) Anatomische Untersuchungen uber die Blutgefasse der Hoden und Nebenhoden von verschiedenen Saugetieren. *Japan J. Med. Sci.* **2**, 135–58.

Hasumi, S. (1929b) Anatomische Untersuchungen uber das Lymphgefasssystem des mannlichen Urogenitalsystems. *Japan J. Med. Sci.* **2**, 159–86.

Hofmann, R. (1960) Die Fefässarchitektur des Bullenhodens, zugleich ein Versuch ihrer funktionellen Deutung. *Zentralblatt Veterinarmed.* **7**, 59–93.

Huggins, C. B. and Entz, F. H. (1931) Absorption from normal tunica vaginalis testis, hydrocele and spermatocele. *J. Urol.* **25**, 447–55.

Hundeiker, M. (1965) Uber besondere Gefasse des Epiorchium. Untersuchungen am Hoden von *Bos taurus*. *Gegenbaurs Jahrb.* **107**, 53–7.

Hundeiker, M. (1966a) Die Vascularisation des Epiorchium beim Menschen. *Gegenbaurs Jahrb.* **108**, 624–8.

Hundeiker, M. (1966b) Die Kapillararchitektur im Stierhoden. *Angiologica* **3**, 343–8.

Hundeiker, M. (1969) Untersuchungen zur Darstellung der Lymphgefasse im Hodenparenchym beim Stier mit der Injektionsmethode. *Andrologie* **1**, 113–7.

Hundeiker, M. (1971a) Die Capillaren im Hodenparenchym. *Arch. klin. exp. Derm.* **239**, 426–35.

Hundeiker, M. (1971b) Untersuchungen über die Vaskularisation des Hodens. *Fortschr. Med.* **89**, 1403–6.

Hundeiker, M. (1972) Vasculäre Regulationseinrichtungen am Hoden. *Arch. Derm. Forsch.* **245**, 229–44.

Hundeiker, M. and Keller, L. (1963) Die Gefäßarchitektur des menschlichen Hodens. *Morphol. Jb.* **105**, 26–73.

Ichev, K. and Bakardjive, A. (1971) The blood supply of the male sexual gland. *Nauchni Tr. Vissh. Med. Inst. Sofia* **50**, 9.

Jantosovicova, J. (1969a) Contribution to the study of the veinal system of the testis and epididymis of rams, boars and stallions. *Folia Vet.* **13**, 2, 13–20.

Jantosovicova, J. (1969b) To the question of anastomoses of the arteries of ram, boar and stallion testis and epididymis. *Folia Vet.* **13**, 2, 21–6.

Jantosovicova, J. (1969c) Intraorgánová sústava tepien semenníka barana, kanca a žrebca. *Folia. vet.* **13**, 3–4, 26–31.

Jarisch, A. (1888–9) Ueber die Schlagadern des menschlichen Hodens. *Ber. Naturw. Med. Ver. Innsbruck* **18**, 32–79.

Joffre, J. and Joffre, M. (1973) Seasonal changes in the testicular blood flow of seasonally breeding mammals: dormouse (*Glis glis*), ferret (*Mustella furo*) and fox (*Vulpes vulpes*). *J. Reprod. Fertil.* **34**, 227–3.

Joffre, M. and Joffre, J. (1971) Debit sanguin testiculaire chez le rat: mise en évidence sur les courbes de clearance du xenon 133 d'une composante liée aux graisses epididymaires. *Compt. Rend. Acad. Sci.* **273**, 496–9.

Joranson, Y., Emmel, V. E. and Pilka, H. J. (1929) Factors controlling the arterial supply of the testis under experimental conditions. *Anat. Rec.* **41**, 157–76.

Kirby, A. (1953) Observations on the blood supply of the bull testis. *Brit. Vet. J.* **109**, 464–72.

Kirby, A. and Harrison, R. G. (1954) A comparison of the vascularization of the testis in Afrikaner and English breeds of bull. *Proc. Soc. Study Fertil.* **6**, 129–39.

Kormano, M. (1967) An angiographic study of the testicular vasculature in the postnatal rat. *Z. Anat. Entwicklungsgeschichte* **126**, 138–53.

Kormano, M. (1970a) Anomalous course of the testicular artery in the rat. *Anat. Anz.* **126**, 505–7.

Kormano, M. (1970b) Effects of serotonin and angiotensin on testicular blood vessels in the rat. *Angiologica* **7**, 291–5.

Kormano, M. (1973) Blood supply to testes and excurrent ducts. *Advan. Biosci.* **10**, 73–82.

Kormano, M. and Koskimies, A. I. (1971) Prenatal development of the testicular artery in the rat. *Z. Anat. Entwicklungsgeschichte* **134**, 73–80.

Kormano, M. and Suoranta, H. (1971a) An angiographic study of the arterial pattern of the human testis. *Anat. Anz.* **128**, 69–76.

Kormano, M. and Suoranta, H. (1971b) Microvascular organization of the adult human testis. *Anat. Rec.* **170**, 31–40.

Kormano, M., Karhunen, P. and Kahanpää, K. (1968) Effect of long-term 5-hydroxytryptamine treatment on the rat testis. *Ann. Med. exp. Fenn* **46**, 474–8.

Larson, L. L. and Foote, R. H. (1974) Testicular blood flow rates in prepubertal and adult rabbits measured by [85]krypton. *Proc. Soc. Exp. Biol. Med.* **147**, 151–3.

Lasserre, R. and Armingaud, F. (1934) Anatomie des vaisseaux testiculaires chez le cheval et applications à la pathologie chirurgicale. *Rev. Vet* (Toulouse) **86**, 13–38.

Laulanie, M. (1884) Sur le renflement érectile terminal de l'artère spermatique dans le foetus. *Compt. Rend. Soc. Biol.* **36**, 624–6.

Laux, W. (1944) Uber die funktionelle Bedeutung der unterschiedlichen Gefassversorgung von Hoden und Nebenhoden des Menschen. *Z. Anat. Entwicklungsgeschichte* **113**, 267–80.

Lindner, H. R. (1963) Partition of androgen between the lymph and venous blood of the testis in the ram. *J. Endocrinol.* **25**, 483–94.

Lindner, H. R. (1967) Participation of lymph in the transport of gonadal hormones. In *Proceedings of the 2nd International Congress Hormonal Steroids, May 1966*, pp. 821–7. Amsterdam: Excerpta Med. Found. (Intern. Congr. Ser. 132).

Linzell, J. L. and Setchell, B. P. (1969) Metabolism, sperm and fluid production of the isolated perfused testis of the sheep and goat. *J. Physiol.* (Lond.) **201**, 129–43.

Ludwig, K. and Tomsa, W. (1961) Die Anfänge der Lymphgefässe in Hoden. *Sitzber. Akad. Wiss. Wien. Math.-Naturw. Kl. Abt.* II **44**, 155–6.

Ludwig, K. and Tomsa, W. (1862) Die Lymphwege des Hodens und ihr Verhaltniss zu den Blut und Samengefassen. *Sitzber. Akad. Wiss. Wien, Math.-Naturw. Kl Abt.* II **46**, 221–37.

Mason, K. E., Shaver, S. L., Hodge, H. C. and Maynard, E. A. (1951) Extensive intratubular and extratubular edema of the rat testis compatible with spermatogenic function. *Anat. Rec.* **109**, 402–3.

Mihalkovics, V. von (1873) Beitrage zur Anatomie und Histologie des Hodens. *Ber. Verhandl. K. Sachs. Ges. Wiss.* **24**, 217–56.

Mitchell, W. A. and Bacon, R. L. (1957) The blood supply of the testis and adnexa of the mouse. *Anat. Rec.* **127**, 475.

Morris, B. and McIntosh, G. H. (1970) The lymphatic drainage of the testis and scrotal serous cavity of the ram. In *Progress in Lymphology II*, pp. 173–6, eds. M. Viamonte, P. R. Kochler, M. Witte and C. Witte, Stuttgart: Verlag.

Morris, B. and McIntosh, G. H. (1971) Techniques for the collection of lymph with special reference to the testis and ovary. *Acta Endocrinol.* Suppl. **158**, 145–68.

Müller, I. (1957) Kanalchen- und Capillararchitektonik des Rattenhodens. *Z. Zellforsch. Mikroskop. Anat.* **45**, 522–37.

Nishida, T. (1964) Comparative and topographical anatomy of the fowl XLII. Blood vascular system of the male reproductive organs. [English summary.] *Nippon Juigaku Zasshi* **26**, 211–21.

O'Steen, W. K. (1963) Serotonin and histamine: Effects of a single injection on the mouse testis and prostate gland. *Proc. Soc. Exptl. Biol. Med.* **113**, 161–3.

Pellanda, C. (1903) La circulation artérielle du testicule. *Intern. Monatsschr. Anat. Physiol.* **20**, 240–66.

Pettersson, S., Soderholm, B., Persson, J. E., Eriksson, S. and Fritjofsson, A. (1973) Testicular blood flow in man measured with venous occlusion plethysmography and xenon[133]. *Scand. J. Urol. Nephrol.* **7**, 115–19.

Picque, R. and Worms, G. (1909) Les voies anastomotiques de la circulation artérielle testiculo-epididymaire. *J. Anat.* (Paris) **45**, 51–64.

Plato, J. (1897) Die interstitiellen Zellen des Hodens und ihre physiologische Bedeutung. *Arch. Mikroskop. Anat. Entwicklungsmech* **48**, 280–304.

Popesko, P. (1965) Vaskularizacia bycieho semennika Pars convoluta et pars marginalis A. spermaticae internae. *Folia Vet.* **9**, 137–46.

Regaud, C. (1897a) Les vaisseaux lymphatiques du testicule. *Compt. Rend. Soc. Biol.* **49**, 659–61.

Regaud, C. (1897b) Les vaisseaux lymphatiques du testicule et les faux endothéliums de la surface des tubes séminifères. Thèse pour Docteur en Médecine, University of Lyon.

Repciuc, E. and Andronescu, A. (1967) Vergleichend-anatomische Betrachtungen uber die muskulos-elastischen Kissen der A. testicularis. *Acta Anat.* **66**, 118–32.

Rinker, J. R. and Allen, L. (1951) A lymphatic defect in hydrocele. *Am. Surgeon* **17**, 681.

Sand, R. S., Dutt, R. H. and Preston, D. F. (1970) Measuring blood flow through ram testes using xenon washout. *J. Anim. Sci.* **31**, 230.

Sand, R. S., Dutt, R. H. and Preston, D. F. (1971) Effect of local heating on ram testis blood flow. *J. Anim. Sci.* **32**, 391.

Sasano, N. and Ichijo, S. (1969) Vascular patterns of the human testis with special reference to its senile changes. *Tohoku J. Exp. Med.* **99**, 269–80.

Sebileau, P. and Arrous, M. (1892) La circulation du testicule. *Compt. Rend. Soc. Biol.* **44**, 53–5.

Schweizer, R. (1929) Uber die Bedeutung der Vascularisation, des Binnendruckes und der Zwischenzellen fur die Biologie des Hodens. *Z. Anat. Entwicklungsgeschichte* **89**, 775–98.

Setchell, B. P. (1970) Testicular blood supply, lymphatic drainage and secretion of fluid. In *The Testis*, eds. A. D. Johnson, W. R. Gomes and N. L. VanDemark, **I**, 101–239, New York: Academic Press.

Setchell, B. P. (1977a) Reproduction in male marsupials. In *The Biology of Marsupials*, eds. D. P. Gilmore and B. Stonehouse, 411–57. London: Macmillan.

Setchell, B. P. (1977b) Male reproductive organs and semen. In *Reproduction in Domestic Animals*, eds. H. H. Cole and P. T. Cupps, 229–56. New York: Academic Press.

Setchell, B. P. and Hinks, N. T. (1969) Absence of countercurrent exchange of oxygen, carbon dioxide, ions or glucose between the arterial and venous blood in the spermatic cords of rams and two marsupials (*Macropus eugenii* and *Megaleia rufa*). *J. Reprod. Fertil.* **20**, 179–81.

Setchell, B. P. and Linzell, J. L. (1974) Soluble indicator techniques for tissue blood flow measurement using $^{86}$Rb-rubidium chloride, urea, antipyrine (phenazone) derivatives or $^3$H-water. *Clin. Exp. Pharmacol. Physiol.* Suppl. 1, 15–29.

Setchell, B. P. and Thorburn, G. D. (1969) The effect of local heating on blood flow through the testes of some Australian marsupials. *Comp. Biochem. Physiol.* 31, 675–7.

Setchell, B. P. and Thorburn, G. D. (1970) The effect of artificial cryptorchidism on the testis and testicular blood flow in an Australian marsupial *Macropus eugenii*. *Comp. Biochem. Physiol.* 38A, 705–8.

Setchell, B. P. and Waites, G. M. H. (1964) Blood flow and the uptake of glucose and oxygen in the testis and epididymis of the ram. *J. Physiol.* (London) 171, 411–25.

Setchell, B. P. and Waites, G. M. H. (1969) Pulse attenuation and counter-current heat exchange in the internal spermatic artery of some Australian marsupials. *J. Reprod. Fertil.* 20, 165–9.

Setchell, B. P., Waites, G. M. H. and Till, A. R. (1964) Variations in flow of blood within the epididymis and testis of the sheep and rat. *Nature* (London) 203, 317–8.

Setchell, B. P., Waites, G. M. H. and Lindner, H. R. (1965) Effect of undernutrition on testicular blood flow and metabolism and the output of testosterone in the ram. *J. Reprod. Fertil.* 9, 149–62.

Setchell, B. P., Waites, G. M. H. and Thorburn, G. D. (1966) Blood flow in the testis of the conscious ram measured with krypton-85; effects of heat, catecholamines and acetyl choline. *Circulation Research* 18, 755–65.

Short, R. V., Mann, T. and Hay, M. F. (1967) Male reproductive organs of the African elephant, *Loxodonta africana*. *J. Reprod. Fertil.* 13, 517–36.

Tuffli, G. (1928) Die Arterienversorgung von Hoden und Nebenhoden. Inaugural Dissertation, Faculty of Veterinary Medicine, University of Zurich.

Vrzgulova, M., Hajovska, B. and Jantosovicova, J. (1965) Studium vyskytu A. accessoria testis v semennika byka. *Folia Vet.* 9, 165–71.

Waites, G. M. H. and Moule, G. R. (1960) Blood pressure in the internal spermatic artery of the ram. *J. Reprod. Fertil.* 1, 223–9.

Waites, G. M. H. and Moule, G. R. (1961) Relation of vascular heat exchange to temperature regulation in the testis of the ram. *J. Reprod. Fertil.* 2, 213–24.

Waites, G. M. H. and Setchell, B. P. (1964) Effect of local heating on blood flow and metabolism of the testes of the conscious ram. *J. Reprod. Fertil.* 8, 339–49.

Waites, G. M. H., Archer, V. and Langford, G. A. (1975) Regional sensitivity of the testicular artery to noradrenaline in the ram, rabbit, rat and boar. *J. Reprod. Fertil.* 45, 159–63.

Waites, G. M. H., Setchell, B. P. and Quinlan, D. (1973) The effect of local heating of the scrotum, testes and epididymides on cardiac output and regional blood flow. *J. Reprod. Fertil.* 34, 41–9.

Wallace, J. C. and Lascelles, A. K. (1964) Composition of testicular and epididymal lymph in the ram. *J. Reprod. Fertil.* 8, 235–42.

Williams, R. G. (1949) Some responses of living blood vessels and connective tissue to testicular grafts in rabbits. *Anat. Rec.* 104, 147–61.

Wolff, J. and Merker, H. J. (1966) Ultrastruktur und Bildung von Poren im Endothel von porösen und geschlossenen Kapillaren. *Z. Zellforsch. Mikroskop. Anat.* **73**, 174–91.

Wolfram, W. (1942) Zur anatomie der Arteria spermatica interna. *Klin. Wochschr.* **21**, 1126–7.

# 4 Nerves of the testis and scrotum

## 4.1 Anatomy of testicular nerves

### 4.1.1 Origin of testicular nerves

The testis, being a visceral organ, receives only visceral afferent and efferent fibres and lacks somatic nerves. The visceral efferent or autonomic nerves to the testis are predominantly of the sympathetic type and there is little evidence for a parasympathetic supply.

The main nerve supply to the testis is the superior spermatic nerve which arises in all species studied, except the rat, from the spermatic ganglion. This ganglion lies close to the origin of the testicular artery and receives its nervous supply from several of the other prevertebral plexuses, namely the intermesenteric and coeliac plexuses, and in many species, but not man, from the caudal mesenteric plexus as well. The spermatic ganglion is also directly supplied by nerves from the lumbar splanchnic nerves, as are the other prevertebral plexuses, with the coeliac plexus also receiving thoracic splanchnic nerves and branches of the vagus. The indirect supply to the spermatic ganglion via the prevertebral plexuses is more important than the direct lumbar splanchnic supply. In the horse and the sheep, the spermatic ganglion is fused to the caudal mesenteric ganglion.

From the spermatic ganglion, the nerves run with the internal spermatic artery, either as the spermatic plexus surrounding the vessel or as a discrete superior spermatic nerve, as for example in man, sheep and horse. The nerves follow the artery through the inguinal canal to the testis (Figure 4.1; Retzius, 1893; Langley and Anderson, 1895; Kuntz, 1919a; Mitchell, 1935, 1938; Kuntz and Morris, 1946; Gray, 1947; Hodson, 1965, 1970; Shioda and Nishida, 1966).

### 4.1.2 Distribution of nerves within the testis

The nerves entering the testis follow the arteries as they encircle the testis and then branch to supply the interior of the testis. The tunics of the testis also receive branches from the superior spermatic nerve.

#### 4.1.2.1 Internal nerves
The nerves of the testicular parenchyma, like the blood and lymph vessels, are confined to the interstitial spaces. They are not especially prominent in conventional sections, but in sections treated to reveal catecholamine-containing fibres, the nerves can be seen to lie mainly in association with the arteries (Baumgarten and Holstein, 1967; Norberg *et al.*, 1967; Baumgarten *et al.*, 1968). While the fibres definitely do not penetrate the boundary tissue of the tubules, several authors (Shioda and Nishida, 1966; Baumgarten and Holstein, 1967) have reported nerve fibres running close to the myoid cells there. It has also been suggested that the Leydig cells may lie in close association with perivascular nerve fibres (Okkels and

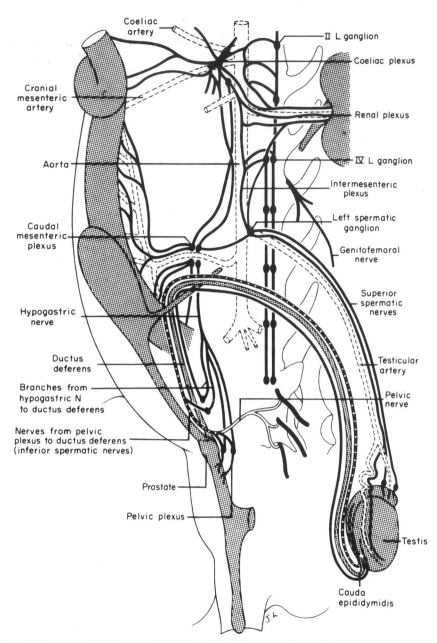

Figure 4.1 A diagram of the nerves to the testis and epididymis in the cat.
Genito-femoral nerve = inguinal nerve; branches from hypogastric nerve to ductus deferens = middle spermatic nerve
Black nerves = sympathetic fibres
White nerves = parasympathetic fibres
Dotted nerves = mixed sympathetic and parasympathetic fibres
Vagal branches and thoracic splanchnic nerves to coeliac plexus and many lumbar splanchnic nerves to all the prevertebral plexuses have been omitted. All nerves contain afferent fibres. (Reproduced from Hodson (1970) who modified a diagram from Langley and Anderson, 1896)

Sand, 1940; van Campenhout, 1949). This certainly seems to be so in man (Baumgarten and Holstein, 1967) and in cat, rat and guinea pig. The Leydig cells near the surface of the testis just under the tunics of these species seem to be more closely associated with nerve fibres than those deep within the testis (Norberg *et al.*, 1967).

The rete testis appears to have a very poor nerve supply, although Kreutz (1964) has reported some fibres entering the testis of the dog along the efferent ducts. These fibres would presumably have come from the superior spermatic nerve, as the head of the epididymis is also supplied by this nerve, in contrast to the tail of the epididymis and the ductus deferens, which receives a mixed adrenergic and cholinergic supply from the inferior spermatic nerve.

**4.1.2.2    Nerve supply to the tunics of the testis**    The tunics of the testis receive an abundant nerve supply, both from the plexus around the blood vessels and direct from the superior spermatic nerve (Risley and Skrepetos, 1964; Norberg *et al.*, 1967; Baumgarten, 1968; Bell and McLean, 1973; Hargrove and Ellis, 1976). The terminal nerve fibres seem to have a reasonably close association with smooth muscle cells in the tunics and there are also numerous specialized nerve endings. These receptors have been described by Corona (1953) as either Pacinian corpuscles or 'intercalated' corpuscles, i.e. midway between encapsulated and free endings. Kreutz (1964) considers that the encapsulated endings are more akin to Meissner's, not Pacinian, corpuscles. Pain receptors are also probably located in the tunics, but these have not been identified anatomically.

## 4.2    Anatomy of scrotal nerves

### 4.2.1    Origin of scrotal nerves

The nerve supply to the scrotum, like its blood supply, is quite distinct from that to the testis. Nerves from three sources reach the scrotum (Figure 4.2). The *pudendal* nerves arise from the ventral branches of the lumbosacral plexus and send two branches to the scrotum, a caudal scrotal nerve and the scrotal branch of the superficial perineal nerve. The *inguinal* nerve (also called the external spermatic or genitofemoral nerve) arises from the lumbar nerves and runs through the inguinal canal on the surface of the cremaster muscle; it supplies this muscle, the tunica vaginalis and the tunica dartos. The *ilio-inguinal* nerve arises from similar or more cranial lumbar nerves; it also runs through the inguinal canal, but then becomes superficial and supplies branches to the dorsal part of the scrotum; in its course, it has connections with both the inguinal and pudendal nerves. The nerves to the scrotum are somatic and visceral and contain sensory, sympathetic (vasomotor, sudomotor and piloerector) and somatic efferent fibres (Langley and Anderson, 1896; Larson and Kitchell, 1958; Linzell, 1959; Hodson, 1970).

### 4.2.2    Thermal receptors in the scrotum

It can be presumed from physiological experiments that the scrotum is abundantly supplied with the free endings which constitute the morphological form of thermal

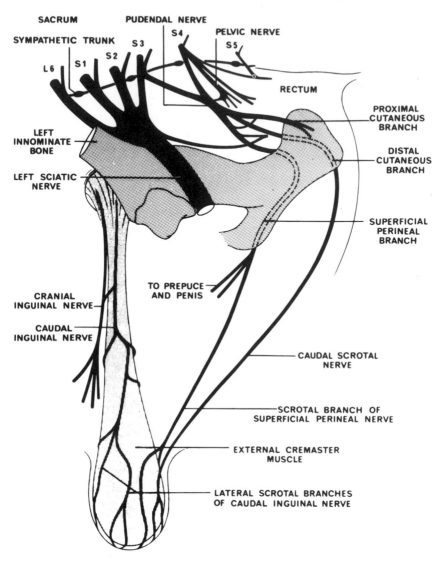

Figure 4.2  A diagram of the nerves of the scrotum of the bull. Cranial and caudal inguinal nerves = inguinal (genito-fermoral) nerve. (Reproduced from Hodson, 1970, who modified a diagram from Larson and Kitchell, 1958)

receptors in other areas of skin. There has been controversy about whether or not a particular sensory attribute can always be linked to a particular receptor type, but now most would agree that there are specific tactile and thermal receptors. Thermal receptors have been defined as those demonstrating: (1) a static discharge along their afferent neurone at a constant skin temperature, the rate of discharge being related to temperature in a constant way; (2) an enhanced rate of discharge with a fall ('cold' receptor) or rise ('warm' receptor) in temperature; (3) insensitivity to non-thermal stimuli; (4) in humans, a threshold sensitivity similar to human perceptual thermal thresholds.

By changing the temperature of rat scrotal skin and recording from afferent neurones attached to single or small groups of receptors in the skin Iggo (1969) and Hellon *et al.* (1975) obtained evidence for 'cold' receptors in the scrotum with maximum sensitivity in the range 23–28°C, and 'warm' receptors which are most sensitive in the range 38–41°C (Figure 4.3). The two receptor types were quite distinct with little or no overlap in the temperature thresholds and both types have minimal activity when skin temperature is close to normal testicular temperature for the rat.

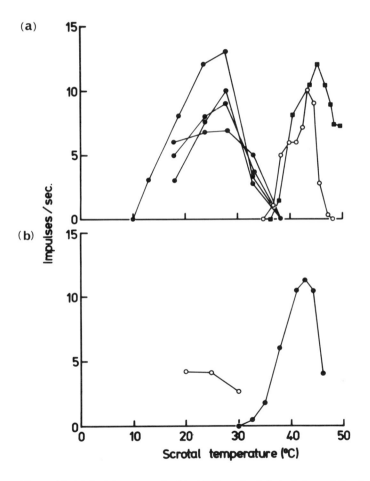

Figure 4.3 (a) Activity curves for 4 'cold' fibres ( ● ), 1 single 'warm' fibre ( ○ ) and a multi-unit 'warm' receptor preparation ( ■ ) of the scrotal nerves of the rats. (Data from Iggo, 1969) (b) Mean impulse activity curves for 6 single cold ( ○ ) and 9 to 14 warm fibres ( ● ) from the scrotal nerves of rats. (Data from Hellon *et al.*, 1975). Note how the activity of the warm receptor reaches a maximum at about 41°C, and the cold receptors at about 25°C, with both reaching a minimum at normal scrotal temperature of about 33°C.

Impulses from these scrotal receptors can be detected in neurones in the dorsal horn of the spinal cord at the level of T13 and L1. In the region where the scrotal nerve enters the cord, 47 per cent of the neurones were responsive to scrotal temperature; half of these were excited by warming and half by cooling. Most of these thermally responding units were not affected by touching the scrotal skin. Both 'warm' and 'cold' neurones responded to steep changes of temperature, some with a short-lived 'dynamic' response which usually returned to initial values within two minutes, some with a 'static' response which was maintained, and some with a combination of these two responses (Figure 4.4). As the receptors in the scrotal skin all showed dynamic responses, it has been suggested that some processing of their output occurs in the dorsal horn (Hellon and Misra, 1973a). Other workers propose that these results can be explained on the basis of the peripheral afferents and that information processing need not take place at the spinal level (Pierau *et al.*, 1975).

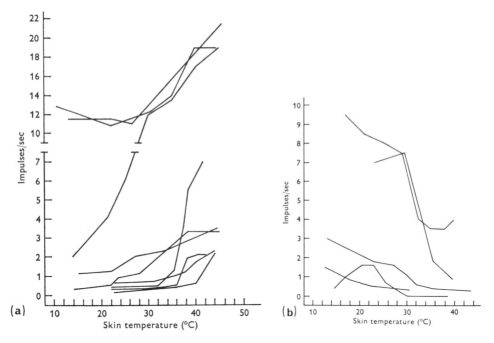

Figure 4.4 Graphs showing the relation between firing rate of neurones in the dorsal horn of the spinal cord and scrotal skin temperature. The neurones shown are those showing only static responses, in the (a) nine neurones showing increased firing rate with rising temperature, and in (b) five neurones showing increased firing with falling temperature. (Reproduced from Hellon and Misra, 1973a)

Neurones responding to changes in scrotal temperature over the range 31–40°C can also be found in the ventrobasal complex of the rat thalamus and the cerebral cortex. Most (82 per cent) of those responding in the thalamus increased their firing rate as scrotal temperature was raised and the rest decreased their firing rate. Most of the individual neurones showed a sudden and maintained change in their activity with changes of scrotal temperature of between 0·5 and 2°C. Mean firing rates changed by factors of about 8 and further increases had no further effect. By contrast, most (83 per cent) of the neurones responding in the cortex decreased their firing rate as scrotal temperature was raised, but like the cells in the thalamus, they changed their firing rates over temperature ranges of between 0·5 and 2°C and then showed no further change (Figures 4.5 and 4.6; Hellon and Misra, 1973b; Hellon *et al.*, 1973).

Almost all the units in the dorsal horn, thalamus and cortex responding specifically to scrotal temperature were equally affected by temperature changes on either side of the scrotum. The receptive fields of these units were bilateral and large, implying a massive convergence of fibres from thermoreceptors on to each central unit.

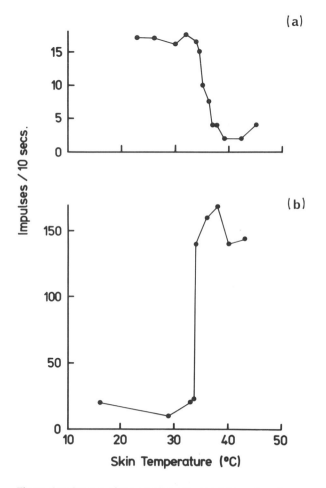

Figure 4.5 Graphs of the relation between firing rates of one cortical neurone (a) and one thalamic neurone (b) and scrotal skin temperature in rats. Note that the neurones change their firing rate sharply but only over a small temperature range; the thalamic neurone increases its firing rate, whereas the cortical neurone decreases its rate as temperature increases. (Reproduced from Hellon and Misra, 1973a, and Hellon *et al.*, 1973b)

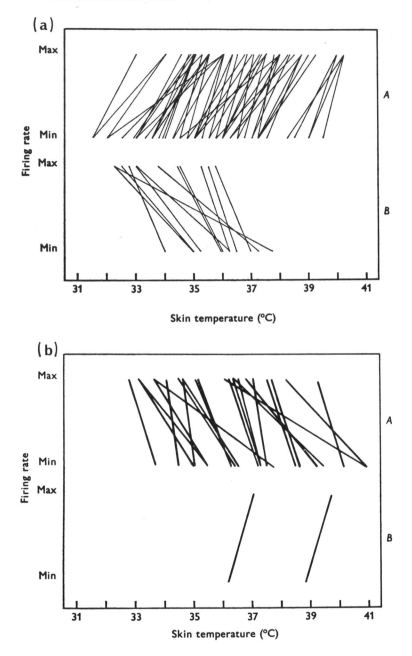

Figure 4.6 Graphs of the operating ranges of thalamic (a) and cortical (b) neurones in rats, showing some cells of both types which increased their firing rate and some which decreased the rate as temperature of the scrotal skin increased. A majority of thalamic neurones increase and a majority of cortical neurones decrease their firing rates as temperature increases. (Reproduced from Hellon and Misra, 1973a, and Hellon *et al.*, 1973b)

In contrast, units sensitive to mechanical stimulation responded only to unilateral stimulation (Hellon and Mitchell, 1975).

## 4.3    Functions of testicular nerves

Several possible functions may be attributed to the testicular nerves. Firstly, the nerves are undoubtedly concerned with the control of vascular tone in the testis or possibly in the spermatic cord. Stimulation of the hypothalamus or of the sympathetic chain of rabbits caused sharp decreases in oxygen tension in the testis (Cross and Silver, 1962a), presumably by decreased blood flow. Stimulation of nerves in the spermatic cord in the isolated perfused testis of the ram caused a pronounced vasoconstriction. However, there appears to be comparatively little resting sympathetic vascular tone (see Section 3.4). It seems that no-one has considered whether the effects of nerve stimulation are on the blood vessels in the spermatic cord or on those in the testis itself.

Secondly, the testicular capsule may be under nervous control because of the many nerve endings found in the capsule. The rate of spontaneous contractions of the testicular capsule can be stimulated by various adrenergic and cholinergic compounds, but apparently no studies have been made on the effects of stimulation of the spermatic nerves (see Section 1.3.3).

Thirdly, it would also appear possible that the movements of the seminiferous tubules themselves may respond to nervous stimulation if the anatomical relation between the myoid cells and nerves is substantiated; the tubules respond *in vitro* to oxytocin but again the effects of nerve stimulation do not appear to have been examined (see Section 1.3.1).

Fourthly, the close relationship of nerve endings and some of the Leydig cells might suggest that their secretion could have some nervous control (see Section 4.1.2.1), as well as the better known control by LH (see Section 10.1). During sexual activity there are increases in the testosterone concentration in serum (see Section 6.1.4.5), which are not necessarily related to equivalent changes in serum LH; nervous factors could be involved under these circumstances.

Fifthly, the sensory nerves are almost certainly involved in the sensation of pain in the testis after trauma (Woollard and Carmichael, 1933). It is not clear whether the pain originates from receptors in the tunica vaginalis and scrotum, rather than in the testis itself, as anaesthesia of the scrotal nerves can abolish pain in the testis produced by compression. The testis is also a site of referred pain from the kidney and upper urinary tract, for example with ureteric obstruction due to a renal calculus. This may be due to the common origin of the nerves to the testis and to the other organs, or because the inguinal nerve lies very near the ureter and may be locally stimulated (Hodson, 1970).

Finally, there is the curious observation by Radnot (1964) that unilateral castration or testicular X-irradiation leads to a decrease in intra-ocular pressure in the contralateral eye. This implies that testicular afferents may also be important.

## 4.4        Effect of neurectomy on the testis

There is now abundant evidence in many species that removal of either parts of the lumbar sympathetic chain and/or the thoracic splanchnic nerves, or the coeliac plexus, or the caudal mesenteric ganglion and spermatic plexus, causes degeneration of testicular cells and vasodilatation of the blood vessels of the testis and epididymis (Kuntz, 1919b; Takahashi, 1922; Marconi, 1923; King and Langworthy, 1940; Bandmann, 1949, 1950; Khodorovsky, 1965; Vague *et al.*, 1965). However, the degeneration only appears after some days, and may be secondary to the vascular effects of neurectomy; these effects need to be defined more precisely by physiological experiments.

## 4.5        Function of the scrotal nerves

The scrotal nerves fulfil two important functions. They conduct impulses from the receptors in the scrotal skin to the central nervous system. They also carry the nervous impulses which help to control the scrotal sweat glands, the tunica dartos and cremaster muscle.

If the scrotum of a ram is heated, there are central physiological responses as well as local effects on the sweat glands (see Chapter 5). Both types of response are blocked by local anaesthesia of the pudendal nerve and therefore this nerve would seem to be the principal route for the afferent neurones from the receptors in the scrotal skin. The efferent fibres to the tunica dartos arise from the lumbar sympathetic outflow as can be demonstrated by unilateral stimulation and the resultant ipsilateral contraction. These fibres also appear to run in the pudendal nerve as the ram scrotum became fully extended when this nerve was locally anaesthetized, even when the animal was standing in a cold environment (Fowler, 1969). The pudendal nerve in the ram also carries most of the sensory fibres from the scrotal skin, as all sensation there appears to be lost after local anaesthesia of this nerve (Waites, 1962; Waites and Voglmayr, 1963).

The efferent fibres to the sweat glands appear to reside in the sympathetic adrenergic nerves. Sympathetic denervation considerably reduces sweat gland activity, but so does adrenalectomy (see Section 5.3.3.1). There would therefore seem to be a dual control—nervous and hormonal.

## References to Chapter 4

Bandmann, F. (1949) Uber die Beeinflussung der Hodenfunktion durch Resektion des lumbalen Grenzstranges. *Chirurg* **20**, 132–6.

Bandmann, F. (1950) Weitere Beobachtungen uber die Hodenfunktion nach lumbaler Grenzstrangresektion. *Bruns' Beitr. Klin. Chir.* **181**, 419–30.

Baumgarten, H. G. and Holstein, A. F. (1967) Catecholaminhaltige Nervenfasern im Hoden des Menschen. *Z. Zellforsch. Mikroskop. Anat.* **79**, 389–95.

Baumgarten, H. G. and Holstein, A.-F. (1968) Adrenerge Innervation im Hoden und Nebenhoden vom Schwan (*Cygnus olor*). *Z. Zellforsch. Microskop. Anat.* **91**, 402–10.

Baumgarten, H. G., Falck, B., Holstein, A. F., Owman, C. and Owman, T. (1968) Adrenergic innervation of the human testis epididymis, ductus deferens and prostate: A fluorescence microscopic and fluorimetric study. *Z. Zellforsch. Mikroskop. Anat.* **90**, 81–95.

Bell, C. and McLean, J. R. (1973) The autonomic innervation of the rat testicular capsule. *J. Reprod. Fertil.* **32**, 253–8.

Campenhout, E. van (1949) Les relations nerveuses de la glande interstitielle des glandes génitales chez les mammifères. *Rev. Can. Biol.* **8**, 374–429.

Corona, G. L. (1953) L'innervazione della vaginale propria del testicolo. *Z. Anat. Entwicklungsgeschichte* **117**, 306–14.

Coujard, R. (1954) Contribution a l'etude des voies nerveuses sympathiques du testicule. *Arch. Anat. Microscop. Morphol. Exptl.* **43**, 321–64.

Cross, B. A. and Silver, I. A. (1962a) Neurovascular control of oxygen tension in the testis and epididymis. *J. Reprod. Fertil.* **3**, 377–95.

Cross, B. A. and Silver, I. A. (1962b) Some factors affecting oxygen tension in the brain and other organs. *Proc. Roy. Soc.* (London) B. **156**, 483–99.

Dayan, A. D. (1970) Variation between species in the innervation of intra-testicular blood vessels. *Experimentia* **26**, 1359–60.

Fowler, D. G. (1969) The relationship between air temperature scrotal surface area and testis temperature in rams. *Aust. J. Exptl. Agric. Anim. Husb.* **9**, 258–61.

Gray, D. J. (1947) The intrinsic nerves of the testis. *Anat. Rec.* **98**, 325–35.

Hanzalova, D. (1970a) Innervation of the scrotum of bulls. I. Sensitive cells in the scrotum. *Folia vet.* **14**, 1–2, 27–34.

Hanzalova, D. (1970b) The innervation of the scrotum in bulls. II. The innervation of the epidermis, corium and subcutis of the scrotum. *Folia. vet.* **14**, 3–4, 19–28.

Hargrove, J. L. and Ellis, L. C. (1976) Autonomic nerves versus prostaglandins in the control of rat and rabbit testicular capsular contractions *in vivo* and *in vitro. Biol. Reprod.* **14**, 651–7.

Hellon, R. F. and Misra, N. K. (1973a) Neurones in the dorsal horn of the rat responding to scrotal skin temperature changes. *J. Physiol.* (London) **232**, 375–88.

Hellon, R. F. and Misra, N. K. (1973b) Neurones in the ventrobasal complex of the rat thalamus responding to scrotal skin temperature changes. *J. Physiol.* **232**, 389–99.

Hellon, R. F. and Mitchell, D. (1975) Convergence in a thermal afferent pathway in the rat. *J. Physiol.* **248**, 359–76.

Hellon, R. F., Misra, N. K. and Provins, K. A. (1973) Neurones in the somatosensory cortex of the rat responding to scrotal skin temperature changes. *J. Physiol.* **232**, 401–11.

Hellon, R. F., Hensel, H. and Schäfer, K. (1975) Thermal receptors in the scrotum of the rat. *J. Physiol.* **248**, 349–57.

Hodson, N. (1964) Role of the hypogastric nerves in seminal emission in the rabbit. *J. Reprod. Fertil.* **7**, 113–22.

Hodson, N. (1965) Sympathetic nerves and reproductive organs in the male rabbit. *J. Reprod. Fertil.* **10**, 209–20.

Hodson, N. (1970) The nerves of the testis, epididymis and scrotum. In *The Testis*, eds. A. D. Johnson, W. R. Gomes and N. L. Vandemark, I, 47–99, New York: Academic Press.

Iggo, A. (1969) Cutaneous thermoreceptors in primates and sub-primates. *J. Physiol.* **200**, 403–30.

Khodorovsky, G. I. (1965) Effects of removal of lateral sympathetic trunks on the structure and function of testes and prostate. *Sechenov Physiol. J. USSR* (English Transl.) **51**, 1123–7.

King, A. N. and Langworthy, O. R. (1940) Testicular degeneration following interruption of the sympathetic pathways. *J. Urol.* **44**, 74–82.

Kreutz, W. (1964) Uber das Vorkommen korpuskulärer Nervenendigungen in der Tunica albuginea testis des Menschen. *Anat. Anz.* **115**, 27–34.

Kuntz, A. (1919a) The innervation of the gonads in the dog. *Anat. Rec.* **17**, 203–19.

Kuntz, A. (1919b) Experimental degeneration in the testis in the dog. *Anat. Rec.* **17**, 221–34.

Kuntz, A. and Morris, R. E. (1946) Components and distribution of the spermatic nerves and the nerves of the vas deferens. *J. Comp. Neurol.* **85**, 33–44.

Langford, G. A. and Silver, A. (1974) Histochemical localization of acetylcholinesterase-containing nerve fibres in the testis. *J. Physiol.* **242**, 9P–10P.

Langley, J. N. and Anderson, H. K. (1895) The innervation of the pelvic and adjoining viscera. Parts III and IV. *J. Physiol.* **19**, 85–121 and 122–30.

Langley, J. N. and Anderson, H. K. (1896) The innervation of the pelvic and adjoining viscera. Part VI. *J. Physiol.* **20**, 372–406.

Larson, L. L. and Kitchell, R. L. (1958) Neural mechanisms in sexual behaviour. II. Gross neuroanatomical and correlative neurophysiological studies of the external genitalia of the bull and the ram. *Am. J. Vet. Res.* **19**, 853–65.

Linzell, J. L. (1959) The innervation of the mammary glands in the sheep and goat with some observations on the lumbo-sacral autonomic nerves. *Q. J. Exptl. Physiol.* **44**, 160–76.

Lombard-Des Gouttes, M-N., Falck, B., Owman, Ch., Rosengren, E., Sjöberg, N-O. and Walles, B. (1974) On the question of content and distribution of amines in the rat testis during development. *Endocrinology* **95**, 1746–9.

Marconi, P. (1923) Atrophie testiculaire par lésions nerveuses. *Compt. Rend. Soc. Biol.* **88**, 356–8.

Mitchell, G. A. G. (1935) The innervation of the kidney, ureter, testicle and epididymis. *J. Anat.* **70**, 10–32.

Mitchell, G. A. G. (1938) The innervation of the ovary, uterine tube, testis and epididymis. *J. Anat.* **72**, 508–17.

Norberg, K-A., Risley, P. L. and Ungerstedt, U. (1967) Adrenergic innervation of the male reproductive ducts in some mammals. 1. The distribution of adrenergic nerves. *Z. Zellforsch. Mikroskop. Anat.* **76**, 278–86.

Okkels, H. and Sand, K. (1940) Morphological relationship between testicular nerves and Leydig cells in man. *J. Endocrinol.* **2**, 38–46.

Peters, H. (1957) Uber die feinere Innervation des Hodens inbesondere des interstitiellen Gewebes und der Hodenkanälchen beim Menschen. *Acta. Neuroveg.* **15**, 235–42.

Pierau, F-K., Torrey, P. and Carpenter, D. O. (1975) Afferent nerve fiber activity responding to temperature changes of scrotal skin of the rat. *J. Neurophysiol.* **38**, 601–12.

Pines, L. and Maiman, R. (1928) Uber die Innervation der Hoden der Saugetiere. *Z. Mikroskop. Anat. Forsch.* **12**, 199–218.

Radnot, M. (1964) Effects of testicular extirpation upon intraocular pressure. *Ann. New York Acad. Sci.* **117**, 614–17.

Retzius, G. (1893) Uber die Nerven der Ovarien und Hoden. *Biol. Untersuch.* [Neue Folge] **5**, 31–4.

Rikimaru, A. and Suzuki, T. (1972) Mechanical responses of the isolated rabbit testis to electrical stimulation and to autonomic drugs. *Tohoku J. exp. Med.* **108**, 283–9.

Risley, P. L. and Skrepetos, C. N. (1964) Histochemical distribution of cholinesterases in the testis, epididymis and vas deferens of the rat. *Anat. Rec.* **148**, 231–49.

Shioda, T. and Nishida, S. (1966) Innervation of the bull testis. *Japan. J. Vet. Sci.* **28**, 251–7.

Shirai, M., Mitsukawa, S. and Matsuda, S. (1975) Impetus to transferring non-motile sperm in the seminiferous tubules into the epididymis. *Tohoku J. exp. Med.* **115**, 95–6.

Stach, W. (1963) Zur Innervation de Leydigschen Zwischenzellen im Hoden. *Z. Mikroskop. Anat. Forsch.* **69**, 569–84.

Takahashi, N. (1922) Hodenatrophie nach Exstirpation des abdominalen Grentzstranges. *Arch. Ges. Physiol.* **196**, 237–42.

Vague, J., Bernard, P., Pache, R., Lieutaud, R. and Mattei, A. (1965) Changes in spermatogenesis and nervous system lesions in man. *European Rev. Endocrinol.* **1**, 127–76.

Waites, G. M. H. (1962) The effect of heating the scrotum of the ram on respiration and body temperature. *Quart. J. Exptl. Physiol.* **47**, 314–23.

Waites, G. M. H. and Voglmayr, J. K. (1963) The functional activity and control of the apocrine sweat glands of the scrotum of the ram. *Aust. J. Agric. Res.* **14**, 839–51.

Woollard, H. H. and Carmichael, E. P. (1933) The testis and referred pain. *Brain* **56**, 293–303.

# 5 The scrotum and thermoregulation

## 5.1 The scrotum

### 5.1.1 Origin and occurrence

A scrotum is not found in vertebrates other than mammals, nor is it present in the Monotremata, which have testes retained within the abdomen. In most marsupials and many eutherian mammals, however, a scrotum develops into a prominent out-pouching of specialized perineal skin containing evaginations of the peritoneum, the tunica vaginalis (Klaatsch, 1890; Weber, 1928; Esser, 1932; van der Broek, 1933). It is usually thought that the scrotum evolved to reduce the temperature of the testis below that of the body (Moore, 1923, 1926; Wislocki, 1933a; Cowles, 1965), possibly in order to reduce the spontaneous mutation-rate (Ehrenberg *et al.*, 1957).

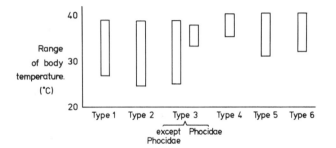

Figure 5.1 Range of body temperatures of animals with different degrees of testicular descent, as defined in Figure 1.1. Note that there are species in all groups with body temperatures approaching 40°C, although some species in groups 1, 2 and 3 have lower body temperatures. The Phocidae have been separated from other species with Type 3 testes because of the peculiar vascular anatomy in these animals (see Section 3.1.2). (Reproduced from Carrick and Setchell, 1977)

More recent evidence suggests that some mammals with abdominal testes have body temperatures as high as mammals with scrotal testes (Figure 5.1). An alternative hypothesis was put forward by Portmann (1952) who suggested that the scrotum evolved as a form of sexual decoration. Ruibal (1957) and Cowles (1958) have argued against this theory. It has also been suggested (Heller, 1929; Bedford, 1977) that it is the epididymis which descends to reach a lower temperature, carrying the testis with it. However, appreciable migration of the testis and epididymis occurs in many species without either organ attaining a lower temperature. It was therefore suggested (Carrick and Setchell, 1977) that some physiological consequence of the elongated testicular artery, which results from testicular migration, confers an evolutionary advantage and the ultimate result of this is a pendulous scrotum.

In eutherian mammals with scrotal testes, the scrotum is formed from the genital swellings which first appear on either side of the urethral plate in the perineal region of early embryos. After the sex of the embryo has been determined by development of the indifferent gonad into an ovary or testis, these swellings give rise to either the paired labia majora or the single scrotum. During the last one-third of gestation, the scrotum becomes a pronounced pair of outpouchings which move towards the anus before fusing together behind the developing phallus (Politzer, 1932).

By contrast, the well developed scrotum found in most marsupials develops in front of the penis (Kaiser, 1931). It is particularly prominent in the kangaroos and wallabies (Figure 5.2). The marsupial mole (*Notoryctes typhlops*) is an exception

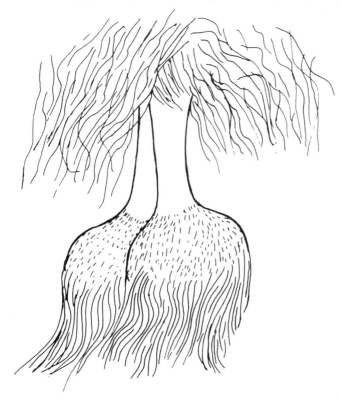

Figure 5.2 Drawing from a photograph of the scrotum of a tammar wallaby (*Macropus eugenii*). Note the very narrow neck, which can be extended as shown when the testis is hot and the animal is at rest; the testis can be drawn up close to the body wall in cold conditions or when the animal is active.

being without a scrotum, the testes remaining between the muscular planes of the abdomen (Stirling, 1891). In the marsupial 'wolf' (*Thylacinus cynocephalus*), the area around the attachment of the scrotum is indented and surrounded anteriorly and laterally by a flap of skin claimed by Pocock (1926) to prevent violent swinging of the scrotum when *Thylacinus* was 'in swift pursuit of its prey'!

The scrota of eutherian mammals range from the pouched scrota of rodents, cats, dogs, horses and pigs, through the more pronounced form of primates, to the exaggerated and pendulous sacs carried by ungulates, including deer, rams, goats and bulls (Figure 1.1)

The testes enter the scrotum midway through pregnancy in ruminants, just before birth in primates or in the weeks after birth in other species. In some primates, the scrotum is often well developed at birth and contains a fully descended testis (Engle, 1932; Wells, 1943; Wyndham, 1943), which later may return into the inguinal canal while the scrotum regresses (Wislocki, 1933b) until puberty.

In seasonally breeding animals, the testes leave the scrotum to return into the inguinal canal or abdomen out of season, e.g. most rodents and wild ungulates (see Section 1.4).

The formation of the scrotum in the fetus and its full development at puberty are believed to be under the control of gonadotrophin-induced androgen secretion (Hamilton, 1936; Andrews, 1940; Wells, 1943). The ability of the rat scrotum to contract with cold, which normally develops at nine weeks of age, was absent in castrated rats; this suggests that the development of the tunica dartos muscle is also under the control of androgen secretion (Andrews, 1940).

### 5.1.2   Structure of the scrotum
The skin is thin, often bare, with very little subcutaneous fat, and with a sheet of smooth muscle, the tunica dartos, lying under the distal portion of the scrotum (Schweizer, 1929; Gutzschenbauch, 1936).

In some animals, such as the rat, there are comparatively few sweat glands, but in sheep and the macropod marsupials there are abundant sweat glands in the scrotal skin, larger and more abundant than those elsewhere in the body (Figure 5.3; Waites

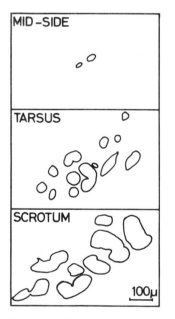

Figure 5.3 Drawing from photographs of sweat glands in midside, tarsal and scrotal skin of a Merino sheep. Note the greater size of the scrotal sweat glands and the greater density of sweat glands in scrotal and tarsal skin compared with midside skin. (Based on a photograph in Waites and Voglmayr, 1962)

and Voglmayr, 1962; Mykytowycz and Nay, 1964; Lyne and Hollis, 1968).

Scrotal skin is often darkly pigmented. In black or brown mice, the scrotum is the first area where pigment appears and in the thirteen-lined ground squirrel (*Citellus tridecemlineatus*), pigmentation of the scrotum is a secondary sexual characteristic. It develops in breeding males coming out of hibernation in the spring, when the scrotum becomes 'deep black', but regresses in June, finally disappearing in August when spermatogenesis ceases. Some marsupials, but by no means all, have a heavily pigmented tunica vaginalis. The depth of the pigmentation in *Didelphis* is increased by androgen administration, but its significance is obscure (Wells, 1935; Finkel, 1945).

## 5.2    Temperature of the testis and of the body

In a thermoneutral environment, the deep body temperature of mammals is normally maintained somewhere in the range 33 to 39°C. There are well defined and often steep temperature gradients from the body core to peripheral areas, which in the case of the extremities are often at a temperature close to that of ambient air.

The scrotal testis is exceptional because it is an abdominal organ which is maintained at a temperature (Moore, 1923, 1926; Moore and Quick, 1923, 1924; Wislocki, 1933a; Phillips and McKenzie, 1934; Waites, 1970, 1976) between 2 and 7°C lower than core temperature. Furthermore, the animal has powerful physiological mechanisms for maintaining this difference (Section 5.3.3) and the testis is damaged if the temperature is allowed to rise (Section 11.1). The temperature of the scrotal testis is determined by the balance between the heat carried to the testis by the arterial blood, the heat generated by metabolism in the testis and the heat lost through the scrotal skin.

Much less information is available for animals with inguinal or abdominal testes but in a number of species, testis temperature has been found to be similar to body temperature (Carrick and Setchell, 1977). However, in some dolphins and porpoises, Kate Parry (personal communication) pointed out that there is in the abdominal wall next to the testis a venous plexus which may bring cooler blood into the abdominal cavity. It has not yet been possible to make measurements to confirm or deny this intriguing possibility.

In seals, the testis is appreciably cooler than the core although it lies under a thick layer of blubber (Bryden, 1967); S. M. McGinnis (personal communication) has suggested that this is probably because cooler venous blood from the back limbs passes through a venous plexus in the blubber overlying the testis.

Kormano (1967a) has studied the development of the rectal–testis temperature gradient in young rats between 20 to 35 days old. The gradient increased from 1·05°C at 20 days to 2·04°C at 30 days and attained adult levels of about 3·5°C at 35 days. During this time there was a striking increase in the capillary network on the testis surface (Section 3.3.3) and attainment of full scrotal development.

The interesting suggestion was made that the length of the internal spermatic artery determined the size of the body-testis temperature gradient among different species (Harrison and Weiner, 1949) and between individuals of the same species, (Kirby and Harrison, 1953). However, this does not appear to be the case, because Waites and Moule (1961) found that the lengths of the artery in a group of rams varied from 151 to 324cm, whereas the rectal-testis temperature differences were all within the range 4·9 to 5·9°C.

When the temperature of the testis itself cannot be measured, scrotal temperature is often measured either by inserting probes subcutaneously or by measuring skin surface temperature. These temperatures follow testis temperature closely, but are usually up to 1°C cooler (Moule and Knapp, 1950; Waites and Moule, 1961).

The scrotum of bulls and rams have decreasing gradients of temperature from the inguinal to the distal regions, and in rams the posterior surface of the scrotum is about 0·5°C warmer than the anterior surface (Riemerschmid and Quinlan, 1941; Fowler, 1968a).

The 'normal' temperature of the human testis is difficult to decide. Most measurements have been made in naked men (Tessler and Krahn, 1966; Stephenson and O'Shaughnessy, 1968; Agger, 1971; Zorgniotti and Macleod, 1973) and this is certainly not a normal condition for most of us. Obviously clothing must have a great effect, but very few reliable measurements in clothed men have been published.

### 5.2.1    Vascular countercurrent heat exchange

The exchange of heat between counterflowing streams of blood is a mechanism available to the extremities for conserving heat in cold conditions. In man, blood in the long femoral and brachial arteries supplying the limbs loses heat to nearby veins. The effect is heightened when hands and feet are artificially cooled (Bazett *et al.*, 1948). A similar heat exchange was either demonstrated or inferred in the limbs of sloths, sea birds and in the fins of whales (Scholander and Schevill, 1955; Scholander and Krog, 1957; Steen and Steen, 1965). In other situations, there are arteries which are coiled to varying degrees and surrounded by venous networks. The cerebral blood vessels of some species have a knotted arrangement of this sort called a carotid rete mirabile. In the sheep this arrangement has been shown to result in heat exchange so that the cerebral arterial blood temperature is influenced by the venous blood temperature (Baker and Heywood, 1968). It is clear that increased surface area and greater proximity of the vessels will improve the efficiency of the heat exchange.

The internal spermatic artery on emergence from the inguinal canal becomes elongated, coiled or branched to varying degrees. Cross-sections of the vessels in the spermatic cord reveal that the veins of the pampiniform plexus closely surround the internal spermatic artery in many mammals. In the dog and sheep, the tunica adventitia of the veins merges with that of the artery (Section 3.1).

Harrison and Weiner (1949) inferred that heat exchange was taking place in the spermatic cord and subsequently, temperature gradients in the blood flowing in the vessels of the spermatic cord were measured by means of small thermocouples in the tips of hypodermic needles or catheters in dogs (Dahl and Herrick, 1959) and rams (Figure 5.4; Waites and Moule, 1961). The latter study also revealed the close relationship between testicular venous temperature and the subcutaneous

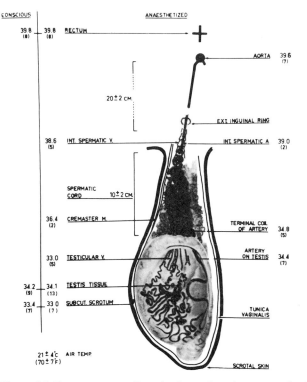

Figure 5.4 Temperature gradients in the testis and scrotum of a Merino ram. Note how the arterial blood is precooled before it reaches the testis by countercurrent exchange of heat in the spermatic cord with the testicular venous blood. Blood in veins on the surface of the testis has been cooled by loss of heat through the scrotal skin, but is warmed to body temperature before it reaches the inguinal ring. (Reproduced from Waites and Moule, 1961)

temperature of the scrotum. The temperature of the venous blood draining the testis is always similar to the temperature beneath the scrotal skin and governs the temperature of the arterial blood entering the testis. In this way, countercurrent heat exchange in the spermatic cord ensures that the temperature deep within the testis is uniform throughout the organ and rapidly follows any variation in scrotal temperature (Figure 5.5). Although blood temperatures have only been measured in species with large vessels, it seems most likely that in all species the blood in the internal spermatic artery is cooled from the abdominal to a value close to testicular temperatures.

Further heat loss may take place from the testicular artery as it winds over the testis surface. This may be particularly important in relatively cold conditions. The heat loss would presumably be most effective in species in which the testicular artery runs a long superficial course around the testis just beneath the tunica albuginea.

It is important to emphasize that the extent of countercurrent heat exchange in the spermatic cord depends solely on the magnitude of the temperature gradient between the body and the scrotum and is in no way autoregulatory. Indeed, vascular exchange can only serve to cool the testis when the returning venous blood is cooler

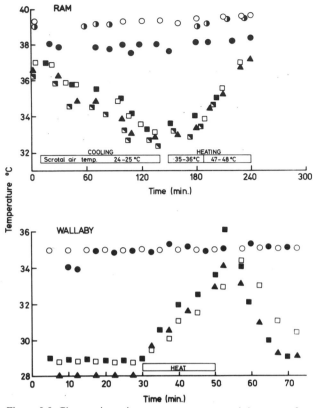

Figure 5.5 Changes in various temperatures around the testes of a ram and of a wallaby during cooling and heating of the testis.

O: rectal; ◑: in aorta; ●: in internal spermatic vein, at inguinal canal; ■: in testicular artery on surface of testis; □: in testis; ▲: in vein on surface of testis; ◪: subcutaneous scrotal.

Note how the temperatures in the testis, in the artery and vein on its surface and under the scrotal skin all change together, while the temperature in the vein above the spermatic cord stay close to body temperature. (Data from Waites and Moule, 1961, and Setchell and Waites, 1969)

than the arterial inflow and this relationship can be maintained only if heat is being lost through the scrotum. In the rat, a normal rectal–testis temperature difference is still maintained by the scrotum even when blood flow in the spermatic vessels is severely reduced (Kormano, 1976b); this would probably not be so in animals with larger testes.

## 5.3    Functional significance of the scrotum

Two aspects must be considered when assessing the scrotum as a thermoregulatory structure: the local control of temperature and the reflex control of general body temperature.

### 5.3.1      Local control of temperature

**5.3.1.1      Blood flow**   The scrotum has a rich blood and lymphatic system which Esser (1932) believed to be associated with the need for fluid secretion through the skin. The smooth muscle of cutaneous arterioles in the scrotum is innervated by postganglionic sympathetic neurones originating in the lumbar sympathetic chain (Section 4.2). Stimulation of the lumbar sympathetic chain or the scrotal nerves causes vasoconstriction in the scrotum (Langley and Anderson, 1895). There was evidence of vasodilatation and increased fluid secretion in the scrotum following removal of the lumbar sympathetic between L2–L5 in sheep (Waites and Voglmayr, 1963) and in analogous clinical cases in man (Monro, 1959). Although there was evidence for residual innervation of blood vessels in both sheep and man, it is clear that the blood vessels of the scrotum are well innervated by the sympathetic nervous system.

It has been estimated that less than 1ml of blood need pass through 100g of human skin per minute to supply the metabolic needs of the tissue, and that blood flow above this rate subserves the function of heat loss to the environment. This is achieved both by raising skin temperature so that conduction, convection and radiation are more effective and by supplying extra fluid for filtration and ultimate evaporation from the skin surface. The body extremities—fingers, toes, hands, feet, ears, tails and peripheral parts of limbs—have labile vasomotor control over skin arterioles so that blood flow can be altered in response to temperature.

There is now good evidence that blood flow in the scrotum can change within wide limits. Elevating scrotal skin temperature, without changing general body temperature, has been achieved in anaesthetized rats by immersion in water and in conscious rams by circulating warmed and dried air around the scrotum. In both species, blood flow increased sharply with temperature so that when the difference between scrotal temperature and the animal's body temperature was reduced by half, blood flow was approximately doubled. In rams, the part of the scrotum furthest from the abdomen tended to have the higher blood flow at all temperatures (Waites et al., 1973; Fowler and Setchell, 1971).

By analogy with other skin areas, it may be assumed that vasodilatation and elevated blood flow in the scrotum can be brought about by the following sequence: direct action on vascular smooth muscle, followed by local reflex removal of sympathetic vasoconstrictor tone derived from adrenergic postganglionic fibres supplying $\alpha$-receptors on the vascular smooth muscle, and then, if there is a rise in general body temperature, there may be total removal of vasoconstrictor tone by an action of the hypothalamus on the medulla oblongata. Possible involvement by local tissue vasodilator hormones cannot be excluded as it should be remembered that bradykinin a potent vasodilator, is released by human eccrine sweat glands.

The skin of the scrotum in sheep contains arterio-venous anastomoses—simple direct communications between arteries and veins (Molyneux, 1965). When glass microspheres were injected into the arteries supplying the scrotum of sheep, spheres of up to 200$\mu$m diameter could be recovered in the venous outflow (G. M. H. Waites, personal communication). Arterio-venous shunts in rabbit ears are assumed to aid in altering the pattern of blood flow through superficial capillaries and in this way to help to maintain maximum heat exchange from the skin surface.

The effect of cooling the scrotum on blood flow has been less fully investigated. Moderately cooling the rat scrotum to 27°C does not significantly reduce blood flow (Waites et al., 1973). It could be assumed that more severe cooling would result in local and reflex vasoconstriction. In extreme cold, bulls in Canada have suffered from scrotal frost-bite presumably compounded by vascular ischaemia (Faulkner et al., 1967).

**5.3.1.2    Surface area and the tunica dartos muscle**  'The scrotum varies greatly in its appearance and size; for under the influence of cold, it is small, contracted and wrinkled; under heat, it is smooth on its surface, and greatly extended' (Cooper, 1830). The smooth muscle sheets of the tunica dartos muscle insert into the skin in such a way that their contractions throw the skin into folds. As the greater thickness of this muscle occurs under the distal part of the scrotum, the greatest rugation occurs there, with the effect of raising the testes towards the abdomen. This contraction can occur in response to a variety of stimuli: local cold or touch, emotion and (presumably) circulating catecholamines. In common with other smooth muscles, the tunica dartos is capable of sustained contraction in the face of long continued cold. (This contraction has often before been attributed to the cremaster muscle, but this is unlikely because the cremaster is a striated muscle and is therefore incapable of sustained contraction.)

The tunica dartos muscles of cat, dog and rabbit are under tonic control from nerves in the lumbar sympathetic outflow and stimulation of these nerves on one side caused ipsilateral contraction (Langley and Anderson, 1895; Lieben, 1908). Reflex adjustment of tone is initiated by receptors in the overlying skin. In rams, the degree of contraction of these muscles is being constantly adjusted in response to changes in skin temperature near the normal testicular temperature (Phillips and McKenzie, 1934; Waites and Moule, 1961). The scrotum starts to extend when its skin temperature rises from about 34 to above 35°C and, when the tunica dartos muscles are fully relaxed, has a surface area of up to 20 per cent greater than normal. When the superficial (superior) perineal nerves were anaesthetized, the scrotum became fully extended, even when the ram was standing in a cold environment (Fowler, 1969). A similar degree of extension was observed for a bull standing in a hot environment (Riemerschmid and Quinlan, 1941).

**5.3.1.3    Evaporation of sweat**  Elevation of scrotal temperature increases moisture loss by simple diffusion maintained by raised blood flow. In some species, e.g. rodents, this is all that is possible. In others, however, the sweat glands of the scrotum can produce considerable volumes of sweat. An inflexion in the rising curve of scrotal skin temperature of heated bull calves (Beakley and Findlay, 1955) and rams (Waites and Voglmayr, 1963) indicates the start of active sweating and the concomitant benefits of evaporative cooling. In rams, the apocrine sweat glands in the scrotum discharge simultaneously, the first expulsion occurring when skin temperature is at about 35·5°C. Thereafter, periodic discharges occur at a frequency of up to 10 per hour. These discharges stopped immediately when the superior perineal nerves were anaesthetized, indicating that they are initiated from receptors in the scrotal skin. The glands themselves remained competent to discharge and could be provoked to do so by intravenous injections of adrenaline or nor-adrenaline (Figure 5.6). As sympathetic denervation reduced sweat gland activity, it is assumed that the efferent side of the reflex is mediated by sympathetic adrenergic nerves possibly releasing noradrenaline to activate $\alpha$-receptors on myoepithelial cells around the sweat glands.

An interesting interaction between sweat and the keratin of wool fibres occurs on the general skin surface of sheep and on the scrotum. If wool was present, a sweat expulsion resulted in a rise of surface temperature of several degrees Celsius, caused by exothermic release of heat, before the evaporation of the sweat could occur and

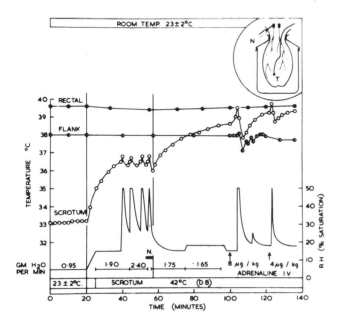

Figure 5.6 The effect of increasing the temperature of the air circulating around the scrotum of a Merino ram on shaved scrotal skin and body temperatures and on sweat production from the scrotum, as indicated by changes in the relative humidity (RH) of the air being circulated round the scrotum.

Note that when scrotal skin temperature reaches a value of about 36.5°C synchronized sweat discharges occur at a rate of about 12 per hour. Each discharge causes a fall in skin temperature. The sweat discharges are stopped by local anaesthesis (N) of the scrotal nerves, but can still be provoked, while the local anaesthesia is still effective, by intravenous injections of adrenaline (as indicated). Adrenaline, unlike scrotal heating, caused sweat discharge also from flank skin, as judged from the drop in temperature. At the higher temperatures, even though sweat discharges do not occur when the scrotum is anaesthetized, the background sweat production (measured by the amount of water trapped from the air circulating around the scrotum) is higher than at room temperature but appreciably less than the rates achieved when sweat discharges are allowed to occur. (Reproduced from Waites and Voglmayr, 1963)

bring about cooling. This would seem to indicate that wool on the skin brings a thermal disadvantage; however the wool on the scrotum protects the underlying skin from radiation, thus delaying the appearance of the first discharge and conserving the sweating mechanism. As apocrine glands cannot discharge with the frequency of eccrine glands, this would seem to be an overriding advantage (see Waites and Voglmayr, 1963).

The large apocrine sweat glands of the scrotum of Merino rams can produce more sweat per unit area than those in midside skin (Waites and Voglmayr, 1963) enough to dissipate more than five times the heat production of the testis at 39°C (Hales and Hutchinson, 1971). In summer there are two- to three-fold increases in sweat production by the scrotum compared with winter (see Waites, 1970). There is also a considerable difference in sweat production by the scrotal skin of two strains of Merino rams, one selected for wrinkly skin and the other selected for plain skin. Rams of the wrinkly strain became infertile during exposure to heat more readily than rams of the plain-skinned strain. This is presumably related to the fact that the scrotal skin of the wrinkly rams produced less sweat and the temperature of their testes rose to higher levels under comparable heat loads (Figure 5.7; Fowler and Waites, 1971). Impaired

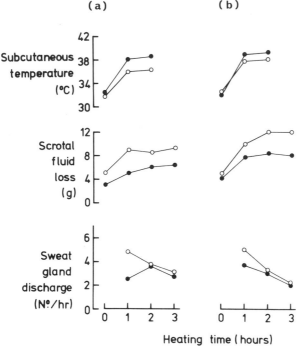

Figure 5.7  A graph of the effect of scrotal heating (by circulating air at (a) 37·6 or (b) 42·5°C through a chamber enclosing the scrotum) on frequency of discharge of the scrotal sweat glands, the scrotal fluid loss and subcutaneous scrotal temperature. ●, rams of a strain selected for wrinkly skin. ○, rams of a strain selected for plain skin. Note that the plain-skinned rams maintain a lower scrotal temperature (and therefore testicular temperature) by a greater frequency of sweat discharge and a greater production of sweat from the scrotum. (Data from Fowler and Waites, 1971)

sweat production in men with lesions in the spinal cord has been suggested as a contributory factor in their infertility (Bors *et al.*, 1950).

Sweat has been seen dripping from the scrota of kangaroos (*Macropus robustus* and *Megaleia rufa*) in inland Australia during very hot weather (Sadleir, 1965). However, these animals also spread saliva over the scrotal skin by licking the scrotum when placed in a hot environment (Dawson *et al.*, 1969). This may be part of a mechanism for general temperature regulation, as a number of animals, including marsupials (kangaroos and wallabies), rodents, bats, cats and elephants, spread saliva on the body in heat. Rats spread saliva particularly on the vascularized surfaces of the scrotum and tail, and when deprived of their salivary glands, suffer from hyperthermia (Hainsworth, 1967).

### 5.3.2     General temperature regulation
Evidence from work on sheep, pigs and rats suggests that reflex responses originating from temperature receptors (see Section 4.2) in the scrotum can influence general thermoregulatory mechanisms and that the scrotum in these species may act as an 'early warning' system.

### 5.3.2.1     Thermal polypnoea
When the temperature of the scrotum of fully-fleeced rams rises above a threshold of 35–36°C, the rate of respiration begins to increase. If the scrotum continues to warm up, respiration rate continues to rise until a fully polypnoeic response develops when the scrotum is at 38–40°C. Warming an equivalent area of flank skin was without a corresponding effect. Local anaesthesia of the scrotum abolished the response; stepwise increases of skin temperature evoked stepwise increases in respiratory rate. The effect is attributed to the recruitment of more and more 'warm' receptors in a richly endowed region. The thermal polypnoea resulted in respiratory evaporative cooling which brought about sharp reductions of deep body temperature (Figures 5.6 and 5.8). In extreme cases, rectal and carotid temperatures decreased by up to 2°C in an hour (Waites, 1962).

In pigs in a warm environment (30°C), heating the scrotum to 42°C also caused a polypnoea, the extent of which depended on trunk skin temperature (Figure 5.9); the polypnoea was sufficient to reduce the rate of rise of body temperature, but not enough to produce a fall in temperature (Ingram and Legge, 1972).

### 5.3.2.2     Control of metabolic rate
Shorn rams with midside skin temperatures of 30–34°C, i.e. 3–7°C lower than in fully-fleeced rams, no longer panted when the scrotum was heated (Figure 5.8). Even so, their body temperature tended to fall, suggesting heat loss by another mechanism. This was shown to be a reduction of $O_2$ consumption from an elevated level brought about by the removal of wool. It is uncertain whether or not cutaneous evaporation was contributing and the vasomotor responses judged from the skin temperature were irregular (Hales and Hutchinson, 1971). Cooling the scrotum after a period of heating, often evoked shivering and a pronounced increase in metabolic rate. Thus, 'cold' receptors in the scrotum, in association with reduced deep body temperature, can evoke appropriate heat gain responses (Waites, 1962; Hales and Hutchinson, 1971).

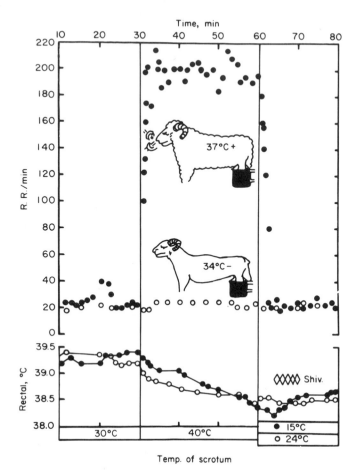

Figure 5.8 The effect of sudden immersion of the scrotum of a ram in water at 40°C (i.e. only slightly hotter than body temperature) on respiratory rate and rectal temperature before (●) and after (O) removal of the fleece. Note that the heat stimulus evokes an immediate polypnoea when the ram is covered with a fleece (when its skin temperature is 37°C or higher) and this causes a pronounced fall in body temperature. When the ram is shorn, the scrotal heating still produces a fall in body temperature but has no effect on respiration rate. The shorn ram began to shiver (◇◇) when his scrotum was cooled to 24°C following the heating. (Reproduced from Waites, 1970)

Heating the scrotum of pigs in a cold environment (15°C) caused a fall in body temperature, the arrest of shivering and a decline in oxygen consumption (Ingram and Legge, 1972).

**5.3.2.3    Reflex sweating**    Hayashi (1968) showed that sweat appeared on the scrotum of rams 5–30 minutes after heating the upper half of the body. As body temperature remained unchanged the response would appear to be due to a peripheral reflex.

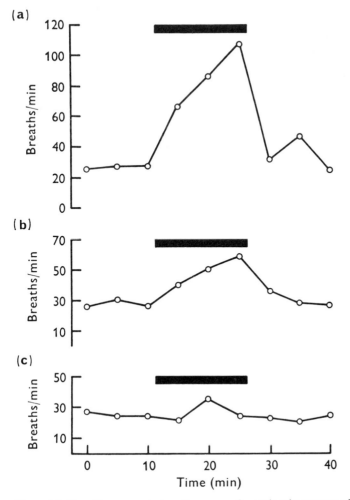

Figure 5.9 The effect on respiratory frequency of warming the scrotum of a pig to 40°C (black bar), at an ambient temperature of 30°C but with the trunk skin temperature controlled by a coat to 38°C (a), 36°C (b) and 34°C (c).
Note the pronounced response in respiratory rate to scrotal heating when trunk skin is also warm (cf. the ram in Figure 5.8). (Reproduced from Ingram and Legge, 1972)

**5.3.2.4     Vasomotor reactions**   In anaesthetized rats, arterial blood pressure and heart rate increased when scrotal temperature was raised to 40°C, but not if the temperature was increased only to 37·5°C (Neya and Pierau, 1976).

## References to Chapter 5

Agger, P. (1971) Scrotal and testicular temperature: its relation to sperm count before and after operation for varicocele. *Fertil. Steril.* **22**, 286–97.

Andrews, F. N. (1940) Thermo-regulatory function of rat scrotum. I. Normal development and effect of castration. *Proc. Soc. Exptl. Biol. Med.* **45**, 867–9.

Badenoch, A. W. (1945) Descent of the testis in relation to temperature. *Brit. Med. J.* **II**, 601–3.

Baker, M. A. and Hayward, J. N. (1968) The influence of the nasal mucosa and the carotid rete upon hypothalamic temperature in sheep. *J. Physiol.* (London) **198**, 561–79.

Bazett, H. C., Love, L., Newton, M., Eisenberg, L., Day, R. and Forster, R. (1948) Temperature changes in blood flowing in arteries and veins in man. *J. Appl. Physiol.* **1**, 3–19.

Beakley, W. R. and Findlay, J. D. (1955) The effect of environmental temperature and humidity on the temperature of the skin of the scrotum of Ayrshire calves. *J. Agric. Sci.* **45**, 365–72.

Bedford, J. M. (1977) Evolution of the scrotum; the epididymis as the prime mover. In *Reproduction and Evolution*, eds. J. H. Calaby and C. H. Tyndale-Biscoe, 171–82. Australian Academy of Science, Canberra.

Bors, E., Engle, E. T., Rosenquist, R. C. and Holliger, V. H. (1950) Fertility in paraplegic males; preliminary report of endocrine studies. *J. clin. Endocrinol.* **10**, 381–98.

Broek, A. J. P. van der (1933) Gonaden und Ausführungsgänge. *Handb. vergl. Anat.* **6**, 1–154.

Bryden, M. M. (1967) Testicular temperature in the southern elephant seal, *Mirounga leonina* (Linn). *J. Reprod. Fertil.* **13**, 583–4.

Carrick, F. N. and Setchell, B. P. (1977) The evolution of the scrotum. In *Reproduction and Evolution*, eds. J. H. Calaby and C. H. Tyndale-Biscoe, 165–70. Australian Academy of Science, Canberra.

Chomiak, M., Lewandowski, M., Kostyra, J., Paroszkiewicz, M., Szewczyk, K., Welento, J. and Bobkiewicz, A. (1954) On the influence of temperature on the descent of the testes. *Ann. Univ. Marie Curie Sklodowska*, Lublin, sec. DD9, 223–34.

Cooper, A. (1830) *Observations on the structure and diseases of the testis*, London: Longman, Rees, Orme, Brown & Green.

Cowles, R. B. (1958) The evolutionary significance of the scrotum. *Evolution* **12**, 417–8.

Cowles, R. B. (1965) Hyperthermia, aspermia, mutation rates and evolution. *Quart. Rev. Biol.* **40**, 341–67.

Dahl, E. V. and Herrick, J. F. (1959) A vascular mechanism for maintaining testicular temperature by counter-current exchange. *Surg. Gynecol. Obstet.* **108**, 697–705.

Dawson, T. J., Denny, M. J. S. and Hulbert, A. J. (1969) Thermal balance of the macropodid marsupial *Macropus eugenii* Desmarest. *Comp. Biochem. Physiol.* **31**, 645–53.

Ehrenberg, L., Ehrenstein, G. von and Hedgram, A. (1957) Gonad temperature and spontaneous mutation-rate in man. *Nature* **180**, 1433–4.

Engle, E. T. (1932) Experimentally induced descent of the testis in the Macacus monkey by hormones from the anterior pituitary and pregnancy urine. *Endocrinology* **16**, 513–20.

Esser, P. H. (1932) Uber die Funktion und den Bau des Scrotums. *Z. Mikroskop.-Anat. Forsch.* **31**, 108–74.

Faulkner, L. C., Hopwood, M. L., Masken, J. F., Kingman, H. E. and Stoddart, H. L. (1967) Scrotal frost-bite in bulls. *J. Amer. Vet. Med. Assn.* **151**, 602–5.

Fiedler, von W. (1959) Uber Differenzierungen der Scrotalhaut von Perodicticus potto |Muller 1766| im Vergleich mit anderen Prosimiae. *Acta Anat.* **37**, 80–105.

Finkel, M. P. (1945) The relation of sex hormones to pigmentation and to testis descent in the opossum and ground squirrel. *Am. J. Anat.* **76**, 93–151.

Foote, W. C., Pope, A. L., Nichols, R. E. and Casida, L. E. (1957) The effects of variations in ambient temperature and humidity on rectal and testis temperature of sheared and unsheared rams. *J. Anim. Sci.* **16**, 144–50.

Fowler, D. G. (1968a) Skin folds and Merino breeding. 5. Variations in scrotal, testis and rectal temperatures as affected by site of measurements, acclimatization to heat and degree of skin fold. *Aust. J. Exptl. Agric. Anim. Husb.* **8**, 125–32.

Fowler, D. G. (1968b) Skin folds and Merino breeding. 7. The relations of heat applied to the testis and scrotal thermoregulation to fertility in the Merino ram. *Aust. J. Exptl. Agric. Anim. Husb.* **8**, 142–8.

Fowler, D. G. (1969) The relationship between air temperature, scrotal surface area and testis temperature in rams. *Aust. J. Exptl. Agric. Animal. Husb.* **9**, 258–61.

Fowler, D. G. and Dun, R. B. (1966) Skin folds and Merino breeding. 4. The susceptibility of rams selected for a high degree of skin wrinkle to heat induced infertility. *Aust. J. Exptl. Agric. Anim. Husb.* **6**, 121–7.

Fowler, D. G. and Kennedy, J. P. (1968) Skin folds and Merino breeding. 6. The effects of varying heat exposures and degree of skin fold on rectal, scrotal and testis temperatures. *Aust. J. Exptl. Agric. Anim. Husb.* **8**, 133–41.

Fowler, D. G. and Setchell, B. P. (1971) Selecting Merino rams for ability to withstand infertility caused by heat. 2. Effect of heat on scrotal and testicular blood flow. *Aust. J. Exptl. Agric. Anim. Husb.* **11**, 143–7.

Fowler, D. G. and Waites, G. M. H. (1971) Selecting Merino rams for ability to withstand infertility caused by heat. 1. Anatomy and functional activity of the scrotum. *Aust. J. Exptl. Agric. Anim. Husb.* **11**, 137–42.

Gutzschebauch, A. (1936) Der Hoden der Haussäugetiere und seine Hüllen in biologischer und ardiagnostischer Hinsicht. *Z. Anat. Entwicklungsgeschichte* **105**, 433–58.

Hainsworth, F. R. (1967) Saliva spreading, activity and body temperature regulation in the rat. *Am. J. Physiol.* **212**, 1288–92.

Hales, J. R. S. and Hutchinson, J. C. D. (1971) Metabolic, respiratory and vasomotor responses to heating the scrotum of the ram. *J. Physiol.* **212**, 353–75.

Hamilton, J. B. (1936) Endocrine control of the scrotum and a 'sexual skin' in the male rat. *Proc. Soc. Exptl. Biol. Med.* **35**, 386–7.

Hamilton, J. B. (1938) The effect of male hormone upon the descent of the testis. *Anat. Rec.* **70**, 533–41.

Hanzalova, D. (1971) The histological structure of the wall of bull scrotum. *Folia. vet.* **15**, 1–2, 63–73.

Harrenstein, K. J. (1928) Uber die Funktion des Skrotums und die Behandlung der Retentio testis beim Menschen. *Zbl. Chir.* **55**, 1734–9.

Harrison, R. G. and Weiner, J. S. (1948) Abdomino-testicular temperature gradients. *J. Physiol.* (London) **107**, 48P–49P.

Harrison, R. G. and Weiner, J. S. (1949) Vascular patterns of the mammalian testis and their functional significance. *J. Exptl. Biol.* **26**, 304–16.

Hartig, F. (1954) Uber die Hodenhullen und ihre funktionelle Bedeutung. *Tierarztliche Umschau* (Munchen) **9**, 150–3.

Hasler, M. J. and Nalbandov, A. V. (1974) Body and peritesticular temperatures of musk shrews (*Suncus murinus*). *J. Reprod. Fertil.* **36**, 397–9.

Hayashi, H. (1968) Functional activity of the sweat glands in the hairy skin of the sheep. *Tohoku J. Exptl. Med.* **94**, 361–75.

Heller, R. E. (1929) New evidence for the function of the scrotum. *Physiol. Zool.* **2**, 9–17.

Ingram, D. L. and Legge, K. F. (1972) The influence of deep body and skin temperatures on thermoregulatory responses in heating of the scrotum in pigs. *J. Physiol.* **224**, 477–87.

Kaiser, W. (1931) Die Entwicklung des Scrotums bei *Didelphis aurita* Wied. *Morphol. Jb.* **68**, 391–433.

Kirby, A. and Harrison, R. G. (1954) A comparison of the vascularization of the testis in Afrikaner and English breeds of bull. *Proc. Soc. Stud. Fert.* **6**, 129–39.

Klaatsch, H. (1890) Uber den Descensus testiculorum. *Morphol. Jb.* **16**, 587–646.

Kormano, M. (1967a) Development of the rectum-testis temperature difference in the post-natal rat. *J. Reprod. Fertil.* **14**, 427–37.

Kormano, M. (1967b) Effect of circulatory disturbance of the testis on the rectum-testis temperature difference in the rat. *Acta Physiol. Scand.* **69**, 209–12.

Langley, J. N. and Anderson, H. K. (1895) The innervation of the pelvic and adjoining viscera. Parts III and IV. *J. Physiol.* **19**, 84–121 and 122–30.

Lieben, S. (1908) Zur Physiologie der Tunica dartos. *Arch. Ges. Physiol.* **124**, 336–52.

Lyne, A. G. and Hollis, D. E. (1968) The skin of the sheep: A comparison of body regions. *Aust. J. Biol. Sci.* **21**, 499–27.

Macdonald, J. and Harrison, R. G. (1954) Effect of low temperatures on rat spermatogenesis. *Fertil. Steril.* **5**, 205–16.

McKenzie, F. F. and Berliner, V. (1937) The reproductive capacity of the ram. Missouri Univ., Agr. Expt. Sta., *Res. Bull* **265**.

Molyneux, G. S. (1965) Observations on the structure, distribution and significance of arteriovenous anastomoses in sheep skin. In *Biology of the Skin and Hair Growth*, eds. A. G. Lyne and B. F. Short, 591–602, Sydney: Angus and Robertson.

Monro, P. A. G. (1959) *Sympathectomy*. London: Oxford University Press.

Moore, C. R. (1923) On the relationship of the germinal epithelium to the position of the testis. *Anat. Rec.* **25**, 142–3.

Moore, C. R. (1924) Properties of the gonads as controllers of somatic and psychical characteristics. VIII. Heat application and testicular degeneration; the function of the scrotum. *Am. J. Anat.* **34**, 337–58.

Moore, C. R. (1926) The biology of the mammalian testis and scrotum. *Quart. Rev. Biol.* **1**, 4–50.

Moore, C. R. and Quick, W. J. (1923) A comparison of scrotal and peritoneal temperatures. *Anat. Rec.* **26**, 344.

Moore, C. R. and Quick, W. J. (1924) The scrotum as a temperature regulator for the testes. *Am. J. Physiol.* **69**, 70–9.

Moule, G. R. and Knapp, B. (1950) Observations on intra-testicular temperatures of Merino rams. *Aust. J. Agric. Res.* **1**, 456–64.

Mykytowycz, R. and Nay, T. (1974) Studies of the cutaneous glands and hair follicles of some species of Macropodidae. *CSIRO Wildlife Res.* **9**, 200–17.

Neya, T. and Pierau, F-K. (1976) Vasomotor response to thermal stimulation of the scrotal skin in rats. *Pflügers Arch.* **363**, 15–18.

Phillips, R. W. and McKenzie, F. F. (1934) The thermoregulatory function and mechanism of the scrotum. Missouri Univ., Agr. Expt. Sta., *Res. Bul.* **217**, 1–73.

Pocock, R. I. (1926) The external characteristics of *Thylacinus, Sarcophilus* and some related marsupials. *Proc. Zool. Soc.* (London) 1037–84.

Politzer, G. (1932) Uber die Entwicklung des Dammes beim Menschen; Nebst Bemerkungen uber die Bildung der ausseren Geschlechtsteile und uber die Fehlbildungen der Kloake und des Dammes. *Z. Anat. Entwicklsgeschichte* **97**, 622–60.

Portmann, A. (1952) *Animal Forms and Patterns*, London: Faber & Faber.

Riemerschmid, G. and Quinlan, J. (1941) Further observations on the scrotal skin temperature of the bull, with some remarks on the intra-testicular temperature. *Onderstepoort J. Vet. Res.* **17**, 123–40.

Ruibal, R. (1957) The evolution of the scrotum. *Evolution* **11**, 376–7.

Sadleir, R. M. F. S. (1965) Reproduction in two species of kangaroo (*Macropus robustus* and *Megaleia rufa*) in the arid Pilbara region of Western Australia. *Proc. Zool. Soc.* (London) **145**, 239–61.

Scholander, P. F. and Krog, J. (1957) Countercurrent heat exchange and vascular bundles in sloths. *J. Appl. Physiol.* **10**, 405–11.

Scholander, P. F. and Schevill, W. E. (1955) Countercurrent vascular heat exchange in the fins of whales. *J. Appl. Physiol.* **8**, 279–82.

Schweizer, R. (1929) Uber die Bedeutung der Vascularization des Binnendruckes und der Zwischenzellen fur die Biologie des Hodens. *Z. Anat. Entwicklungsgeschichte* **89**, 775–96.

Setchell, B. P. and Waites, G. M. H. (1969) Pulse attenuation and countercurrent heat exchanges in the internal spermatic artery of some Australian marsupials. *J. Reprod. Fertil.* **20**, 165–9.

Steen, I. and Steen, J. B. (1965) The importance of the legs in the thermoregulation of birds. *Acta Physiol. Scand.* **63**, 286–91.

Stephenson, J. D. and O'Shaughnessy, E. J. (1968) Hypospermia and its relationship to varicocele and intrascrotal temperature. *Fertil. Steril.* **19**, 110–17.

Stirling, E. C. (1891) Description of a new genus and species of marsupial. *Trans. R. Soc.* (S. Aust.) **14**, 154–87.

Tessler, A. N. and Krahn, H. P. (1966) Varicocele and testicular temperature. *Fertil. Steril.* **17**, 201–3.

Waites, G. M. H. (1961) Polypnoea evoked by heating the scrotum of the ram. *Nature* **190**, 172–3.

Waites, G. M. H. (1962) The effect of heating the scrotum of the ram on respiration and body temperature. *Q. J. Exptl. Physiol.* **47**, 314–23.

Waites, G. M. H. (1970) Temperature regulation and the testis. In *The Testis,* eds. A. D. Johnson, W. R. Gomes and N. L. VanDemark, **I**, 241–79, London: Academic Press.

Waites, G. M. H. (1976) Temperature regulation and fertility in male and female mammals. *Israel J. med. Sci.* **9**, 982–93.

Waites, G. M. H. and Moule, G. R. (1961) Relation of vascular heat exchange to temperature regulation in the testis of the ram. *J. Reprod. Fertil.* **2**, 213–24.

Waites, G. M. H. and Voglmayr, J. K. (1962) Apocrine sweat glands of the scrotum of the ram. *Nature* (London) **196**, 965–7.

Waites, G. M. H. and Voglmayr, J. K. (1963) The functional activity and control of the apocrine sweat glands of the scrotum of the ram. *Aust. J. Agric. Res.* **14**, 839–51.

Waites, G. M. H., Setchell, B. P. and Quinlan, D. (1973) The effect of local heating of the scrotum, testes and epididymides of rats on cardiac output and regional blood flow. *J. Reprod. Fertil.* **34**, 41–9.

Weber, M. (1928) *Die Saugetiere*, Jena: G. Fischer.

Wells, L. J. (1935) Seasonal sexual rhythm and its experimental modification in the male of the thirteen-lined ground squirrel (*Citellus tridecemlineatus*). *Anat. Rec.* **62**, 409–47.

Wells, L. J. (1943) Descent of the testis: Anatomical and hormonal considerations. *Surgery* **14**, 436–72.

Wislocki, G. B. (1933a) Location of the testes and body temperature in mammals. *Quart. Rev. Biol.* **8**, 385–96.

Wislocki, G. B. (1933b) Observations on the descent of the testes in the macaque and in the chimpanzee. *Anat. Rec.* **57**, 133–48.

Wyndham, N. R. (1943) A morphological study of testicular descent. *J. Anat.* **77**, 179–88.

Zorgniotti, A. W. and MacLeod, J. (1973) Studies in temperature, human semen quality and varicocele. *Fertil. Steril.* **24**, 854–63.

# 6 Endocrinology of the testis

The endocrine function of the testis, as manifested by the development of the secondary sexual characteristics and libido, has been known since the earliest times (see Chapter 1) but the identity of the active substance was not known.

With the developing interest in steroid compounds in the 1920s and 1930s, a series of steroids, some with androgenic activity (the ability to restore the sexual characteristics in castrated animals), were isolated from urine and from testes, often using many kilogrammes of tissue obtained at an abattoir (Butenandt, 1931; David *et al.*, 1935; Ruzicka and Prelog, 1943; Brady, 1951; Savard *et al.*, 1956; Neher and Wettstein, 1960a, b). The final links in the chain were forged when it was shown that some of the many steroid compounds found in the testis were released into the blood during its passage through the testis (Section 6.1.2) and that several of these compounds were potent androgens (Dorfmann and Shipley, 1956).

## 6.1    Steroid hormones

### 6.1.1    Pathways of steroid synthesis
Many features of steroid synthesis are common to all steroid forming tissues but this discussion is concerned, as far as possible, with results obtained from the testis.

**6.1.1.1    Source of testicular steroids**    Two major sources of testicular steroids have been suggested: cholesterol, formed elsewhere in the body and transported to the testis in the blood, and acetate, either derived from the blood as such or formed as acetyl-coenzyme A during the metabolism of glucose. The relative importance of these two sources appears to depend on the species studied. Plasma cholesterol contributes about 13 per cent to testicular androgens in the guinea pig (Werbin and Chaikoff, 1961) but about 40 per cent in the rat (Morris and Chaikoff, 1959). *In vitro* studies with rat testis revealed no incorporation of $^{14}$C-acetate into testosterone (Brady, 1951). Slices of rabbit testis incorporated acetate rapidly into testosterone (Brady, 1951; Hall and Eik-Nes, 1962) but they also converted $^{3}$H-cholesterol into testosterone (Hall, 1966). Acetate appears to be converted first to cholesterol (Hall, 1970) which is present mainly in the free not esterified form in the interstitial tissue (Section 9.1.3).

**6.1.1.2    Conversion of acetate to cholesterol**    Cholesterol is an important lipid and is of much wider importance than just the synthesis of steroids. Its formation from acetate is a process common to many other tissues both endocrine and non-endocrine and there is no suggestion that the pathway is different in the testis.

In the synthesis of cholesterol from acetate, both carbon atoms are used and the probable pathway is shown in Figure 6.1. In short, two acetyl-CoA molecules combine to form acetoacetyl-CoA which then combines with another molecule of acetyl-

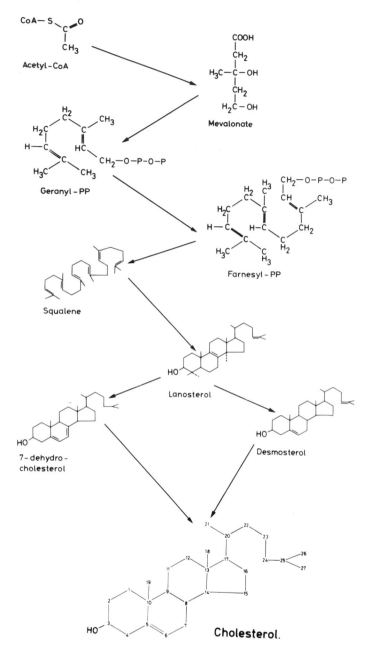

Figure 6.1 A much-simplified summary of the pathway of formation of cholesterol from acetate, with many intermediate compounds omitted. From squalene onwards, and in succeeding figures, no hydrogen atoms attached to carbon atoms are shown. The individual carbon atoms in the cholesterol molecule are shown as the numbers by which they are usually known.

CoA to form mevalonic acid, a six-carbon compound. By a series of conversions and condensations, the carbon chain is elongated until squalene, a 30-carbon compound is formed. This compound can be folded to resemble a steroid nucleus and by a series of ring closures can be transformed to the cyclic compound lanosterol, which is, in turn, converted to cholesterol (Gaylor and Tsai, 1964; Tsai *et al.*, 1964; Ying *et al.*, 1965; Nightingale *et al.*, 1967).

**6.1.1.3    Conversion of cholesterol to pregnenolone**    Cholesterol differs from the endocrine steroids in having a long side chain attached to the C-17 atom of the steroid nucleus and this side chain must be split off as the first stage of the transformation. This leads to the formation of $\Delta^5$-pregnenolone (Figure 6.2) which is a key

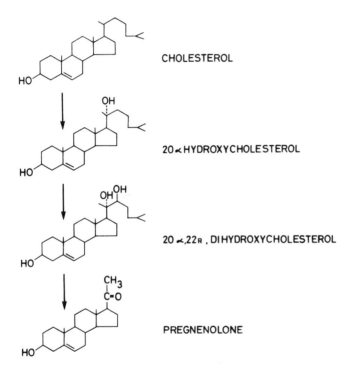

CHOLESTEROL

20∝HYDROXYCHOLESTEROL

20∝,22R , DIHYDROXYCHOLESTEROL

PREGNENOLONE

Figure 6.2  The probable pathway of formation of pregnenolone from cholesterol.

steroid in all subsequent conversions, not only in the testis but in other endocrine organs as well (Samuels, 1960). The conversion requires NADPH and oxygen and occurs in the mitochondria.

Successive oxidations of cholesterol to 20α-hydroxycholesterol occur, a second hydroxyl group is added to carbon 22 forming 20α-, 20R-dihydroxycholesterol and then the side chain is cleaved between atoms 20 and 22 under the influence of $C_{20}$-$C_{22}$ lyase to yield pregnenolone (Burstein and Gut, 1971).

This part of the pathway for the formation of testosterone is important because it is here that gonadotrophins appear to exert their control (Section 10.1.2.3).


**6.1.1.4      Conversion of pregnenolone to testosterone and other androgens**    The basic ring of pregnenolone has a double bond between carbons 5 and 6 ($\Delta^5$) whereas testosterone has a double bond between carbons 4 and 5 ($\Delta^4$). The conversion therefore involves transposition of this bond under the influence of an isomerase and a 3$\beta$-hydroxysteroid dehydrogenase. It also requires successively the introduction of a hydroxyl group with an $\alpha$-configuration on the C-17 carbon, removal of the remaining 2 carbons of the side chain on carbon C-17, and at the same time the hydroxyl group there is oxidized to a keto group. Finally this keto group is reduced to a hydroxyl group with the $\beta$-configuration.

The isomerization from the $\Delta^5$ to the corresponding $\Delta^4$ steroid can theoretically take place at any stage and therefore there are four possible pathways (Figure 6.3; Hall, 1970; Eik-Nes, 1975).

*In vitro* it would appear that the $\Delta^4$ pathway, pregnenolone → progesterone → 17$\alpha$-hydroxyprogesterone → androstenedione → testosterone is quantitatively the most important. This is so for rat-testicular microsomes (Tamaoki and Shikita, 1966), interstitial tissue and isolated tubules (Bell *et al.*, 1968, 1971; Bell, 1972). However *in vivo* significant amounts of dehydroepianodrosterone are secreted by the rat, dog, boar and human testis and 17$\alpha$-hydroxypregnenolone by the dog testes (see Section 6.1.2) so the $\Delta^5$ pathway, pregnenolone → 17$\alpha$-hydroxypregnenolone → dehydroepiandrosterone → androstenediol or androstenedione → testosterone would also seem to be operative *in vivo* (Tamaoki *et al.*, 1969). There may also be important species differences; the $\Delta^4$ pathway may be more important in the rat and the $\Delta^5$ pathway in the human (Yanahara and Troen, 1972).

A pathway involving steroid sulphates has also been suggested as an important route for testosterone production from pregnenolone sulphate. Testosterone sulphate may also be formed directly via $\Delta^5$-androstendiol disulphate, although this compound has not yet been isolated from testis tissue. The key enzyme for the formation of free steroids from the sulphates, a sulphatase, is inhibited by 5$\alpha$-androstane-3$\alpha$, 17$\beta$-diol and by 5$\alpha$-androstane-3$\beta$, 17$\beta$-diol, both of which are formed when testicular tissue is incubated with $\Delta^5$-androstenediol-3$\beta$-yl-sulphate (Notation and Ungar, 1969a, b; Payne and Jaffe, 1970; see also Eik-Nes, 1975).

5$\alpha$-androstane-3$\alpha$,17$\beta$-diol (DIOL) is also formed from testosterone via 5$\alpha$-dihydro-testosterone (DHT) (see Figure 6.4) inside the seminiferous tubules.

This conversion is greatest in rats at about the beginning of meiosis (Rivarola and Podesta, 1972; Rivarola *et al.*, 1972, 1973; Folman *et al.*, 1972, 1973; Podesta and Rivarola, 1974). It is interesting that DHT and DIOL penetrate into the seminiferous tubules much more slowly than testosterone (see Section 8.7.4).

Oestrogens can also be formed by testicular tissue, probably by the pathways indicated in Figure 6.5. The activity of these pathways has been shown by incubations with various radioactive precursors but no indication has been given of which pathway is the most important. The testes of stallions and boars are particularly

# CHOLESTEROL

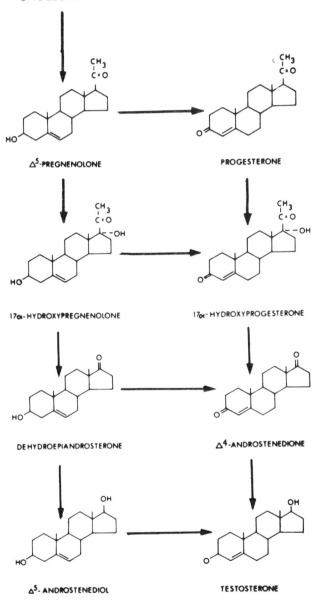

Δ⁵-PREGNENOLONE        PROGESTERONE

17α-HYDROXYPREGNENOLONE     17α-HYDROXYPROGESTERONE

DEHYDROEPIANDROSTERONE     Δ⁴-ANDROSTENEDIONE

Δ⁵-ANDROSTENEDIOL        TESTOSTERONE

Figure 6.3 The various pathways from pregnenolone to testosterone. The righthand column shows the 'Δ⁴-pathway', the lefthand column the 'Δ⁵-pathway'. Each of the conversions from the left to the right column involves an isomerase and 3β-hydroxysteroid dehydrogenase. Each of the pairs of conversions, shown vertically, of the compounds shown on the same horizontal level, involve the same enzyme, namely 17α-hydroxylase, 17–20 lyase and 17β-hydroxysteroid dehydrogenase respectively.

active in producing oestrogens, and oestrogen sulphates are also present in high concentrations in these species (Raeside, 1965, 1969; Bedrak and Samuels, 1969; Oh and Tamaoki, 1970).

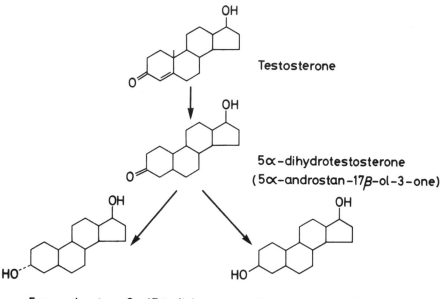

Figure 6.4  The formation of DHT and diols from testosterone, under the influence of 5α-reductase and 3α-reductase respectively. These conversions occur in many accessory reproductive tissues, as well as in the seminiferous tubules.

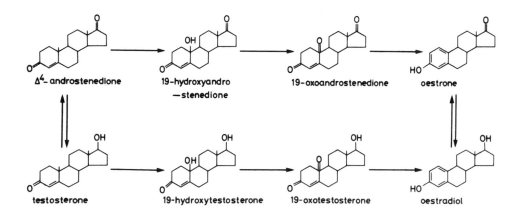

Figure 6.5  Pathways of formation of oestrogens from androgens.

## 6.1.2    Steroid hormones secreted by the testis

**6.1.2.1    Testosterone**    Testosterone was first isolated from more than 100kg of bull testes in 1935 (David *et al.*, 1935) and synthesized in the same year (Ruzicka and Wettstein, 1935). Of the 20 or more steroids which have been isolated from testicular tissue, testosterone is one of the most potent androgens, and it appears to be the major androgen in the blood from the internal spermatic vein of a variety of species (West *et al.*, 1952; Lucas *et al.*, 1957; Hollander and Hollander, 1958; Lindner, 1959, 1961a, c; Suzuki and Eto, 1962; Eik-Nes, 1962; Ibayashi *et al.*, 1965; Ewing and Eik-Nes, 1966; Bardin and Petersen, 1967; Baulieu *et al.*, 1967; Lipsett, 1970; Bullock and New, 1971, Ewing *et al.*, 1975; Amann and Ganjam, 1976; Hammond *et al.*, 1977).

In the bull, the androgenic potency of blood plasma from the internal spermatic vein corresponded well with the predicted potency estimated from the concentration of testosterone and androstenedione. When the extracts were separated on paper chromatography, only the areas where testosterone and androstenedione were present showed androgenic activity (Savard *et al.*, 1961). Testosterone is also present in high concentrations in testicular lymph (Lindner, 1963), and in rete testis fluid (see Section 8.2).

**6.1.2.2    Androstenedione and androstenediol**    These two steroids are the two possible immediate precursors of testosterone (see Figure 6.3). Androstenedione has been reported to occur in blood from the internal spermatic veins of men, bulls, rams, boars, stallions, dogs and rats (Lindner, 1961a, c; Savard *et al.*, 1961; Eik-Nes, 1967a, b; Gandy and Peterson, 1968; Resko *et al.*, 1968; Amann and Ganjam, 1976; Hammond *et al.*, 1977). Its rate of production in mature animals of most of these species is usually about 10 per cent of that of testosterone, but because the production of androstenedione is less affected by age, it contributes a greater fraction of the androgen produced by the testes of young animals.

Androstenedione is converted by red blood cells of cattle, and to a lesser extent, by those of sheep, to another steroid *d*-epitestosterone, which has almost no androgenic potency (Lindner, 1961b).

Androstenediol has been reported in blood from the spermatic veins of dogs in concentrations between one-half and one-twentieth of the testosterone concentration (Eik-Nes, 1970; Yamaji *et al.*, 1968, Amann and Ganjam, 1976).

**6.1.2.3    Dehydroepiandrosterone (DHEA)**    The boar and dog testis secrete appreciable quantities of this steroid both free and as the sulphate (Baulieu *et al.*, 1967; Ibayashi *et al.*, 1965; Eik-Nes, 1971). In man, there is an appreciable secretion of DHEA into the blood stream but much more of it comes from the adrenal than from the testis; only about 6 per cent of the total secretion of almost 7mg per day is produced by the testis (Gandy and Peterson, 1968). It has also been reported that in man orchidectomy scarcely altered the plasma concentration of DHEA, whereas castration of boars caused a pronounced reduction.

**6.1.2.4    17α-hydroxyprogesterone and 17α-hydroxypregnenolone**    Both these compounds are possible precursors of testosterone and both enter the blood stream of humans at a rate of about 2mg per day. However, more than 90 per cent of the hydroxyprogesterone is secreted by the testis, but only about 30–40 per cent of the hydroxypregnenolone, the rest coming from the adrenals (Bermudez *et al.*, 1969; Strott *et al.*, 1969; Lipsett, 1970; Hammond *et al.*, 1977).

**6.1.2.5    Pregnenolone and progesterone**  Some progesterone appears to be secreted by the testes of dogs and bulls (Eik-Nes, 1970; Amann and Ganjam, 1976). The concentration of pregnenolone in blood from the internal spermatic veins of perfused dog testes is higher than in peripheral blood. This indicates that some of this steroid is secreted by the testis, but insufficient data was given to gauge the importance of the secretion. It may simply be due to the 'leakiness' of an endocrine gland as defined by Short (1960), but it is interesting that the secretion of pregnenolone rises during barbiturate anaesthesia while the production of testosterone falls (Tcholakian and Eik-Nes, 1971). Both pregnenolone and progesterone are present in higher concentrations in spermatic venous than peripheral venous blood in humans (Hammond, *et al.*, 1977).

**6.1.2.6    5α-dihydrotestosterone (DHT) and androstanediols**  The DHT secretion by the human testis is so low that many authors have not been able to detect it, but measurable concentrations have been reported in internal spermatic venous plasma of dogs, rabbits and humans (Folman *et al.*, 1972; Haltmeyer and Eik-Nes, 1972; Ewing *et al.*, 1975; Hammond *et al.*, 1977).

3α- and 3β-androstanediol appear in small amounts in testicular venous blood of rabbits and bulls; the total concentration of the diols reaches about 25 per cent of the testosterone concentration in rabbits and about 10 per cent in bulls (Ewing *et al.*, 1975; Amann and Ganjam, 1976, see also Moger, 1977).

**6.1.2.7    16-unsaturated $C_{19}$ steroids**  These compounds, 5α-androst-16-en-3-one (5α-androstenone), 5α-androst-16-en-3α-ol (an-α) and 5α-androst-16-en-3β-ol (an-β), are produced by the testes of boars. In fact, the concentrations of 5α-androstenone in peripheral and spermatic vein plasma were comparable to those of testosterone and the concentrations of all three $\Delta^{16}$-unsaturated steroids in the testis were much higher than that of testosterone. In blood, the two alcohols occur principally as sulphates, the ketone as the 'free' steroid. 5α-androstenone is concentrated in the body fat, giving it a characteristic 'boar taint' when cooked, and 5α-androstenone, an-α and an-β are taken up by the salivary glands (Patterson, 1968b) and secreted in the saliva where they act as pheromones (Reid *et al.*, 1974). An-α is also found in human urine, both male and female after puberty, although the concentration decreases in old age. None of the three compounds have any detectable androgenic activity in bioassays.

The $\Delta^{16}$-unsaturated steroids are formed from pregnenolone, and progesterone, via 4,16-androstadien-3-one, but not via testosterone or dehydroepiandrosterone (Patterson, 1968a; Gower, 1972).

**6.1.2.8    Oestrogens**  There are appreciable amounts of oestrogens in urine from human males. It has been suggested that in men most of the oestradiol is not secreted as such by the testis but is formed elsewhere in the body from testosterone. Some oestrone may also form in this way from androstenedione but some is probably secreted by the testis (Lipsett, 1970). However, this is based on rather indirect evidence and earlier interpretations of similar evidence suggested that the majority of the oestrogens were secreted as such by the testes (Fishman *et al.*, 1967). Several authors (Kelch *et al.*, 1972; Longcope *et al.*, 1972; Scholler *et al.*, 1973; Baird *et al.*, 1973) have now shown by measurements in spermatic venous and peripheral plasma that oestradiol is secreted by human testes, probably about 20 per cent of total production. Substantial amounts of oestradiol are produced also by the testis of dogs and monkeys (Kelch *et al.*, 1972), bulls (Amann and Ganjam, 1976), rams and boars, especially the older individuals (R. B. Heap and B. P. Setchell, unpublished observations). There is also evidence for secretion of oestrogens by perfused stallion testes (Nyman *et al.*, 1959). In the rat, the testes contribute 20 per cent of total oestradiol secretion (de Jong *et al.*, 1973, 1974).

### 6.1.3        Source of testicular steroids

**6.1.3.1        Cellular origin**   The original description of Leydig (1850) of the cells named after him did not envisage an endocrine function for these cells. It was only at the turn of the century that Bouin and Ancel (1903) noticed that the male characteristics of an animal persisted in cryptorchidism, and in other conditions involving atrophy of the seminiferous tubules, and suggested that the interstitial tissue and in particular the Leydig cells were the source of the androgens. Further support for this view came from studies with testicular grafts (see Section 7.6.1.5) and from studies with X-irradiated testes (see Section 11.1).

More direct evidence has recently been added. Christensen and Mason (1965), using a technique the essentials of which were described by de Graaf in 1668, separated the seminiferous tubules from the interstitial tissue and showed that the tubular tissue converted progesterone to androgens but at a very much lower rate than the interstitial tissue. Separated interstitial tissue converted $^3$H-cholesterol to testosterone but tubular tissue did not do so (Hall *et al.*, 1969). When $^3$H-cholesterol was injected into a rat, appreciable amounts appeared in the interstitial tissue both as cholesterol and as other derived steroids, but very little radioactivity appeared in the tubules (Parvinen *et al.*, 1970).

When interstitial tissue was incubated *in vitro*, its content of testosterone increased with time, whereas the amount of testosterone in isolated tubules decreased during incubation or was unchanged (Cooke *et al.*, 1972). Furthermore, mitochondria isolated from separated interstitial tissue produced about 50 times as much testosterone as mitochondria from separated tubules (van der Vusse *et al.*, 1973). However, the oestradiol content of isolated tubules, although lower to start with than that in interstitial tissue, increased during incubation, while that in interstitial tissue did not change (de Jong *et al.*, 1974). This suggests that the conversion of testosterone to oestradiol may occur in the tubules (Figure 6.6).

Sertoli cells cultured from the testes of pre-pubertal rats formed appreciable amounts of oestradiol-17β and oestrone when testosterone, 19-hydroxytestosterone, androstenedione or 19-hydroxyandrostenedione were added to the medium. This aromatization was stimulated more than 10-fold by FSH, cyclic-AMP or cholera toxin; it was greatest in cells from 5 day old rats and had fallen to undetectable levels in cells cultured from 30 to 40 day old rats (Armstrong *et al.*, 1975; Dorrington and Armstrong, 1975; Dorrington *et al.*, 1976).

Furthermore, the tubules may have a capacity for converting progesterone and pregnenolone not only to 17α-hydroxyprogesterone, androstenedione and

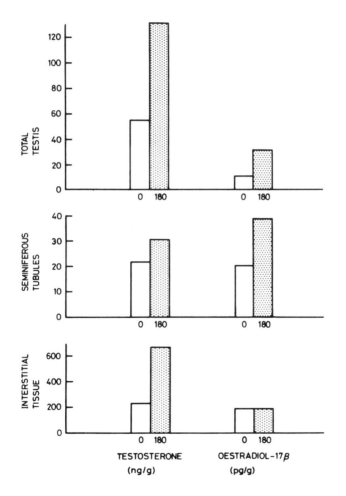

Figure 6.6 The concentrations of testosterone and oestradiol-17β in the rat testis, and isolated seminiferous tubules and interstitial tissue before and after 180 min incubation. Note the large increase in testosterone concentration in the interstitial tissue and whole testis during incubation, and compare this with the small non-significant increase in testosterone in the tubules. (Other experiments by Cooke *et al.* (1972) using similar preparations showed small decreases or no change.) However the concentrations of oestradiol in the seminiferous tubules and whole testis rise significantly during incubation, with no change in the concentration in the interstitial tissue. (Data from de Jong *et al.*, 1974)

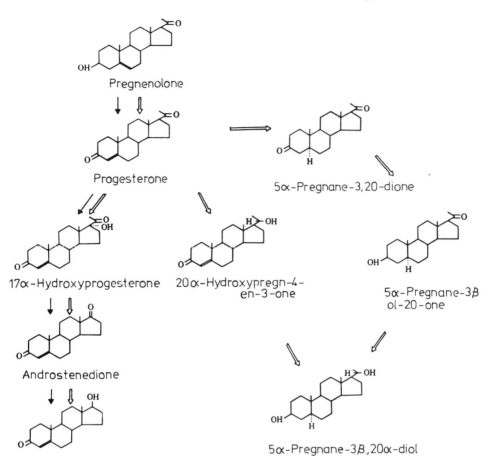

Figure 6.7 Pathways of conversion of progesterone in the interstitial tissue (solid arrows) and seminiferous tubules (open arrows) of the rat testis, according to Bell *et al.* (1971), from whose paper this figure was adapted.

testosterone but also to other compounds not formed by the interstitial tissue. These compounds include 20α-hydroxypregn-4-en-3-one, 5α-pregnane-3,20-dione, 5α-pregnane-3β,20α-diol and 5α-pregnane-3β-ol-20-one (Figure 6.7; Bell *et al.*, 1968, 1971). The significance, or indeed the specificity, of these additional transformations is not known, but I do not think that they should be referred to as steroidogenesis, which should be used to mean synthesis from cholesterol. This still appears to be a function of the interstitial tissue, presumably of the Leydig cells.

Other evidence for the location of the main testosterone synthesis in Leydig cells comes from studies on the histochemical localization of 3β-hydroxysteroid dehydrogenase in these cells (see Section 9.2.5) or with fluorescent or peroxidase-labelled antibodies against testosterone coupled to bovine serum albumin (Woods and Domm, 1966; Bubenik *et al.*, 1975).

# Table 6.1

## The concentration of testosterone in various fluids from the testes of various species

| | Man | Monkey | Dog | Rat | Rabbit | Ram | Bull | Boar |
|---|---|---|---|---|---|---|---|---|
| Weight of one testis (g) | 25 | 25 | 15 | 1·5 | 4·0 | 300 | 400 | 500 |
| Blood flow through one testis (ml per min)* | 3·5 | 4·0 | 3·0 | 0·32 | 1·1 | 27 | 20 | 20 |
| Concentration (ng per ml) of testosterone in: | | | | | | | | |
| General arterial blood plasma | 5·0 | 4·2 | 2·5 | 2·4 | 2·2 | 4·8 | 4·0 | 4·5 |
| Testicular arterial blood plasma | 11 | 4·9 | 3·8 | 6·7 | 6·9 | 5·4 | 5·6 | 5·4 |
| Internal spermatic venous plasma | 440 | 50 | 144 | 50 | 140 | 70 | 185 | 110 |
| Rete testis fluid | — | 2·5 | — | 25 | 70 | 20 | 33 | 17 |
| Daily production of testosterone by both testes (mg per day) | 2·6 | 0·31 | 0·73 | 0·024 | 0·25 | 3·5 | 7·0 | 3·6 |

Data from: Bayard et al., 1975; Einer-Jensen and Waites, 1977; Free, 1977; Setchell, 1970; Pettersson et al., 1973 (see Chapter 3); Comhaire and Vermeulen, 1976; Cooper and Waites, 1974; Cooper et al., 1976; Ganjam and Amann, 1976; Harris and Bartke, 1974; Setchell et al., 1971; Voglmayr et al., 1966; Waites and Einer-Jensen, 1974 (see Chapter 8); Amann and Ganjam, 1976; Carlson et al., 1971; Ewing et al., 1975; Gustafson and Shemesh, 1976; Hammond et al., 1977; Kelch et al., 1972; Lindner, 1961a, c, 1963; Setchell et al., 1965; and B. P. Setchell and S. J. Main, unpublished observations.

* Values given are for total venous outflow, except for the monkey where capillary flow is quoted. In species where both measurements have been made, capillary flow is about 70 per cent of total flow (see Section 3.4). The calculations of daily production assume a haematocrit of between 30 and 40 per cent depending on species, and that the concentration of testosterone in the red blood cells does not change during the passage of the blood through the testis.

**6.1.3.2    Origin within the cells**    Different steps of the process of production of androgens in the interstitial cells occur in different parts of the cell (Samuels and Eik-Nes, 1968). Cholesterol esters are converted to cholesterol in the lipid droplets, but the cholesterol must then be transported into the mitochondria. There it is converted to 20$\alpha$,22R-dihydroxycholesterol and the side chain is removed to give pregnenolone (Toren et al., 1964; Moyle et al., 1973). This substance leaves the mitochondria and the rest of the steps leading to the production of testosterone take place in the smooth endoplasmic reticulum (Shikita et al., 1964; Shikita and Tamaoki, 1965; Tamaoki, 1973), which is very abundant in the Leydig cells (see Section 1.3.2).

**6.1.3.3    Rate of production and route of secretion**    Discussion here will be confined to the principal androgens, testosterone and androstenedione. The rates of production have been estimated directly in a few species by multiplying the testicular blood flow by the increase in concentration of the androgen as blood passes through the testis (Table 6.1). A number of indirect estimates have been based on the doses of testosterone needed to maintain the size of the accessory organs or the concentration of certain compounds in the semen. Indirect estimates can also be derived from the metabolic clearance rate (Lipsett, 1970), but these estimates also include androgens produced in other organs of the body besides the testes.

Variations in rate of production can also be estimated in some species from the rate of excretion of a specific metabolite in the urine. However, this method is of very dubious value as changes in rate of production of testosterone could easily lead to changes in the fraction appearing as the chosen metabolite.

An overwhelming proportion of the androgen passes from the testes to the rest of the body via the blood stream, rather than the lymph, although testicular lymph contains much more testosterone than arterial blood (Lindner, 1963, 1967, 1969). This is not, as some authors have suggested, because of the slightly lower concentration of testosterone in testicular lymph than in blood from the internal spermatic vein (4–6$\mu$g/100ml compared with 6–10$\mu$g/100ml in normal rams), but because of the hundred-fold difference in flow rate—more than 1000ml blood per hour compared with about 10ml lymph per hour (see Chapter 3). However, blood from the deferential vein contains more testosterone than arterial blood (see Section 6.1.6.3) and blood from arteries leaving the spermatic cord may contain more testosterone than arterial blood elsewhere in the body, because of vein to artery transfer in the cord (see Section 3.2.2).

A significant proportion of testosterone in blood is bound to albumin (Eik-Nes et al., 1954; Eik-Nes, 1970) and also to a cortisol-binding globulin found in the plasma of a number of species. Human plasma also contains a testosterone-binding globulin which appears to be distinct from cortisol-binding globulin (Pearlman and Crepy, 1967; Rosner and Deakins, 1968; Vermeulen, 1973). As a result, in human males, about 2 per cent of the testosterone is free, about 1 per cent is bound to cortisol-binding globulin, about 40 per cent is albumin bound and about 60 per cent is bound to the testosterone-binding globulin. This protein is found in the plasma of several other species but is not present in the plasma of rat or boar (Corval and Bardin,

1973) and the proportion of free testosterone is therefore much higher in rats than in man.

The testosterone-binding globulin is formed in the liver and in the rabbit is chemically and immunlogically indistinguishable from the androgen-binding protein formed inside the tubules (see Section 8.2.1).

### 6.1.4    Metabolism and excretion of androgens

In men, practically 100 per cent of injected radioactive testosterone appears in the urine within 5 days. Before excretion by the kidney, testosterone is metabolized and conjugated either with glucuronic acid or sulphuric acid in the liver. In other species, much less information is available but it is likely that substantial amounts are excreted in the bile and faeces (Baulieu and Robel, 1970).

The clearance of testosterone by the liver is not as complete as it is with many other steroids. In man, about 60 per cent of the circulating testosterone is removed during one passage through the liver; the binding to proteins in the plasma probably explains the low value. Androstenedione is cleared more effectively; its metabolic clearance rate actually exceeds hepatic blood flow so some extrahepatic metabolism must also occur (Horton and Tait, 1966).

Of the principal compounds formed from testosterone, the glucuronides and sulphates of 5$\alpha$- and 5$\beta$-androsterone (5$\beta$-androsterone is also known as etiocholanolone) and the glucuronides of testosterone and 5$\alpha$- and 5$\beta$-androstanediol are the main compounds found in the urine (Baulieu and Robel, 1970). Some of the differences between the amounts of the various metabolites in urine will be due to their handling by kidney. The renal clearance of 5$\beta$-androsterone glucuronide exceeds glomerular filtration rate, so it is one of that small group of compounds actively excreted by the tubular cells of the kidney. The clearance of 5$\alpha$-androsterone glucuronide is only about half glomerular filtration, so that about half the filtered steroid is reabsorbed in the tubules. The clearance of androsterone sulphate is even less (Bongiovanni and Eberlein, 1957; Kellie and Smith, 1967).

In the older literature, the estimation of urinary 17-ketosteroids was often done as an index of androgen secretion. These compounds include 5$\alpha$- and 5$\beta$-androsterone, but only about 30 per cent are of testicular origin and the majority are probably metabolic products of adrenal steroids (see Dorfman and Shipley, 1956). The estimation is not now in common use.

### 6.1.5    Factors affecting androgen production

**6.1.5.1    Species**  The concentration of testosterone in the testis and the peripheral plasma seems to bear little relation to the position of the testis (Figure 6.8). If countercurrent exchange of testosterone in the spermatic cord of the scrotal testis (see Section 3.2.2) were important, one might expect to find higher concentrations in the testis (in relation to peripheral levels) in animals with well developed scrota. This does not appear to be so.

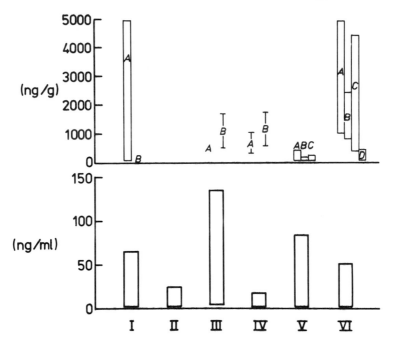

Figure 6.8 The variation in concentration of testosterone in the peripheral plasma and testis of various species of mammals, grouped according to the position of their testes (see Figure 1.1). Reproduced from Carrick and Setchell (1977), where the references for the individual species are given. IA, African elephant. IB, Hyrax. IIIA, Pipistrelle bat. IIIB, Noctule bat. IVA, Mole. IVB, House shrew (*Suncus*). VA, Rat. VB, Dog. VC, Boar. VIA, Red Deer. VIB, Roebuck. VIC, Bull. VID, Ram.

**6.1.5.2    Gonadotrophins**    It is very likely that the major causes of variation in androgen production are changes in gonadotrophin secretion. There is so far no definite evidence that any other factor directly affects the secretion of androgens with the possible exception of reduced blood flow which has been shown, under rather unphysiological conditions, to reduce androgen output by the perfused testis (Eik-Nes, 1964a).

Human chorionic gonadotrophin (hCG) and ovine and bovine luteinizing hormone (LH), also known as interstitial cell stimulating hormone (ICSH), have been shown to stimulate testosterone secretion *in vivo* (see Figure 6.9; Brinck-Johnsen and Eik-Nes, 1957; Lindner, 1961a; Saez and Bertrand, 1968; Laatikainen *et al.*, 1971; El Safoury and Bartke, 1974; Moger and Armstrong, 1974) and in infused testes in anaesthetized dogs (Eik-Nes, 1964b, 1967a, b; Eik-Nes and Hall, 1965). Intravenous administration of hCG stimulated oestradiol secretion as well as testosterone production by the rat testis but prolonged administration of hCG caused a rise in the testis content of both oestradiol and testosterone, and in the secretion of testosterone but not that of oestradiol (Figure 6.10). Conversely, hypophysectomy reduced testis content of testosterone and oestradiol, and the secretion of testosterone, but did not affect the secretion of oestradiol (de Jong *et al.*, 1973, 1974). hCG also stimulates the secretion of $\Delta^{16}$-unsaturated steroids by the boar testis (Andresen, 1975). Ovine

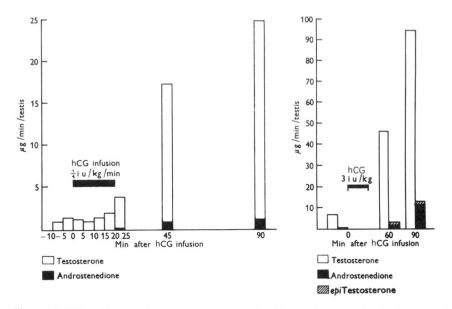

Figure 6.9 Effect of human chorionic gonadotrophin (hCG) on androgen production by one testis of an anaesthetized bull. (Reproduced from Lindner, 1961a)

LH also stimulates the synthesis of androgens by rat and rabbit testis *in vitro* (Hall and Eik-Nes, 1962, 1963a, b; Hall and Young, 1968). Pregnant mare serum gonadotrophin (PMSG) also appears to stimulate steroidogenesis (Lindner, 1961a; Eik-Nes and Hall, 1965a). FSH may act synergistically with LH and permit a higher rate of testosterone synthesis (Connell and Eik-Nes, 1968; Johnson and Ewing, 1971; Seilicovich and Rosner, 1974) and FSH increases the responsiveness of the testis of young rats to LH (Odell *et al.*, 1973; Sizonenko *et al.*, 1973; Odell and Swerdloff, 1976). Prolactin also appears to potentiate the effects of LH (see Bartke, 1976).

One surprising feature of the effect of hCG on the secretion of testosterone by the testis is the speed with which the secretion increases. Lindner (1961a) and Eik-Nes (1967a, b) have observed significant increases in testosterone secretion within minutes of the injection of hCG and it is therefore likely that there is release of preformed hormone as well as stimulated synthesis.

LH is secreted in 'bursts', so that the concentration in plasma shows sudden rises, each with a quick return to baseline. Every peak of LH is followed by a similar peak of testosterone, with a delay of between 15 and 30 minutes (see Figure 6.11; Katongole *et al.*, 1971, 1974; Gombe *et al.*, 1973; Purvis *et al.*, 1974; Sanford *et al.*, 1974; Schanbacher and Ford, 1976; Davies *et al.*, 1977).

The concentration of LH in the plasma of bulls can be increased by single injections of prostaglandin $F_{2\alpha}$, which also produce rises in testosterone concentration (Haynes, *et al.*, 1975; Kiser, *et al.*, 1976). Prostaglandins $E_2$ and $F_{2\alpha}$ in high doses over several days depressed the plasma concentration of testosterone in rats and mice (Bartke *et al.*, 1973;

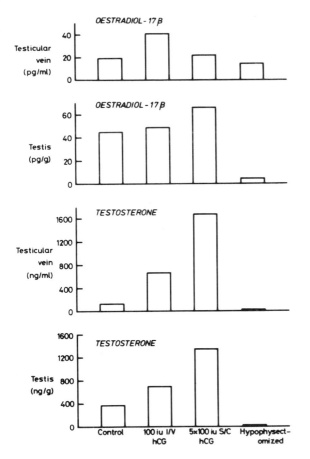

Figure 6.10 Effect of gonadotrophins on the secretion of oestradiol-17β and testosterone by the rat testis. Rats were either injected intravenously with human chorionic gonadotrophin (hCG) immediately before the 30 minute collection of testicular venous blood, or given daily subcutaneous injections of hCG for 5 days before sampling or hypophysectomized 6 days before sampling. Concentrations in arterial blood are not shown here but were probably no greater than 3pg/ml for oestradiol and 5ng/ml for testosterone in unstimulated rats. (Data from de Jong *et al.*, 1973, 1974)

Saksena, *et al.*, 1973) possibly by their action in reducing testicular blood flow (see Section 3.4). However prostaglandins A, and A$_2$, which are vasodilator rather than vasoconstrictor in other tissues, also decrease serum testosterone in rats and mice (Saksena *et al.*, 1974).

The details of the way LH stimulates steroidogenesis by the interstitial cells are dealt with in Chapter 10 and will not be further considered here.

**6.1.5.3    Age**    It is obvious that one of the principal sources of variation in androgen secretion by the testis is the age of the animal, but details of the quantitative and qualitative variations are only now being established.

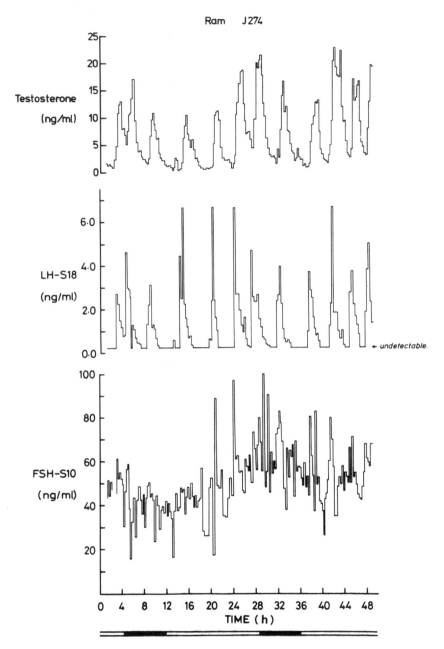

Figure 6.11 The concentrations of testosterone, luteinizing hormone (LH) and follicle stimulating hormone (FSH) in blood from the posterior vena cava of a Clun Forest ram in Cambridge in June. The blood was withdrawn continuously at a rate of 0·25ml/min, and 20 minute collections made with a fraction collector. The bar at the bottom indicates when the lights were switched on. Note the sharp peaks in LH, each followed about 30–45min later by a rather flatter peak of testosterone. FSH showed random variation with a proportionally higher baseline level. (Data from Davies *et al.*, 1977)

(a) Fetal testis

The evidence for androgen secretion by the fetal testis is entirely circumstantial as no actual measurements of androgens in fetal spermatic venous blood have been made. However fetal testis of several species do secrete testosterone in tissue culture (Acevedo *et al.*, 1961, 1963; Bloch, 1964, 1967; Rice *et al.*, 1966; Zaaijer *et al.*, 1966; Ortiz *et al.*, 1966; Price *et al.*, 1967; McArthur *et al.*, 1967; Bloch *et al.*, 1971; Moon *et al.*, 1973; Abramovich *et al.*, 1974; Sitteri and Wilson, 1974; Stewart and Raeside, 1976).

Various enzymes associated with steroid conversions have been demonstrated in the testis of fetal mice, rats and humans (see Section 9.2.5).

Attal (1969) has demonstrated significant concentrations of testosterone, androstenedione, oestrone and oestradiol in the testes of fetal sheep and these levels reach maxima at about 90 days of pregnancy (Figure 6.12). In human fetal testes, the peak of testosterone concentration is reached at about the time of gonadal differentiation, about 13 weeks p.c. (Huhtaniemi *et al.*, 1970; Reyes *et al.*, 1973; Diez d'Aux and Murphy, 1974). In pig fetal testis, appreciable concentrations of testosterone are found by 30 days p.c. but the concentrations continue to rise until just before birth (Raeside and Sigman, 1975; Segal and Raeside, 1975; Booth, 1975). In rats, the peak is towards the end of pregnancy, and the concentration falls abruptly after birth (Warren *et al.*, 1973). In rabbits, the concentration is highest between days 20 and 23 of pregnancy and then falls (Veyssiere *et al.*, 1976).

The concentration of testosterone in fetal blood has been shown in several species to be higher in the middle of pregnancy than either near term, or after birth before puberty (Figure 6.13; Kim *et al.*, 1972; Abramovich and Rowe, 1973; Resko *et al.*, 1973; Reyes *et al.*, 1974; Meussy-Dessolle, 1974, Strott *et al.*, 1974; Veyssiere *et al.*, 1976).

(b) Newborn

Behavioural changes in adulthood have been produced by short treatments of newborn animals with hormones (see Section 6.1.6.4) and it is relevant that the testes of newborn rats (Resko *et al.*, 1968) and sheep (Attal *et al.*, 1972) contain more androgen than slightly older animals. This is reflected by the concentrations of testosterone in peripheral blood (Resko *et al.*, 1968). Furthermore the testes of newborn rats avidly convert progesterone to testosterone, but as the animal grows older, although the amount of progesterone converted does not change, the proportion appearing as testosterone falls and a greater proportion appears as androstenedione (Steinberger and Ficher, 1968).

(c) Puberty

As might be expected, the production of testosterone rises sharply at the time of puberty, although even in prepubertal boys there is a higher concentration of testosterone in their blood than in girls of the same age (Frasier and Horton, 1966; Winter *et al.*, 1976). Associated with the rise at puberty, there is comparatively little change or a decrease in the secretion of androstenedione so that the ratio of testosterone to androstenedione in blood from the internal spermatic vein rises from

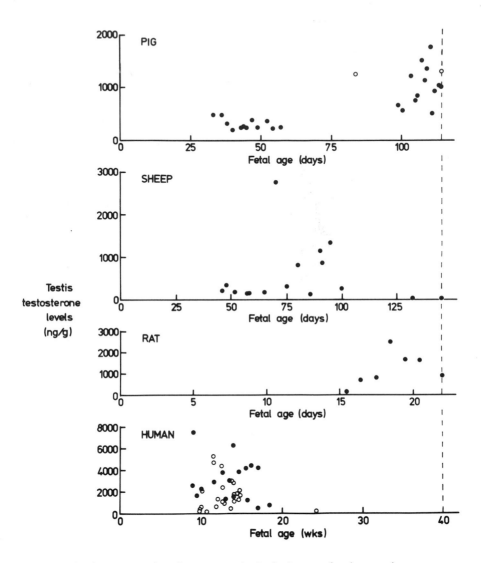

Figure 6.12 The concentration of testosterone in the fetal testes of various species.
Pig:●, data from Raeside and Sigman (1975) and Segal and Raeside (1975);○, data from Booth (1975)
Sheep: data from Attal (1969) and Attal *et al.* (1972)
Rat: data from Warren *et al.* (1973)
Human:●, data from Diez d'Aux and Pearson Murphy (1974);○, data from Reyes *et al.* (1973) The vertical dashed line indicates the approximate age at birth.

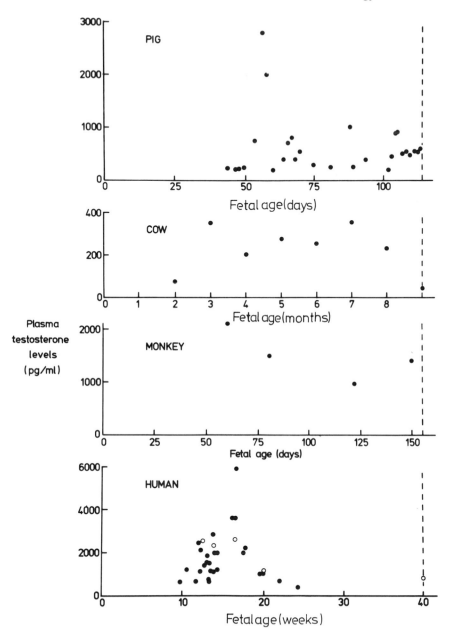

Figure 6.13 The concentration of testosterone in the fetal plasma of various species
Pig: data from Meussy-Dessolle (1974)
Cow: data from Kim *et al.* (1972)
Monkey: data from Resko *et al.* (1973)
Human:●, data from Reyes *et al.* (1974);○data from Abramovich and Rowe (1973) The vertical
dashed line indicates the approximate age at birth.

about 0·1 to about 20 in the bull (Lindner, 1961a, 1969); similar ratios can be found in the bull testis (Lindner and Mann, 1960). In the ram, there is usually more testosterone than androstenedione in the testis but the ratio rises as the animal matures (Lindner, 1961c; Skinner *et al.*, 1968; Attal *et al.*, 1972). Likewise in the rat, there is always more testosterone than androstenedione in the peripheral blood and testes but testosterone becomes preponderant at about the time of puberty (Resko *et al.*, 1968).

It has been demonstrated in a variety of species that the androgenic function of the testis, measured indirectly by semen analysis, develops before spermatogenesis is fully established (Mann, 1964; Mann *et al.*, 1967; Skinner *et al.*, 1968).

### (d) Old age

The total concentration of testosterone in the peripheral serum of men decreases with age (Figure 6.14), and, as the metabolic clearance rate also falls in old age, the secretion of testosterone is probably even more reduced than appears from the circulating concentration (Kent and Acone, 1966). This change in metabolic clearance rate may be because a smaller proportion of the circulating testosterone is free in the old men; the protein bound testosterone is presumably broken down less readily. Moreover, as

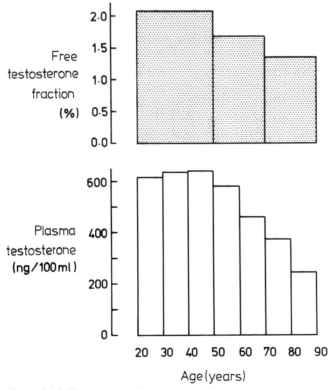

Figure 6.14 The concentration in men of different ages of total testosterone and the fraction of this which is 'free', i.e. not bound to protein. (Data from Vermeulen *et al.*, 1972)

the free testosterone is probably the biologically active hormone, the change in bound to free ratio with age would have the effect of intensifying the effect of the fall in concentration (Vermeulen *et al.*, 1971, 1972; Rubens *et al.*, 1974; Mazzi *et al.*, 1974). Serum testosterone levels are higher in rats at 4 months of age than at 18 months (Chan *et al.*, 1977).

**6.1.5.4    Season**    Many species show marked variation in the amount of testosterone secreted by the testis at different times of the year. Much of the evidence for this is indirect, depending on either the histological appearance of the interstitial cells or on the size of the accessory glands. In general, these data suggested that androgen secretion begins either slightly before or at about the same time as spermatogenesis. (Section 1.4.)

More recently, these observations have been extended and made more quantitative in the roe deer (Short and Mann, 1966) the red deer (Lincoln, 1971), hare (Lincoln, 1974) and hyrax (Millar and Glover, 1973) by estimating the amount of testosterone in the testes. Testosterone becomes detectable in the testes as spermatogenesis begins at the end of June in the red deer and in mid-February in the roebuck. The testosterone content rises to a peak in the 'rut' (as the period of maximal sexual activity is known) in September–October and July–August respectively and the maximal amounts found are about 50- and 10-times those present at the beginning of spermatogenesis. The rut lasts only about a month and then the testosterone content declines rapidly to very low values. Spermatogenesis continues for only a short time after the amount of testosterone in the testes has fallen again. Similar changes in testis size have been described in the white-tailed deer in Minnesota and in this species testis size is closely correlated with serum testosterone levels (McMillin *et al.*, 1974).

In noctule bats (*Nyctalus noctula*), the concentration of testosterone in the testes and blood begins to rise at about the same time as spermatogenesis begins, but maximum concentrations are not reached until the testis weight has begun to decline. Then the testosterone levels remains at levels slightly below the maximal values; the testes regress but the epididymides and accessory organs remain enlarged and the bats are capable of mating (Racey, 1974).

In an American bat (*Myotis lucifugus*), plasma testosterone concentrations did not begin to rise until the latter part of spermatogenesis; maximum levels were reached in mid-August when the spermatozoa were passing into the epididymis and the testis was beginning to shrink. By the beginning of the breeding season, plasma testosterone levels have fallen again to quite low levels (Gustafson and Shemesh, 1976).

In other seasonally breeding mammals, in which spermatogenesis and the breeding season coincide (e.g. fox, hyrax, horse, stoat and ferret) much higher plasma concentrations of testosterone are found during the breeding season that at other times of the year (Joffre and Joffre, 1975; Neaves, 1973; Berndtson *et al.*, 1974a; Gulamhusein and Tam, 1974; Neal *et al.*, 1977). Moving hamsters from a 14 hr light—10 hr dark environment to continuous darkness was associated with a marked fall in plasma testosterone concentrations (Desjardins *et al.*, 1971).

An annual cycle in plasma testosterone has also been described in monkeys (Plant

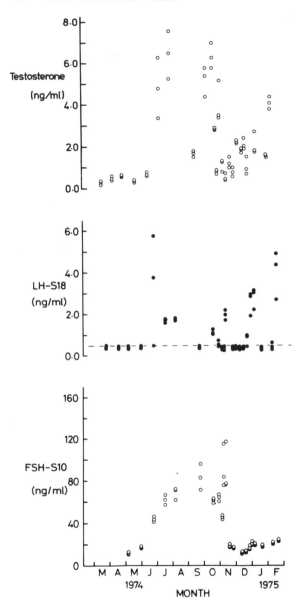

Figure 6.15 Seasonal variation in the concentrations of testosterone, luteinizing hormone (LH) and follicle stimulating hormone (FSH) in jugular venous blood of a Clun Forest ram in Cambridge. On each
occasion of sampling, a set of 3 samples were taken at 5 minute intervals, and individual values are
shown here, emphasizing the short-term temporal variation in LH and testosterone. The levels of both
gonadotrophins begin to rise about the longest day and testosterone slightly afterwards; FSH decreases
again in November, whereas LH and testosterone show some high values as late as February. (Data
from Davies *et al.*, 1977)

*et al.*, 1974; Robinson *et al.*, 1975; Gordon *et al.*, 1976; Michael and Bonsall, 1977). In sheep also there are substantial seasonal changes in testosterone concentrations in serum. In England these are lowest in May or thereabouts, rise to a peak in October and begin to decline again in December. This is associated with similar changes in serum LH and FSH (Figure 6.15; Katongole *et al.*, 1974; Davies *et al.*, 1977). The testes also enlarge from about 200g each at the beginning of the season to about 350g at its height, and then regress. (Fig. 1.10; see also Sections 1.4 and 7.6.1.2.)

**6.1.5.5    Time of day**    There is some dispute about the existence of a diurnal or circadian rhythm in androgen secretion in man. A number of authors have reported a significant fluctuation with the lowest concentrations of testosterone at around midnight and the highest values in the mornings (Southren *et al.*, 1965; Dray *et al.*, 1965; Resko and Eik-Nes, 1966; Saxena *et al.*, 1968; Faiman and Winter, 1971; Okamoto *et al.*, 1971; Rose *et al.*, 1972b; Piro *et al.*, 1973; Judd *et al.*, 1973; Rose *et al.*, 1974). On the other hand, other authors have found no such variation (Figure 6.16) although there was, in some instances, a diurnal fluctuation in cortisol in the same subjects (Hudson *et al.*, 1967; Gordon *et al.*, 1968; Boon *et al.*, 1972; Alford *et al.*, 1973; Serio *et al.*, 1974). There is a diurnal variation in the rhesus monkey (Goodman *et al.*, 1974; Mukku *et al.*, 1976), pigs (Ellendorff *et al.*, 1975) and horses (Kirkpatrick *et al.*, 1976).

In other species such as the bull and the ram, there are large irregular fluctuations in the concentrations of both LH and testosterone (see Figure 6.8) but no consistent diurnal rhythm.

**6.1.5.6    Copulation**    There is a suggestion in a number of species—rat, rabbit, bull, man, monkey and elephant (Endröczi and Lissak, 1962; Saginor and Horton, 1968; Haltmeyer and Eik-Nes, 1969; Herz *et al.*, 1969; Katongole *et al.*, 1971; Jainudeen *et al.*, 1972; Rose *et al.*, 1972a; Purvis and Haynes, 1974; Agmo, 1976; Younglai *et al.*, 1976; Kamel *et al.*, 1977; but cf. Fox *et al.*, 1972; Gombe *et al.*, 1973; Stearns *et al.*, 1973) but not ram or boar (Ellendorff *et al.*, 1975; Illius *et al.*, 1976a, b)—that copulation may be associated with rises in the concentrations of testosterone in the blood and in the testis. In humans, this rise is not produced by masturbation so presumably it is associated not with ejaculation but with sexual excitement (Purvis *et al.*, 1976).

The rise in testosterone is not always associated with or preceded by rises in serum LH concentrations; and in the pig, there are rises in LH without equivalent changes in testosterone (Ellendorf *et al.*, 1975). This raises the possibility that other factors, for example nervous stimulation, may also cause rises in secretion of testosterone.

**6.1.5.7    Nutrition**    Many studies have been made on the effects of undernutrition on reproduction using indirect methods for assessing endocrine function. In general they suggested that androgen production was reduced. This conclusion has been substantiated and quantitated in two more recent studies in which testosterone output was measured directly in groups of underfed and well fed adult rams (Setchell *et al.*, 1965) or in a pair of identical twin bull calves (Mann *et al.*, 1967). Both studies also

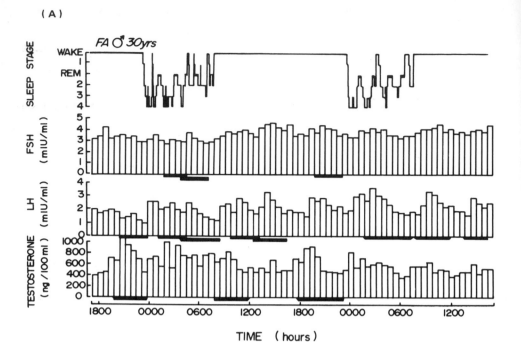

(A)

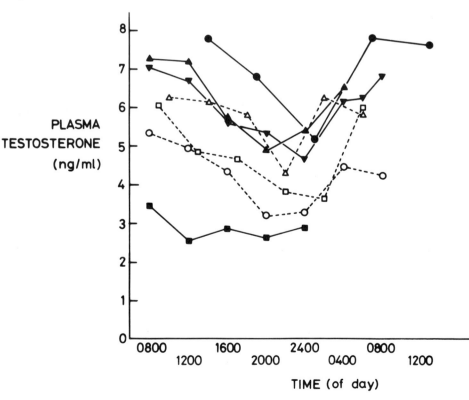

(B)

concluded that the androgenic function of the testis was more affected than sper-matogenesis. Undernutrition also reduces the testosterone concentration in the testes of hyraxes (Millar and Fairall, 1976). The effects of undernutrition appear to be produced by a reduced production of gonadotrophins (Howland, 1975) but reduced sensitivity of the testis to gonadotrophic stimulation may also occur (see also Section 11.2.4).

**6.1.5.8    Temperature and cryptorchidism**    Many observations on indirect in-dicators of androgen secretion suggested that cryptorchidism or heating the testis by insulation of the scrotum had little effect on androgen production. However, Clegg (1960) noted an initial decrease, then an increase above normal followed by a final decrease in fructose content of the coagulating glands of bilaterally cryptorchid rats. Androgen content of the cryptorchid testis was shown to be decreased (Hanes and Hooker, 1937; Gunn et al., 1961; Skinner and Rowson, 1968a) and cryptorchid testes secreted less testosterone than the scrotal testes of the same animals. Furthermore testosterone secretion was decreased by exposing the testis to a temperature of 39·5°C for 1 hour, although no immediate effects were seen when the temperature was varied in the range 35·5 to 39·5°C (Eik-Nes, 1966).

The concentration of testosterone in the serum of bulls exposed to high temperatures fell briefly and then returned to normal although spermatogenesis was affected for the whole period of heat exposure (Rhynes and Ewing, 1973), although the concentration of testosterone in the plasma and testes of rams fell when they were exposed to temperatures between 28 and 32°C (Gomes et al., 1971). The concentra-tion of testosterone in the sera of rats whose testes had been locally heated at 43°C for 30min did not show any significant changes in a 7-week period following heating (Main, Davies and Setchell, unpublished observations) and the concentration of testosterone in ram rete testis fluid did not change when the testes were heated to 41°C for 3 hours (Setchell et al., 1971; see also Section 11.1).

Several of the enzymes involved in steroidogenesis are affected by cryptorchidism or heat (Llaurado and Dominguez, 1963; Kormano et al., 1964; Inano and Tamaoki, 1968; LeVier and Spaziani, 1969).

An interesting possibility of temperature induced changes in the steroids secreted by the gonads was suggested by the observation of Hill (1937) who found that ovaries grafted into the ears of castrated male mice secreted enough androgens to maintain

---

Figure 6.16 (A)    A graph of the concentrations of testosterone, follicle stimulating hormone (FSH) and luteinizing hormone (LH) in blood samples drawn continuously from a human forearm vein for con-secutive 20 min. periods. The heavy lines indicate significant short term fluctuations, defined as 'abrupt or progressive rises in hormone concentration followed by progressive declines over at least two con-secutive samples, where the peak level was significantly different (p< 0·01) from both preceding and following lowest levels.' Note that there appears to be no significant diurnal rhythms. (Reproduced from Alford et al., 1973b)

(B) A summary of the results of several groups of workers who have demonstrated a significant diurnal rhythm in blood testosterone.●data of Dray et al., 1965;■, data of Resko and Eik-Nes, 1966;□data of Southern et al., 1965;○data of Saxena et al., 1969;▲data of Nieschlag and Ismail, 1970;▼data of Okamoto et al., 1971;△data of Faiman and Winter, 1971.

the accessory glands if the mice were kept at 22°C. When the temperature was increased to 33°C, no androgen secretion could be detected. No one has apparently measured oestrogen secretion by testes kept at ovarian temperature.

**6.1.5.9    Anaesthetics and narcotics**    Ether anaesthesia in rats caused a rapid decline in the production of testosterone (Bardin and Petersen, 1967; Fariss et al., 1969). In testes of barbiturate-anaesthetized dogs infused at constant rate with the animals own blood, the production of testosterone declined with time while the release of pregnenolone doubled in the first 75 minutes (Tcholakian and Eik-Nes, 1971). Both these procedures were possibly associated with a considerable amount of stress and, in the latter preparation, unphysiological conditions. In rams there was no difference between testosterone secretion when they were lightly anaesthetized with pentobarbitone and supported in a sling in an approximately normal posture or when conscious and standing quietly in a pen, but the secretion rate was lower in the days following anaesthesia (Setchell et al., 1965). In humans anaesthetized with barbiturate after premedication and maintained by a muscle relaxant and halothane–nitrous oxide–oxygen, there was no change in plasma testosterone during the operations which lasted about one hour. However, there was a sharp fall afterwards, beginning between 1·5 and 4·5h after the operation, reaching a minimum of about 15 per cent of pretreatment values on days 1 and 2 and returning to normal between 6 and 9 days later (Carstensen et al., 1973a, b; see also Oyama and Kudo, 1972). A number of narcotics including morphine, heroin, codeine, and etorphine, have also been found to reduce plasma testosterone concentrations in rats, mice and humans (Azizi et al., 1973; Cicero et al., 1975a, b, 1976a, b, 1977; Mendelson et al., 1975a, b, 1976; Thomas and Dombrosky, 1975; Mirin et al., 1976; Cicero, 1977). One report suggests that marihuana smoking reduces plasma testosterone levels (Kolodny et al., 1974; but cf. Mendelson et al., 1974). Both acute and chronic intake of alcohol reduce plasma testosterone levels and cause atrophy of the testes in humans, rats and mice (see Section 11.3.6).

**6.1.6    Functions of androgens**

**6.1.6.1    Sex differentiation**    The development of the genital system in a male embryo occurs earlier than the corresponding changes in the female and it takes place under the influence of the fetal testis. Practically nothing is known about the hormonal or chemotactic factors controlling the migration of the germ cells to join the rest of the testis near the mesonephros, or of the factors controlling development of the testis itself. Such evidence as is available would suggest that recognized androgens play little or no part.

If we turn to the rest of the genital tract, it appears that the fetal testis controls the development of the male system from the Wolffian ducts, the regression of the Müllerian ducts (which would otherwise give rise to the female organs) and also the differentiation of the derivatives of the urogenital sinus into male organs. If fetal male rabbits are castrated, the Wolffian ducts fail to develop, the Müllerian ducts persist and the external genitalia are female. It is known that the fetal testis secretes testosterone (see Section 6.1.4.2) and if androgens are injected into castrated fetuses, the Wolffian ducts develop, the external genitalia become male but the Müllerian ducts still persist. Development of the Wolffian duct and the external genitalia can be blocked in males with the anti-androgen cyproterone acetate and stimulated in females with testosterone. However regression of the Müllerian ducts is unaffected by treatment with cyproterone—again suggesting that androgens are not involved in this process.

Moreover if male fetal rabbits were unilaterally castrated, the external genitalia still developed normally but the Müllerian duct persisted on the castrated side. This suggests that blood-borne androgens caused the development of the genitalia but that the effects on the Wolffian and Müllerian ducts were due to locally restricted hormones from the testis (see Section 2.3 for references).

**6.1.6.2    Spermatogenesis**    Two lines of evidence suggest that testosterone is important in spermatogenesis in the seminiferous tubules. First, testosterone given in large doses to rats prevents the regression of the seminiferous tubules which occurs after hypophysectomy (Walsh *et al.*, 1933, 1934; Nelson and Gallagher, 1936; Nelson and Merckel, 1937; Wells, 1942; Clermont and Harvey, 1967) or injections of oestrogens (Steinberger and Duckett, 1965). It cannot re-initiate spermatogenesis once atrophy has occurred. In squirrels and monkeys, testosterone appears to be able to prevent regression and also to re-initiate spermatogenesis (Wells, 1942; Smith, 1944). Second, small doses of testosterone administered to intact adult male rats causes regression of the testes, presumably by reducing gonadotrophin secretion. Larger doses do not have this effect (Moore and Price, 1937, 1938; Selye and Friedmann, 1941; Zahler, 1944; Ludwig, 1950; Berndtson *et al.*, 1974b; Weddington *et al.*, 1976) and it has been assumed that this is because these doses recreate the normal concentration of testosterone around the seminiferous tubules. Because of the close proximity of the Leydig cells to the seminiferous tubules, much smaller amounts of naturally produced testosterone are equivalent to the much larger amounts injected elsewhere in the body. Testosterone injected directly into the testis (Dvoskin, 1944) or introduced in a silastic implant in the testis (Ahmad *et al.*, 1973) also maintains spermatogenesis but it is difficult to make this technique quantitative and impossible to ensure even treatment throughout the testis. There is also evidence for a local effect of testosterone in young boys with unilateral Leydig cell tumours; spermatogenesis begins in the affected testis but not in the other one (Schmidt and Tonutti, 1956; Root *et al.*, 1972).

There is also some suggestion that other androgens can substitute for this action of testosterone. In fact, several steroids such as pregnenolone, 17α-hydroxy-pregnenolone, progesterone or 17α-hydroxyprogesterone which are less active androgens than testosterone on the accessory glands, are just as active in maintaining spermatogenesis (Nelson and Merckel, 1937; Selye and Friedmann, 1941; Masson, 1945; Steinberger *et al.*, 1975). These steroids may be converted to testosterone inside the tubules, because the administration of some of them has been shown to increase the concentration of testosterone in rete testis fluid but not in testicular venous blood (Harris and Bartke, 1975). Spermatogenesis can also be maintained with 5α-dihydrotestosterone and 5α-androstane-3α,17β-diol, which are formed from testosterone in the tubules, but cannot be converted back into testosterone

(Chowdhury and Steinberger, 1975; Chemes *et al.*, 1976). In immature rats treated for 10 days with a wide range of doses of testosterone propionate, 5$\alpha$-dihydrotestosterone propionate or 5$\alpha$-androstane-3$\alpha$, 17$\beta$-diol, low doses reduced testis weight, due to suppression of gonadotrophin secretion, but with higher doses, testis weight was normal, presumably because of a direct action of the steroid on the testis. However the testis weight was related to the concentration of 5$\alpha$-dihydrotestosterone in the testis, not to testosterone or 5$\alpha$-androstane-3$\alpha$,17$\beta$-diol concentrations, suggesting that 5$\alpha$-dihydrotestosterone is the active mediator of androgen action in the testis (Weddington *et al.*, 1976; Purvis *et al.*, 1977).

The site of action of testosterone seems to be primarily in enabling diakinesis and the reduction division of meiosis to occur. Hypophysectomy in rats causes arrest of spermatogenesis at the primary spermatocyte stage and the administration of testosterone enables meiosis to proceed (Clermont and Harvey, 1967; Steinberger, 1971; Steinberger and Steinberger, 1974). An involvement of testosterone in the conversion of gonocytes to type A spermatogonia has been suggested on the rather tenuous grounds that spermatogenesis *in vitro* can be initiated less easily from rat testes containing only gonocytes and that testes of rats of that age contain significant concentrations of testosterone (Steinberger, 1971).

**6.1.6.3    Accessory organs**    Castration affects all the accessory organs of reproduction. The survival of spermatozoa in the epididymis is reduced by castration and this effect can be counteracted by injections of testosterone (Orgebin-Crist *et al.*, 1975). Castration has the very obvious effect of reducing the size of the accessory organs of reproduction such as the prostate and the seminal vesicles; this reduction can be prevented and the size returned to normal by injection of testosterone, and the changes were the basis for many of the older biological assays for androgenic activity (see Parkes, 1966). Testosterone is also the most important determinant of the rate of formation of certain of the specific secretions of some of the accessory organs, for example: fructose by the coagulating gland and dorsolateral prostate of rats, the ampullae and seminal vesicles of man, bull, ram, guinea pig, hedgehog and mole, and the glandula vesicularis and prostate of rabbits; citric acid by the prostate of human, hedgehog and mole, the ventral prostate and seminal vesicles of rat, and the seminal vesicles of man, bull, ram, boar and guinea pig. Other specific secretions also under the control of testosterone are inositol from the seminal vesicles and ergothioneine from the ampullary glands of the stallion (see Mann, 1964, 1967, 1969, 1975).

It is interesting that while most of these actions are produced by blood-borne hormone, the action on the epididymis may be from the luminal surface since unilateral ligation of the efferent ducts leads to ipsilateral reduction in the peaks of blood flow found in certain parts of the epididymis (Brown and Waites, 1972), and in the secretion of inositol and carnitine in the corpus epididymidis (B. T. Hinton and B. P. Setchell, unpublished observations). Androgens in the luminal fluid may also affect other accessory organs, as unilateral ligation of the ductus deferens causes regression of the ipsilateral ampullae (Skinner and Rowson, 1968b). The prostate, especially that of the dog, may in addition be affected by the high concentration of androgens in the deferential vein, as this blood can reach the prostate directly (Pierrepoint *et al.*, 1974, 1975a, b).

**6.1.6.4    Behaviour**    The differences in behaviour produced by castration have been known since antiquity and were probably the main reason for the introduction of castration of male domestic animals. The most striking differences in sexual behaviour are produced by prepuberal castration following which the general form and behaviour of the animal remains infantile; normal sexual behaviour may never be started. In contrast, sexual activity may persist for a variable time after castration of adults, the extent probably depending on the degree of involvement of psychological elements in libido. Sexual activity after castration can be restored by injections of androgens. However, in intact males, libido is poorly correlated with androgen levels in blood and is unaffected by administration of androgens (Stone, 1939; Beach, 1949, 1952, 1974, 1975; Kinsey *et al.*, 1948; Riss and Young, 1954; Johnson and Masters, 1961; Lisk and Suydam, 1967; Hart, 1968; Resko and Phoenix, 1972; Phoenix *et al.*, 1973; Phoenix, 1974, 1977; Robinson *et al.*, 1975). Likewise impotence in man is commonly associated with normal concentrations of testosterone in blood (Hudson *et al.*, 1967). In newborn rats, the presence of androgens (or more probably of oestrogens derived from androgens) is essential for the development of a male pattern of hypothalamic function in adult life. Androgens in newborn rats also induce male behaviour patterns in later life (see Section 2.3.3).

Castration also produces more general changes in behaviour, such as reduced aggression, and even reduces general activity in rats (Wang *et al.*, 1925; Richter and Wislocki, 1928; Richter, 1933; Pedersen-Bjergaard and Madsen, 1938). Androgens, probably of adrenal origin, are thought to be important in determining libido in females (Carter *et al.*, 1947).

**6.1.6.5    Regulation of gonadotrophin secretion**    Castration leads to an increase in weight of the pituitary and to changes in its histological appearance (Fichera, 1905, Severinghaus *et al.*, 1932; Severinghaus, 1937; but cf Moore, 1922); it is also followed by an increase in the secretion of both LH and FSH while testosterone and several other androgens including testosterone, $5\alpha$-dihydrotestosterone and $5\alpha$-androstan-$3\alpha$-, $17\beta$-diol inhibit secretion of LH by the pituitary (Figure 6.17). They also inhibit FSH secretion, but only at equivalent or greater doses than those that affect LH (Gay and Dever, 1971; Dufy-Barbe and Franchimont, 1972; Lee *et al.*, 1972; Demoulin *et al.*, 1973; Ojeda and Ramirez, 1973; Sherins and Loriaux, 1973; Swerdloff and Walsh, 1973; Swerdloff *et al.*, 1973; Walsh *et al.*, 1973; Zanisi *et al.*, 1973; Martini *et al.*, 1974; Stewart-Bentley *et al.*, 1974; Verjans *et al.*, 1974; Moger, 1976a; Verjans and Eik-Nes, 1976). Removal of free testosterone from the circulating blood by active or passive immunization against testosterone conjugated with albumin leads to a greater rise in LH than FSH, compared with the effects of castration (Hillier *et al.*, 1975; Nieschlag *et al.*, 1975; Main *et al.*, 1977, see Figure 6.20) and cyproterone, an anti-androgen, causes a greater rise in LH than in FSH (see Section 6.1.8).

The relation between LH and the testicular androgens thus conforms to the classical negative feedback by a hormone on the secretion of its trophic hormone.

(A)

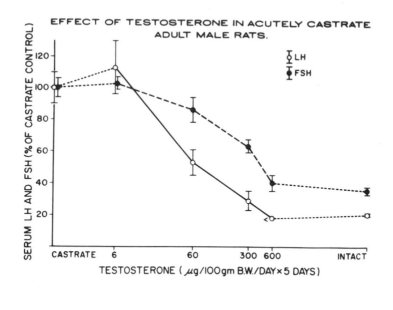

(B)

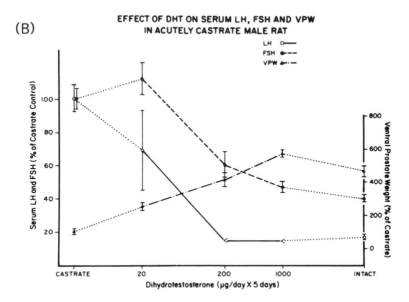

Figure 6.17  The effect of 5 daily doses of (A) testosterone or (B) dihydrotestosterone on serum LH and FSH in acutely castrated adult male rats. Note that both hormones reduce LH at lower doses than are needed to reduce FSH, but both gonadotrophins can be reduced to intact levels with sufficiently high doses of either hormone. (Reproduced from Swerdloff and Walsh, 1973 and Swerdloff et al., 1973)

While androgens do have some control over the secretion of FSH, other factors (see Section 6.2) must be invoked to explain some of the observations.

**6.1.6.6    Anabolism**  Most androgens have anabolic or myotrophic activity which is not necessarily related to their androgenic potency. In fact the most active androgens have only moderate anabolic activity and vice versa. It has been claimed that administration of testosterone to a castrate leads to nitrogen retention and muscular growth (Kochakian and Murlin, 1935; Kochakian, 1946, 1949, 1950, 1976) but more recent work has raised some doubts about this effect (Hervey and Hutchinson, 1973). One muscle, the levator ani muscle, has been the subject of many investigations in rats (Eisenberg and Gordon, 1950). It has been shown that administration of testosterone caused marked hypertrophy of the individual muscle cells rather than cell multiplication (Venable, 1966) in contrast to the prostate where cell proliferation is a marked response to androgens. However this muscle seems to be quite atypical and should be considered as an accessory sexual organ rather than a representative muscle (Hayes, 1965).

**6.1.6.7    Miscellaneous**  Androgens have a number of incidental effects, often confined to one species or family of animal. The development of the scrotum in those animals where it is found, is dependent on androgen secretion (see Chapter 5). The growth of horns in sheep is controlled by androgens, the entire males growing much larger horns than castrate males or females. In other animals (e.g. cattle) horn growth is not affected by androgens (see Parkes, 1966). In deer, the change from 'velvet' to 'hard horn' antlers depends on an increase in androgen secretion; the shedding of the hard horn antlers at the end of the rutting season is produced by a sudden fall-off in androgen production which can be simulated by castration. The growth of the 'velvet' antlers is not androgen dependent (Lincoln et al., 1970; Lincoln, 1971). Hair patterns in certain deer and primates are also androgen determined. Beard growth, the shape of the hair line and the occurrence of baldness in humans is also dependent on androgens, although once developed, these characteristics are not much affected by castration (Parkes, 1966).

Sudden growth of the larynx causing 'breaking' of the voice is best known in humans although it occurs to some extent in other species. The human also shows more prominently than most species the continued growth of the long bones in castrates which usually results in them being taller and particularly longer limbed than entire males. However this phenomenon can also be seen in castrate male goats. Various odour glands in such species as goats and guinea pigs, and the sebaceous glands on the sternum of the brush-tailed possum are androgen dependent (Parkes, 1966).

The kidneys of male animals are consistently heavier than those of females or castrate males and contain a higher concentration of certain enzymes (Kochakian, 1947). The kidneys of castrated male rats also show a reduced compensatory hypertrophy after unilateral nephrectomy.

The liver and heart are also sensitive to testosterone (Korenchevsky et al., 1941). Testosterone produces uterine enlargement in female animals (Velardo et al., 1956;

Velardo, 1959) and can cause virilization if administered for long periods (Parkes, 1966).

Androgens have a pronounced effect on the thymus. Castration increases thymus weight (Dougherty, 1952; Eidinger and Garrett, 1972; Castro, 1974a), increases the immunological response of the animal (Castro, 1974b) and prevents involution of the thymus following stress (Selye, 1952) and undernutrition (Dougherty, 1952). Injections of testosterone cause involution of lymphoid tissue and immunosuppression (Korenchevsky et al., 1932; Plagge, 1941) and prevent the relative thymic enlargement seen after castration (Castro, 1974c) (see Section 10.5 for evidence for an effect of the thymus on the testis).

Certain specialized cells in the submaxillary salivary glands of mice are also under the control of androgens (Junquiera et al., 1949), and so are the concentrations in this gland of nerve growth factor and epidermal growth factor (see Levi-Montalcini and Angeletti, 1968).

### 6.1.7    Mechanism of action of androgens

Although androgens exert an important effect on the seminiferous tubules, and there are receptors for androgens in this tissue (see below), most of our knowledge about the mechanism of action of androgens comes from studies on the ventral prostate of the rat. These studies are not immediately relevant to this book, but will be summarized here on the assumption that the action of androgens in the testis is basically similar.

The first step in any androgen action is the entry of the hormone into the cell. There is evidence for selective uptake and retention of androgens by androgen-sensitive tissues and there are receptor-proteins in both the cytoplasm and nuclei of the cells of these tissues (Figure 6.18). These receptor-proteins have a much greater affinity for DHT than for testosterone, and therefore it would appear that testosterone enters the cells, and then is transformed to DHT which binds to the cytoplasmic receptor. By decreasing the intracellular concentration of free DHT, the receptors would increase the production of DHT. The cytoplasmic receptors in the prostate appear to be of two types with slightly different chemical characteristics and neither binds testosterone to any appreciable extent. One of these receptors (the $\beta$-receptor) seems to be involved in translocating the DHT into the nucleus where the receptor-hormone complex is transformed into a characteristically nuclear complex. Isolated nuclei appear to be unable to form the complex in the absence of cytoplasm, so it appears to be a two-stage process. The nuclei retain the hormone-receptor complex long after the activity has disappeared from the cytoplasm, and this may involve the reaction of the $\beta$-receptor with 'acceptor' molecules in the nucleus. The next stage of androgen action probably involves stimulation of a nucleolar $Mg^{2+}$-stimulated RNA polymerase, followed by increases in the activity of a Mn-activated enzyme, also in the nucleolus and finally increased RNA synthesis (see Bruchovsky and Wilson, 1968a, b; Anderson and Liao, 1968; Liao and Fang, 1970; Williams-Ashman, 1975; Wilson, 1975; Mainwaring, 1976).

Androgen-receptors in testis tissue have recently been described; their characteristics are essentially similar to those from the prostate except that the testis

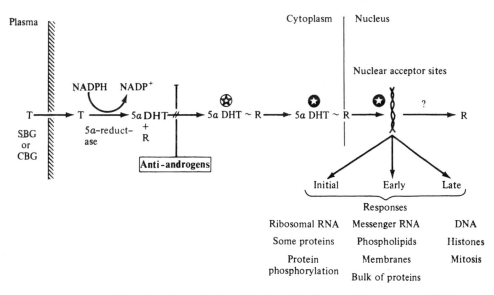

Figure 6.18 Diagram illustrating the mechanism of action of androgens. T, testosterone; SBG, sex steroid-binding β-globulin; CBG, corticosteroid-binding α₂-globulin; 5αDHT, 5α-dihydrotestosterone; ✪ and ✪ indicate changes in configuration of receptor complex during activation. (Based on a drawing by Mainwaring, 1977)

receptors appear to bind testosterone and DHT with approximately the same affinity. These receptors appear to be located in germ cells, interstitial cells and Sertoli cells (Mulder *et al.*, 1974a, 1975a, b; Hansson *et al.*, 1974a; Smith *et al.*, 1975a, b; McLean *et al.*, 1976; Wilson and Smith, 1975; Wilson and French, 1976). Interstitial tissue also contains oestradiol receptors, both in the cytoplasm and nucleus but their significance is obscure (Brinkmann *et al.*, 1972; van Beurden-Kamers *et al.*, 1974; Mulder *et al.*, 1974b; de Boer *et al.*, 1976). The androgen-receptors in the tubular cells are distinct from the androgen-binding protein which is secreted by the Sertoli cells of several species into the rete testis fluid (see Section 8.2.1). This protein is chemically similar to testosterone-binding globulin but its physiological significance is not clear. It may be concerned with increasing the entry of testosterone into the tubules or in transporting large amounts of androgen into the epididymis.

### 6.1.8    Anti-androgens
Some compounds with anti-androgenic activity are modified steroids, but several non-steroidal substances are also active. Among the steroidal anti-androgens, the most studied are cyproterone and cyproterone acetate. Cyproterone acetate is about three times as potent an anti-androgen as cyproterone, but cyproterone acetate also has progestational activity whereas cyproterone does not (see Neumann and Steinbeck, 1974). Their anti-androgenic effect is produced by competition with testosterone or DHT for binding sites in the receptor cells (Fang and Liao, 1969; Baulieu and Jung, 1970).

Both compounds block the effects of androgens on the accessory organs in adult males and interfere with development of the Wolffian duct and internal genitalia in male fetuses (see Neumann *et al.*, 1970; Neumann and Steinbeck, 1974). Cyproterone causes an increase in plasma LH and sometimes in plasma FSH, presumably by interfering with feed-back by androgens at the hypothalamus (Berswordt-Wallrabe and Neumann, 1967, 1968; Rausch-Strooman *et al.*, 1970; Vosbeck and Keller, 1971; Walsh *et al.*, 1972). Cyproterone acetate may cause a slight decrease in gonadotrophin output by the pituitary because of its progestational activity or gonadotrophin levels remain normal (Viguier-Martinez and Pelletier, 1972; Morse *et al.*, 1973; Brotherton, 1974). These different effects on the pituitary are important in explaining the effects of the two anti-androgens on the testis, although, in general, studies with these drugs confirm that androgens are necessary for spermatogenesis. Cyproterone acetate causes a progressive inhibition of spermatogenesis due to its anti-androgenic action (Ott, 1968; Markevitz *et al.*, 1969; Steinbeck *et al.*, 1971; Städtler, 1972) and in some cases a decreased activity of the Leydig cells (Schoones *et al.*, 1971; Sorcini *et al.*, 1971; Brotherton, 1974). Cyproterone causes a temporary reduction in spermatogenesis after about three weeks of treatment, but then the increased production of testosterone by the testis in response to the increased secretion of gonadotrophins overcomes the anti-androgenic effect and spermatogenesis returns to normal. The effectiveness of the drugs as anti-androgens in these experiments was checked by measuring the atrophy of the accessory organs (Steinbeck *et al.*, 1971).

Another steroidal compound with anti-androgenic activity is BOMT (6α-bromo-17β-hydroxy-17α-methyl-4-oxa-5α-androstan-3-one). This compound has no androgenic, oestrogenic or progestational activity but is a potent anti-androgen (Boris *et al.*, 1970); it competes effectively for the specific, high-affinity binding sites for DHT in the rat prostate (Mangan and Mainwaring, 1972) and depresses testis weight (Boris *et al.*, 1970).

SKF 7690 (17α-methyl-B-nor testosterone) has also been reported to have anti-androgenic activity (Saunders *et al.*, 1964). It appears to produce only a slight decrease in the weight of the testes of rats (Mahesh *et al.*, 1966), but inhibits spermatogenesis in hamsters (Lubicz-Nawrocki and Glover, 1973). It also acts by competing with DHT at receptor sites (Tveter, 1971). As no measurements were made of the effects of BOMT or SKF 7690 on serum gonadotrophins, it is not clear whether they are antigonadotrophic or have a direct effect on the testis.

R-2956 (17β-hydroxy-2, 2, 17α-trimethylestra-4,9,11-trien-3-one) and spironolactone have also been shown to have an anti-androgenic effect on the prostrate (Azadian-Boulanger *et al.*, 1974; Rasmussen *et al.*, 1972) again by competing with DHT (Bonne *et al.*, 1974; Bonne and Raynaud, 1974); R-2956 has no effect on testis weight (Azadian-Boulander *et al.*, 1974) but no studies to have been done on the effects of spironolactone on the testis.

Two closely related non-steroid anti-androgens—Ro 2-7239 (2-acetyl-7-oxo-1,2,3,4,4a,4b,5,6,7,9,10,10a-dodecahydro-phenanthrene) and Ro 5-2537 (which has a 1-ethinyl-1-hydroxyethyl group in place of the acetyl at the 2 position)—have apparently different effects on testis weight. Ro 5-2537 has little effect on the testis, in doses which cause pronounced reductions in seminal vesicle and prostate weights (Boris, 1965). Ro 2-7239 causes a moderate decrease in testis weight as well as prostate and seminal vesicle weight (Boris, 1962, but cf. Eviatar *et al.*, 1961 for smaller doses) but again no measurements were made of the concentrations of gonadotrophins in the serum.

Flutamide (4′-nitro-3′.trifluoro-methylisobutyranilide, Sch 13521) is a potent, non-steroidal anti-androgen without progestational or oestrogenic effects (Neri *et al.*, 1972). It also interferes with the binding and retention of DHT by the rat prostate (Liao *et al.*, 1974; Peets *et al.*, 1974) and causes atrophy of the accessory organs, but has no effect on the testis (Neri *et al.*, 1972). Flutamide is rapidly metabolized to $\alpha,\alpha,\alpha$,trifluoro-2-methyl-4′-nitro-*m*-lacto-luidide, which may be the active form (Katchen and Buxbaum, 1975; Katchen *et al.*, 1976).

A study has also been made of the *in vivo* effects on the testis of six *in vitro* inhibitors of testicular microsomal $C_{21}$-$3\beta$-hydroxysteroid-$\Delta^5$-dehydrogenase. Four of these compounds given subcutaneously for 6 weeks reduced serum testosterone and DHT although only one (19-norspiroxenone) also reduced fertility, the size of the testis and the concentration of both testosterone and DHT there; it also reduced serum and pituitary LH and FSH. It is interesting that this compound is also a potent anti-oestrogen (Goldman *et al.*, 1976).

## 6.2    Inhibin

The secretion by the testis of another hormone which controls the production of FSH by the pituitary (Figure 6.19) has been suspected for many years, and the name 'inhibin' was coined in 1932 by McCullagh for this substance. Its existence was suggested in many experiments dating back to 1923, in which the germinal cells of the testis were damaged supposedly without affecting the Leydig cells and androgen production. This usually led to changes in the pituitary and a rise in the concentration of serum FSH. When the germinal epithelium was disrupted after efferent duct ligation, the rise in FSH was associated with unchanged serum concentrations of LH and testosterone (Setchell *et al.*, 1977a, b). However, with other supposedly specific lesions in the germinal epithelium, there were rises in both serum FSH and LH with normal or slightly reduced levels of serum testosterone, for example after X-irradiation, local heating of the testis or cryptorchidectomy, and antispermatogenic drugs such as busulphan or WIN 18336. Suggestive evidence for inhibin has also been obtained in experiments with ORF 1616 (Figure 6.20), nitrofurans, or cadmium, during vitamin E deficiency, testicular ischaemia, immunological aspermatogenesis and in many clinical conditions in man involving the germinal cells more or less specifically. However, no change in serum FSH was seen in vitamin A deficient rats, in rats given hydroxyurea in their drinking water or in androgen-insensitive male pseudohermaphrodite rats (Stanley-Gumbreck rats), in all of which germinal cells were depleted; serum FSH was also normal in some clinical conditions in men with reduced numbers of spermatids (see Setchell and Main, 1974; Setchell *et al.*, 1977b). Inhibin is therefore probably produced by the Sertoli cells but the route

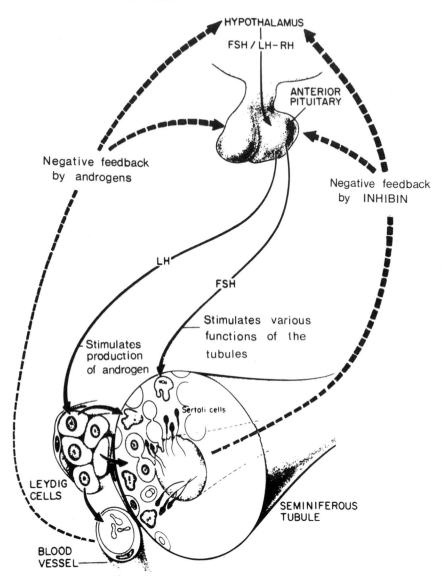

Figure 6.19 A diagram illustrating the probable pituitary controls acting on the testis and the negative feedback by androgens and inhibin on LH and FSH production respectively. LH is thought to act on the Leydig cells causing them to secrete androgens, which enter the tubules and also the blood vessels. The blood-borne hormones control the secondary sexual characteristics and also feed back on both hypothalamus and pituitary to control LH production, and at high levels also FSH production. FSH acts on the Sertoli cells. Inhibin is probably formed inside the tubules and it may feed back on the hypothalamus and/or pituitary to control FSH production, and at high levels also LH production. (Based on a diagram by Fawcett, 1975)

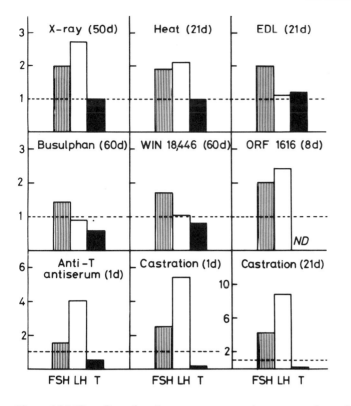

Figure 6.20 The effect of various treatments on the concentrations of follicle-stimulating hormone (FSH, shaded columns), luteinizing hormone (LH, open columns) and testosterone (dark columns) in rats. Values given are those for the treated animals as a percentage of sham-operated or saline-injected controls, at the time of the maximal change in FSH. Note that efferent duct ligation (EDL) leads to a substantial rise in FSH with no change in LH or testosterone; that heat, X-irradiation and ORF 1616 cause a rise in both FSH and LH without any change in testosterone; and busulphan and WIN 18,446 cause a rise in FSH and a fall in testosterone with no change in LH. However in all these instances, the rise in FSH is similar to or proportionally greater than that in LH, whereas after castration or the injection of antiserum against a testosterone-3 oxime conjugate with bovine serum albumin, there is proportionally a much greater effect on LH. Therefore it is necessary to postulate a lack of 'inhibin' to explain the selective rise in FSH after selective damage to the germinal epithelium (see Setchell *et al.*, 1977b; and Section 6.1.6.5).

of secretion is not known. Sertoli cells in culture inhibit FSH secretion by pituitary cells cultured with them or in media in which the Sertoli cells had been growing (Steinberger and Steinberger, 1976).

Inhibin-like activity has been found in rete testis fluid and seminal plasma as well as in extracts of testes and spermatozoa (Figure 6.21), but the biological significance of these various preparations is uncertain. Although inhibin has not yet been isolated and purified, active fractions from all four sources appear to be peptides, not steroids. There is some disagreement about the size of the molecule; under some conditions, activity can be found in rete testis fluid corresponding to molecular weights

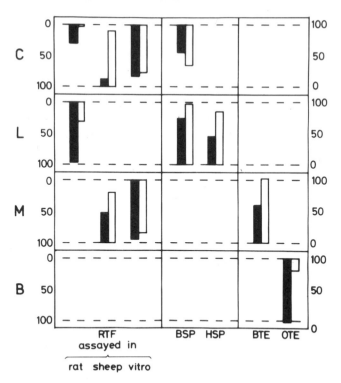

Figure 6.21 The effect of various 'inhibin' preparations on FSH and LH. The effect is expressed either as the percentage inhibition of the post-castration rise in gonadotrophins *in vivo* or of the LH-RH stimulated release of gonadotrophins *in vitro*, plotted down from the upper dotted line (scale on left) or as the effect on the actual concentrations of gonadotrophins in serum of treated animals, plotted up from the lower dotted line (scale on right). Note that most of the preparations affect FSH more than LH, unlike all steroids tested which either affect LH but not FSH, or LH and FSH similarly. The values shown are either the maximal effect, where this is known, or, more usually, the effect of an arbitrarily chosen dose.

C, Cambridge: see Setchell and Jacks; 1974, Davies *et al.*, 1976; Setchell *et al.*, 1977a. L, Liège; see Franchimont, 1972; Franchimont *et al.*, 1975a, b. M, Melbourne: see Lee *et al.*, 1974; Baker *et al.*, 1976. B, Bangalore: see Nandini *et al.*, 1976.

RTF, rete testis fluid; BSP, bovine seminal plasma; HSP, human seminal plasma; BTE, bovine testicular extract; OTE, ovine testicular extract.

    The RTF assayed in rats in Cambridge was concentrated 10x with an Amicon filter UM2; that assayed in Liège was a fraction (D II of RTF$_3$) of concentrated fluid separated on Sephadex G-200 and then DEAE cellulose. The Cambridge *in vitro* assay used incubated rat pituitary halves; the Melbourne *in vitro* assay used cultured pituitary cells. The BSP assayed in Cambridge was precipitated with ethanol and the precipitate washed with acetone before assay; that assayed in Liège was treated in the same way and a fraction (AcII) prepared with Sephadex G-100.

of less than 5000 Daltons as well as about 20,000 and 80,000 Daltons. In seminal plasma and testis extracts, most of the activity behaves like material of less than 5000 or about 20,000 Daltons molecular weight, but the active fraction extracted from spermatozoa seems to be a much smaller peptide, about 1000 to 2000 Daltons (Setchell and Sirinathsinghji, 1972; Franchimont, 1972; Lee *et al.*, 1974; Lugaro *et al.*, 1974; Setchell and Jacks, 1974; Franchimont *et al.*, 1975a, b; Baker *et al.*, 1976; Davies *et al.*, 1976; Keogh *et al.*, 1976; Moodbidri *et al.*, 1976; Nandini *et al.*, 1976). This apparent disparity in molecular size may be due to polymerization of the active component, to association of a small active molecule with larger proteins, or the breakdown of a large active molecule into smaller still active fragments.

Antibodies have been raised in rabbits against partially purified inhibin and injected into rats. The antibodies appear to block the action of the rats own inhibin, and the plasma FSH rises, both in male and female animals (Franchimont *et al.*, 1975b).

The testis also produces appreciable amounts of (+)-1,4-diphenylbutane-2,3-diol (DPB) (Eik-Nes, 1967b); this compound had no effect on serum LH or FSH when administered subcutaneously to rats, but elevated serum FSH when implanted in the median eminence (Iturriza *et al.*, 1977).

## References to Chapter 6

Aakvaag, A., Hagen, A. A. and Eik-Nes, K. B. (1964) Biosynthesis *in vivo* of testosterone and $\Delta^4$-androstenedione from dehydroepiandrosterone sodium sulfate by the canine testis and ovary. *Biochim. Biophys. Acta* **86**, 622–7.

Abramovich, D. R. and Rowe, R. (1973) Foetal plasma testosterone levels at mid-pregnancy and at term: relationship to foetal sex. *J. Endocrinol.* **56**, 621–2.

Abramovich, D. R., Baker, T. G. and Neal, P. (1974) Effect of human chorionic gonadotrophin on testosterone secretion by the foetal human testis in organ culture. *J. Endocrinol.* **60**, 179–85.

Acevedo, H. F., Axelrod, L. R., Ishikawa, E. and Takaki, F. (1961) Steroidogenesis in the human fetal testis. The conversion of pregnenolone-$7\alpha$-$^3$H to dehydroepiandrosterone, testosterone and 4-androstene-3,17-dione. *J. Clin. Endocrinol. Metab.* **21**, 1611–3.

Acevedo, H. F., Axelrod, L. R., Ishikawa, E. and Takaki, F. (1963) Studies in fetal metabolism. II. Metabolism of progesterone-4-$C^{14}$ and pregnenolone-$7\alpha$-$H^3$ in human fetal testes. *J. Clin. Endocrinol. Metab.* **23**, 885–90.

Agmo, A. (1976) Serum luteinizing hormone and testosterone after sexual stimulation in male rabbits. *Acta Physiol. Scand.* **96**, 140–2.

Ahmad, N. and Sowell, J. G. (1976) *In vitro* conversion of testosterone to various metabolites by the teased testicular tissue of normal and posthypophysectomized regressed rats. *Biol. Reprod.* **14**, 561–5.

Ahmad, N., Haltmeyer, G. C. and Eik-Nes, K. B. (1973) Maintenance of spermatogenesis in rats with intratesticular implants containing testosterone or dihydrotestosterone (DHT). *Biol. Reprod.* **8**, 411–19.

Ahmad, N., Haltmeyer, G. C. and Eik-Nes, K. B. (1975) Maintenance of spermatogenesis with testosterone or dihydrotestosterone in hypophysectomized rats. *J. Reprod. Fertil.* **44**, 103–7.

Alford, F. P., Baker, H. W. G., Patel, Y. C., Rennie, G. C., Youatt, G., Burger, H. G. and Hudson, B. (1973a) Temporal patterns of circulating hormones as assessed by continuous blood sampling. *J. Clin. Endocinol. Metab.* **36**, 108–16.

Alford, F. P., Baker, F. W. G., Burger, H. G., Kretser, D. M. de, Hudson, B., Johns, M. W., Masterton, J. P., Patel, Y. C. and Rennie, G. C. (1973b) Temporal patterns of integrated plasma hormone levels during sleep and wakefulness II. Follicle-stimulating hormone, luteinizing hormone, testosterone and estradiol. *J. clin. Endocrinol. Metabol.* **37**, 848–54.

Amann, R. P. and Ganjam, V. K. (1976) Steroid production by the bovine testis and steroid transfer across the pampiniform plexus. *Biol. Reprod.* **15**, 695–703.

Anderson, K. M. and Liao, S. (1968) Selective retention of dihydrotestosterone by prostatic nuclei. *Nature* (London) **219**, 277–9.

Andresen, Ø. (1975) 5α-Androstenone in peripheral plasma of pigs, diurnal variation in boars, effect of intravenous HCG administration and castration. *Acta Endocrinol.* **78**, 385–91.

Armstrong, D. T., Moon, Y. S., Fritz, I. B. and Dorrington, J. H. (1975) Synthesis of estradiol-17β by Sertoli cells in culture; stimulation by FSH and dibutyryl cyclic AMP. *Curr. Top. Molec. Endoc.* **2**, 85–96.

Attal, J. (1969) Levels of testosterone, androstenedione, estrone, and estradiol-17β in the testes of fetal sheep. *Endocrinology* **85**, 280–9.

Attal, J., Andre, D. and Engels, J-A. (1972) Testicular androgen and oestrogen levels during the postnatal period in rams. *J. Reprod. Fertil.* **28**, 207–12.

Azadian-Boulanger, G., Bonne, C., Secchi, J. and Raynaud, J. P. (1974) Antiandrogenic activity of R 2956 (17β-hydroxy-2,2,17α-trimethylestra-4,9,11-trien-3-one). I. Endocrinological profile. *J.Pharmacol.* (Paris) **5**, 509–20.

Azizi, F., Vagenakis, A. G., Longcope, C., Ingbar, S. H. and Braverman, L. D. (1973) Decreased serum testosterone concentration in male heroin and methadone addicts. *Steroids* **22**, 467–72.

Baird, D. T., Galbraith, A., Fraser, I. S. and Newsam, J. E. (1973) The concentration of oestrone and oestradiol-17β in spermatic venous blood in man. *J. Endocrinol.* **57**, 285–8.

Baker, H. W. G., Bremner, W. J., Burger, H. G., Kretser, D. M. de, Dulmanis, A., Eddie, L. W., Hudson, B., Keogh, E. J., Lee, V. W. K. and Rennie, G. C. (1976) Testicular control of FSH secretion. *Rec. Prog. Horm. Res.* **32**, 429–69.

Bardin, C. W. and Peterson, R. E. (1967) Studies of androgen production by the rat: testosterone and androstenedione content of blood. *Endocrinology* **80**, 38–44.

Bartke, A. (1976) Pituitary–testis relationships. Role of prolactin in the regulation of testicular function. *Prog. Reprod. Biol.* **1**, 136–52.

Bartke, A. and Lloyd, C. W. (1970) Influence of prolactin and pituitary isografts on spermatogenesis in dwarf mice and hypophysectomized rats. *J. Endocrinol.* **46**, 321–9.

Bartke, A., Musto, N., Caldwell, B. V. and Behrman, H. R. (1973) Effects of a cholesterol esterase inhibitor and of prostaglandin $F_{2\alpha}$ on testis cholesterol and on plasma testerone in mice. *Prostaglandins* **3**, 97–104.

Baulieu, E-E. and Jung, I. (1970) A prostatic cytosol receptor. *Biochem. Biophys. Res. Comm.* **38**, 599–606.

Baulieu, E-E. and Robel, P. (1970) Catabolism of testosterone and androstenedione. In *The Androgens of the Testis*, ed. K. B. Eik-Nes, 49–91, New York: Marcel Dekker.

Baulieu, E. E., Fabre-Jung, I. and Huis in't Veld, L. G. (1967) Dehydroepiandrosterone sulfate: a secretory product of the boar testis. *Endocrinology* **81**, 34–8.

Beach, F. A. (1947) A review of physiological and psychological studies of sexual behaviour in mammals. *Physiol. Rev.* **27**, 240–307.

Beach, F. A. (1949) *Hormones and Behaviour*, New York: P. Hoeber.

Beach, F. A. (1952) Sex and species differences in the behavioural effects of gonadal hormones. *Ciba. Colloq. Endocrinol.* **3**, 3–14.

Beach, F. A. (1974) Behavioural endocrinology and the study of reproduction. *Biol. Reprod.* **10**, 2–18.

Beach, F. A. (1975) Hormonal modification of sexually dimorphic behaviour. *Psychoendocrinology* **1**, 3–23.

Bedrak, E. and Samuels, L. T. (1969) Steroid biosynthesis by the equine testis. *Endocrinology* **85**, 1186–95.

Bell, J. B. G. (1972) *In vitro* testosterone production from endogenous precursors by the seminiferous tubules and interstitium of the human testis. *Steroid Lipid Res.* **3**, 315–20.

Bell, J. B. G., Vinson, G. P., Hopkin, D. J. and Lacy, D. (1968) Pathways for androgen biosynthesis from [7α-³H] pregnenolone and [4-¹⁴C] progesterone by rat testis interstitium *in vitro. Biochim. Biophys. Acta* **164**, 412–20.

Bell, J. B. G., Vinson, G. P. and Lacy, D. (1971) Studies on the structure and function of the mammalian testis. III. *In vitro* steroidogenesis by the seminiferous tubules of rat testis. *Proc. Roy. Soc.* (London) B. **176**, 433–43.

Berndtson, W. E., Pickett, B. W. and Nett, T. M. (1974a) Reproductive physiology of the stallion. IV. Seasonal changes in the testosterone concentration of peripheral plasma. *J. Reprod. Fertil.* **39**, 115–18.

Berndtson, W. E., Desjardins, C. and Ewing, L. L. (1974b) Inhibition and maintenance of spermatogenesis in rats implanted with polydimethylsiloxane capsules containing various androgens. *J. Endocrinol.* **62**, 1–11.

Berswordt-Wallrabe, R. von and Neumann, F. (1967) Influence of a testosterone antagonist (cyproterone) on pituitary and serum FSH content in juvenile male rats. *Neuroendocrinology* **2**, 107–12.

Berswordt-Wallrabe, R. von and Neumann, F. (1968) Influence of a testosterone antagonist (Cyproterone) on pituitary and serum ICSH content in juvenile male rats. *Neuroendocrinology* **3**, 332–6.

Beurden-Lamers, W. M. O. van, Brinkmann, A. O., Mulder, E. and Molen, H. J. van der (1974) High-affinity binding of oestradiol-17β by cytosols from testis interstitial tissue, pituitary, adrenal, liver and accessory sex glands of the male rat. *Biochem. J.* **140**, 495–502.

Bloch, E. (1964) Metabolism of 4-¹⁴C-progesterone by human fetal testis and ovaries. *Endocrinology* **74**, 833–45.

Bloch, E. (1967) The conversion of 7-³H-pregnenolone and 4-¹⁴C-progesterone to testosterone and androstenedione by mammalian fetal testis. *Steroids* **9**, 415–30.

Bloch, E. and Benirschke, K. (1965) *In vitro* steroid biosynthesis by fetal, newborn and adult armadillo adrenals and by fetal armadillo testes. *Endocrinology* **76**, 43–51.

Bloch, E., Lew, M. and Klein, M. (1971) Studies on the inhibition of fetal androgen formation. Inhibition of testosterone synthesis in rat and rabbit fetal testes with observations on reproductive tract development. *Endocrinology* **89**, 16–31.

Boer, W. de, Mulder, E. and Molen, H. J. van der (1976) Effects of oestradiol-17β, hypophysectomy and age of cytoplasmic oestradiol-17β receptor sites in rat testis interstitial tissue. *J. Endocrinol.* **70**, 397–407.

Bongiovanni, A. M. and Eberlein, W. R. (1957) The renal clearance of neutral 17-ketosteroids in man. *J. Clin. Endocr.* **17**, 238–49.

Bonne, C. and Raynaud, J. P. (1974) Mode of spironolactone anti-androgenic action: inhibition of androstanolone binding to rat prostate androgen receptor. *Mol. Cell. Endocrinol.* **2**, 59–67.

Bonne, C., Raynaud, J. P., Chantier, M. D. and Courtois, P. (1974) Antiandrogenic activity of R 2956 (17β-hydroxy-2,2,17α-trimethylestra-4,8,11-trien-3-one). II. Mechanism of action. *J. Pharmacol.* (Paris) **5**, 521–32.

Boon, D. A., Keenan, R. E. and Slaunwhite, W. R. (1972) Plasma testosterone in men: variation but not circadian rhythm. *Steroids* **20**, 269–78.

Booth, W. D. (1975) Changes with age in the occurrence of $C_{19}$ steroids in the testis and submaxillary gland of the boar. *J. Reprod. Fertil.* **42**, 459–72.

Boris, A. (1962) Further studies on the endocrinology of a dodecahydro-phenanthrene derivative. *Acat. Endocrinol.* **41**, 280–6.

Boris, A. (1965) Endocrine studies of a nonsteroid anti-androgen and progestin. *Endocrinology* **76**, 1062–7.

Boris, A. and Stevenson, R. H. (1966) Further studies on a nonsteroidal anti-androgen. *Endocrinology* **78**, 549–55.

Boris, A., DeMartino, L. and Trmal, T. (1970) Some endocrine studies of a new anti-androgen, 6α-bromo-17β-hydroxy-17α-methyl-4-oxa-5α-androstan-3-one (BOMT). *Endocrinology* **88**, 1086–91.

Bouin, P. and Ancel, P. (1903) Recherches sur les cellules interstitielles du testicule des mammifères. *Arch. Zool. Exptl. Gen.* **1**, 437–523.

Brady, R. O. (1951) Biosynthesis of radioactive testosterone *in vitro*. *J. Biol. Chem.* **193**, 145–8.

Brinck-Johnsen, T. and Eik-Nes, K. B. (1957) Effect of human chorionic gonadotrophin on the secretion of testosterone and 4-androstene-3, 17-dione by the canine testis. *Endocrinology* **61**, 676–83.

Brinckmann, A. O., Mulder, E., Lamers-Stahlhofen, G. J. M., Mechielsen, M. J. and Molen, H. J. van der (1972) An oestradiol receptor in rat testis interstitial tissue. *FEBS Letters* **26**, 301–5.

Brotherton, J. (1974) Effect of oral cyproterone acetate on urinary and serum FSH and LH levels in adult males being treated for hypersexuality. *J. Reprod. Fertil.* **36**, 177–87.

Brown, P. D. C. and Waites, G. M. H. (1972) Regional blood flow in the epididymis of the rat and rabbit; effect of efferent duct ligation and orchidectomy. *J. Reprod. Fertil.* **28**, 221–3.

Bruchovsky, N. and Wilson, J. D. (1968a) The conversion of testosterone to 5α-androstan-17β-ol-3-one by rat prostate *in vivo* and *in vitro*. *J. Biol. Chem.* **243**, 2012–21.

Bruchovsky, N. and Wilson, J. D. (1968b) The intranuclear binding of testosterone and 5α-androstane-17β-ol-3-one by rat prostate. *J. Biol. Chem.* **243**, 5953–60.

Bubenik, G. A., Brown, G. M. and Grota, L. J. (1975) Localization of immunoreactive androgen in testicular tissue. *Endocrinology* **96**, 63–9.

Bullock, L. P. and New, M. I. (1971) Testosterone and cortisol concentration in spermatic adrenal and systemic venous blood in adult male guinea pigs. *Endocrinology* **88**, 523–526.

Burstein, S. and Gut, M. (1971) Biosynthesis of pregnenolone. *Rec. Progr. Horm. Res. 27*, 303–45.

Butenandt, A. (1931) Uber die chemische Untersuchung der Sexualhormone. *Z. Angew. Chem.* **44**, 905–8.

Butenandt, A. and Harrisch, G. (1935) Uber Testosteron-Unwandlung des Dehydroandrostendione in Androstendiol und Testosteron; ein Weg sur. *Z. Physiol. Chem.* **237**, 89–97.

Carlson, I. H., Stratman, F. and Hauser, E. (1971) Spermatic vein testosterone in boars during puberty. *J. Reprod. Fertil.* **27**, 177–80.

Carrick, F. N. and Setchell, B. P. (1977) The evolution of the scrotum. In *Reproduction and Evolution*, eds. J. H. Calaby and C. H. Tyndale-Biscoe, 165–70. Australian Academy of Science, Canberra.

Carstensen, H., Amer, B., Amer, I. and Wide, L. (1973a) The post-operative decrease of plasma testosterone in man after major surgery, in relation to plasma FSH and LH. *J. Ster. Biochem.* **4**, 45–55.

Carstensen, H., Amer, I., Wide, L. and Amer, B. (1973b) Plasma testosterone, LH and FSH during the first 24 hours after surgical operations. *J. Ster. Biochem.* **4**, 605–11.

Castro, J. E. (1974a) Orchidectomy and the immune response. I. Effect of orchidectomy on lymphoid tissues. *Proc. Roy. Soc.* (London) B. **185**, 425–36.

Castro, J. E. (1974b) Orchidectomy and the immune response. II. Response of orchidectomized mice to antigens. *Proc. Roy. Soc.* (London) B. **185**, 437–51.

Castro, J. E. (1974c) The hormonal mechanism of immunopotentiation in mice after orchidectomy. *J. Endocrinol.* **62**, 311–18.

Chan, S. W. C., Leathem, J. H. and Esashi, T. (1977) Testicular metabolism and serum testosterone in ageing male rats. *Endocrinology* **101**, 128–33.

Chemes, H. E., Podesta, E. and Rivarola, M. A. (1976) Action of testosterone, dihydrotestosterone and $5\alpha$-androstane-$3\alpha$, $17\beta$-diol on the spermatogenesis of immature rats. *Biol. Reprod.* **14**, 332–8.

Chowdhury, A. K. and Steinberger, E. (1975) Effect of $5\alpha$ reduced androgens on sex accessory organs, initiation and maintenance of spermatogenesis in the rat. *Biol. Reprod.* **12**, 609–17.

Christensen, A. K. and Mason, N. R. (1965) Comparative ability of seminiferous tubules and interstitial tissue of rat testes to synthesize androgens from progesterone-4-$^{14}$C *in vitro*. *Endocrinology* **76**, 646–56.

Cicero, T. J., Bell, R. D., Wiest, W. G., Allison, J. H., Polakoski, K. and Robins, E. (1975a) Function of the male sex organs in heroin and methadone users. *N. Engl. J. Med.* **292**, 882–7.

Cicero, T. J., Meyer, E. R., Wiest, W. G., Olney, J. W. and Bell, R. D. (1975b) Effects of chronic morphine administration of the reproductive system of the male rat. *J. Pharmacol. Exp. Ther.* **192**, 542–8.

Cicero, T. J., Meyer, E. R., Bell, R. D. and Koch, G. A. (1976a) Effects of morphine and methadone on testosterone, luteinizing hormone and the secondary sex organs of the male rat. *Endocrinology* **98**, 365–70.

Cicero, T. J., Wilcox, C. E., Bell, R. D. and Meyer, E. R. (1976b) Acute reductions in serum testosterone levels by narcotics in the male rat: Stereospecificity, blockade by naloxone and tolerance. *J. Pharmacol. Exp. Ther.* **198**, 340–6.

Cicero, T. J., Bell, R. D., Meyer, E. R. and Schweitzer, J. (1977) Narcotics and the hypothalamic-pituitary-gonadal axis: Acute effects on luteinizing hormone, testosterone and androgen-dependent systems. *J. Pharmacol. Exp. Ther.* **201**, 76–83.

Clegg, E. J. (1960) Some effects of artificial cryptorchidism on the accessory reproductive organs of the rat. *J. Endocrinol.* **20**, 210–19.

Clermont, Y. and Harvey, S. C. (1967) Effects of hormones on spermatogenesis in the rat. *Ciba Colloq. Endocrinol.* **16**, 173–89.

Connell, G. M. and Eik-Nes, K. B. (1966) Testosterone biosynthesis by rabbit testis slices: Glucose-U-$^{14}$C as a carbon source for testicular steroids and testicular proteins. *Proc. Natl. Acad. Sci.* (U.S.) **55**, 410–6.

Connell, G. M. and Eik-Nes, K. B. (1968) Testosterone production by rabbit testis slices. *Steroids* **12**, 507–16.

Cooke, B. A., Jong, F. H. de, Molen, H. J. van der and Rommerts, F. F. G. (1972) Endogenous testosterone concentrations in rat testis interstitial tissue and seminiferous tubules during *in vitro* incubation. *Nature New Biol.* **237**, 255–6.

Corvol, P. and Bardin, C. W. (1973) Species distribution of testosterone-binding globulin. *Biol. Reprod.* **8**, 277–82.

Crafts, R., Llerena, L. A., Guevara, A., Lobotsky, J. and Lloyd, C. W. (1968) Plasma androgens and 17-hydroxycorticosteroids throughout the day in submarine personnel. *Steroids* **12**, 151–63.

Crim, L. W. and Geschwind, I. I. (1972) Testosterone concentration in spermatic vein plasma of the developing ram. *Biol. Reprod.* **7**, 42–6.

Cutuly, E., McCullagh, D. R. and Cutuly, E. (1938) Effects of androgenic sterols in hypophysectomized and in castrated rats. *Am. J. Physiol.* **121**, 786–93.

Danzo, B. J. and Etter, B. C. (1975) Steroid-binding proteins in rabbit plasma: separation of testosterone-binding globulin (TeBG) from corticosteroid-binding globulin (CBG), preliminary characterization of TeBG and changes in TeBG concentration during sexual maturation. *Mol. Cell. Endocrinol.* **2**, 351–68.

David, K., Dingemanse, E., Freud, J. and Laqueur, E. (1935) Ueber krystallinisches mannliches Hormon aus Hoden (Testosteron), wirksamer als aus Harn oder aus Cholesterin bereitetes Androsteron. *Z. Physiol. Chem.* **233**, 281–2.

Davies, R. V., Main, S. J., Young, M. G. W. L. and Setchell, B. P. (1976) Bio-assay of inhibin-like activity in rete testis fluid and its partial purification. *J. Endocrinol.* **68**, 26P.

Davies, R. V., Main, S. J. and Setchell, B. P. (1977) Seasonal changes in plasma follicle-stimulating hormone luteinizing hormone and testosterone in rams. *J. Endocrinol.* **72**, 12P.

Demoulin, A., Thieblot, P. and Franchimont, P. (1973) Influence de différents steroides et de la prostaglandin E$_1$ sur les taux des gonadotrophines sériques chez le rat mâle castré. *C. R. Seances Soc. Biol. Ses Fil.* **167**, 1684–7.

Desjardins, C., Ewing, L. L. and Johnson, B. H. (1971) Effects of light deprivation upon the spermatogenic and steroidogenic elements of hamster testes. *Endocrinology* **89**, 791–800.

Desjardins, C., Ewing, L. L. and Irby, D. C. (1973) Response of the rabbit seminiferous epithelium to testosterone administered in polydimethylsiloxane capsules. *Endocrinology* **93**, 450–60.

Diez D'Aux, R. C. and Pearson Murphy, B. E. (1974) Androgens in the human fetus. *J. Steroid Biochem.* **5**, 207–10.

Dixon, R., Vincent, V. and Kase, N. (1965) Biosynthesis of steroid sulfates by normal human testis. *Steroids* **6**, 757–69.

Dorfman, R. I. (1968) Antiandrogens. In *Testosterone*, ed. J. Tamm, 130, New York: Hafner.

Dorfman, R. I. and Shipley, R. A. (1956) *Androgens*, New York: Wiley.

Dorfman, R. I., Forchielli, E. and Gut, M. (1963) Androgen biosynthesis and related studies. *Recent Prog. Hormone Res.* **19**, 251–67.

Dorfman, R. I., Menon, K. M. J., Sharma, D. C., Joshi, S. and Forchielli, E. (1967) Steroid hormone biosynthesis in rat, rabbit and capuchine testis. *Ciba Found. Colloq. Endocrinol.* **16**, 91–104.

Dorrington, J. H. and Armstrong, D. T. (1975) Follicle-stimulating hormone stimulates estradiol-17$\beta$ synthesis in cultured Sertoli cells. *Proc. Nat. Acad. Sci. U.S.A.* **72**, 2677–81.

Dorrington, J. H., Fritz, I. B. and Armstrong, D. T. (1976) Site at which FSH regulates estradiol-17$\beta$ biosynthesis in Sertoli cell preparations in culture. *Molec. Cell. Endocrinol.* **6**, 117–22.

Dougherty, T. F. (1952) Effects of hormones on lymphatic tissue. *Physiol. Rev.* **32**, 379–401.

Dray, F., Reinberg, A. and Sebaoun, J. (1965) Rythme biologique de la testosterone libre du plasma chez l'homme adulte sain: existence d'une variation circadienne. *C. r. hebd. Seanc. Acad. Sci.* (Paris) **261**, 573–7.

Dufy-Barbe, L. and Franchimont, P. (1972) Influence des différents steroides gonadiques sur le taux de la FSH et de la LH chez le rat castré. *C.R. Seances Soc. Biol. Ses Fil.* **166**, 960–4.

Dvoskin, S. (1944) Local maintenance of spermatogenesis by intracellularly implanted pellets of testosterone in hypophysectomized rats. *Am. J. Anat.* **75**, 289–327.

Eidinger, D. and Garrett, T. J. (1972) Studies of the regulatory effects of the sex hormones on antibody formation and stem cell differentiation. *J. Exp. Med.* **136**, 1098–116.

Eik-Nes, K. B. (1962) Secretion of testosterone in anesthetized dogs. *Endocrinology* **71**, 101–6.

Eik-Nes, K. B. (1964a) On the relationship between testicular blood flow and secretion of testosterone in anesthetized dogs stimulated with human chorionic gonadotrophin. *Can. J. Physiol. Pharmacol.* **42**, 671–7.

Eik-Nes, K. B. (1964b) Effects of gonadotrophins on secretion of steroids by the testis and ovary. *Physiological Rev.* **44**, 609–44.

Eik-Nes, K. B. (1966) Secretion of testosterone by the eutopic and the cryptorchid testes in the same dog. *Cand. J. Physiol. Pharmacol.* **44**, 629–33.

Eik-Nes, K. B. (1967a) Factors controlling the secretion of testicular steroids in the dog. *J. Reprod. Fertil.* Suppl. **2**, 125–41.

Eik-Nes, K. B. (1967b) Factors influencing the secretion of testosterone in the anaesthetized dog. *Ciba. Colloq. Endocrinol.* **16**, 120–36.

Eik-Nes, K. B. (1969) An effect of isoproterenol on rates of synthesis and secretion of testosterone. *Am. J. Physiol.* **217**, 1764–70.

Eik-Nes, K. B. (1970) Synthesis and secretion of androstenedione and testosterone. In *The Androgens of the Testis*, ed. K. B. Eik-Nes, 1–47, New York: Marcel Dekker.

Eik-Nes, K. B. (1971) Production and secretion of testicular steroids. *Recent Progr. Hormone Res.* **27**, 517–35.

Eik-Nes, K. B. (1975) Biosynthesis and secretion of testicular steroids. Handbk. Physiol. Section 7, Endocrinology. V. Male Reproductive System, 95–115.

Eik-Nes, K. B. and Hall, P. F. (1965a) The action of pregnant mare serum on the production of testosterone *in vivo* and *in vitro. J. Reprod. Fertil.* **9**, 233–41.

Eik-Nes, K. B. and Hall, P. F. (1965b) Secretion of steroid hormones *in vivo. Vitamins Hormones* **23**, 153–208.

Eik-Nes, K. B. and Kekre, M. (1963) Metabolism *in vivo* of steroids by the canine testes. *Biochim. Biophys. Acta* **78**, 457–63.

Eik-Nes, K. B. and Kekre, M. (1964) Metabolism *in vivo* of progesterone to testosterone in the dog. *Biochim. Biophys. Acta* **82**, 121–4.

Eik-Nes, K. B., Schellman, J. A., Lumry, R. and Samuels, L. T. (1954) The binding of steroids to protein. I. Solubility determinations. *J. Biol. Chem.* **206**, 411–19.

Eisenberg, E. and Gordan, G. S. (1950) The levator ani muscle of the rat as an index of myotrophic activity of steroidal hormones. *J. Pharmacol. Exp. Therap.* **99**, 38–44.

Ellendorff, F., Parvizi, N., Pomerantz, D. K., Hartjen, A., König, A., Smidt, D. and Elsaesser, F. (1975) Plasma luteinizing hormone and testosterone in the adult male pig: 24 hour fluctuations and the effect of copulation. *J. Endocrinol.* **67**, 403–10.

Ellis, B. W., Evans, P. F., Phillips, P. D., Murray, M. A. F., Jacobs, H. S., James, V. H. T. and Dudley, H. A. F. (1976) Effects of surgery on plasma testosterone, luteinizing hormone and follicle-stimulating hormone: a comparison of pre- and postoperative patterns of secretion. *J. Endocr.* **69**, 25P.

Endröczi, E. and Lissak, K. (1962) Role of reflexogenic factors in testicular hormone secretion. Effect of copulation on the testicular hormone production of the rabbit. *Acta Physiol. Acad. Sci. Hung.* **21**, 203–6.

Eviatar, A., Danon, A. and Sulman, F. G. (1961) The mechanism of the 'push and pull' principle. V. Effect of the antiandrogen RO 2-7239 on the endocrine system. *Arch. Int. Pharmacodyn.* **133**, 75–88.

Ewing, L. L. and Eik-Nes, K. B. (1966) On the formation of testosterone by the perfused rabbit testis. *Can. J. Biochem.* **44**, 1327–44.

Ewing, L., Brown, B., Irby, D. C. and Jardine, I. (1975) Testosterone and $5\alpha$-reduced androgen secretion by rabbit testes-epididymides perfused *in vitro. Endocrinology* **96**, 610–17.

Faiman, C. and Winter, J. S. D. (1971) Diurnal cycles in plasma FSH, testosterone and cortisol in man. *J. Clin. Endocrinol. Metab.* **33**, 186–92.

Falvo, R. E., Buhl, A. E., Reimers, T. J. Foxcroft, G. R., Hunzicker-Dunn, M. and Dziuk, P. J. (1975) Diurnal fluctuations of testosterone and LH in the ram: effect of HCG and gonadotrophin-releasing hormone. *J. Reprod. Fertil.* **42**, 503–10.

Fang, S. and Liao, S. (1969) Antagonistic action of anti-androgens on the formation of a specific dihydrotestosterone-receptor protein complex in rat ventral prostate. *Mol. Pharmacol.* **5**, 428–31.

Fang, S., Anderson, K. M. and Liao, S. (1969) Receptor proteins for androgens. On the role of specific protein in selective retention of $17\beta$-hydroxy-$5\alpha$-androstan-3-one by rat ventral prostate *in vivo* and *in vitro. J. Biol. Chem.* **244**, 6584–95.

Fariss, B. L., Hurley, T. J., Hane, S. and Forsham, P. H. (1969) Reduction of testicular testosterone in rats by ether anesthesia. *Endocrinology* **84**, 940–2.

Fawcett, D. W. (1975) Gametogenesis in the male: prospects for its control. In *The Developmental Biology of Reproduction*, eds. D. L. Markest and J. Papaconstantinou, 25–53, New York: Academic Press.

Fazekas, A. G. and Sandor, T. (1972) Metabolism of androgens by isolated human hair follicles. *J. Steroid Biochem.* **3**, 485–91.

Fichera, S. (1905) Sur l'hypertrophie de la glande pituitaire consécutive à la castration. *Arch. Ital. Biol.* **43**, 403–26.

Fishman, L. M., Sarfaty, G. A., Wilson, H. and Lipsett, M. B. (1967) The role of the testis in oestrogen production. *Ciba Colloq. Endocrinol.* **16**, 156–66.

Folman, Y., Haltmayer, G. C. and Eik-Nes, K. B. (1972) Production and secretion of 5α-dihydrotestosterone by the dog testis. *Am. J. Physiol.* **222**, 653–6.

Folman, Y., Ahmad, N., Sowell, J. G. and Eik-Nes, K. B. (1973) Formation *in vitro* of 5α-dihydrotestosterone and other 5α-reduced metabolites of $^3$H-testosterone by the seminiferous tubules and interstitial tissue from immature and mature rats testes. *Endocrinology* **92**, 41–7.

Fox, C. A., Ismail, A. A. A., Love, D. N., Kirkham, K. E. and Loraine, J. A. (1972) Studies on the relationship between plasma testosterone levels and human sexual activity. *J. Endocrinol.* **52**, 51–8.

Franchimont, P. (1972) Human gonadotrophin secretion. *J. Roy. Coll. Physicians* (London) **6**, 283–98.

Franchimont, P., Chair, S. and Demoulin, A. (1975a) Hypothalamus–pituitary–testis interaction. *J. Reprod. Fertil.* **44**, 335–50.

Franchimont, P., Chari, S., Hagelstein, M. T. and Duraiswami, S. (1975b) Existence of a follicle-stimulating hormone inhibiting factor, 'inhibin', in bull seminal plasma. *Nature* (London) **257**, 402–4.

Frasier, S. D. and Horton, R. (1966) Androgens in the peripheral plasma of prepubertal children and adults. *Steroids* **8**, 777–84.

Gallagher, T. F. and Koch, F. C. (1929) The testicular hormone. *J. Biol. Chem.* **84**, 495–500.

Gandy, H. M. and Peterson, R. E. (1968) Measurement of testosterone and 17-ketosteroids in plasma by the double isotope dilution derivative technique. *J. Clin. Endocr.* **28**, 949–77.

Gay, V. L. and Dever, N. W. (1971) Effects of testosterone propionate and estradiol benzoate alone or in combination on serum LH and FSH in orchidectomized rats. *Endocrinology* **89**, 161–8.

Gaylor, J. L. and Tsai, S. C. (1964) Testicular sterols. II. Conversion of lanosterol to cholesterol and steroid hormones by cell-free preparations of rat testicular tissue. *Biochim. Biophys. Acta.* **84**, 739–48.

Gidari, A. S., Lane, S. E. and Levere, R. D. (1976a) Cyproterone-mediated stimulation of δ-aminolevulinic acid synthetase in chick embryo liver cells. *Endocrinology* **99**, 130–6.

Gidari, A. S., Lane, S. E. and Levere, R. D. (1976b) Induction of δ-aminolevulinic acid synthetase by flutamide, a non-steroidal antiandrogen. *Biochim. Biophys. Acta* **451**, 326–31.

Goldman, A. S., Shapiro, B. H. and Root, A. W. (1976) Effects of new multi-site hormone blockers on the fertility of male rats. *J. Endocrinol.* **69**, 11–21.

Gombe, S., Hall, W. C., McEntee, K., Hansel, W. and Pickett, B. W. (1973) Regulation of blood levels of LH in bulls: influence of age, breed, sexual stimulation and temporal fluctuations. *J. Reprod. Fertil.* **35**, 493–503.

Gomes, W. R., Butler, W. R. and Johnson, A. D. (1971) Effect of elevated ambient temperature on testis and blood levels and *in vitro* biosynthesis of testosterone in the ram. *J. Anim. Sci.* **33**, 804–7.

Gomez, E. C. and Hsia, S. L. (1968) *In vitro* metabolism of testosterone-4-$^{14}$C and $\Delta^4$-androstene-3, 17-dione-4-$^{14}$C in human skin. *Biochemistry* **7**, 24–32.

Goodman, R. L., Hotchkiss, J., Karsch, P. J. and Knobil, E. (1974) Diurnal variations in serum testosterone concentrations in the adult male rhesus monkey. *Biol. Reprod.* **11**, 624–30.

Gordon, R. D., Spinks, J., Dulmanis, A., Hudson, B., Halberg, F. and Bartter, F. C. (1968) Amplitude and phase relations of several circadian rhythms in human plasma and urine. Demonstration of rhythm for tetrahydrocortisol and tetrahydrocorticosterone. *Clin. Sci.* **35**, 307–24.

Gordon, T. M., Rose, R. M. and Bernstein, I. S. (1976) Seasonal rhythm in plasma testosterone levels in the rhesus monkey (*Macaca mulatta*): a three-year study. *Horm. Behav.* **7**, 229–44.

Gower, D. B. (1972) 16-unsaturated $C_{19}$ steroids: A review of their chemistry, biochemistry and possible physiological role. *J. Steroid Biochem.* **3**, 45–103.

Gower, D. B. and Haslewood, G. A. D. (1961) Biosynthesis of androst-16-en-3α-ol from acetate by testicular slices. *J. Endocrinol.* **23**, 253–60.

Grosso, L. I. and Ungar, F. (1964) Conversion of pregnenolone and 4-androstene 3β, 17β-diol to testosterone by mouse testes *in vitro*. *Steroids* **3**, 67–75.

Guillot-Manteghetti, M., Jallageas, M. and Assenmacher, I. (1976) Effects of chronic testosterone treatment on plasma testosterone and weight of accessory reproductive organs in intact or gonadectomized rats. *J. Physiol.* (Paris) **72**, 833–40.

Gulamhusein, A. P. and Tam, W. H. (1974) Reproduction in the male stoat, *Mustela erminea*. *J. Reprod. Fertil.* **41**, 303–12.

Gunn, S. A., Gould, T. C. and Anderson, W. A. D. (1961) Seasonal variation in endocrine response to cryptorchidism. Acta Endocrinol. **37**, 589–96.

Gustafson, A. W. and Shemesh, M. (1976) Changes in plasma testosterone levels during the annual reproductive cycle of the hibernating bat, *Myotis lucifugus lucifugus* with a survey of plasma testosterone levels in adult male vertebrates. *Biol. Reprod.* **15**, 9–24.

Hall, P. F. (1965) Influence of temperature upon the biosynthesis of testosterone by rabbit testis *in vitro*. *Endocrinology* **76**, 396–402.

Hall, P. F. (1966) On the stimulation of testicular steroidogenesis in the rabbit by interstitial cell-stimulating hormone. *Endocrinology* **78**, 690–8.

Hall, P. F. (1970) Endocrinology of the testis. In *The Testis*, eds. A. D. Johnson, W. R. Gomes and N. L. VanDemark, II, 1–71, New York: Academic Press.

Hall, P. F. and Eik-Nes, K. B. (1962) The action of gonadotrophic hormones upon rabbit testes *in vitro*. *Biochim. Biophys. Acta* **63**, 411–23.

Hall, P. F. and Eik-Nes, K. B. (1963a) The influence of gonadotrophins *in vivo* upon the biosynthesis of androgens by homogenate of rat testis. *Biochim. Biophys. Acta* **71**, 438–47.

Hall, P. F. and Eik-Nes, K. B. (1963b) The effect of interstitial cell-stimulating hormone on the production of pregnenolone by rabbit testis in the presence of an inhibitor of 17α-hydroxylase. *Biochim. Biophys. Acta* **86**, 604–9.

Hall, P. F. and Young, D. G. (1968) Site of action of trophic hormones upon the biosynthetic pathways to steroid hormones. *Endocrinology* **82**, 559–68.

Hall, P. F., Nishizawa, E. E. and Eik-Nes, K. B. (1963) Biosynthesis of testosterone by rabbit testis: Homogenate v slices. *Proc. Soc. Exptl. Biol. Med.* **114**, 791–4.

Hall, P. F., Sozer, C. C. and Eik-Nes, K. B. (1964) Formation of dehydroepiandrosterone during *in vivo* and *in vitro* biosynthesis of testosterone by testicular tissue. *Endocrinology* **74**, 35–43.

Hall, P. F., Irby, D. C. and Kretser, D. M. de (1969) Conversion of cholesterol to androgens by rat testes: comparison of interstitial cells and seminiferous tubules. *Endocrinology* **84**, 488–96.

Haltmeyer, G. C. and Eik-Nes, K. B. (1969) Plasma levels of testosterone in male rabbits following copulation. *J. Reprod. Fertil.* **19**, 273–7.

Hammond, G. L., Ruokonen, A., Kontturi, M., Koskela, E. and Vikho, R. (1977) The simultaneous radioimmunoassay of seven steroids in human spermatic and peripheral venous blood. *J. Clin. Endocrinol. Metab.* **45**, 16–24.

Hanes, F. M. and Hooker, C. W. (1937) Hormone production in the undescended testis. *Proc. Soc. Exptl. Biol. Med.* **35**, 549–50.

Hasimoto, I. and Suzuki, Y. (1966) Androgens in testicular venous blood in the rat, with special reference to pubertal changes in the secretory pattern. *Endocrinol. Japon.* **13**, 326–37.

Hansson, V., Djoseland, O., Reusch, E., Attramadal, A. and Torgersen, O. (1973) An androgen-binding protein in the testis cytosol fraction of adult rats. Comparison with the androgen binding protein in the epididymis. *Steroids* **21**, 457–74.

Hansson, V., McLean, W. S., Smith, A. A., Tindall, D. J., Weddington, S. C., Nayfeh, S. N., French, F. S. and Ritzen, E. M. (1974a) Androgen receptors in rat testis. *Steroids* **23**, 823–32.

Hansson, V., Trygstad, O., French, F. S., McLean, W. S., Smith, A. A., Tindall, D. J., Weddington, S. C., Petrusz, P., Nayfeh, S. N. and Ritzen, E. M. (1974b) Androgen transport and receptor mechanisms in testis and epididymis. *Nature* **250**, 387–91.

Hansson, V., Ritzen, M. E., French, F. S., Weddington, S. C. and Nayfeh, S. N. (1975) Testicular androgen-binding protein (ABP): comparison of ABP in rabbit testis and epididymis with a similar androgen-binding protein (TeBG) in rabbit serum. *Mol. Cell. Endocrinol.* **3**, 1–20.

Harris, M. E. and Bartke, A. (1975) Maintenance of rete testis fluid testosterone and dihydrotestosterone levels by pregnenolone and other $C_{21}$ steroids in hypophysectomized rats. *Endocrinology* **96**, 1396–402.

Hart, B. L. (1968) Role of prior experience in the effects of castration on sexual behaviour of male dogs. *J. Comp. Physiol. Phychol.* **66**, 719–25.

Hayes, K. J. (1965) The so-called 'Levator ani' of the rat. *Acta Endocrinol.* **48**, 337–47.

Haynes, N. B., Hafs, H. D., Waters, R. J., Manns, J. G. and Riley, A. (1975) Stimulatory effect of prostaglandin $F_{2\alpha}$ on the plasma concentration of testosterone in bulls. *J. Endocr.* **66**, 329–38.

Heller, C. G. and Nelson, W. O. (1948) The testis–pituitary relationship in man. *Rec. Prog. Horm. Res.* **3**, 229–43.

Heller, C. G., Morse, H. C., Su, M. and Rowley, M. J. (1970) Role of FSH, ICSH and endogenous testosterone during testicular suppression by exogenous testosterone in normal men. In *The Human Testis*, eds. E. Rosemberg and C. A. Paulsen, 249–59, New York: Plenum.

Hervey, G. R. and Hutchinson, I. (1973) The effects of testosterone on body weight and composition in the rat. *J. Endocrinol.* **57**, xxiv–xxv.

Herz, Z., Folman, Y. and Drori, D. (1969) Testosterone content of the testes of mated and unmated rats. *J. Endocrinol.* **44**, 127–8.

Hill, R. T. (1937) Ovaries secrete male hormones. III. Temperature control of male hormone output by grafted ovaries. *Endocrinology* **21**, 633–6.

Hillier, S. G., Groom, G. V., Boyns, A. R. and Cameron, E. H. D. (1975) The active immunisation of intact adult rats against steroid-protein conjugates: Effects on circulating hormone levels and related physiological processes. Steroid Immunoassay, Proc. 5th Tenovus Workshop, 1974, pp. 97–110.

Hoffman, K. and Nieschlag, E. (1977) Circadian rhythm of plasma testosterone in the male Djungarian Hamster (Phodopus sungorus). *Acta Endocrin.* **86**, 193–9.

Hollander, N. and Hollander, V. P. (1958) The microdetermination of testosterone in human spermatic vein blood. *J. Clin. Endocr.* **18**, 966–71.

Horton, R. and Tait, J. F. (1966) Androstenedione production and interconversion rates measured in peripheral blood and studies on the possible site of its conversion to testosterone. *J. Clin. Invest.* **45**, 301–13.

Horton, R. and Tait, J. F. (1967) *In vivo* conversion of dehydroisoandrosterone to plasma androstenedione and testosterone in man. *J. Clin. Endocrinol.* **27**, 79–88.

Howland, B. E. (1975) The influence of feed restriction and subsequent re-feeding on gonadotrophin secretion and serum testosterone levels in male rats. *J. Reprod. Fertil.* **44**, 429–36.

Hudson, B., Coghlan, J. P. and Dulmanis, A. (1967) Testicular function in man. *Ciba Found. Colloqu. Endocrinol.* **16**, 140–53.

Huggins, C. B. and Moulder, P. V. (1945) Estrogen production by Sertoli-cell tumors of the testis. *Cancer Res.* **5**, 510–14.

Huhtaniemi, I., Ikonen, M. and Vihko, R. (1970) Presence of testosterone and other neutral steroids in human fetal testes. *Biochem. Biophys. Res. Comm.* **38**, 715–20.

Ibayashi, H., Nakamura, M., Uchikawa, T., Murakawa, S., Yoshida, S., Nakao, K. and Okinaka, S. (1965) $C_{19}$ steroids in canine spermatic venous blood following gonadotrophin administration. *Endocrinology* **76**, 347–52.

Illius, A. W., Haynes, N. B. and Lamming, G. E. (1976a) Effects of ewe proximity on peripheral plasma testosterone levels and behaviour in the ram. *J. Reprod. Fertil.* **48**, 25–32.

Illius, A. W., Haynes, N. B., Purvis, K. and Lamming, G. E. (1976b) Plasma concentrations of testosterone in the developing ram in different social environments. *J. Reprod. Fertil.* **48**, 17–24.

Inano, H. and Tamaoki, B. (1968) Effect of experimental bilateral cryptorchidism on testicular enzymes related to androgen formation. *Endocrinology* **83**, 1074–82.

Iturriza, F., Carlini, M. R., Piva, F. and Martini, L. (1977) Neuroendocrine effects of a non-steroidal compound of testicular origin. *Experimentia.* **33**, 396–8.

Jaffe, R. B. and Payne, A. H. (1971) Gonadal steroid sulfates and sulfatase. IV. Comparative studies on steroid sulfokinase in the human fetal testis and adrenal. *J. Clin. Endocrinol. Metab.* **33**, 592–6.

Jainudeen, M. R., Katongole, C. B. and Short, R. V. (1972) Plasma testosterone levels in relation to musth and sexual activity in the male Asiatic elephant. *Elephas maximus. J. Reproduct. Fertil.* **29**, 99–104.

Janszen, F. H. A., Cooke, B. A., Van Driel, M. J. A. and Molen, H. J. van der (1976) Purification and characterization of Leydig cells from rat testes. *J. Endocrinol.* **70**, 345–59.

Joffre, M. and Joffre, J. (1975) Variations de la testosteronemie au cours de la période prepubère du Renardeau et au cours du cycle génital saisonier du Renard mâle adulte (*Vulpes vulpes*) en captivité. *C.R. Acad. Sci. Ser.* D. **281**, 819–21.

Johnsen, S. G. (1964) Studies on the testicular-hypophyseal feed-back mechanism in man. *Acta Endocrinol.* (Copenhagen) **90**, 99–124.

Johnsen, S. G. (1972) Studies on the pituitary–testicular axis in male hypogonadism, particularly in infertile men with 'cryptogenetic' hypospermatogenesis. In *Gonadotropins*, eds. B. B. Saxena, C. G. Beling and H. M. Gandy, 593–608, New York: Wiley.

Johnson, B. H. and Ewing, L. L. (1971) FSH and the regulation of testosterone secretion in rabbit testes. *Science* (New York) **173**, 635–7.

Jong, F. H. de and Sharpe, R. M. (1976) Evidence for inhibin-like activity in bovine follicular fluid. *Nature* **263**, 71–2.

Jong, F. H. de, Hey, A. H. and Molen, H. J. van der (1973) Effect of gonadotrophins on the secretion of oestradiol-17$\beta$ and testosterone by the rat testis. *J. Endocrinol.* **57**, 277–84.

Jong, F. H. de, Hey, A. H. and Molen, H. J. van der (1974) Oestradiol-17$\beta$ and testosterone in rat testis tissue: effect of gonadotrophins, localization and production *in vitro. J. Endocrinol.* **60**, 409–19.

Judd, H. L., Parker, D. C., Rakaff, J. S., Hopper, B. R. and Yen, S. S. C. (1973) Elucidation of mechanism(s) of the nocturnal rise of testosterone in men. *J. Clin. Endocrinol. Metab.* **38**, 134–41.

Jungmann, R. A. (1968) Androgen biosynthesis. I. Enzymatic cleavage of the cholesterol side-chain to dehydroepiandrosterone and 2 methylheptan-6-one. *Biochim. Biophys. Acta* **164**, 110–23.

Junquiera, L. C. U., Fajer, A., Rabinowitch, M. and Frankenthal, L. (1949) Biochemical and histochemical observations on the sexual dimorphism of mice submaxillary glands. *J. cell. comp. Physiol.* **34**, 129–58.

Kamel, F., Wright, W. W., Mock, E. J. and Frankel, A. I. (1977) The influence of mating and related stimuli on plasma levels of luteinizing hormone, follicle stimulating hormone, prolactin and testosterone in the male rat. *Endocrinology* **101**, 421–9.

Katchen, B. and Buxbaum, S. (1975) Disposition of a new, nonsteroid, antiandrogen, $\alpha,\alpha,\alpha$-trifluoro-2-methyl-4′-nitro-*m*-propionotoluidide (flutamide), in men following a single oral 200mg dose. *J. Clin. Endocrinol. Metab.* **41**, 373–9.

Katchen, B., Dancik, S. and Millington, G. (1976) Percutaneous penetration and metabolism of topical [$^{14}$C] flutamide in men. *J. Invest. Dermatology* **66**, 379–82.

Katongole, C. B., Naftolin, F. and Short, R. V. (1971) Relationship between blood levels of luteinizing hormone and testosterone in bulls, and the effects of sexual stimulation. *J. Endocrinol.* **50**, 457–66.

Katongole, C. B., Naftolin, F. and Short, R. V. (1974) Seasonal variations in blood luteinizing hormone and testosterone levels in rams. *J. Endocrinol.* **60**, 101–6.

Kelch, R. P., Jenner, M. R., Weinstein, R., Kaplan, S. I. and Grumbach, M. M. (1972) Estradiol and testosterone secretion by human, simian and canine testis in males with hypogonadism and in the male pseudohermaphrodite with feminizing testis syndrome. *J. clin. Invest.* **51**, 824–30.

Kellie, A. E. and Smith, E. R. (1967) Renal clearance of 17-oxo steroid conjugates found in human peripheral plasma. *Biochem. J.* **66**, 490–5.

Kent, J. R. and Acone, A. B. (1966) Plasma testosterone levels and ageing in males. In *Androgens in Normal and Pathological Conditions*, eds. A. Vermeulen and D. Exely, 31–6, Excerpta Med. Found. Intern. Congr. Ser. 101, Amsterdam.

Keogh, E. J., Lee, V. W. K., Rennie, G. C., Burger, H. G., Hudson, B. and de Kretser, D. M. (1976) Selective suppression of FSH by testicular extracts. *Endocrinology* **98**, 997–1004.

Kim, C. K., Yen, S. S. and Benirschke, K. (1972) Serum testosterone in fetal cattle. *Gen. Comp. Endocrinol.* **18**, 404–6.

Kinsey, A. C., Pomeroy, W. B. and Martin, C. E. (1948) Sexual behaviour in the human male. Philadelphia: Saunders.

Kirkpatrick, J. F., Vail, R., Devous, S., Schwend, S., Baker, C. B. and Wiesner, L. (1976) Diurnal variation of plasma testosterone in wild stallions. *Biol. Reprod.* **15**, 98–101.

Kiser, T. E., Hafs, H. D. and Oxender, W. D. (1976) Increased blood LH and testosterone after administration of prostaglandin $F_{2\alpha}$ in bulls. *Prostaglandins* **11**, 545–53.

Knorr, D. W., Vanha-Perttula, T. and Lipsett, M. B. (1970) Structure and function of rat testis through pubescence. *Endocrinology* **86**, 1298–304.

Kochakian, C. D. (1946) The protein anabolic effects of steroid hormones. *Vitam. Horm.* **4**, 225–310.

Kochakian, C. D. (1947) The role of hydrolytic enzymes in some of the metabolic activities of steroid hormones. *Rec. Progr. Horm. Res.* **1**, 177–214.

Kochakian, C. D. (1949) Renotrophic, androgenic and somatotrophic properties of further steroids. *Am. J. Physiol.* **158**, 51–6.

Kochakian, C. D. (1950) Comparison of protein anabolic property of various androgens in the castrated rat. *Am. J. Physiol.* **160**, 53–61.

Kochakian, C. D. (ed.) (1976) *Anabolic-Androgenic Steroids*, Berlin: Springer-Verlag.

Kochakian, C. D. and Murlin, J. R. (1935) The effect of male hormones on the protein and energy metabolism of castrate dogs. *J. Nutr.* **10**, 437–59.

Kochakian, C. D., Hill, J. and Aonuma, S. (1963) Regulation of protein biosynthesis in mouse kidney by androgens. *Endocrinology* **72**, 354–63.

Kochakian, C. D., Tomana, M. and Strickland, B. (1974) Role of cytosol and polysomes in the stimulation by androgen of protein biosynthesis in the mouse kidney. *Mol. cell. Endocrinol.* **1**, 129–38.

Kolodny, R. C., Masters, W. H., Kolodner, R. M. and Toro, G. (1974) Depression of plasma testosterone levels after chronic intensive marihuana use. *New Engl. J. Med.* **290**, 872–4.

Kormano, M., Harkonen, M. and Kontinen, E. (1964) Effect of experimental cryptorchidism on the histochemically demonstrable dehydrogenases in the rat testis. *Endocrinology* **74**, 44–51.

Korenchevsky, V. M., Dennison, M. and Schalit, R. (1932) The response of castrated male rats to the injection of testicular hormone. *Biochem. J.* **26**, 1306–14.

Korenchevsky, V., Hall, K., Burbank, R. C. and Cohen, J. (1941) Hepatotrophic and cardiotrophic properties of sex hormones. *Brit. Med. J.* **I**, 396–9.

Kretser, D. M. de, Catt, K. J., Dufau, M. L. and Hudson, B. (1971) Studies on rat testicular cells in tissue culture. *J. Reprod. Fertil.* **24**, 311–18.

Laatikainen, T., Laitinen, E. A. and Vihko, R. (1969) Secretion of neutral steroid sulfates by the human testis. *J. Clin. Endocrinol. Metab.* **29**, 219–24.

Laatikainen, T., Laitinen, E. A. and Vihko, R. (1971) Secretion of free and sulfate-conjugated neutral steroids by the human testis. Effect of administration of human chorionic gonadotrophin. *J. Clin. Endocrinol. Metabl.* **32**, 59–64.

Lacy, D. (1973) Androgen dependency of spermatogenesis and the physiological significance of steroid metabolism *in vitro* by the seminiferous tubules. In *Endocrine Function of the Human Testis*, eds. V. H. T. James, M. Serio and L. Martini, **1**, 493–532, New York: Academic Press.

Leach, R. B., Maddock, W. O., Tokuyama, I., Paulsen, C. A. and Nelson, W. O. (1956) Clinical studies of testicular hormone production. *Recent Progr. Hormone Res.* **12**, 377–98.

Leathem, J. H. (1944) Influence of testosterone propionate on the adrenals and testes of hypophysectomized rats. *Anat. Rec.* **89**, 155–61.

Leathem, J. H. and Brent, B. J. (1943) Influence of preneninolone and pregnenolone on spermatogenesis in hypophysectomized adult rats. *Proc. Soc. Exp. Bio. Med.* **52**, 341–3.

Lebeau, M-C. and Baulieu, E. E. (1966) Contribution du cholesterol et du sulfate de cholesterol circulants a la biosynthese des hormones testiculaires et corticosurrenaliennes. *Compt. Rend.* **263**, 158–61.

Lee, P. A., Jaffe, R. B., Midgley, A. R., Kohen, F. and Niswender, B. D. (1972) Regulation of human gonadotropins. VIII. Suppression of serum LH and FSH in adult males following exogenous testosterone administration. *J. Clin. Endocrinol. Metab.* **35**, 636–41.

Lee, V. W. K., Keogh, E. J., de Kretser, D. M. and Hudson, B. (1974) Selective suppression of FSH by testis extracts. *IRCS Med. Sci.* **2**, 1406.

Lee, V. W. K., Kretser, D. M. de, Hudson, B. and Wang, C. (1975) Variations in serum FSH, LH and testosterone levels in male rats from birth to sexual maturity. *J. Reprod. Fertil.* **42**, 121–8.

Levier, R. R. and Spaziani, E. (1969) The influence of temperature on steroidogenesis in the rat testis. *J. Exptl. Zool.* **169**, 113–20.

Levi-Montalcini, R. and Angeletti, P. U. (1968) Nerve growth factor. *Physiol. Rev.* **48**, 534–69.

Liao, S. and Fang, S. (1970) Receptor-proteins for androgens and the mode of action of androgens on gene transcription in ventral prostate. *Vitam. Horm.* **27**, 17–90.

Liao, S., Howell, D. K. and Chang, T-M. (1974) Action of a nonsteroidal antiandrogen, flutamide, on the receptor-binding and nuclear retention of 5α-dihydrotestosterone in rat ventral prostate. *Endocrinology* **94**, 1205–9.

Lincoln, G. A. (1971a) The seasonal reproductive changes in the red deer stag (*Cervus elaphus*). *J. Zool.* (London) **163**, 105–23.

Lincoln, G. A. (1971b) Puberty in a seasonally breeding male, the red deer stag (*Cervus elephas* L.) *J. Reprod. Fertil.* **25**, 41–54.

Lincoln, G. A. (1974) Reproduction and 'March madness' in the brown hare (*Lepus europaeus*). *J. Zool.* (London) **174**, 1–14.

Lincoln, G. A. (1976) Seasonal variation in the episodic secretion of luteinizing hormone and testosterone in the ram. *J. Endocrinol.* **69**, 213–26.

Lincoln, G. A., Youngson, R. W. and Short, R. V. (1970) The social and sexual behaviour of the red deer stag. *J. Reprod. Fertil.* Suppl. **11**, 71–103.

Lindner, H. R. (1959) Androgens in the bovine testis and spermatic vein blood. *Nature* **183**, 1605–6.

Lindner, H. R. (1961a) Androgens and related compounds in the spermatic vein blood of domestic animals. 1. Neutral steroids secreted by the bull testis. *J. Endocrinol.* **23**, 139–59.

Lindner, H. R. (1961b) Androgens and related compounds in the spermatic vein blood of domestic animals. II. Species-linked differences in the metabolism of androstenedione in blood. *J. Endocrinol.* **23**, 161–6.

Lindner, H. R. (1961c) Androgens and related compounds in the spermatic vein blood of domestic animals. IV. Testicular androgens in the ram, boar and stallion. *J. Endocrinol.* **23**, 171–8.

Lindner, H. R. (1963) Partition of androgen between the lymph and venous blood of the testis in the ram. *J. Endocrinol.* **25**, 483–94.

Lindner, H. R. (1967) Participation of lymph in the transport of gonadal hormones. *Excerpta Med. Found. Intern. Congr.* Ser. **132**, 821–7.

Lindner, H. R. (1969) The androgenic secretion of the testis in domestic ungulates. In *The Gonads*, ed. K. W. McKerns, 615–48, New York: Appleton-Century-Crofts.

Lindner, H. R. and Mann, T. (1960) Relationship between the content of androgenic steroids in the testes and the secretory activity of the seminal vesicles in the bull. *J. Endrocrinol.* **21**, 341–60.

Lipsett, M. B. (1970) Steroid secretion by the human testis. *Adv. Exptl. Med. Biol.* **10**, 407–18.

Lipsett, M. B. and Tullner, W. W. (1965) Testosterone synthesis by the fetal rabbit gonad. *Endocrinology* **77**, 273–7.

Lisano, M. E., Beverly, J. R., Sorensen, A. M. Jr. and Fleeger, J. L. (1972) *In vivo* incorporation of $^{14}$C from acetate-1-$^{14}$C into testicular steroids in the conscious, standing ram. *Steroids* **19**, 159–76.

Lisk, R. D. and Suydam, A. J. (1967) Sexual behaviour patterns in the prepubertally castrated rat. *Anat. Rec.* **157**, 181–90.

Llaurado, J. G. and Dominguez, O. V. (1963) Effect of cryptorchidism on testicular enzymes involved in androgen biosynthesis. *Endocrinology* **72**, 292–5.

Longcope, C., Widrich, W. and Sawin, C. T. (1972) The secretion of estrone and estradiol-17$\beta$ by human testis. *Steroids* **20**, 439–48.

Lubicz-Nawrocki, C. M. and Glover, T. D. (1973) The influence of 17$\alpha$-methyl-B-nortestosterone (SK and F 7690) on the fertilizing ability of spermatozoa in hamsters. *J. Reprod. Fertil.* **34**, 331–9.

Lucas, W. M., Whitmore, W. F. and West, C. D. (1957) Identification of testosterone in human spermatic vein blood. *J. Clin. Endocrinol.* **17**, 465–72.

Lucis, O. J., Raheja, M. C., Millard, O. and Morse, W. T. (1967) Utilization of circulating precursors for the synthesis of testosterone by human testis *in vivo. Can. J. Biochem.* **45**, 1213–8.

Ludwig, D. J. (1950) The effect of androgen on spermatogenesis. *Endocrinology* **46**, 453–81.

Lugaro, G., Casellato, M. M., Mazzola, G., Fachini, G. and Carrea, G. (1974) Evidence for the existence in spermatozoa of a factor inhibiting the follicle stimulating hormone releasing hormone synthesis. *Neuroendocrinology*, **15**, 62–8.

Lynn, W. S., Jr. and Brown, R. H. (1958) The conversion of progesterone to androgens by testes. *J. Biol. Chem.* **232**, 1015–30.

McArthur, E., Short, R. V. and O'Donnell, V. J. (1967) Formation of steroids by equine fetal testis. *J. Endocrinol.* **38**, 331–6.

McCullagh, D. R. (1932) Dual endocrine activity of testes. *Science* **76**, 19–20.

McCullagh, E. P. and Schaffenburg, C. A. (1952) Role of the seminiferous tubules in the production of hormones. *Ann. N.Y. Acad. Sci.* **55**, 674–84.

McLean, W. S., Smith, A. A., Hansson, V., Naess, O., Nayfeh, S. N. and French, F. S. (1976) Further characterization of the androgen receptor in rat testis. *Mol. Cell. Endocrinol.* **4**, 239–56.

McMillin, J. M., Seal, U. S., Keenlyne, K. D., Erickson, A. W. and Jones, J. E. (1974) Annual testosterone rhythm in the adult white-tailed deer (*Odocoileus virginianus borealis*). *Endocrinology* **94**, 1034–40.

Mahesh, V. B., Zarate, A., Roper, B. K. and Greenblatt, R. B. (1966) Studies on the action of 17α-methyl-B-nortestosterone as an antiandrogen. *Steroids* **8**, 297–308.

Main, S. J., Davies, R. V., Young, M. G. W. L. and Setchell, B. P. (1976) Serum and pituitary gonadotrophins after destruction of germinal cells in the testis by X-irradiation or heat. *J. Endocrinol.* **69**, 23P.

Main, S. J., Davies, R. V. and Setchell, B. P. (1977) The effect of injections of an antiserum to testosterone on serum gonadotrophins and testosterone in rats. *J. Endocrinol.* **72**, 22P.

Mainwaring, W. I. P. (1977) The mechanism of action of androgens. *Monogr. Endocrinol.* **10**, Berlin: Springer Verlag.

Mainwaring, W. I. P., Mangan, F. R., Feherty, P. A. and Freifeld, M. (1974) An investigation into the anti-androgenic properties of the non-steroidal compound Sch 13 521 (4'-nitroso-3'-trifluoromethyl iso butyrylanilide). *Mol. Cell. Endocrinol.* **1**, 113–28.

Mangan, F. R. and Mainwaring, W. I. P. (1972) An explanation of the antiandrogenic properties of 6α-bromo-17β-hydroxy-17α-methyl-4-oxa-5α-androstane-3-one. *Steroids* **20**, 331–43.

Mann, T. (1964) *Biochemistry of the Semen and of the Male Reproductive Tract*, London: Methuen.

Mann, T. (1967) Appraisal of endocrine testicular activity by chemical analysis of semen and male accessory secretion. *Ciba. Colloq. Endocrinol.* **16**, 233–44.

Mann, T. (1969) Physiology of semen and of the male reproductive tract. In *Reproduction in Domestic Animals*, eds. H. H. Cole and P. T. Cupps, 2nd edition, 277–312, New York: Academic Press.

Mann, T. (1975) Biochemistry of semen. *Handb. Physiol.* Section 7, **5**, 461–71.

Mann, T., Rowson, L. E. A., Short, R. V. and Skinner, J. D. (1967) The relationship between nutrition and androgenic activity in pubescent twin calves, and the effect of orchitis. *J. Endocrinol.* **38**, 455–68.

Markewitz, M., Veenema, R. J., Fingerhut, B., Neehme-Haily, D. and Sommers, S. C. (1969) Cyproterone acetate, effect on histology and nucleic acid synthesis in the testes of patients with prostatic carcinoma. *Invest. Urol.* **6**, 638–49.

Martini, L., Massa, R., Motta, M. and Zanisi, M. (1974) Recent views on the mechanisms controlling LH and FSH secretion. In *Male Fertility and Sterility*, eds. R. E. Mancini and L. Martini, 359–88, New York: Academic Press.

Masson, G. (1945) Spermatogenic activity of various steroids. *Am. J. Med. Sci.* **209**, 324–7.

Mason, N. R. and Samuels, L. T. (1961) Incorporation of acetate-1-$C^{14}$ into testosterone and $3\beta$-hydroxysterols by the canine testis. *Endocrinology* **68**, 899–906.

Masters, W. H. and Johnson, V. E. (1966) *Human Sexual Response*, Boston: Little, Brown.

Matsumoto, K. and Yamada, M. (1973) $5\alpha$-reduction of testosterone *in vitro* by rat seminiferous tubules and whole testes at different stages of development. *Endocrinology* **93**, 253–5.

Mauvais-Jarvis, P., Floch, H. H. and Bercovici, D-P. (1968) Studies on testosterone metabolism in human subjects with normal and pathological sexual differentiation. *J. Clin. Endocrinol. Metab.* **28**, 460–71.

Mazzi, C., Riva, L. P. and Bernasconi, D. (1974) Gonadotrophins and plasma testosterone in senescence. In *The Endocrine Function of the Human Testis*, eds. V. H. T. James, M. Serio and L. Martini, **II**, 51–66, New York: Academic Press.

Mendelson, J. H., Kuehnle, J. C., Ellingboe, J. and Babor, T. F. (1974) Plasma testosterone levels before, during and after chronic marihuana smoking. *N. Engl. J. Med.* **291**, 1105 1–55.

Mendelson, J. H., Mendelson, J. E. and Patch, V. D. (1975a) Plasma testosterone levels in heroin addiction and during methadone maintenance. *J. Pharmacol. Exp. Ther.* **192**, 211–17.

Mendelson, J. H., Meyer, R. E., Ellingboe, J., Mirin, S. M. and McDougle, M. (1975b) Effects of heroin and methadone on plasma cortisol and testosterone. *J. Pharmacol. Exp. Ther.* **195**, 296–302.

Mendelson, J. H., Intrurrisi, C. E., Renault, P. and Sinay, E. C. (1976) Effects of acetylmethadol on plasma testosterone. *Clin. Pharmacol. Ther.* **19**, 371–4.

Menon, K. M. J., Dorfman, R. I. and Forchielli, E. (1965) The obligatory nature of cholesterol in the biosynthesis of testosterone in rabbit testis. *Steroids* Suppl. **2**, 165–75.

Meussy-Dessolle, M. (1974) Evolution du taux de testosterone plasmatique au cours de la vie foetale chez le Proc domestique (*Sus scrofa* L.). *C.R. Acad. Sc. Paris.* **278**, 1257–60.

Michael, R. P. and Bonsall, R. W. (1977) A 3-year study of an annual rhythm in plasma androgen levels in male rhesus monkeys (*Macaca mulatta*) in a constant laboratory environment. *J. Reprod. Fertil.* **49**, 129–31.

Millar, R. and Fairall, N. (1976) Hypothalamic, pituitary and gonadal hormone production in relation to nutrition in the male hyrax (*Procavia capensis*). *J. Reprod. Fertil.* **47**, 339–41.

Millar, R. P. and Glover, T. D. (1973) Regulation of seasonal sexual activity in an ascrotal mammal, the rock hyrax (*Procavia capensis*). *J. Reprod. Fertil.* Suppl. **19**, 203–20.

Mirin, S. M., Mendelson, J. H., Ellingboe, J. and Meyer, R. E. (1976) Acute effects of heroin and naltrexone on testosterone and gonadotropin secretion: A pilot study. *Psychoneuroendocrinology* **1**, 359–69.

Mizuno, M., Lobotsky, J., Lloyd, C. W., Kobayashi, T. and Murasawa, Y. (1968) Plasma androstenedione and testosterone during pregnancy and in the newborn. *J. Clin. Endocrinol.* **28**, 1133–42.

Mock, E. J., Kamel, F., Wright, W. W. and Frankel, A. I. (1975) Seasonal rhythm in plasma testosterone and luteinizing hormone of the male laboratory rat. *Nature* **256**, 61–3.

Moger, W. H. (1976a) Effect of testosterone implants on serum gonadotropin concentrations in the male rat. *Biol. Reprod.* **14**, 665–9.

Moger, W. H. (1976b) Serum testosterone response to acute LH treatment in estradiol treated rats. *Biol. Reprod.* **14**, 115–17.

Moger, W. H. (1977) Serum 5α-androstane-3α,17β-diol, androsterone and testosterone concentrations in the male rat. Influence of age and gonadotropin stimulation. *Endocrinology* **100**, 1027–32.

Moger, W. H. and Armstrong, D. T. (1974) Changes in serum testosterone levels following acute LH treatment in immature and mature rats. *Biol. Reprod.* **11**, 1–6.

Molen, H. J. van der, Vusse, G. J. van der, Cooke, B. A. and Jong, F. H. de (1975) Cellular and subcellular compartmentalization of steroid metabolism in the rat testis. *J. Reprod. Fertil.* **44**, 351–62.

Moodbidri, S. B., Joshi, L. R. and Sheth, A. R. (1976) Isolation of an inhibin-like substance from ram testis. *IRCS Med. Sci.* **4**, 217.

Moon, Y. S., Hardy, M. H. and Raeside, J. I. (1973) Biological evidence for androgen secretion by the fetal pig testes in organ culture. *Biol. Reprod.* **9**, 330–7.

Moor, B. C. and Younglai, E. V. (1975) Variations in peripheral levels of LH and testosterone in adult male rabbits. *J. Reprod. Fertil.* **42**, 259–66.

Moore, C. R. (1922) On the physiological properties of the gonads as controllers of somatic and psychical characteristics. V. The effects of gonadectomy in the guinea pig on growth, bone lengths and weight of organs of internal secretion. *Biol. Bull.* **43**, 285–312.

Moore, C. R. and Price, D. (1937) Some effects of synthetically prepared male hormone (androsterone) in the rat. *Endocrinology* **21**, 313–29.

Moore, C. R. and Price, D. (1938) Some effects of testosterone and testosterone propionate in the rat. *Anat. Rec.* **71**, 59–78.

Morehead, J. R. and Morgan, C. F. (1967) Hormone production by experimental cryptorchid testes as indicated by radiographic studies of the seminal vesicles and coagulating glands. *Fertil. Steril.* **18**, 232–7.

Morris, M. D. and Chaikoff, I. L. (1959) The origin of cholesterol in liver, small intestines, adrenal gland, and testis of the rat: dietary versus endogenous contributions. *J. Biol. Chem.* **234**, 1095–7.

Morse, H. C., Leach, D. R., Rowley, M. J. and Heller, C. G. (1973) Effect of cyproterone acetate on sperm concentration, seminal fluid volume, testicular cytology and levels of plasma and urinary ICSH, FSH and testosterone in normal men. *J. Reprod. Fertil.* **32**, 365–78.

Mottram, J. C. and Cramer, W. (1932) On the general effects of exposure to radium on metabolism and tumour growth in the rat and the special effects on testis and pituitary. *Q. J. Exptl. Physiol.* **13**, 209–29.

Moyle, W. R., Jungas, R. L. and Greep, R. O. (1973) Metabolism of free and esterified cholesterol by Leydig cell tumour mitochondria. *Biochem. J.* **134**, 415–24.

Mukku, V., Prahalda, S. and Moudgal, N. R. (1976) Effect of constant light on nychthermal variations in serum testosterone in male *Macaca radiata*. *Nature* (London) **260**, 778–80.

Mulder, E., Peters, M. J., Beurden, W. M. O. van and Molen, H. J. van der (1974a) A receptor for testosterone in mature rat testes. *FEBS Letters* **47**, 209–11.

Mulder, E., Beurden-Lamers, W. M. O. van, Boer, W. de, Brinkmann, A. O. and Molen, H. J. van der (1974b) Testicular estradiol receptors in the rat. *Curr. Topic. Molec. Endocrinol.* **1**, 343–55.

Mulder, E., Peters, M. J., Vries, J. de and Molen, H. J. van der (1975a) Characterization of a nuclear receptor for testosterone in the seminiferous tubules of mature rat. *Mol. Cell. Endocrinol.* **2**, 171–82.

Mulder, E., Peters, M. J. and Molen, H. J. van der (1975b) Androgen receptors in testis tissue enriched in Sertoli cells. *Curr. Top. Mol. Endocrinol.* **2**, 287–91.

Murakami, M. (1970) Regressive changes in fine structure of the testicular interstitial tissue of the senile and experimentally treated rats. *Advances Andrology* **1**, 103–5.

Nandini, S. G., Lipner, H. and Moudgal, N. R. (1976) A model system for studying inhibin. *Endocrinology* **98**, 1460–5.

Nayfeh, S. N. and Baggett, B. (1966) Metabolism of progesterone by rat testicular homogenates. I. Isolation and identification of metabolites. *Endocrinology* **78**, 460–70.

Nayfeh, S. N., Barefoot, S. W., Jr. and Baggett, B. (1966) Metabolism of progesterone by rat testicular homogenates. II. Changes with age. *Endocrinology* **78**, 1041–8.

Neal, J., Murphy, B. D., Moger, W. H. and Oliphant, L. W. (1977) Reproduction in the male ferret: gonadal activity during the annual cycle; recrudescence and maturation. *Biol. Reprod.* **17**, 380–5.

Neaves, W. B. (1973) Changes in testicular Leydig cells and in plasma testosterone levels among seasonally breeding rock hyrax. *Biol. Reprod.* **8**, 451–66.

Neher, R. and Wettstein, A. (1960a) Steroide und andere Inhaltsstoffe aus Stierhoden. *Helv. Chim. Acta.* **43**, 1628–39.

Neher, R. and Wettstein, A. (1960b) Occurrence of $\Delta^5$-$3\beta$-hydroxysteroids in adrenal and testicular tissue. *Acta Endocrinol.* **35**, 1–7.

Nelson, W. O. and Gallagher, T. F. (1936) Some effects of androgenic substances in the rat. *Science* **84**, 230–2.

Nelson, W. O. and Merckel, C. (1937) Maintenance of spermatogenesis in testis of the hypophysectomized rat with sterol derivatives. *Proc. Soc. Exptl. Biol. Med.* **36**, 825–8.

Neri, R., Florance, K., Koziol, P. and Cleave, S. van (1972) A biological profile of a non-steroidal antiandrogen, SCH 13521 (4′-nitro-3′-trifluoromethylisobutyranilide). *Endocrinology* **91**, 427–37.

Neumann, F. and Berswordt-Wallrabe, R. von (1966) Effects of the androgen antagonist cyproterone acetate on the testicular structure, spermatogenesis and accessory sexual glands of testosterone-treated adult hypophysectomized rats. *J. Endocrinol.* **35**, 363–71.

Neumann, F. and Elger, W. (1966) Permanent changes in gonadal function and sexual behaviour as a result of early feminization of male rats by treatment with an antiandrogenic steroid. *Endokrinologie* **50**, 209–25.

Neumann, F. and Steinbeck, H. (1974) Antiandrogens. *Handb. Exptl. Pharmacol.* **35,** 235–484.

Neumann, F., Elger, W., Berswordt-Wallrabe, R. von and Kramer, M. (1966) Beeinflussung der Regelmechanismen des Hypophysenzwischenhirnsystems von Ratten durch einen Testosteron-Antagonisten, Cyproteron (1, 2a-Methylen-6-chlor-4,6-pregnadien-17a-ol-3,20-dion). *Naunyn-Schmiedebergs Arch. Pharmakol. exp. Path.* **255,** 221–35.

Neumann, F., Berswordt-Wallrabe, R. von, Elger, W., Steinbeck, H., Hahn, J. D. and Kramer, M. (1970) Aspects of androgen-dependent events as studied by antiandrogens. *Rec. Prog. Hor. Res.* **26,** 337–405.

Nieschlag, E. and Ismail, A. A. A. (1970) Bestimmung der Tagesschwankungen des Plasma-Testosterons normaler Männer durch kompetitive Proteinbindung. *Klin. Wschr.* **48,** 53–4.

Nieschlag, E., Usadel, K. H. and Kley, H. K. (1975) Active immunization with steroids as an approach to investigating testicular and adrenal feedback control. *J. Steroid Biochem.* **6,** 537–40.

Nightingale, M. S., Tsai, S-C. and Gaylor, J. L. (1967) Testicular sterols. VI. Incorporation of mevalonate into squalene and sterols by cell-free preparations of testicular tissue. *J. Biol. Chem.* **242,** 341–9.

Notation, A. D. and Ungar, F. (1969a) Rat testis steroid sulfatase. II. Kinetic study. *Steroids* **14,** 151–9.

Notation, A. D. and Ungar, F. (1969b) Regulation of rat testis steroid sulfatase. A kinetic study. *Biochemistry* **8,** 501–6.

Noumura, T., Weisz, J. and Lloyd, C. W. (1966) *In vitro* conversion of 7-$^3$H-progesterone to androgens by the rat testis during the second half of fetal life. *Endocrinology* **78,** 245–53.

Nyman, M. A., Geiger, J. and Goldzieher, J. W. (1959) Biosynthesis of estrogen by the perfused stallion testis. *J. Biol. Chem.* **234,** 16–8.

Odell, W. R. and Swerdloff, R. S. (1976) Etiologies of sexual maturation: a model based on the sexually maturing rat. *Rec. Progr. Hor. Res.* **32,** 245–77.

Odell, W. D., Swerdloff, R. S., Jacobs, H. S. and Hescox, M. A. (1973) FSH induction of sensitivity to LH: one cause of sexual maturation in the male rat. *Endocrinology* **92,** 160–5.

Oh, R. and Tamaoki, B-I. (1970) Steroidogenesis in equine testis. *Acta Endocrinol.* **64** 1–16.

Ojeda, S. R. and Ramirez, V. D. (1973) Short-term steroid treatment on plasma LH and FSH in castrated rats from birth to puberty. *Neuroendocrinology* **13,** 100–14.

Okamoto, M., Setaishi, C., Nakagawa, K., Horiuchi, Y., Moriya, K. and Itoh, S. (1971) Diurnals variations in the levels of plasma and urinary androgens. *J. Clin. Endocrinol. Metab.* **32,** 846–51.

Orgebin-Crist, M-C., Danzo, B. J. and Davies, J. (1975) Endocrine control of the development and maintenance of sperm fertilizing ability in the epididymis. *Handbk. Physiol. Sect. 7,* **5,** 319–338.

Ortiz, E., Price, D. and Zaaijer, J. J. P. (1966) Organ culture studies of hormone secretion in endocrine glands of fetal guinea pigs. II. Secretion of androgenic hormone in adrenals and testes during early stages of development. *Konikl. Ned. Akad. Wetenschap. Proc.* C**69,** 400–8.

Ott, F. (1968) Hypersexualitat, Antiandrogene und Hodenfunktion. *Praxis* **57,** 218–20.

Oyama, T. and Kudo, T. (1972) Effect of thiopentone-nitrous oxide anaesthesia and surgery on plasma testosterone levels in human males. *Brit. J. Anaesth.* **44**, 704–6.

Parkes, A. S. (1966) The internal secretions of the testis. In *Marshall's Physiology of Reproduction*, ed. A. S. Parkes, **III**, 412–569, London: Longmans.

Parvinen, M., Hurme, P. and Niemi, M. (1970) Penetration of exogenous testosterone, pregnenolone, progesterone and cholesterol into the seminiferous tubules of the rat. *Endocrinology* **87**, 1082–4.

Patterson, R. L. S. (1968a) 5α-androst-16-ene-3-one: compound responsible for taint in boar fat. *J. Sci. Fd. Agric.* **19**, 31–8.

Patterson, R. L. S. (1968b) Identification of 3α-hydroxy-5α-androst-16-ene as the musk odour component of boar submaxillary salivary gland and its relationship to the sex odour taint in pork meat. *J. Sci. Fd. Agric.* **19**, 434–8.

Payne, A. H. and Jaffe, R. B. (1970) Comparative roles of dehydroepiandrosterone sulfate and androstenediol sulfate as precursors of testicular androgens. *Endocrinology* **87**, 316–22.

Payne, A. H. and Mason, M. (1965) Conversion of dehydroepiandrosterone sulfate to androst-5-enediol-3-sulfate by soluble extracts of rat testis. *Steroids* **6**, 323–37.

Payne, A. H., Jaffe, R. B. and Abell, M. R. (1971) Gonadal steroid sulfates and sulfatase. III. Correlation of human testicular sulfatase, 3β-hydroxysteroid dehydrogenase-isomerase, histologic structure and serum testosterone. *J. Clin. Endocrinol. Metab.* **33**, 582–91.

Payne, A. H., Kawano, A. and Jaffe, R. B. (1973) Formation of dihydrotestosterone and other 5α-reduced metabolites by isolated seminiferous tubules and suspension of interstitial cells in a human testis. *J. Clin. Endocrinol. Metab.* **37**, 448–53.

Pearlman, P. L. (1950) The functional significance of testis cholesterol in the rat: effects of hypophysectomy and cryptorchidism. *Endocrinology* **46**, 341–6.

Pearlman, W. H. and Crepy, O. (1967) Steroid-protein interaction with particular reference to testosterone by human serum. *J. Biol. Chem.* **242**, 182–9.

Pedersen-Bjergaard, K. and Madsen, G. B. (1938) Effect of oestrogenic and androgenic hormone on spontaneous muscular activity of gonadectomized male and female rats. *Acta Path. Microbiol. Scand.* Suppl. **37**, 431–7.

Peets, E. A., Henson, M. F. and Neri, R. (1974) On the mechanism of the anti-androgenic action of flutamide (α-α-α-trifluoro-2-methyl-4'-nitro-*m*-propionotoluidide) in the rat. *Endocrinology* **94**, 532–40.

Phoenix, C. H. (1974) Effects of dihydrotestosterone on sexual behaviour of castrated male rhesus monkeys. *Physiol. Behav.*, **12**, 1045–55.

Phoenix, C. H. (1977) Factors influencing sexual performance in male rhesus monkeys. *J. comp. Physiol. Psychol.* **91**, 697–710.

Phoenix, C. H., Slob, A. K. and Goy, R. W. (1973) Effects of castration and replacement therapy on sexual behaviour of adult male rhesuses. *J. comp. physiol. Psychol.* **84**, 472–81.

Pierrepoint, C. G., Davies, P. and Wilson, D. W. (1974) The role of the epididymis and ductus deferens in the direct and unilateral control of the prostate and seminal vesicles of the rat. *J. Reprod. Fertil.* **41**, 413–23.

Pierrepoint, C. G., Davies, P., Lewis, M. H. and Moffat, D. B. (1975a) Examination of the hypothesis that a direct control system exists for the prostate and seminal vesicles. *J. Reprod. Fertil.* **44**, 395–409.

Pierrepoint, C. G., Davies, P., Millington, D. and John, B. M. (1975b) Evidence that the deferential vein acts as a local transport system for androgens in the rat and the dog. *J. Reprod. Fertil.* **43**, 293–303.

Piro, C., Fraioli, F., Sciarra, F. and Conti, C. (1973) Circadian rhythm of plasma testosterone, cortisol and gonadotrophins in normal male subjects. *J. Steroid Biochem.* **4**, 321–9.

Plagge, J. C. (1941) Thymus gland in relation to sex hormones and reproductive processes in the albino rat. *J. Morphol.* **68**, 519–45.

Plant, T. M., Zumpe, D., Sauls, M. and Michael, R. P. (1974) An annual rhythm in the plasma testosterone of adult male rhesus monkeys maintained in the laboratory. *J. Endocrinol.* **62**, 403–4.

Podesta, E. J. and Rivarola, M. A. (1974) Concentration of androgens in whole testis, seminiferous tubules and interstitial tissue of rats at different stages of development. *Endocrinology* **95**, 455–61.

Price, D., Ortiz, E. and Zaaijer, J. J. P. (1967) Organ culture studies of hormone secretion in endocrine glands of fetal guinea pigs. III. The relation of testicular hormone to sex differentiation of the reproductive ducts. *Anat. Rec.* **157**, 27–42.

Purvis, K. and Haynes, N. B. (1974) Short-term effects of copulation, human chorionic gonadotrophin injection and non-tactile association with a female on testosterone levels in the male rat. *J. Endocrinol.* **60**, 429–39.

Purvis, K., Illius, A. W. and Haynes, N. B. (1974) Plasma testosterone concentrations in the ram. *J. Endocrinol.* **61**, 241–53.

Purvis, K., Landgren, B-M., Cekan, Z. and Diczfalusy, E. (1976) Endocrine effects of masturbation in men. *J. Endocrinol.* **70**, 439–44.

Purvis, K., Calandra, R., Haug, E. and Hansson, V. (1977) 5α reduced androgens and testicular function in the immature rat: effects of 5α-androstan-17β-ol-3-one (DHT) propionate and 5α-androstan-3α, 17β-diol. *Mol. cell. Endocr.* **7**, 203–19.

Racey, P. A. (1974) The reproductive cycle in male noctule bats, *Nyctalus noctula. J. Reprod. Fertil.* **41**, 169–82.

Raeside, J. L. (1965) Urinary excretion of dehydroepiandrosterone and oestrogens by the boat. *Acta Endocrinol.* **50**, 611–20.

Raeside, J. I. (1969) The isolation of estrone sulphate and estradiol-17β sulphate from stallion testis. *Can. J. Biochem.* **47**, 811–15.

Raeside, J. I. and Sigman, D. M. (1975) Testosterone levels in early fetal testes of domestic pigs. *Biol. Reprod.* **13**, 318–21.

Raheja, M. C. and Lucis, O. (1970) Role of dehydroepiandrosterone and its sulphate in synthesis of testosterone by human testes *in vivo* and *in vitro. J. Endocrinol.* **46**, 21–8.

Rampini, E., Voigt, W., Davis, B. P., Moretti, G. and Hsia, S. L. (1971) Metabolism of testosterone-4-[14]C by rat skin: variations during the hair cycle. *Endocrinology* **89**, 1506–14.

Rasmusson, G. H., Chen, A., Reynolds, G. F., Patanelli, D. J., Patchett, A. A. and Arth, G. E. (1972) Antiandrogens. 2′, 3′α-tetrahydrofuran-2′-spiro-17-(1,2α-methylene-4-androsten-3-ones). *J. Med. Chem.* **15**, 1165–8.

Rausch-Strooman, J. G., Petry, R., Hocevar, V., Mauss, J. and Senge, T. (1970) Influence of an anti-androgen (cyproterone) on the gonadotrophic function of the pituitary gland, on the gonads and on metabolism in normal men. *Acta Endocrinol.* **63**, 595–608.

Reed, H. C. B., Melrose, D. R. and Patterson, R. L. S. (1974) Androgen steroids as an aid to the detection of oestrus in pig artificial insemination. *Brit. Vet. J.* **130**, 61–7.

Resko, J. A. (1967) Plasma androgen levels of rhesus monkey: effects of age and season. *Endocrinology* **81**, 1203–12.

Resko, J. A. (1970a) Androgen secretion by the fetal and neonatal rhesus monkey. *Endocrinology* **87**, 680–7.

Resko, J. A. (1970b) Androgens in systemic plasma of male guinea pigs during development and after castration in adulthood. *Endocrinology* **86**, 1444–7.

Resko, J. A. and Eik-Nes, K. B. (1966) Diurnal testosterone levels in peripheral plasma of human male subjects. *J. Clin. Endocrinol. Metab.* **26**, 573–5.

Resko, J. A. and Phoenix, C. H. (1972) Sexual behaviour and testosterone concentrations in the plasma of the rhesus monkey before and after castration. *Endocrinology* **91**, 499–503.

Resko, J. A., Feder, H. H. and Goy, R. W. (1968) Androgen concentrations in plasma and testis of developing rats. *J. Endocrinol.* **40**, 485–91.

Resko, J. A., Malley, A., Begley, D. and Hess, D. L. (1973) Radioimmunoassay of testosterone during fetal development of the rhesus monkey. *Endocrinology* **93**, 156–61.

Reyes, F. I., Winter, J. S. D. and Faiman, C. (1973) Studies on human sexual development. I. Fetal gonadal and adrenal sex steroids. *J. Clin. Endocrinol. Metab.* **37**, 74–8.

Reyes, F. I., Boroditsky, R. S., Winter, J. S. D. and Faiman, C. (1974) Studies on human sexual development. II. Fetal and maternal serum gonadotropin and sex steroid concentrations. *J. Clin. Endocrinol. Metab.* **38**, 612–17.

Rhynes, W. E. and Ewing, L. L. (1973) Testicular endocrine function in Hereford bulls exposed to high ambient temperature. *Endocrinology* **92**, 509–15.

Rice, B. F., Johanson, C. A. and Sternberg, W. H. (1966) Formation of steroid hormones from acetate-1-$^{14}$C by a human fetal testis preparation grown in organ culture. *Steroids* **7**, 79–90.

Richter, C. P. (1933) The effect of hypophyseal injections and implants on the activity of hypophysectomized rats. *Endocrinology* **21**, 481–8.

Richter, C. P. and Wislocki, G. B. (1929) Activity studies on castrated male and female rats with testicular grafts, in correlation with histological studies of the grafts. *Am. J. Physiol.* **86**, 651–60.

Rigaudiere, N., Pelardy, G., Robert, A. and Delost, P. (1976) Changes in the concentrations of testosterone and androstenedione in the plasma and testis of the guinea-pig from birth to death. *J. Reprod. Fertil.* **48**, 291–300.

Riss, W. and Young, W. C. (1954) The failure of large quantities of testosterone propionate to activate low drive male guinea pigs. *Endocrinology* **54**, 232–5.

Ritzen, E. M., Dobbins, M. C., Tindall, D. J., French, F. S. and Nayfeh, S. N. (1973) Characterization of an androgen binding protein (ABP) in rat testis and epididymis. *Steroids* **21**, 593–607.

Ritzen, E. M., Hagenas, L., Hansson, V., Weddington, S. C., French, F. S. and Nayfeh, S. N. (1975) Androgen-binding and transport in testis and epididymis. *Vit. Horm.* **33**, 283–95.

Rivarola, M. and Podesta, E. J. (1972) Metabolism of testosterone $^{14}$C by seminiferous tubules of mature rats: Formation of androstan-3$\alpha$ 17$\beta$-diol-$^{14}$C. *Endocrinology* **90**, 618–23.

Rivarola, M. A., Podesta, E. J. and Chemes, H. E. (1972) *In vitro* testosterone-$^{14}$C metabolism by rat seminiferous tubules at different stages of development: Formation of 5α-androstandiol at meiosis. *Endocrinology* **91**, 537–42.

Rivarola, M. A., Podesta, E. J., Chemes, H. E. and Aguilar, D. (1973) *In vitro* metabolism of testosterone by whole human testis, isolated seminiferous tubules and interstitial tissue. *J. Clin. Endocrinol. Metab.* **37**, 454–60.

Roberts, K. D., Bandi, L., Calvin, H. I., Drucker, W. D. and Lieberman, S. (1964) Evidence that steroid sulfates serve as biosynthetic intermediates. IV. Conversion of cholesterol sulfate *in vivo* to urinary $C_{19}$ and $C_{21}$ steroid sulfates. *Biochemistry* **3**, 1983–8.

Robinson, J. A., Scheffler, G., Eisele, S. G. and Goy, R. W. (1975) Effects of age and season on sexual behaviour and plasma testosterone and dihydrotestosterone concentrations of laboratory-housed male rhesus monkeys (*Macaca mulatta*). *Biol. Reprod.* **13**, 203–10.

Root, A. T., Steinberger, E., Smith, K., Steinberger, A., Russ, D., Somers, L. and Rosenfield, R. (1972) Isosexual pseudoprecocity in a 6-year old boy with testicular interstitial cell adenoma. *J. Pediatr.* **80**, 264–8.

Rose, R. M., Gordon, T. P. and Bernstein, I. S. (1972a) Plasma testosterone levels in the male rhesus: influences of sexual and social stimuli. *Science* **178**, 643–5.

Rose, R. M., Kreuz, L. E., Holaday, J. W., Sulak, K. J. and Johnson, C. E. (1972b) Diurnal variation of plasma testosterone and cortisol. *J. Endocrinol.* **54**, 177–8.

Rosner, W. and Deakins, S. M. (1968) Testosterone-binding globulins in human plasma: studies on sex distribution and specificity. *J. Clin. Invest.* **47**, 2109–16.

Rowe, P. H., Lincoln, G. A., Racey, P. A., Lehane, J., Stephenson, M. J., Shenton, J. C. and Glover, T. (1974) Temporal variations of testosterone levels in the peripheral blood plasma of men. *J. Endocrinol.* **61**, 63–73.

Rubens, R., Dhont, M. and Vermeulen, A. (1974) Further studies on Leydig cell function in old age. *J. Clin. Endocrinol. Metab.* **39**, 40–5.

Ruzicka, L. and Prelog, V. (1943) Untersuchungen von Extracten aus Testes. 1. Mitteilung. Zur Kenntnis der Lipoide aus Schweintestes. *Helv. Chim. Acta.* **26**, 975–95.

Ruzicka, L. and Wettstein, A. (1935) Uber die kunstliche Herstellung des Testikel-hormons Testosteron (Androsten-3-on-17-ol). *Helv. Chem. Acta* **18**, 1264–75.

Sacerdote, F. L., Burgos, M. H. and Bari, D. R. (1972) Effects of testosterone propionate on some ultrastructural and chemical characteristics of rat testis mitochondria. *Endocrinology* **91**, 1020–4.

Saez, J. M. and Bertrand, J. (1968) Studies on testicular function in children: plasma concentrations of testosterone, dehydro-epiandrosterone and its sulfate before and after stimulation with human chorionic gonadotrophin. *Steroids* **12**, 749–62.

Saez, J. M., Saez, S. and Migeon, C. J. (1967) Identification and measurement of testosterone in the sulfate fraction of plasma of normal subjects and patients with gonadal and adrenal disorders. *Steroids* **9**, 1–14.

Safoury, S. El and Bartke, A. (1974) Effects of follicle-stimulating hormone and luteinizing hormone on plasma testosterone levels in hypophysectomized and in intact immature and adult male rats. *J. Endocrinol.* **61**, 193–8.

Saginor, M. and Horton, R. (1968) Reflex release of gonadotropin and increased plasma testosterone concentration in male rabbits during copulation. *Endocrinology* **82**, 627–30.

Saksena, S. K., El Safoury, S. and Bartke, A. (1973) Prostaglandins $E_2$ and $F_{2\alpha}$ decrease plasma testosterone levels in male rats. *Prostaglandins* **4**, 235–42.

Saksena, S. K., Lau, I. F. and Bartke, A. (1974) Prostaglandins A₁ and A₂ decrease testosterone levels in mice and rats. *Endocrinology* **95**, 311–14.

Samuels, L. T. (1960) Metabolism of steroid hormones. In *Metabolic Pathways*, ed. D. M. Greenberg, **I**, 2nd edition, 431–80, New York: Academic Press.

Samuels, L. T. and Eik-Nes, K. B. (1968) Metabolism of steroid hormones. In *Metabolic Pathways*, ed. D. M. Greenberg, **II**, 2nd edition, 169–200, New York: Academic Press.

Sanborn, B. M., Elkington, J. H., Tcholakian, R. K. and Steinberger, E. (1975) Some properties of androgen-binding activity in rat testis. *Mol. Cell. Endocrinol.* **3**, 129–42.

Sanford, L. M., Winter, J. S. D., Palmer, W. M. and Howland, B. E. (1974) The profile of LH and testerone secretion in the ram. *Endocrinology* **95**, 627–31.

Saunders, H. L., Holden, K. and Kerwin, J. F. (1964) The anti-androgenic activity of 17α-methyl-B-nortestosterone (SK & F 7690). *Steroids* **3**, 687–98.

Saunders, H. L., Tomaszewski, J., Pauls, J. and Zuccarello, W. (1969) Antifertility effect of 17α-methyl-B-nortestosterone (SK & F 7690). *Endocrinology* **85**, 960–3.

Savard, K. and Goldzieher, J. W. (1960) Biosynthesis of steroids in stallion testis tissue. *Endocrinology* **66**, 617–24.

Savard, K., Dorfman, R. I., Baggett, B. and Engel, L. L. (1956) Biosynthesis of androgens from progesterone by human testicular tissue *in vitro*. *J. Clin. Endocrinol.* **16**, 1629–31.

Savard, K., Mason, N. R., Ingram, J. T. and Gassner, F. X. (1961) The androgens of bovine spermatic venous blood. *Endocrinology* **69**, 324–30.

Saxena, B. B., Leyerdecker, G., Chen, W., Gandy, H. M. and Peterson, R. E. (1969) Radioimmunoassay of follicle-stimulating (FSH) and luteinizing (LH) hormones by chromatoelectrophoresis. *Acta Endocrinol.* Suppl. **142**, 185–203.

Schanbacher, B. D. and Ford, J. J. (1976) Seasonal profiles of plasma luteinizing hormone, testosterone and estradiol in the ram. *Endocrinology* **99**, 752–7.

Schmidt, G. W. and Tonutti, E. (1956) Pseudo pubertas praecox und unvollständige Pubertas praecox bei einem Leydig-Zell-Tumor des Hodens. *Helv. pediatr. Acta.* **11**, 436–54.

Schoonees, R., Schalch, D. S. and Murray, G. P. (1971) The hormonal effects of anti-androgen (SH 714) treatment in man. *Invest. Urol.* **8**, 635–9.

Segal, D. H. and Raeside, J. I. (1975) Androgens in testes and adrenal glands of the fetal pig. *J. Steroid Biochem.* **6**, 1439–44.

Seilicovich, A. and Rosner, J. M. (1974) Effect of FSH on the *in vivo* uptake and metabolism of ³H-testosterone by the immature rat testis. *Steroids Lipids Res.* **5**, 1–9.

Selye, H. (1937) Studies on adaptation. *Endocrinology* **21**, 169–88.

Selye, H. and Friedman, S. (1941) The action of various steroid hormones on the testis. *Endocrinology* **28**, 129–40.

Serio, M., Crosignani, P. G., Romano, S., Trojsi, L., Forti, G., Fiorelli, G. and Pazzagli, M. (1974) Temporal variations of testosterone and gonadotrophins in man. In *The Endocrine Function of the Human Testis*, eds. V. H. T. James, M. Serio and L. Martini, **II**, 13–39, New York: Academic Press.

Setchell, B. P. (1974) Secretions of the testis and epididymis. *J. Reprod. Fertil.* **37**, 165–77.

Setchell, B. P. and Jacks, F. (1974) Inhibin-like activity in rete testis fluid. *J. Endocrinol.* **62**, 675–6.

Setchell, B. P. and Main, S. J. (1974) Bibliography with review on inhibin. *Bibliog. Reprod.* **24**, 245–52 and 362–7.

Setchell, B. P. and Sirinathsinghji, D. J. (1972) Antigonadotrophic activity in rete testis fluid a possible 'inhibin'. *J. Endocrinol.* **53**, lx–lxi.

Setchell, B. P., Waites, G. M. H. and Lindner, H. R. (1965) Effect of undernutrition on testicular blood flow and metabolism and the output of testosterone in the ram. *J. Reprod. Fertil.* **9**, 149–62.

Setchell, B. P., Voglmayr, J. K. and Hinks, N. T. (1971) The effect of local heating on the flow and composition of rete testis fluid in the conscious ram. *J. Reprod. Fertil.* **24**, 81–9.

Setchell, B. P., Main, S. J. and Davies, R. V. (1977a) The effect of ligation of the efferent ducts of the testis on serum gonadotrophins and testosterone in rats. *J. Endocrinol.* **72**, 13P–14P.

Setchell, B. P., Davies, R. V. and Main, S. J. (1977b) Inhibin. In *The Testis, IV*, eds. A. D. Johnson and W. R. Gomes, 189–238, New York: Academic Press.

Severinghaus, A. E. (1937) Cellular changes in the anterior hypophysis with special reference to its secretory activities. *Physiol. Rev.* **17**, 556–88.

Severinghaus, A. E., Engle, E. T. and Smith, P. E. (1932) Anterior pituitary changes referable to the reproductive hormones, and the influence of the thyroid and the adrenals on genital function. In *Sex and Internal Secretions*, ed. E. Allen, 805–27, Baltimore: Williams and Wilkins.

Sharp, F., Hay, J. B. and Hodgkins, M. B. (1976) Metabolism of androgens *in vitro* by human foetal skin. *J. Endocrinol.* **70**, 491–9.

Sherins, R. J. and Loriaux, D. L. (1973) Studies on the role of sex steroids in the feedback control of FSH concentrations in men. *J. Clin. Endocrinol. Metab.* **36**, 886–93.

Shikita, M. and Tamaoki, B. (1965) Testosterone formation by subcellular particles of rat testes. *Endocrinology* **76**, 563–9.

Shikita, M., Kakizaki, H. and Tamaoki, B. (1964) The pathway of formation of testosterone from 3β-hydroxypregn-5-en-20-one by rat testicular microsomes. *Steroids* **4**, 521–31.

Short, R. V. (1960) The secretion of sex hormones by the adrenal gland. *Biochem. Soc. Symp.* **18**, 59–84.

Short, R. V. and Mann, T. (1966) The sexual cycle of a seasonally breeding mammal, the roebuck (*Capreolus capreolus*). *J. Reprod. Fertility* **12**, 337–51.

Siiteri, K. and Wilson, J. D. (1974) Testosterone formation and metabolism during male sexual differentiation in the human embryo. *J. Clin. Endocrinol. Metab.* **38**, 113–25.

Sizonenko, P. C., Cueudet, A. and Paumier, L. (1973) FSH. I. Evidence for its mediating role on testosterone secretion in cryptorchidism. *J. Clin. Endocrinol. Metab.* **37**, 68–73.

Skinner, J. D. and Rowson, L. E. A. (1968a) Some effects of unilateral cryptorchidism and vasectomy on sexual development of the pubescent ram and bull. *J. Endocrinol.* **42**, 311–21.

Skinner, J. D. and Rowson, L. E. A. (1968b) Effects of testosterone injected unilaterally down the vas deferens on the accessory glands of the ram. *J. Endocrinol.* **42**, 355–6.

Skinner, J. D., Booth, W. D., Rowson, L. E. A. and Karg, H. (1968) The postnatal development of the reproductive tract of the Suffolk ram, and changes in the gonadotrophin content of the pituitary. *J. Reprod. Fertil.* **16**, 463–77.

Slaunwhite, W. R., Jr. and Burgett, M. J. (1965) *In vitro* testosterone synthesis by rat testicular tissue. *Steroids* **6**, 721–35.

Slaunwhite, W. R., Jr. and Samuels, L. T. (1956) Progesterone as a precursor of testicular androgens. *J. Biol. Chem.* **220**, 341–52.

Smith, A. A., McLean, W. S., Hansson, V., Nayfeh, S. N. and French, F. S. (1975a) Androgen receptor in nuclei of rat testis. *Steroids* **25**, 569–86.

Smith, A. A., McLean, W. S., Nayfeh, S. N., French, F. S., Hansson, V. and Ritzen, M. (1975b) Androgen receptors in rat testis. *Curr. Top. Mol. Endocrinol.* **2**, 257–80.

Smith, P. E. (1944) Maintenance and restoration of spermatogenesis in hypophysectomized rhesus monkeys by androgen administration. *Yale J. Biol. Med.* **17**, 281–7.

Snipes, C. A., Forest, M. G. and Migeon, C. J. (1969) Plasma androgen concentration in several species of Old and New World monkeys. *Endocrinology* **85**, 941–5.

Sorcini, G., Sciarra, F., diSilverio, F. and Fraili, F. (1971) Further studies on plasma androgens and gonadotrophins after cyproterone acetate. *Folia Endocr.* **24**, 196–201.

Southren, A. L., Tochimoto, S., Carmody, N. C. and Isurugi, K. (1965) Plasma production rates of testosterone in normal adult men and women and in patients with the syndrome of feminizing testes. *J. Clin. Endocrinol.* **25**, 1441–50.

Srere, P. A., Chaikoff, I. L., Treitman, S. S. and Burstein, L. S. (1950) The extrahepatic synthesis of cholesterol. *J. Biol. Chem.* **182**, 629–34.

Städtler, F. (1972) Histologische Befunde am menschlichen Hodenbiopsen vor und unter Antiandrogenbehandlung. In *Life Science Monographs* **2**, 101–112, ed. G. Raspe, Oxford: Pergamon.

Städtler, F. and Horn, H. J. (1972) Histometrische und ferment histochemische Untersuchungen an menschlichen Hodenbiopsien vor und unter Antiandrogenbehandlung *Verhandl. Dtsch. Ges. Path.* **56**, 580–3.

Stearne, E. L., Winter, J. S. D. and Faiman, C. (1973) Effects of coitus on gonadotropin, prolactin and sex steroid levels in man. *J. Clin. Endocrinol. Metab.* **37**, 687–91.

Steinbeck, H., Mehring, M. and Neumann, F. (1971) Comparison of the effects of cyproterone, cyproterone acetate and oestradiol on testicular function, accessory sexual glands and fertility in a long-term study on rats. *J. Reprod. Fertil.* **26**, 65–76.

Steinberger, A. and Steinberger, E. (1976) Secretion of an FSH-inhibiting factor by cultured Sertoli cells. *Endocrinology* **99**, 918–21.

Steinberger, E. (1971) Hormonal control of mammalian spermatogenesis. *Physiol. Rev.* **52**, 1–22.

Steinberger, E. and Duckett, G. E. (1965) Effect of estrogen or testosterone on initiation and maintenance of spermatogenesis in the rat. *Endocrinology* **76**, 1184–9.

Steinberger, E. and Ficher, M. (1968) Conversion of progesterone to testosterone by testicular tissue at different stages of maturation. *Steroids* **11**, 351–68.

Steinberger, E. and Ficher, M. (1971) Formation and metabolism of testosterone in testicular tissue of immature rats. *Endocrinology* **89**, 679–84.

Steinberger, E. and Steinberger, A. (1974) Hormonal control of testicular function in mammals. *Handbk. Physiol.* Sect. 7, vol. **IV**, Part 2, 325–45.

Steinberger, E., Root, A., Ficher, M. and Smith, K. D. (1973) The role of androgens in the initiation of spermatogenesis in man. *J. clin. Endocrinol. Metab.* **37**, 746–51.

Steinberger, E., Chowdhury, A. K., Tcholakian, R. K. and Roll, H. (1975) Effects of $C_{21}$ steroids on sex accessory organs and testes of mature hypophysectomized rats. *Endocrinology* **96**, 1319–23.

Stewart, D. W. and Raeside, J. I. (1976) Testosterone secretion by the early fetal pig testes in organ culture. *Biol. Reprod.* **15**, 25–8.

Stewart-Bentley, M., Odell, W. and Horton, R. (1974) The feedback control of luteinizing hormone in normal adult men. *J. Clin. Endocrinol. Metab.* **38**, 545–53.

Stone, C. P. (1939) Copulatory activity in adult male rats following castration and injections of testosterone propionate. *Endocrinology* **24**, 165–74.

Straus, J. S. and Pochi, P. E. (1969) Recent advances in androgen metabolism and their relation to the skin. *Arch. Dermatol.* **100**, 621–36.

Strickland, A. L., Nayfeh, S. N. and French, F. S. (1970) Conversion of cholesterol to testosterone and androstanediol in testicular homogenates of immature and mature rats. *Steroids* **15**, 373–87.

Strott, C. A., Yoshimi, T. and Lipsett, M. B. (1969) Plasma progesterone and 17-hydroxyprogesterone in normal men and children with congenital adrenal hyperplasia. *J. Clin. Invest.* **48**, 930–9.

Sundby, A., Tollman, R. and Velle, W. (1975) Long-term effect of HCG on plasma testosterone in bulls. *J. Reprod. Fertil.* **45**, 249–54.

Suzuki, Y. and Eto, T. (1962) Androgens in testicular venous blood in the adult rat. *Endocrinol. Japon.* **9**, 277–83.

Swerdloff, R. S. and Walsh, P. C. (1973) Testosterone and oestradiol suppression of LH and FSH in adult male rats: Duration of castration, duration of treatment and combined treatment. *Acta Endocrinol.* (Copenhagen) **73**, 11–21.

Swerdloff, R. S., Grover, P. K., Jacobs, H. S. and Bain, J. (1973) Search for a substance which selectively inhibits FSH: Effects of steroids and prostaglandins on serum FSH and LH levels. *Steroids* **21**, 703–22.

Takashima, I., Adachi, K. and Montagna, W. (1970) Studies of common baldness in the stumptailed macaque. IV. *In vivo* metabolism of testosterone in the hair follicle. *J. Invest. Dermatol.* **55**, 329–34.

Tamaoki, B-I. (1973) Steroidogenesis and cell structure—biochemical pursuit of sites of steroid biosynthesis. *J. steroid Biochem.* **4**, 89–118.

Tamaoki, B-I. and Shikita, M. (1966) A dynamic study on the pathway of testosterone formation from progesterone and pregnenolone by rat testicular microsomes. *European J. Steroids* **1**, 351–60.

Tamaoki, B-I., Inano, H. and Nakano, H. (1969) *In vitro* synthesis and conversion of androgens in testicular tissue. In *The Gonads*, ed. K. W. McKerns, 547–613, New York: Appleton Century Crofts.

Tcholakian, R. K. and Eik-Nes, K. B. (1971) $\Delta^5$-pregnenolone and testosterone in spermatic venous blood of anaesthetized dogs. *Am. J. Physiol.* **221**, 1824–6.

Thomas, J. A. and Dombrosky, J. T. (1975) Effects of methadone on the male reproductive system. *Arch. Int. Pharmacodyn. Ther.* **215**, 215–21.

Toren, D., Menon, K. M. J., Forchielli, E. and Dorfmann, R. I. (1964) *In vitro* enzymatic cleavage of the cholesterol side-chain in rat testis preparations. *Steroids* **3**, 381–90.

Tsai, S. C., Ying, B. P. and Gaylor, J. L. (1964) Testicular sterols. I. Incorporation of mevalonate and acetate into sterols by testicular tissue from rats. *Arch. Biochem. Biophys.* **105**, 329–38.

Tsang, W. N., Lacy, D. and Collins, P. M. (1973) Leydig cell differentiation, steroid metabolism by the interstitium *in vitro* and the growth of the accessory sex organs in the rat. *J. Reprod. Fertil.* **34**, 351–5.

Tveter, K. J. (1971) Effect of 17α-methyl-β-nortestosterone (SK and F 7690) on the binding *in vitro* of 5α-dihydrotestosterone to macromolecular components from the rat ventral prostate. *Acta Endicrinol.* **66**, 352–6.

Velardo, J. T. (1959) Steroid hormones and uterine growth. *Ann. N.Y. Acad. Sci.* **75**, 441–62.

Velardo, J. T., Hisaw, F. L. and Bever, A. T. (1956) Inhibitory action of desoxycorticosterone acetate, cortisone and testosterone on uterine growth induced by estradiol-17β. *Endocrinology* **59**, 165–9.

Venable, J. H. (1966) Constant cell populations in normal, testosterone deprived and testosterone stimulated levator ani muscles. *Am. J. Anat.* **119**, 263–301.

Verjans, H. L. and Eik-Nes, K. B. (1976) Effects of androstenes, 5α-androstanes, 5β-androstanes, oestrenes and oestratriennes on serum gonadotrophin levels and ventral prostate weights in gonadectomized, adult male rats. *Acta Endocrinol.* **83**, 201–10.

Verjans, H. L., Eik-Nes, K. B., Aafjes, J. H., Vels, F. J. M. and Molen, H. J. van der (1974) Effects of testosterone propionate, 5-dihydrotestosterone propionate and oestradiol benzoate on serum levels of LH and FSH in the castrated adult male rat. *Acta Endocrinol.* (Copenhagen) **77**, 643–54.

Vermeulen, A. (1973) The physical state of testosterone in plasma. In *The Endocrine Function of the Human Testis*, eds. V. H. T. James, M. Serio and L. Martini, I, 157–70, New York: Academic Press.

Vermeulen, A., Stoica, T. and Verdonck, L. (1971) The apparent free testosterone concentration, an index of androgenicity. *J. clin. Endocrinol.* **33**, 759–67.

Vermeulen, A., Rubens, R. and Verdonck, L. (1972) Testosterone secretion and metabolism in male senescence. *J. Clin. Endocrinol. Metab.* **34**, 730–5.

Vernon, R. G., Kopec, B. and Fritz, I. B. (1974) Observations on the binding of androgens by rat testis seminiferous tubules and testis extracts. *Mol. Cell. Endocrinol.* **1**, 167–88.

Veyssiere, G., Berger, M., Jean-Faucher, C., de Turckhein, M. and Jean, C. (1976) Levels of testosterone in the plasma, gonads and adrenals during fetal development of the rabbit. *Endocrinology* **99**, 1263–8.

Viguier-Martinez, M. and Pelletier, J. (1972) Etude comparée des effets de la cyprotérone et de l'acetate de cyprotérone sur la LH hypophysaire la LH plasmatique et l'appareil génital du rat mâle prépubière. *C.R. Acad. Sci. Paris.* Ser D **274**, 2696–9.

Vosbeck, K. and Keller, P. J. (1971) The influence of anti-androgens on the excretion of FSH, LH and 17-ketosteroids in males. *Hormone Metab. Res.* **3**, 273–6.

Vusse, G. J. van der, Kalkman, M. L. and Molen, H. J. van der (1973) Endogenous production of steroids from total rat testis and from isolated interstitial tissue and seminiferous tubules. *Biochim, Biophys. Acta.* **297**, 179–85.

Walsh, E. L., Cuyler, W. K. and McCullagh, D. R. (1933) Effect of testicular hormone on hypophysectomized rats. *Proc. Soc. Exptl. Biol. Med.* **30**, 848–50.

Walsh, E. L., Cuyler, W. K. and McCullagh, D. R. (1934) Physiologic maintenance of male sex glands; effect of androtin on hypophysectomized rats. *Am. J. Physiol.* **107**, 508–12.

Walsh, P. C., Swerdloff, R. S. and Odell, W. D. (1972) Cyproterone: Effect on serum gonadotrophins in the male. *Endocrinology* **90**, 1655–9.

Walsh, P. C., Swerdloff, R. S. and Odell, W. D. (1973) Feedback control of FSH in the male: Role of oestrogen. *Acta Endocrinol.* (Copenhagen) **74**, 449–60.

Wang, G. H., Richter, C. P. and Guttmacher, A. F. (1925) Activity studies on male castrated rats with ovarian transplants, and correlation of the activity with the histology of the grafts. *Am. J. Physiol.* **73**, 581–98.

Warren, D. W., Haltmeyer, G. C. and Eik-Nes, K. B. (1972) Synthesis and metabolism of testosterone in the fetal rat testis. *Biol. Reprod.* **7**, 94–9.

Warren, D. W., Haltmeyer, G. C. and Eik-Nes, K. B. (1973) Testosterone in the fetal rat testis. *Biol. Reprod.* **8**, 560–5.

Weddington, S. C., Hansson, V., Purvis, K., Varaas, T., Verjans, H. L., Eik-Nes, K. B., Ryan, W. H., French, F. S. and Ritzen, E. M. (1976) Biphasic effect of testosterone propionate on Sertoli cell secretory function. *Mol. Cell. Endocrinol.* **5**, 137–45.

Wells, L. J. (1942) The response of the testis to androgens following hypophysectomy. *Anat. Rec.* **82**, 565–85.

Weniger, J. P. and Zeis, A. Z. (1970) Transformation de la testosterone en dihydrotestosterone par les canaux de Wolff de l'embryon de Souris. *Compt. Rend.* **271**, 1307–10.

Werbin, H. and Chaikoff, I. L. (1961) Utilization of adrenal gland cholesterol for synthesis of cortisol by the intact normal and the ACTH-treated guinea pig. *Arch. Biochem. Biophys.* **93**, 476–82.

West, C. D., Hollander, V. P., Kritchevsky, T. H. and Dobriner, K. (1952) The isolation and identification of testosterone, 4-androstenedione-3,17, and 7-ketocholesterol from spermatic vein blood. *J. Clin. Endocrinol.* **12**, 915–6.

Whalen, R. E. (1964) Hormone induced changes in the organization of sexual behaviour in the male rat. *J. Comp. Physiol. Psychol.* **57**, 175–82.

Wiele, R. L. van de, MacDonald, P. C., Gurpide, E. and Lieberman, S. (1963) Studies on the secretion and interconversion of the androgens. *Recent Progr. Hormone Res.* **19**, 275–305.

Williams-Ashman, H. G. (1965) Androgenic control of nucleic acid and protein synthesis in male accessory genital organs. *J. Cell. Comp. Physiol.* **66**, 111–24.

Williams-Ashman, H. G. (1975) Metabolic effects of testicular androgens. *Handbk. Physiol.* Sect. 7, vol. **5**, 473–90.

Wilson, E. M. and French, F. S. (1976) Binding properties of androgen receptors. Evidence for identical receptors in rat testis, epididymis and prostate. *J. Biol. Chem.* **251**, 5620–9.

Wilson, E. M. and Smith, A. A. (1975) Localization of androgen receptors in rat testis: biochemical studies. *Curr. Top. Mol. Endocrinol.* **2**, 281–6.

Wilson, J. D. (1962) Localization of the biochemical site of action of testosterone on protein synthesis in the seminal vesicle of the rat. *J. Clin. Invest.* **41**, 153–61.

Wilson, J. D. (1975) Metabolism of testicular androgens. *Handbk. Physiol.* Sect. 7, vol. **5**, 491–508.

Wilson, J. D. and Walker, J. D. (1969) The conversion of testosterone to 5-androstan-17-ol-3-one (dihydrotesterone) by skin slices of man. *J. Clin. Invest.* **48**, 371–9.

Winter, J. S. D., Hughes, I. A., Reyes, F. I. and Faiman, C. (1976) Pituitary-gonadal relations in infancy: 2. Patterns of serum gonadal steroid concentrations in man from birth to two years of age. *J. Clin. Endocrinol. Metab.* **42**, 679–86.

Woods, J. E. and Domm, L. V. (1966) Histochemical identification of the androgen-producing cells in the gonads of the domestic fowl and albino rat. *Gen. Comp. Endocrinol.* **7**, 559–70.

Yamaji, T., Motohashi, K., Tahioka, T. and Ibayashi, H. (1968) Androstenediol in canine spermatic vein blood and its significance in testosterone biosynthesis *in vivo*. *Endocrinology* **83**, 992–8.

Yanahara, T. and Troen, P. (1972) Studies of the human testis. I. Biosynthetic pathways for androgen formation in human testicular tissue *in vitro*. *J. Clin. Endocrinol. Metab.* **34**, 783–92.

Ying, B. P., Chang, Y. J. and Gaylor, J. L. (1965) Testicular sterols. III. Effects of gonadotrophins on the biosynthesis of testicular steroids. *Biochim. Biophys. Acta.* **100**, 256–62.

Young, D. G. and Hall, P. F. (1968) Biosynthesis of cholesterol sulfate by slices of rabbit testis. *Endocrinology* **82**, 291–5.

Young Lai, E. V., Moor, B. C. and Dimond, P. (1976) Effects of sexual activity on luteinizing hormone and testosterone levels in the adult male rabbit. *J. Endocrinol.* **69**, 183–91.

Zaaijer, J. J. P., Price, D. and Ortiz, E. (1966) Organ culture studies of hormone secretion in endocrine glands of fetal guinea pigs. I. Androgenic secretion as demonstrated by a bioindicator method. *Koninkl. Ned. Akad. Wetenschap.*, Proc. C 69, 389–99.

Zahler, H. (1944) Uber die Wirkung verschiedener Gaben von Androgen auf den Rattenhoden. *Virchow's Arch. pathol. Anat.* **312**, 138–64.

Zanisi, M., Motta, M. and Martini, C. (1973) Inhibitory effect of 5α-reduced metabolites of testosterone on gonadotrophin secretion. *J. Endocrinol.* **56**, 315–16.

Zumoff, B., Bradlow, H. L., Finkelstein, J., Boyar, R. M. and Hellman, L. (1976) The influence of age and sex on the metabolism of testosterone. *J. Clin. Endocrinol. Metab.* **42**, 703–6.

# 7 Spermatogenesis

## 7.1 Cell types of germinal line

Four basic cell types can be recognized among the germinal cells of mature animals, and apart from the spermatozoa, they are usually referred to by the names given to them by von La Valette St George (1865, 1876, 1878, 1887): spermatogonia, spermatocytes and spermatids. There are two types of spermatocytes, primary and secondary (Figure 7.1).

### 7.1.1 Spermatogonia

The spermatogonia are large diploid cells which lie against the boundary tissue of the tubule and divide mitotically; morphologically they can be classified into three types: A, Intermediate (In) and B spermatogonia. The A spermatogonia are characterized by a large pale ovoid nucleus usually lying with its long axis parallel to the boundary tissue. The chromatin is homogeneous and dust-like. Several generations of A spermatogonia are produced by successive mitoses; some authors describe unsynchronized divisions of the stem-spermatogonia (see Section 7.3) but the last three or four divisions of the A spermatogonia are synchronized with other events in spermatogenesis.

After several generations of morphologically very similar cells, with considerable degeneration at each division, cells are formed with the chromatin showing conspicuous crusts attached to the nuclear membrane. These are the intermediate or In spermatogonia and these divide again by mitosis to give rise to the B spermatogonia, in which the crust-like accumulations of chromatin are even more pronounced; the nuclei are also appreciably smaller and more nearly spherical (Figure 7.2). These cells, like the A spermatogonia, are still in contact with the boundary tissue of the tubule, although there is a tendency for tongues of Sertoli cell cytoplasm to interpose at their edges. Depending on the species there are one to four successive generations of B spermatogonia, which cannot be distinguished morphologically, but only by the cells with which they are associated. Finally the B spermatogonia divide mitotically to produce pre-leptotene spermatocytes (see Ortavant, 1959; Bishop and Walton, 1960; Roosen-Runge, 1962, 1977; Leblond et al., 1963; Courot et al., 1970; Clermont, 1972; Steinberger and Steinberger, 1975). The duration of DNA synthesis increases progressively from A spermatogonia up to the last B spermatogonia (Monesi, 1962; Hilscher, 1964; Hochereau, 1967).

The spermatogonia arise from the gonocytes of the fetal testis, which cease dividing on about day 18 of fetal life. Many of the gonocytes degenerate in the rat between 3 and 7 days after birth, but the remainder (often as little as one-quarter of those present at birth) then begin to divide to form the first spermatogonia (Clermont and Perey, 1957; Sapsford, 1962; Beaumont and Mandl, 1963; Huckins and Cler-

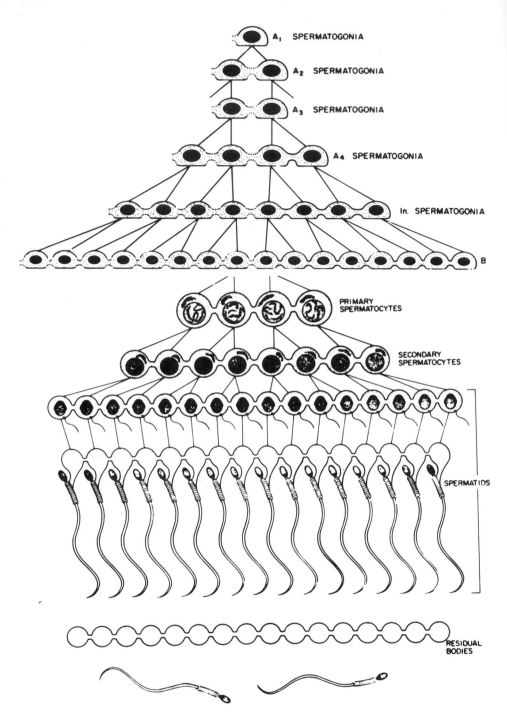

Figure 7.1 A diagram of the various cell types of the germinal line. (B refers to B spermatogonia.) The diagram also illustrates the intercellular bridges formed by these cells, although because of the large numbers of cells involved, only the progeny of 2 of the possible 4 $A_3$ spermatogonia and 2 of the possible 32 B spermatogonia are shown. (Reproduced from Dym and Fawcett, 1971)

RAT

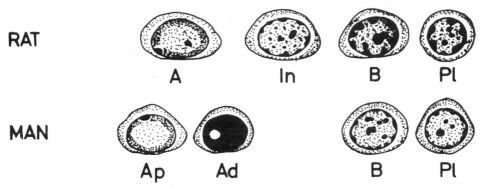

A          In          B         Pl

MAN

Ap    Ad          B         Pl

Figure 7.2 A diagram illustrating the appearance of the different types of spermatogonia and pre-leptotene spermatocytes (P1) found in rat and human testis. Note that in the human there are two types of A spermatogonia, A-pale (Ap) and A-dark (Ad) and no recognisable intermediate spermatogonia. (Based on a drawing by Clermont, 1972)

mont, 1968; Roosen-Runge and Leik, 1968; Hilscher *et al.*, 1972, 1973). Gonocyte degeneration is less obvious in sheep and cattle (Courot, 1962; Attal and Courot, 1963).

### 7.1.2    Spermatocytes

The spermatocytes formed by the last spermatogonial mitosis are now called 'pre-leptotene spermatocytes'; they used to be called 'resting spermatocytes' by morphologists, but since it has been shown that this is the stage when almost all the DNA is formed for the meiotic divisions, this term has rightly fallen out of use. Morphologically these cells resemble the B spermatogonia, but when filaments become distinguishable within the crusts of chromatin in the nucleus, the cells have entered leptotene, the real beginning of the meiotic prophase. By the end of this stage, spiralization and contraction of the chromosomes occur, just after DNA synthesis has stopped in the sex chromosomes, which are the last to finish. Therefore at this time, these cells contain a tetraploid amount of DNA. During the next stage, zygotene, the analogous chromosomes pair off and become thicker, then gather in a bouquet-like arrangement. Synaptinemal complexes appear between pairs of chromosomes and the nucleoli become more easily visible.

The next stage, pachytene, is the longest in the meiotic prophase, and is also the stage at which the cells seem to be particularly susceptible to many forms of damage (see Chapter 11). Each chromosome thickens and begins to show a longitudinal split, except at the level of the centromeres. This is the stage in mammals when crossing over occurs between the paired chromosomes. Finally in diplotene, the nucleus reaches its maximum size and duplication of the chromosomes is complete so that complete tetrads are formed. These then separate during the very rapid diakinesis, which can be seen in rodents but is extremely rapid and therefore seldom observed in rams and bulls. The members of the pairs of homologous chromosomes separate into the two secondary spermatocytes, but as each chromosome has doubled its DNA content, each of these cells contains a diploid amount of DNA. These have round nuclei with granules of DNA connected by a fine network of chromatin. Their life-span is very short, only a few hours, and they divide again, by a process very like mitosis, in which the halves of the doubled chromosomes are separated to produce two haploid cells (see Ortavant, 1959; Roosen-Runge, 1962, 1977; Leblond *et al.*, 1963; Courot *et al.*, 1970; Clermont, 1972; Steinberger and Steinberger, 1975).

### 7.1.3    Spermatids

The cells which result from the second meiotic division are initially small round cells with a pale nucleus and a moderate amount of undistinguished cytoplasm. These cells are then transformed into the very complex spermatozoa, undergoing condensation of the nucleus, formation of the acrosome, virtual elimination of the cytoplasm, development of a tail and the arrangement of its mitochondria into a helix to produce the midpiece (Figure 7.3). This process is known as spermiogenesis or sper-

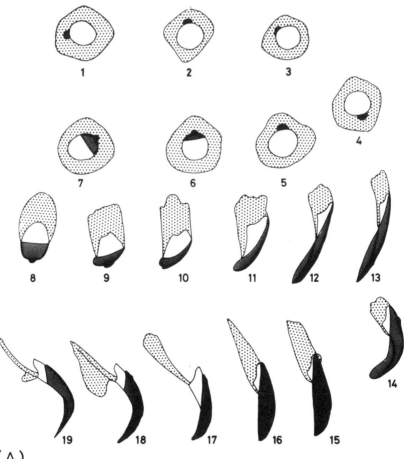

(A)

Figure 7.3 Diagrams of the various stages in the transformation of an early round spermatid into a spermatozoa in (A) rat, (B, opposite) ram and (C, overleaf) man. (Based on drawings by Leblond and Clermont, 1952a and Clermont and Leblond, 1955)

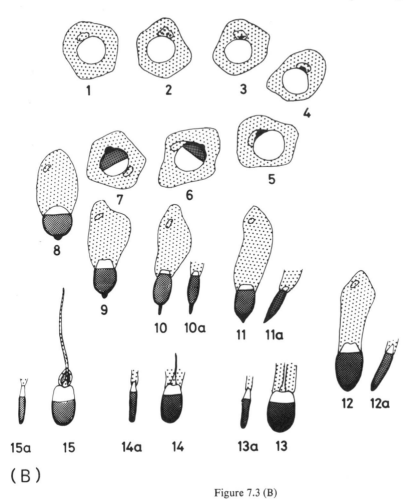

Figure 7.3 (B)

mateleosis. Just how these changes are achieved is not fully understood but as the spermatids develop in close association with the Sertoli cells the latter may well be actively involved. Because of the lack of information of the mechanisms concerned, the following account is almost entirely descriptive but as these changes are so remarkable, they are worth considering in some detail.

**7.1.3.1    Nuclear condensation**    When the spermatid is first formed, its spherical nucleus shows, when stained with the Feulgen technique, several karyosomes dispersed in a fine dusty chromatin. As one would expect, the amount of DNA, whether determined by Feulgen staining or by UV-microspectrophotometry, is nearly one-half of that in the secondary spermatocytes (and of diploid cells from other tissues) and a quarter of that in the primary spermatocytes. Then there is a progressive aggregation of the chromatin granules which finally produces the highly electron-dense nucleus of the late spermatid. As this occurs, the ratio of the amount of DNA

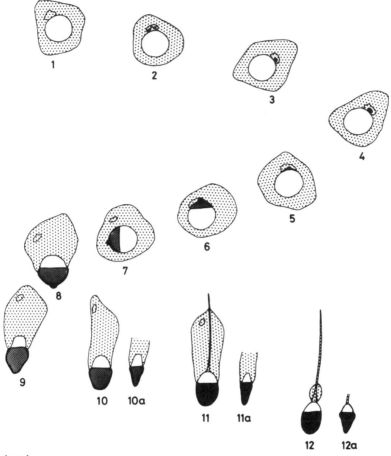

(C)

Figure 7.3 (C)

determined by Feulgen staining to that determined by UV-microspectrophotometry falls, presumably because the condensed chromatin cannot react as readily with the stain (Figure 7.4). In the human, in contrast to most other species, the condensation of the chromatin often leaves nuclear vacuoles which show up in electron micrographs as clear zones of variable size and shape (Fawcett, 1970). During the condensation of the nucleus, the ultrastructure of the chromatin changes from a beaded pattern, typical of histones, to a smooth fibre (Kirzenbaum and Tres, 1975).

The nucleus is also reshaped during condensation of the chromatin to the form characteristic of the species. External forces from the spermatid cytoplasm or even the Sertoli cell have been suggested but it now seems likely that the process is determined by forces generated within the nucleus itself (Fawcett et al., 1971).

**7.1.3.2    The acrosome**    The acrosome of the spermatozoon which plays such an important part in fertilization, is formed during spermiogenesis (Figure 7.5). It

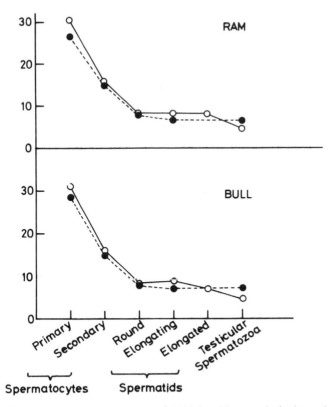

RAM

BULL

Primary
Secondary
Round
Elongating
Elongated
Testicular
Spermatozoa

Spermatocytes      Spermatids

Figure 7.4 Graph of the amount of DNA in arbitrary units in the nucleus of various germ cells determined by UV-microspectrophotometry (●─ ─●) and Feulgen staining (O───O). (Data of Esnault
(1965) quoted by Courot et al., 1970)

originates in the Golgi complex, possibly beginning before meiosis in the spermatocyte, where several of the cisternae appear to synthesize protein granules called
pro-acrosomal granules in their lumina. These structures fuse together, the pro-
acrosomal granules forming a single acrosomal granule, surrounded by an
acrosomal vesicle produced by the fusion of the dilated cisternae in which the
original granules were formed. The granule and vesicle then migrate towards the
nucleus and become attached to it (Figure 7.5.A), while further material is added
from the Golgi complex. The extent of this addition of material can be seen particularly well in freeze-fractured preparations which reveal the inner surface of the
outer membrane of the acrosomal vesicle. Then the acrosome spreads out along the
nuclear membrane, itself becoming flattened as it does so, and indenting the nucleus
(Figure 7.5.B). Finally it envelopes the end of the condensed spermatid nucleus
(Burgos and Fawcett, 1955; Sandoz, 1970; Bedford and Nicander, 1971; Ploen,
1971; Figure 7.5.C to F). The form of the acrosome is quite variable between species
ranging from the neat 'skull-caps' of man, ram and bull to the extravagant forms seen
in some rodents (Fawcett, 1970, 1975a). Once the acrosome is formed, the remaining

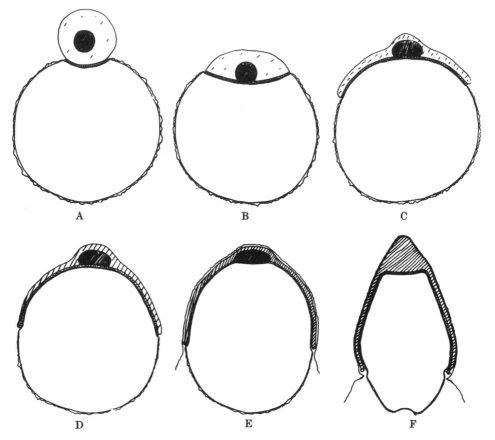

Figure 7.5 Diagram of the development of the acrosome and the change in shape of the spermatid nucleus in the rabbit. (Reproduced from Ploen, 1971)

Golgi complex migrates round to the other side of the nucleus and is eventually included in the residual cytoplasm left behind when the spermatozoon is shed (Fawcett and Phillips, 1969a; Fawcett *et al.*, 1971).

**7.1.3.3    Formation of neck and tail**   At the end of the second meiotic division, the two centrioles formed do not separate but migrate to a position opposite the Golgi apparatus, immediately beneath the cell membrane. One centriole, the distal centriole, is orientated perpendicular to the cell surface and at right angles to the other, the proximal centriole. The distal centriole then gives rise to the tail of the spermatozoon (Figure 7.6), beginning as an outgrowth of the axonemal complex, an

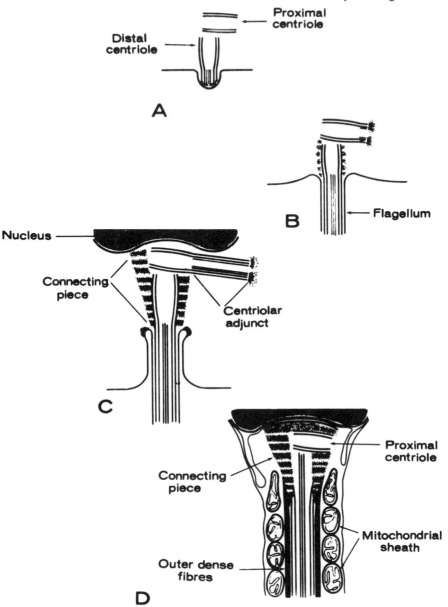

Figure 7.6 Diagram illustrating the development of the tail and neck of the spermatozoa from the two centrioles. A. The distal centriole of the early spermatid gives rise to the axoneme of the flagellum. B. Soon thereafter small aggregations of filamentous material appear along the sides of the distal centriole and at the end of the proximal centriole. C. The proximal centriole gives rise at its end to a transient organelle, the centriolar adjunct. The material accumulating around the distal centriole polymerizes to form the cross-striated columns of the connecting piece. D. In the mature sperm, the centriolar adjunct and distal centriole have disappeared and the cross-striated elements of the connecting piece are continuous with the outer dense fibres of the flagellum. (Reproduced from Fawcett, 1972)

inner pair of single microtubules and nine outer double microtubules. The tail gradually elongates and from the nine outer double microtubules, the nine dense outer fibres develop, petal-shaped in cross-section. These eventually become much larger than the original microtubules and separate from them along most of their length, but their origin is still obvious in mature spermatozoa from the fusion of the two sets of fibres at their distal ends. There are also a variable number of finer satellite fibres set among the dense outer fibres. The general pattern is common to all mammals but the size, density, shape and position of the dense outer and satellite fibres is very variable (Fawcett, 1965, 1970, 1975a).

As the tail develops, it protrudes from the spermatid and at its base the cell membrane is attached to the annulus, a dense ring attached by delicate radial densities to the distal centriole. During elongation of the spermatid, the annulus retains its position relative to the nucleus so that as the cytoplasm moves distally, a space, known as the flagellar canal, is formed between it and the tail; this space is filled by a tongue of Sertoli cell cytoplasm.

There are also changes in the proximal centriole during development of the tail. It gives rise to a thin sheet of dense material which arches over its upper surface and eventually develops into the articular surface of the capitulum or connecting piece which joins the tail to the nucleus. At a slightly later stage, the pair of centrioles and the base of the flagellum move towards the nucleus and become associated with its caudal pole, at a spot known as the implantation fossa, where there is a local thickening on the outer aspect of the nuclear envelope known as the basal plate. So far the proximal centriole has retained its structure of nine evenly spaced sets of triple microtubules. As the nuclear chromatin condenses, dense material is deposited between the triplets. This dense matrix is most pronounced on the nuclear side and extends outward to become confluent with one end of the thin sheet of dense material mentioned earlier as the anlage of the capitulum. The fusion of this dense lamina with protrusions from the centriole continues until the capitulum is complete. The dense material also spreads into the proximal centriole, so that in some species it becomes almost unrecognizable, although in other animals it is still clearly visible in mature spermatozoa. At one end the proximal centriole develops a cylindrical prolongation, known as the centriolar adjunct, similar but not identical to the centriole proper. This adjunct persists until the late stages of spermiogenesis when it disappears suddenly, and its significance is not apparent.

The distal centriole, after giving rise to the flagellum, also undergoes profound changes and ultimately disappears as a distinct entity. Its cylindrical shape is obvious when the sperm flagellum is newly-formed, but as development proceeds, the walls bow out at the proximal end and the central pair of microtubules extend through its cavity to contact the wall of the proximal centriole, whose long axis always coincides with the plane of flattening of the fully condensed nucleus. Dense material, similar to that formed near the proximal centriole, is deposited near the distal centriole. The dense material which appears near the two centrioles soon develops a periodic structure and joins with the cranial ends of the dense outer fibres. Consequently there are nine cross-striated longitudinal columns at the caudal end of the neck, but towards the capitulum, the number is reduced to seven by the coalescence of two on either side to form major columns. The other five continue as minor columns and all converge upon and merge with the capitulum. The actual connection between the capitulum and the sperm nucleus appears to be formed by numerous fine filaments which are oriented perpendicular to the two surfaces, and which cross the apparently clear zone in between (Fawcett and Phillips, 1969b; Woolley and Fawcett, 1973).

**7.1.3.4    Manchette and nuclear ring**    The manchette or caudal sheath is a transient sleevelike structure composed of laterally associated microtubules. Just before the chromatin begins to condense, the manchette appears around the caudal pole of the nucleus, extending back into the cytoplasm behind the nucleus (Figure 7.7). At

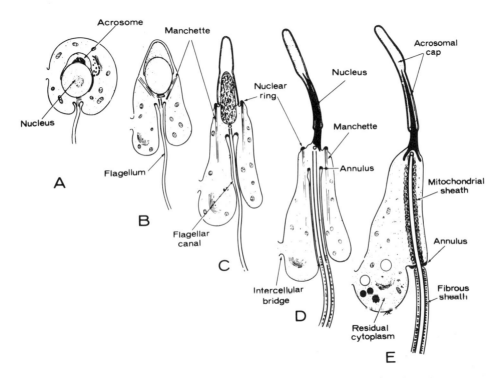

Figure 7.7 Diagram of the development and disappearance of the manchette of a guinea pig spermatozoon. (A) The nucleus and acrosomal cap of the spermatid are centrally located in the spermatid cytoplasm in the Golgi and cap phases of development. (B) With completion of the acrosomal cap, the Golgi migrates into the caudal cytoplasm, microtubules appear tangential to the caudal half of the nucleus and spermatid elongation begins. (C) Caudal displacement of cytoplasm brings the plasma membrane into close apposition to the acrosomal cap. Nuclear condensation begins; the microtubules become arranged in a cylinder originating in a dense nuclear ring at the caudal margin of the acrosomal cap. (D) Nuclear condensation and spermatid elongation continue. The nuclear ring moves back to the level of the base of the flagellum; the flagellar canal lengthens and a fibrous sheath begins to form around axoneme of the free portion of the flagellum. (E) Nuclear ring and manchette disappear; annulus migrates caudally to the anterior margin of the fibrous sheath; mitochondria gather around the base of the flagellum anterior to the annulus and form the mitochondrial helix of the middle piece. (Reproduced from Fawcett et al., 1971)

the same time as the cell membrane makes contact with the other acrosomal membrane, the microtubules of the manchette take up a position tangential to the caudal half of the nucleus and the cell membrane near the posterior margin of the acrosomal cap becomes locally specialized with a layer of dense fibrillar material on its cytoplasmic surface. This encircles the cell and is known as the nuclear ring. The proximal ends of the manchette microtubules are embedded in the nuclear ring and gradually change their orientation, starting off oblique and then becoming parallel to the flagellum. The cylindrical manchette excludes the mitochondria from the region of the centrioles but there are many vesicles and cisternae associated with both the inner and outer aspects of the manchette.

The annulus and the nuclear ring are thus similar in providing points at which the cell membrane is fixed to the axial components of the spermatid. Finally, as nuclear condensation nears completion, the nuclear ring migrates from its initial position at the margin of the acrosomal cap, back to the level of the posterior margin of the nucleus. Then it and the manchette disappear (Burgos and Fawcett, 1955; Courot and Flechon, 1966; Fawcett *et al.*, 1971).

The function of the manchette is obscure. It has been suggested that the microtubules of the manchette may be important in elongation of the spermatid, but the tubules seem to terminate posteriorly far short of the caudal end of the cell. It also seems unlikely that the manchette plays any part in the shaping of the spermatid nucleus particularly as the orientation of the manchette is similar in marsupials, although in these animals the sperm head is flattened in a plane perpendicular to the axis of the tail, i.e. in a direction at right angles to that in eutherian mammals (Sapsford *et al.*, 1967; Phillips, 1970). Instead it seems more likely that the manchette provides a track for the translocation of cytoplasm from the anterior to the posterior part of the cell and in this function the vacuoles may be especially significant (Fawcett *et al.*, 1971).

### 7.1.3.5   Nuage, the chromatoid body and its satellite
Germ cells, from the time they are found in the epithelium of the embryonic gut, contain dense fibrous material, known as 'nuage'. This is either in the form of small discrete bodies in the cytoplasm or 'cementing material' in the interstices of mitochondrial clusters. It is found in gonocytes, most spermatogonia and spermatocytes, but cannot be discerned in B spermatogonia or leptotene or zygotene spermatocytes. During the latter part of pachytene, the nuage dissociates from the mitochondria to form the chromatoid body (Eddy, 1974).

Under the light microscope, the chromatoid body is a homogeneous dense body of irregular shape situated near the nucleus, the acrosomal vesicle and the Golgi complex. A spherical body of slightly different appearance is often found nearby; this is called the satellite of the chromatoid body. Under the electron microscope, the chromatoid body seems to be formed of thin filaments, either consolidated into a compact mass or dense strands of varying thickness that branch and anastomose to form an irregular network. The satellite is also filamentous, organized somewhat more loosely than those of the chromatoid body.

When the acrosomal vesicle has just attached to the nucleus, the chromatoid body and its satellite lie nearby and slender prolongations of its substance extend towards pores in the neighbouring nuclear envelope.

The chromatoid body then migrates to the caudal pole of the nucleus, forms a ring around the flagellum near the annulus, and finally disappears without contributing to any of the structures of the mature spermatozoon (Figure 7.8; Sud, 1961a, b; Eddy,

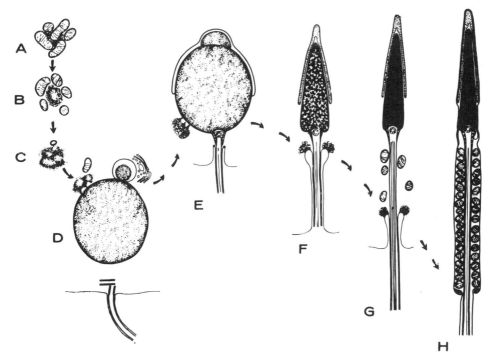

Figure 7.8 Diagram of the development and disappearance of the chromatoid body. A and B. The dense material ('nuage') which gives rise to the chromatoid body is first located in mitochondrial clusters of the spermatocytes. C. The mitochondrial clusters disperse and the dense material aggregates into the single chromatoid body. D and E. This establishes a close relationship to nuclear pores in the spermatid, and then migrates round to the base of the flagellum. F. There it forms a ring around the flagellum adjacent to the annulus. G and H. In the course of the caudal migration of the annulus, the associated chromatoid body disperses. It leaves no residue in the mature sperm. (Reproduced from Fawcett, 1972)

1970; Fawcett *et al.*, 1970). The composition and function of the chromatoid body and its satellite still remain obscure, but rapid movements of these bodies have been observed in isolated tubules (Parvinen and Jokelainen, 1974). Incorporation of [$^3$H] uridine into the chromatoid body *in vitro* could be demonstrated 14h after a 2h incubation with the radioactive precursor, but not immediately after the 2h incubation (Söderström and Parvinen, 1976).

**7.1.3.6    Mitochondria**   When the spermatids are formed they contain quite ordinary-looking mitochondria scattered through the cytoplasm. When the manchette forms, it excludes the mitochondria from the area of the developing flagellum, but when it disappears and the annulus moves away from the nucleus, the mitochondria migrate through the cytoplasm and form a double helical coil around the midpiece of the sperm. The individual mitochondria are arranged end to end so it is difficult to say how many individual organelles are involved, but the number of gyres in the sheath varies from about 10 in man and bull to about 350 in rats (Fawcett, 1970, 1975a). In some species, the mitochondria undergo extensive internal reorganization during spermiogenesis involving separation of the cristae, condensation of the intercristal matrix and the formation of large areas of clear 'pseudo-matrix' (André, 1962; Woolley, 1970). These changes are of unknown significance but resemble the changes seen in isolated mitochondria when the metabolism is altered from a slow respiration to active oxidative phosphorylation.

The annulus remains after migration at the end of the mitochondrial helix, where it forms the point at which the cell membrane is reflected onto the flagellum, and develops a shape characteristic of the species (Fawcett, 1970, 1975a).

**7.1.3.7    Metabolic changes**    During these complex structural changes, the composition of the spermatid changes, with appreciable RNA synthesis in the early stages and replacement of the nuclear histones of the somatic type, rich in lysine, by a new type of histone rich in arginine. These changes are discussed more fully in Sections 9.6 and 9.7.

## 7.2    Cytoplasmic bridges between germinal cells

One characteristic feature of the germinal cell population is that the cytoplasm does not separate completely when the nuclei divide. Instead, the cells remain linked in clones until the sperm are released. This phenomenon was noticed by the early histologists (van La Valette St George, 1865; Sertoli, 1877; von Ebner, 1888) but then largely ignored until it was demonstrated conclusively by electron microscopy (Figure 7.9). The process begins with the spermatogonia and continues as the successive cell types are formed, so that theoretically as many as 512 interconnected spermatids may result (Figures 7.1 and 7.10; Fawcett *et al.*, 1959; Dym and Fawcett, 1971; Moens and Go, 1972; Moens and Hugenholtz, 1975). In fact, the numbers of conjoined cells is usually less than this because of degeneration of individual cells, which breaks the cell chain into fragments.

During division of the nuclei, the intercellular bridges exhibit multiple parallel transverse cisternae which presumably isolate the individual cells temporarily and interrupt cytoplasmic continuity (Figure 7.10). These also prevent the spread of degeneration throughout the entire clone, as degeneration usually occurs during actual division (Dym and Fawcett, 1971).

One clone of cells may form the basis of the unit segments suggested as the basis of the spatial wave of spermatogenesis (see Section 7.4.2). It has been estimated that some 40 clones enter meiosis simultaneously in each square millimetre of seminiferous epithelium, but complete synchrony may not occur in the mitotic divisions during the development of a clone.

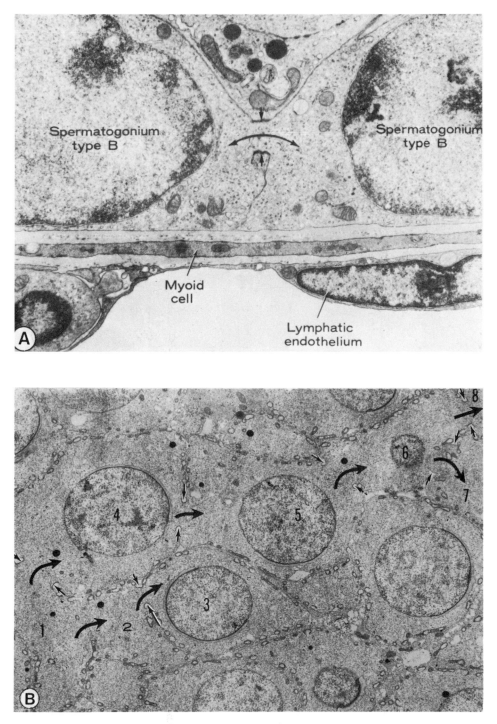

Figure 7.9

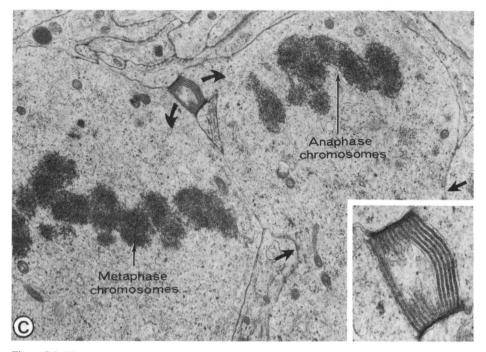

Figure 7.9 Electron micrographs of intercellular bridges. A. Between spermatogonia. B. Between spermatids. Eight different spermatids are joined together by seven bridges in the plane of this thin section. C. Bridge between two dividing spermatogonia showing the multiple transverse cisternae which form a temporary barrier between the two cells. It is interesting that the two cells are at slightly different stages in their division, one at metaphase and the other at anaphase with the constriction between the daughter cells already beginning (arrows). (Reproduced from Dym and Fawcett, 1971)

Linked spermatids may also be the origin of the multinucleated cells which so commonly occur following damage to the germinal epithelium (see Chapter 11).

### 7.3    Release of spermatozoa

When the spermatid has finally assumed all the characteristic features of the spermatozoa, its head is embedded in the luminal surface of the cytoplasm of the Sertoli cell, with its tail protruding into the lumen (Figure 7.11). The cytoplasm which still surrounds the midpiece is drawn out to one side and is also embedded into the Sertoli cell cytoplasm, but at a slightly different site from the head (Fawcett and Phillips, 1969a). The cytoplasmic 'neck' joining the midpiece with the majority of the cytoplasm (the 'residual cytoplasm') is gradually attenuated as the sperm is slowly extruded towards the lumen, until finally the neck breaks, releasing the spermatozoon (with a small remnant of the cytoplasm, the cytoplasmic or kinoplasmic droplet) into the lumen (Fawcett and Phillips, 1969a) and leaving the residual cytoplasm (now known as the residual body) embedded in the cytoplasm of the Sertoli cell (Kingsley-

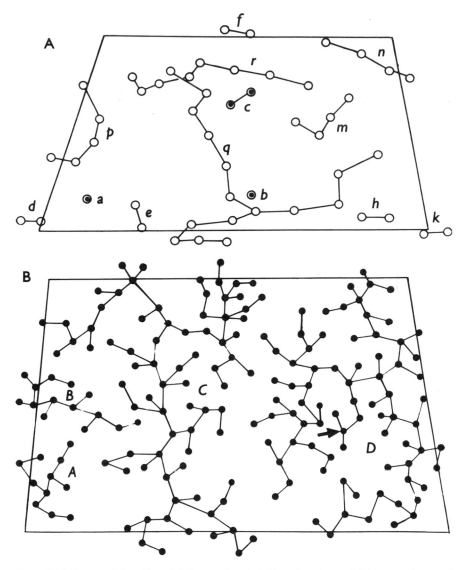

Figure 7.10  A map of the cells and their cytoplasmic bridges, based on serial electron micrographs of a piece of rat seminiferous epithelium (about 0·05mm³) at Stage I of the cycle. Cells are represented by circles and bridges by lines. Only cells within the boundaries were examined in their entirety.
A. Spermatogonia. Each syncitium is lettered. Groups *a*, *b* and *c* are dense type A spermatogonia.
B. Spermatocytes, overlying the spermatogonia shown in A. The arrow marks a branch where one of the terminal cells has failed to divide or a cell was lost. (Reproduced from Moens and Hugenholtz, 1975)

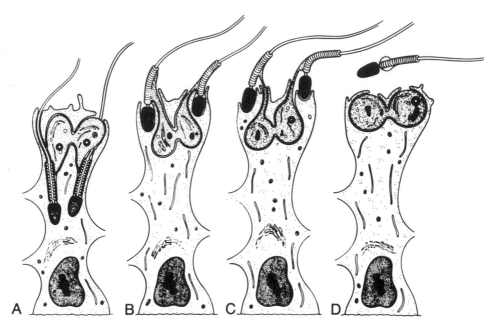

Figure 7.11 Diagram illustrating successive stages in sperm release. A: elongated spermatids deeply embedded in the Sertoli cell with cytoplasm caudal to the middle piece. B: extrusion of axial components of spermatids while lobules of cytoplasm are held in the epithelium. C: continued extrusion of sperm head resulting in attenuation of the slender stalk connecting the cytoplasm to the neck region. D: breaking of the stalk with retraction of its proximal portion to neck region forming the cytoplasmic droplet and leaving behind rounded residual bodies still held in the epithelium. (Reproduced from Fawcett, 1975b)

Smith and Lacy, 1959; Firlit and Davis, 1965; Dietert, 1966; Sapsford *et al.*, 1969b). The spermatid appears to be relatively passive during this process which seems to depend largely on the activity of the Sertoli cells, although specialized anchoring devices on the spermatids have been described recently (Russell and Clermont, 1976; Figure 7.12).

The large numbers of spermatids shed at one time in one section of the seminiferous tubule would therefore require some degree of coordination between a number of Sertoli cells, but the mechanism for this is not known at present.

## 7.4    Cell associations and the spermatogenic cycle

One of the most remarkable facts about spermatogenesis is that in any given species, certain germinal cells are always found in association with one another, and never with certain other types of cell. These associations were first recognized by von Ebner (1871), Benda (1887) and Regaud (1901). They extend for considerable lengths of the tubule in many species, e.g. rat, rabbit, ram, bull, pig and some marsupials, so that whole cross-sections of a tubule contain the same cell association. In

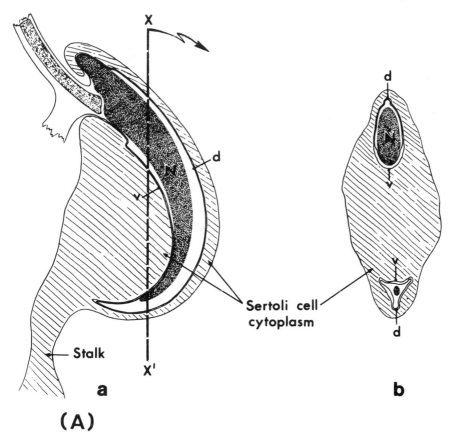

**(A)**

Figure 7.12 A. Diagram illustrating the relationship of a step 19 spermatid, just before release, with an apical cytoplasmic process of a Sertoli cell. This large cytoplasmic process is connected to the rest of the Sertoli cell by a narrow stalk. b. A section through the plane XX'. N, nucleus of spermatid; v, ventral and d. dorsal surface of spermatid head. (Reproduced from Russell and Clermont, 1976)

humans, on the other hand, the individual associations cover much smaller areas of the tubular wall, so that in any single cross-section, several associations can be discerned. These associations include all the germinal cell types in the epithelium, except one very rare type of spermatogonium, the $A_0$ or $A_s$ spermatogonium (see Section 7.5), and the later mitotic divisions of the spermatogonia and the meiotic division are always associated with certain other cell types (see Roosen-Runge, 1962, 1977; Courot et al., 1970; Clermont, 1972).

Two principal classifications of these associations are in current use. One depends on the general features of all cells in the association and usually recognizes eight stages which are conventionally numbered 1 to 8 in arabic numerals, counting from the shedding of the mature spermatozoa (Figure 7.13; Curtis, 1918; Roosen-Runge and Giesel, 1950; Ortavant, 1959; Roosen-Runge, 1962; Courot et al., 1970). The other, while recognizing the constancy of the cell associations, classifies them according to the characteristics of the acrosome of the spermatids, which are revealed by the periodic acid-Schiff stain. The spermatids can be classified into nineteen separate

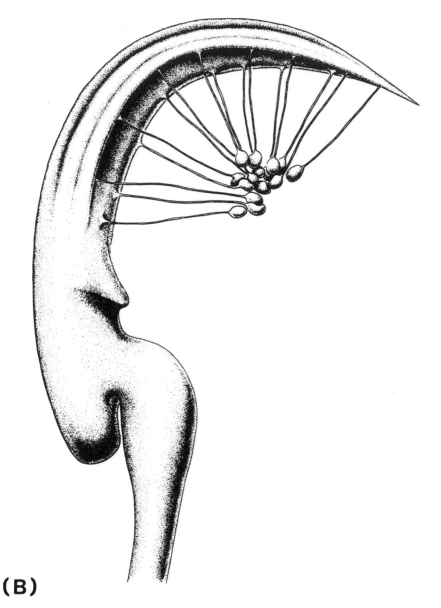

**(B)**

Figure 7.12  B. Three-dimensional drawing of the external form of a step 19 spermatid of the rat, showing the tubulobulbar processes extending from the ventral aspect of the sickle-shaped head. These processes are thought to anchor the head of the spermatid into the apical Sertoli cell cytoplasm. None of the Sertoli cell component of the tubulobulbar complex is shown. Disintegration of the complex would be followed by the release of the spermatozoon into the lumen of the seminiferous tubule. (Reproduced from Russell and Clermont, 1976)

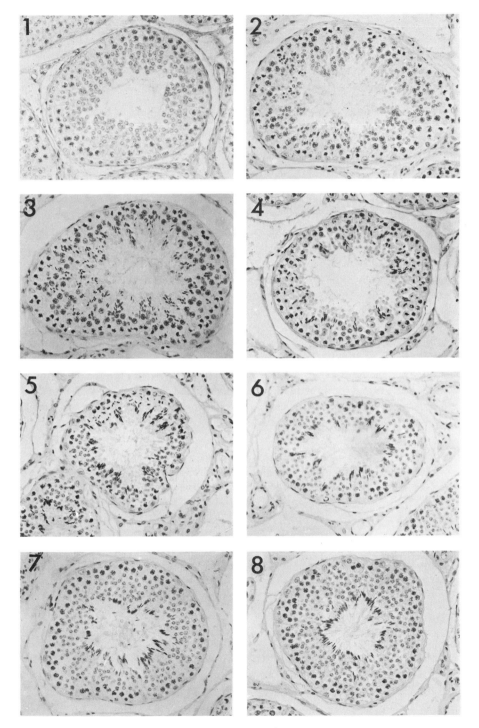

Figure 7.13

steps in the rat, but some cell associations have two generations of spermatids (e.g. step 1 with step 15 and step 6 with step 18, etc.) so that 14 cell associations are recognized. These are usually numbered in the rat I to XIV in Roman numerals, counting from the meiotic division (Figure 7.14; Leblond and Clermont, 1952a, b). In the monkey twelve associations can be recognized (Clermont and Leblond, 1959; Clermont, 1969; Clermont and Antar, 1973) and in man six (Clermont, 1963, 1966). The two schemes can be reconciled (Figure 7.15).

### 7.4.1    The spermatogenic cycle in time

This constancy of the cell association involving four or five generations of cells means that the various steps in spermatogenesis are closely synchronized. Therefore, if an observer could view the same section of tubule over a period of time, it would change from one cell association to the next and so on in a cycle, completing one turn of the cycle when the same cell association reappears. The whole portion of the tubule at one stage would not necessarily change to the next stage at the same time; the 'unit segment' involved (see p. 210) may be associated with one clone of cells derived by the persistent cytoplasmic bridges between daughter cells.

The time that each stage lasts, as a fraction of the whole cycle, can be estimated by classifying a large number of randomly chosen cross-sections of tubules (see Roosen-Runge, 1962; Courot et al., 1970). The assumption involved is that if the choice is truly random, then the chance of finding one particular stage is greater if that stage lasts longer (Figure 7.15). As will be seen in Section 7.4.2, this does not apply to short lengths of a single tubule.

The length of the whole cycle can be determined because if tritiated thymidine is injected into an animal, or directly into its testis or the testicular artery, the most advanced cell types to incorporate radioactivity into DNA are the pre-leptotene or early leptotene spermatocytes (Courot et al., 1970). Therefore if sections of testes are

---

Figure 7.13 Photographs of cross-sections of seminiferous tubules of a ram testis showing the eight stages defined by Curtis (1918), Roosen-Runge and Giesel (1950) and Ortavant (1959).
Stage 1. From the end of the spermatozoa release into the lumen until the beginning of the spermatid nuclei elongation. It is characterized by the presence of spermatids with round nuclei only.
Stage 2. From the elongation of the spermatid nuclei up to the formation of the bundles of spermatids. This is the phase of nuclear elongation of the spermatids.
Stage 3. From the formation of the first elongated spermatid bundles in the Sertoli cell cytoplasm up to the first maturation divisions.
Stage 4. From the appearance of the first divisions to the disappearance of the second maturation divisions.
Stage 5. From the end of the last maturation divisions up to the appearance of dusty chromatin in the nuclei of the young spermatids: during this stage the latter have a small nucleus containing some karyosomes connected by a chromatin network.
Stage 6. From the appearance of the dusty chromatin in the young spermatids up to the migration of the bundles of elongated spermatids toward the lumen of the seminiferous tubules.
Stage 7. From the beginning to the end of the centripetal migration of the elongated spermatids toward the lumen.
Stage 8. From the end of the migration of the spermatids to their complete release as spermatozoa into the lumen.
The tubule labelled Stage 5 is mostly in this stage, but a small segment is still in Stage 4.

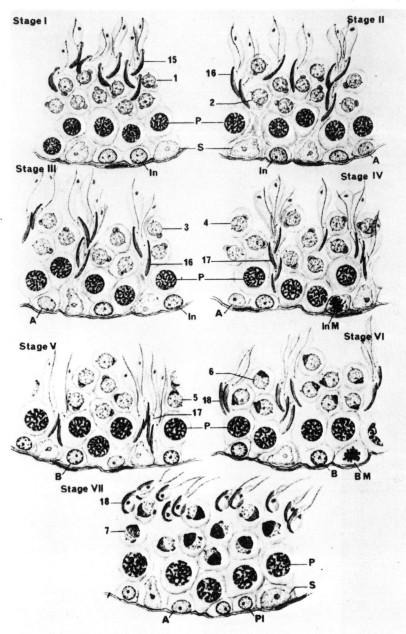

Figure 7.14  (continued overleaf) Diagram of the 14 stages of the cycle of the seminiferous epithelium of the rat based on the morphology of the acrosome, as revealed by the periodic acid—Schiff stain.
S, Sertoli cell. A, A type spermatogonia. In, intermediate type spermatogonia. B, B type spermatogonia. M, mitosis. Pl, pre-leptotene primary spermatocyte. L, leptotene. Z, zygotene. P, pachytene. Di, diplotene primary spermatocyte. IM, primary spermatocyte dividing. II, secondary spermatocyte. IIM, secondary spermatocyte dividing. 1 to 19, successive stages of spermatids. (See also Figure 7.3A)

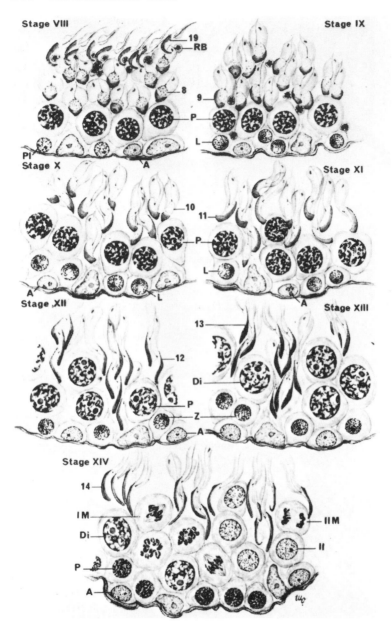

Stage I. From the appearance of new spermatids until the formation of pro-acrosomic granules in their idiosome which is then uniformly stained.

Stage II. From the appearance of the pro-acrosomic granules in the idiosome of the young spermatids until their complete fusion into a single one.

Stage III. From the formation of a single round acrosomic granule resulting from the fusion of the pro-acrosomic granules until its flattening.

examined at various times afterwards, the most advanced cells which are labelled will have been pre-leptotene spermatocytes when the label was injected (Figure 7.16). From this information, and a knowledge of the relative lengths of the various stages of the cycle, a precise estimate can be made of the length of one cycle. The duration of the whole spermatogenic process can also then be calculated, from the length of one cycle and the fact that from the appearance of the A spermatogonia to the release of the spermatozoa is about 4·5 turns of the cycle, and from the beginning of the meiotic prophase to the liberation of the spermatozoa is a little over three complete cycles.

## 7.4.2 The spermatogenic wave in space

Regaud in 1901 made the statement that 'l'onde est dans l'espace ce que le cycle est dans le temps'. Unfortunately this elegantly simple hypothesis has not been substantiated by subsequent observations. There is no doubt that, in several species, successive stages of spermatogenesis are found in a descending order along the tubule, beginning at the rete testis; the stage immediately next to the rete can be any one of the eight or fourteen recognized. However there are commonly modulations in the wave where the order is reversed for a short length, and then reverts to its original order (Figure 7.17). Nevertheless even during these modulations, adjacent stages in space are always adjacent numbers in the cycle (Perey et al., 1961; Hochereau, 1963a). As the tubules are usually two-ended and this process also begins at the other end of the tubule, the two

---

Stage IV. From the hemispherical acrosomic granule flattening over the young spermatid nucleus to the clear-cut differentiation of the head cap.

Stage V. From the beginning of the head cap differentiation through its extent over the nucleus.

Stage VI. The nuclear cap covers from one-fourth to one-third of the head. The old elongated spermatid bundles move toward the tubular lumen.

Stage VII. From the maximum size of the head cap to its orientation toward the basement membrane. The head cap covers half the nucleus. The B type spermatogonia divide giving young pre-leptotene spermatocytes. The old elongated spermatids have lost their bundle arrangement and are found near the lumen.

Stage VIII. From the orientation of the young spermatid acrosomic system toward the basement membrane until the initiation of its nuclear flattening: the spermatozoa are then being released into the tubular lumen.

Stage IX. From the appearance of the flattening and slight assymetry of the spermatid nucleus until its elongation. $A_1$ spermatogonial divisions then occur.

Stage X. From the spermatid nucleus elongation to the protruding of the acrosome at the tip of the nucleus. The caudal end of the acrosome reaches the lower part of the nucleus.

Stage XI. From the protruding of the acrosome at the tip of the nucleus in the form of a round tube to its triangular-shaped appearance. The spermatid nuclei are markedly elongated and curved.

Stage XII. From the appearance of the triangular shape of the acrosomal protrusion until the flattening of the acrosome itself along the dorsal edge of the nucleus. The spermatid nuclei are less curved in shape, narrow and more stainable. They form bundles of spermatids attached to the Sertoli cells.

Stage XIII. From the flattening of the acrosome along the dorsal edge of the nucleus to the longitudinal contraction of the spermatid nuclei.

Stage XIV. From the longitudinal contraction of the nuclei to two-thirds of their former length through the appearance of new young round spermatids; the older primary spermatocytes undergo division to give rise to secondary spermatocytes. These latter soon divide again. A new cycle of the seminiferous epithelium then begins. (Reproduced from Clermont, 1972)

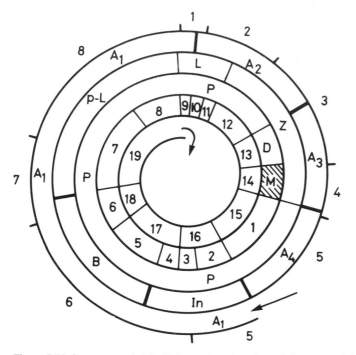

Figure 7.15 Spermatogenic 'clock' for rat based on data of Courot *et al.* (1970) and Clermont (1972) for the frequency of occurrence of each of the associations shown in Figures 7.13 and 7.14. The duration of stages 1 to 8 (Figure 7.13) are shown around the outside, stages I to XIV (Figure 7.14) correspond to the equivalent stage of development of the spermatids. These stages of development are shown as arabic numbers 1 to 19 near the centre; spermatids of stages 15 to 19 coincide with those of stages 1 to 8, as shown.

It must be emphasized that this diagram does not represent a progression of cells around the wall of an individual tubule. The cell types which can be read off along any radius are those found in association with one another, and the angle formed at the centre (as a fraction of a complete revolution) by each cell type or association indicates the fraction of one cycle taken up by this cell type or association. $A_1$, $A_2$, $A_3$, $A_4$, In and B: types of spermatogonia with the heavy dividing lines representing the mitotic divisions at the end of each stage.

p-L, L, Z, P and D: preleptotene, leptotene, zygotene, pachytene and diplotene primary spermatocytes. M; the meiotic divisions and the secondary spermatocytes.

The outer arrow indicates the entry of $A_1$ spermatogonia into the system without implying anything about their origin.

(See Figures 7.20 and 7.21)

waves meet at a site of reversal, somewhere about the centre of tubule (Figure 7.18; Perey *et al.*, 1961). This site appears to have no particular anatomical characteristic and its situation may be determined by chance. Even when the tubule branches, as it does more often in the bull than in the rat, the order is unbroken. The order of stages can be recognized in whole isolated tubules as well as in serial or longitudinal sections.

It is most important to realize that the length of a single tubule in any stage of the cycle bears little relationship to the relative duration of that stage. At first sight this is a surprising finding, as the relative duration is determined by the frequency with

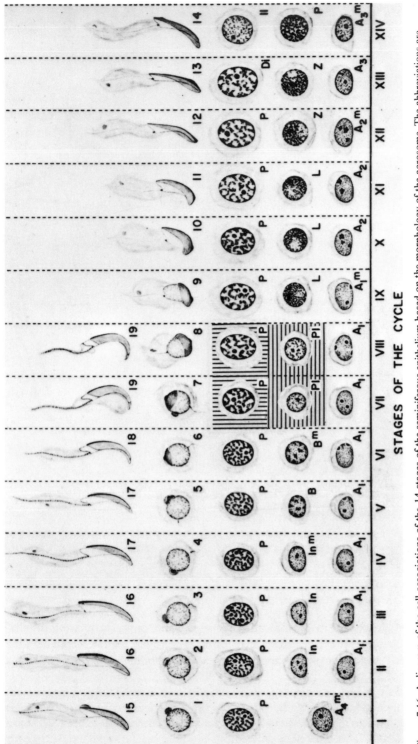

Figure 7.16 A diagram of the cell associations of the 14 stages of the seminiferous epithelium based on the morphology of the acrosome. The abbreviations are as in Figure 7.14. This diagram also illustrates the most-developed cells which would have incorporated ³H-thymidine (injected one hour previously) into DNA (pre-leptotene spermatocytes at stages VII and VIII of the cycle, horizontal shading) and the most-developed cells which could have incorporated ³H-thymidine (injected 12 days previously) into DNA (pachytene spermatocytes at stages VII and VIII, vertical shading). (Modified from a diagram by Dym and Clermont, 1970)

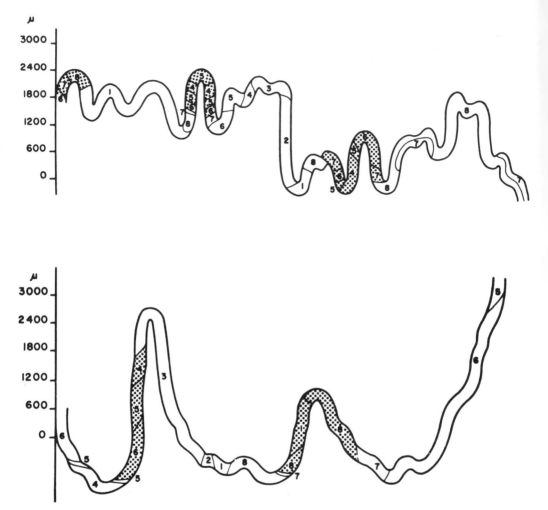

Figure 7.17 The arrangement of stages (classified according to the Curtis scheme) along sections of seminiferous tubules of a bull. Note that in general adjacent stages are also consecutive but there are several 'modulations' (stippled) where the progression is reversed. Nevertheless, even in these modulations, two adjacent patches of tubular wall are always one stage later or one stage earlier. (Reproduced from Hochereau, 1963a)

which that stage occurs in a large number of random cross-sections of tubules. Although each wave and each tubule may contribute to the overall frequency, the lengths of the stages in individual waves can reveal substantial departures from the overall average.

This situation is best explained if each clone of cells formed from a single stem spermatogonia, and linked by cytoplasmic bridges, forms a unit of synchronized cells (Figure 7.19). A series of such adjacent units could be at the same stage but could have entered at different times. Therefore different units would reach the end of the time for

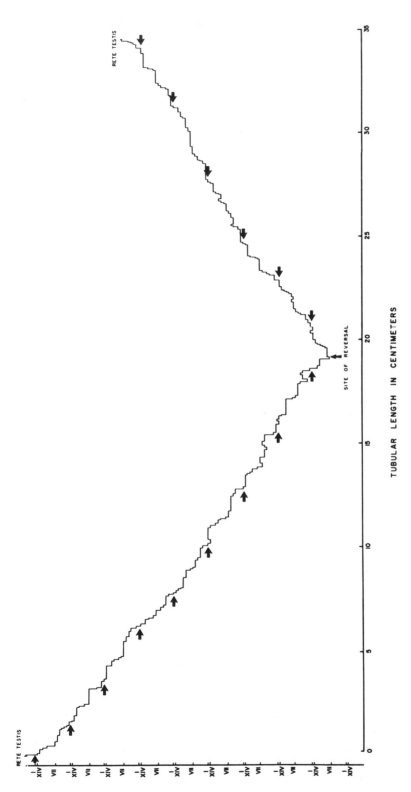

Figure 7.18 The stages (classified according to the Leblond–Clermont scheme) along the entire length of one seminiferous tubule or a rat, beginning and ending at the rete testis. Note that the stages in general descend from the rete at both ends of the tubule. There are 8 complete waves with 5 modulations in one arm of the tubule and 5 complete waves with 4 modulations in the other. (Reproduced from Perey *et al.*, 1961)

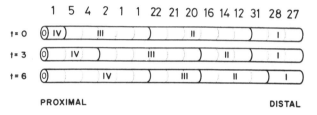

Figure 7.19 A diagram of a seminiferous tubule at 3 times, 0h (t = 0) 3h (t = 3) and 6h (t = 6). The 'unit-segments' are divided by dotted lines, and the boundaries between stages by heavy lines. 'Proximal' and 'distal' refer to the relation with the rete. It is assumed that all of one unit-segment progresses simultaneously from one stage to the next. However not all adjacent unit-segments which happen to be at the same stage at any particular time entered that stage at the same time, and therefore will not leave it together. If it is assumed that Stage I lasts 35 h, Stage II 23h, Stage III 6h and Stage IV 13h, then the observed pattern (an increasing and then decreasing length of tubule in Stage III and a decreasing and then increasing length of tubule in Stage II) would be the result if the various segments entered at the times shown at the top, in hours before time 0, into the stages shown at time 0. Based on a diagram by Perey *et al.* (1961)

that stage at different times, so the length of tubule at that stage would vary. Even if it were a very short-lived stage, if a large number of adjacent units had entered that stage just before the tissue was fixed, then the length of tubule at that stage, as a proportion of the length of that particular wave in that tubule, would be greater than the duration of the stage as a proportion of the duration of a whole cycle (Perey *et al.*, 1961).

This scheme does not explain why only stages with consecutive numbers, either one higher or one lower, are found next to one another. This is presumably a consequence of the way in which spermatogenesis is established at puberty or at the beginning of the breeding season. However the existence of some factor progressing along the tubule and producing the orderly sequence is suggested by experiments in which focal lesions are produced in the tubules. These lesions are followed by a progressive depopulation which spreads along the tubule but only in the direction of the rete (Suoranto, 1971; Setchell *et al.*, 1978; but cf. Hall and Hupp, 1970). There are obviously many questions yet to be answered.

## 7.5     Spermatogonial renewal

In the testis, unlike the ovary, there is apparently continual recruitment of cells to begin meiosis. It is always presumed that these cells come from a population of stem cells which maintain their own numbers while producing cells for the reduction division. However there is a disagreement concerning the exact method by which this is achieved. Two schemes have been proposed. In the first (Figure 7.20) Clermont (1972) has suggested that, in the rat, the $A_4$ spermatogonia divide either to give the two intermediate spermatogonia or to give two $A_1$ spermatogonia. The intermediate spermatogonia go on to yield B spermatogonia and spermatocytes; the $A_1$ revert to the beginning of the spermatogenic process and then divide again to give $A_2$, $A_3$ and again $A_4$ spermatogonia. No one has suggested how the individual $A_4$ spermatogonia

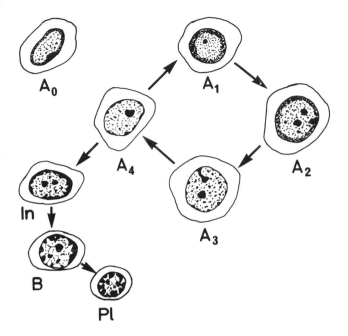

Figure 7.20 Clermont's scheme for spermatogonial renewal. A$_1$ spermatogonia divide to form A$_2$, A$_3$ and then A$_4$ spermatogonia, some of which divide to form A$_1$ spermatogonia, while most of them divide to form Intermediate (In) and B spermatogonia and then pre-leptotene (Pl) spermatocytes. There is a population of reserve or stem A$_0$ spermatogonia which divide only if the other spermatogonia are reduced in number, for example following X-irradiation, when the A$_0$ spermatogonia repopulate the tubules. (Based on drawings by Clermont and Bustos-Obregon, 1968, and Dym and Clermont, 1970)

know which course to take. Another type of spermatogonia has also been described, the A$_0$ spermatogonia, which Clermont believes either do not divide at all, or divide very rarely and their divisions are not synchronized with other events in the spermatogenic cycle (Clermont and Bustos-Obregon, 1968; Clermont and Hermo, 1975). Only when the A$_1$ to A$_4$ spermatogonia are depleted, for example by X-irradiation, do the A$_0$ spermatogonia begin to divide and thus repopulate the tubule (Dym and Clermont, 1970).

   The other scheme (Figure 7.21), proposed by Huckins (1971a, b) and Oakberg (1971a, b), recognizes the same cell types but differs in the role they ascribe to the A$_0$ or, as they call them, A$_s$ spermatogonia. It also denies that A$_4$ spermatogonia can give rise to A$_1$ spermatogonia. Instead it suggests that A$_s$ spermatogonia are always dividing, irregularly but at an appreciable rate, either to give other isolated A$_s$ cells, or pairs or chains of aligned (A$_{al}$) spermatogonia. In Stage V the population of A$_{al}$ has reached a maximum and mitosis ceases; these cells then transform morphologically into A$_1$ spermatogonia which at stage IX enter the first of their mitotic divisions that is synchronized with other events in the cycle. Clermont apparently does not recognize a separate class of A$_{al}$ spermatogonia but considers that they are instead 'tight linear groups of type A$_2$, A$_3$ and A$_4$ cells in G phase' and no

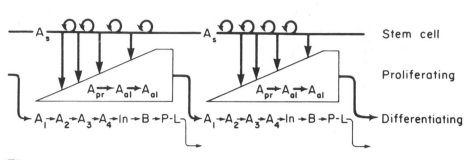

Figure 7.21 Huckins' and Oakberg's scheme for spermatogonial renewal. According to these authors, the $A_0$ or as they call them $A_s$ spermatogonia are continually dividing to maintain their numbers, or to give a pair of spermatogonia ($A_{pr}$). These divide again to give four aligned ($A_{al}$) spermatogonia which may divide again. The divisions of these cells are not linked to the cycle of the seminiferous epithelium, but at a certain point in the cycle, all the groups of aligned spermatogonia become $A_1$ spermatogonia and thenceforth their divisions are synchronized with the cycle. What determines whether an $A_s$ spermatogonia divides to give two more $A_s$ or a pair of $A_{pr}$ spermatogonia is presumably whether or not the cytoplasm divides completely with the nucleus. (Reproduced from Huckins, 1971b)

general agreement has been reached on which of these schemes is correct. Huckins scheme may be supported by Clermont's data showing that the increase in $A_1$ spermatogonia occurs in stages II to V, long after the $A_4$ spermatogonia have divided, and the Huckins–Oakberg scheme is certainly more appealing from a logical point of view. The observations suggesting the existence of a spermatogonial chalone (see Section 7.6.2.2) would be equally relevant for both schemes.

In the bull and ram, it appears that $A_1$ spermatogonia are formed both from $A_0$ and $A_2$ spermatogonia (Hochereau, 1970, 1976).

The existence of single and twin A spermatogonia and chains of between 4 and 13 A spermatogonia has been confirmed by careful examination of some serial electron microscopical sections (Moens and Go, 1972; Moens and Hugenholtz, 1975). These authors also conclude that 'regions of seminiferous epithelium which are identical in terms of spermatid and spermatocyte criteria have, in fact, quantitative and qualitative differences in their spermatogonial population'. They conclude that 'there is a periodic build-up of spermatogonia which then produce several successive quanta of spermatocytes and when the spermatogonia are depleted, the process repeats'. However, they do not make clear what the relation is between their shorter syncytia of spermatogonia and the $A_{al}$ spermatogonia of Huckins and Oakberg. In any case, their conclusions are based on only a very small area of seminiferous tubules and more observations must be made before the significance of these results can be appreciated.

## 7.6     Rate of production of spermatozoa

As the timing of spermatogenesis is apparently constant for any given species, two factors determine the number of spermatozoa produced by the testis. These are the weight of the testes and the sperm production per unit weight of testis. Alternatively it can be considered that sperm production is determined by the number of spermatogonia feeding into the synchronized cell divisions and the extent of cell degeneration at each stage. Factors deciding the numbers of spermatogonia can only be determined when the exact nature of spermatogonial renewal is decided as different control mechanisms would apply if the $A_1$ spermatogonia are formed by de-differentiation of some of the $A_4$ spermatogonia or if they are formed from stem spermatogonia (see Section 7.5). Degeneration is obviously of considerable importance and has been demonstrated at various stages of spermatogenesis (see Ortavant, 1959; Barr et al., 1971; Roosen-Runge, 1977; Russell and Clermont, 1977) by showing that the observed numbers of the later cell types do not correspond to the calculated numbers (see Figure 7.22).

Sperm production rates can be measured:

(1) directly, by cannulating the rete testis (see Section 8.2.1) and counting the spermatozoa in the fluid leaving the testis during a certain time;
(2) by counting the spermatozoa in each testis, at a known time (16–24h) after ligation on one side of the efferent ducts (see Section 8.2.5) (Setchell et al., 1973);
(3) by quantitative histology of the testis and a knowledge of the kinetics of spermatogenesis (Amann, 1970);
(4) by counting the spermatozoa in ejaculates, collected sufficiently frequently that no sperm escape in the urine or degenerate in the epididymis (Amann, 1970);
(5) by counting the spermatozoa in a daily collection of urine, treated so that the sperm do not agglutinate (Lino et al., 1967).

### 7.6.1     Factors affecting sperm production

**7.6.1.1     Species**     The location of the testis seems to bear little relation to the characteristics of spermatogenesis (Figure 7.23), although very little quantitative information is available from species with abdominal or inguinal testes.

**7.6.1.2     Age and season**     Huckins (1965) has reported that the duration of the spermatogonial divisions is shorter in rats at puberty than in adult rats, although there was no difference in the duration of spermiogenesis. This interesting observation does not appear to have been followed up.

The testes continue to enlarge for some time after puberty, presumably by the recruitment of more spermatogonia, and therefore the total production of spermatozoa increases although the production per unit weight of testis rapidly reaches a maximal value (Attal and Courot, 1963; MacMillan and Hafs, 1968; Setchell and

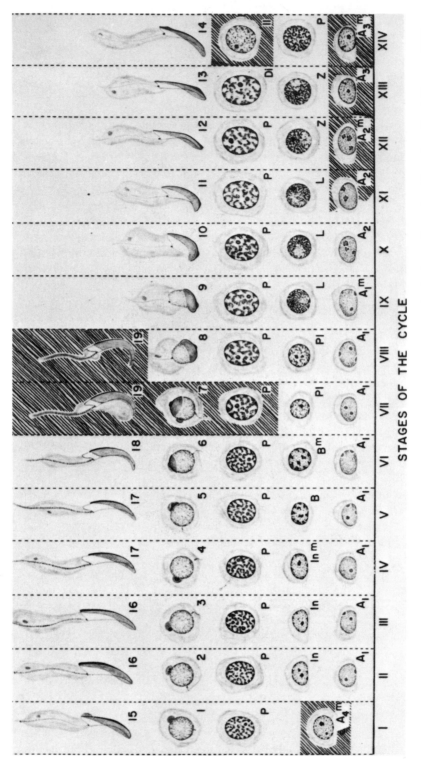

Figure 7.22 A diagram illustrating the cell types (shaded) which are most susceptible to degeneration in the rat (Reproduced from Russell and Clermont, 1976). Degeneration of the A spermatogonia, of pachytene spermatocytes and of cells during meiosis has also been reported in the ram and bull (Ortavant, 1956a; Attal and Courot, 1963; Courot, 1971) but in the ram, under optimal conditions, degeneration of spermatids does appear to be important (Courot, 1971).

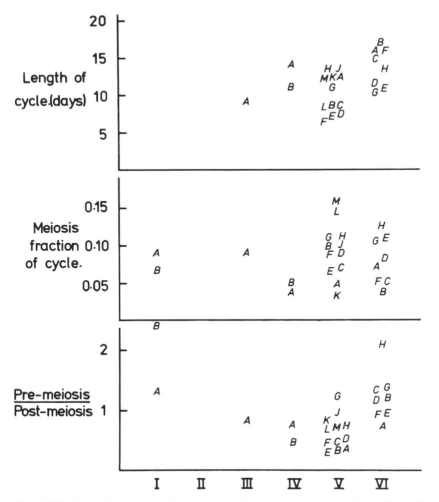

Figure 7.23 Some characteristics of spermatogenesis in mammals, grouped according to the position of their testes (see Figure 1.1).
Reproduced from Carrick and Setchell (1977) where the references for individual species are given.
IA, Indian elephant; IB, African elephant; IIIA, Hedgehog; IVA, House shrew (*Suncus*); IVB, Mole; VA, Rat; VB, Mouse; VC, Hamster; VD, Field vole; VE, Prairie vole; VF, Bank vole; VG, Rabbit; VH, Dog; VJ, Ferret; VK, Coyote; VL, Boar; VM, Stallion; VIA, Tammar wallaby; VIB, Bennett's wallaby; VIC, Brush-tailed possum; VID, Monkey; VIE, Baboon; VIF, Man; VIG, Ram, VIH, Bull.

Waites, 1971, 1972). There is also a reduction with age in the extent of degeneration during spermatogenesis (Bartke, 1970). Similarly the variations in testis weight which occur with season (see Section 1.4) are also associated with changes in total sperm production, in some species, with little change in production per unit weight of testis until spermatogenesis almost ceases (see Amann, 1970). A possible exception to this generalization may be the rabbit, in which Orgebin-Crist (1968) has observed a decrease in sperm production per unit weight of testis with no change in testis weight.

The kinetics of spermatogenesis in sheep are unaffected by artificial variations in day length (Ortavant, 1958). (See also Section 6.1.5.4.)

**7.6.1.3    Hormones and chalones**    Although certain hormones, particularly testosterone and FSH are necessary for the initiation of spermatogenesis and testosterone, at least, is necessary for its maintenance (see Sections 6.1.6.2 and 10.2.2.4) there is no conclusive evidence that the rate of production of spermatozoa, once established is under the control of any hormones (see Setchell *et al.*, 1973). This is not to say that some control does not exist and it is an area where more work is urgently needed.

The effects of unilateral castration are particularly interesting. In rats, there is no change in the concentration of FSH in the serum, although LH is higher than normal up to 30 days after hemicastration (Howland and Skinner, 1975), and there is virtually no compensatory hypertrophy of the remaining testis (Liang and Liang, 1970; Setchell and Waites, 1972), unless the castration is done before 10 days of age (Hochereau-de Reviers, 1971). However, in adult rats, from which the seminal vesicles and coagulating glands had been removed, sperm output fell after hemicastration but returned to pre-castration levels after about 28 days (Mauss and Hackstedt, 1972). In rams (Walton *et al.*, 1978), as in humans (Bramble *et al.*, 1975), FSH rises after unilateral orchidectomy and remains high long after the LH and testosterone concentrations have returned to normal. In seasonally breeding sheep, unilateral castration at the end of one breeding season results in hypertrophy of the remaining testis in the next season. There is little immediate effect if hemicastration is done during the season, but the decrease in size at the end of the season is less (Hochereau-de Reviers *et al.*, 1975). In Australian sheep, in which there is little evidence of seasonal variation in sexual activity, compensatory hypertrophy occurred within 120 days after unilateral castration and was accompanied by a measured increase in sperm production (Voglmayr and Mattner, 1968).

Gonadotrophins and androgens have no effect on the kinetics of spermatogenesis in man (Heller and Clermont, 1964) or rats (Clermont and Harvey, 1965; Desclin and Ortavant, 1963; Go *et al.*, 1971) but may be involved in regulating the extent of degeneration at several critical stages (see Section 10.2.2.4).

High doses of prostaglandins $E_1$, $E_2$, $F_{1\alpha}$ and $F_{2\alpha}$ interfere with spermatogenesis in mice and rats and PG $E_1$ and $E_2$ reduce testis weight (Ericsson, 1973; Memon, 1973; Abbatiello *et al.*, 1976; see also Section 6.1.5.2).

In many tissues, substances have been found which regulate cell division locally. These substances are known as chalones and a chalone in the testes of adult rats has been shown to reduce the incorporation of tritiated thymidine into A spermatogonia of X-irradiated or young (33 day-old) rats (Clermont and Mauger, 1974). This substance may be of importance in controlling the mitotic rate of these cells under normal conditions.

**7.6.1.4    Sexual activity**    The effects of sexual activity *per se* on sperm production seem rather uncertain. In rams and bulls, there is no effect of sexual activity on sperm production rates (Ortavant, 1958; Amann, 1962a). The rabbit appears to be unusual as sexual activity appears to produce larger testes, similar numbers of spermatids per unit weight of testis (Amann, 1970) and therefore probably a greater sperm production rate.

**7.6.1.5    Testicular grafts and temperature**    It is well known that spermatogenesis is disrupted by increasing the temperature of scrotal testes to body temperature, either by making them cryptorchid or by applying heat artificially, while androgen production is comparatively unaffected (see Section 11.1). Similarly when pieces of testes are grafted into another animal, only the interstitial tissue remains functional if the graft is made into a site which is at body temperature (see Moore, 1926). Spermatogenesis can be maintained or initiated if the graft is made into the anterior chamber of the eye (Turner, 1938), into a chamber in the ear of a rabbit (Williams, 1950) or into the testis (Turner and Asakawa, 1963; Luxembourger and Aron, 1966; Aron and Luxembourger, 1968), areas where the temperature is similar to normal scrotal values. It is also interesting that the timing of spermatogenesis is unchanged by many factors such as a short period of heating which cause disruption and even death of some of the cells (see Courot *et al.*, 1970) but is accelerated by a sustained slight increase in testicular temperature (Meistrich *et al.*, 1973).

There does not seem to be any essential differences in the kinetics of spermatogenesis in different species with scrotal or abdominal testis (see Carrick and Setchell, 1977) but only very few quantitative observations have been made on species other than rodents, ungulates and primates (Figure 7.23).

**7.6.2    Spermatogenesis in culture**
Many attempts have been made to maintain the germinal epithelium in organ culture of testicular fragments, originally using plasma clots to support and nourish the tissue. Only limited success was achieved and the germinal cells usually degenerated within a few days (see Steinberger and Steinberger, 1970). When cultures were prepared from the testes of adult rats using a stainless steel grid to support the explants in a liquid medium, tubular architecture, the Sertoli cells and primitive A type spermatogonia (a type of cell possessing intermediate characteristics between gonocytes and type A spermatogonia) were maintained for six months or more. More mature cell types survived only for a limited period, spermatocytes for several weeks and spermatids for several days. Pre-leptotene primary spermatocytes already present in the explant proceeded through meiotic prophase as far as the pachytene stage, but then did not develop any further and subsequently degenerated (A. Steinberger *et al.*, 1964; E. Steinberger *et al.*, 1964; Steinberger and Steinberger, 1970). The development of the cells *in vitro* was confirmed by labelling the spermatogonia and pre-leptotene spermatocytes *in vivo* with tritiated thymidine (Steinberger and Steinberger, 1965).

When testis fragments were taken from younger rats, tubular architecture, Sertoli cells and primitive type A spermatogonia could again be maintained for several months. No permanent damage was caused to the spermatogonia by the culture conditions, because if the explants were implanted into the testes of normal adult rats, spermatogenesis proceeded to completion in the cultured fragments within 8 to 10 weeks (Steinberger et al., 1970; Steinberger, 1975).

The addition of vitamins A, C and E and glutamine to the cultures enabled primitive type A spermatogonia to differentiate to pachytene spermatocytes, but, as with the cultures of adult testes, the cells degenerated at this stage. Subsequent waves of primary spermatocytes failed to appear (Steinberger and Steinberger, 1966a, b, 1967, 1970). The yield of spermatocytes depended on the age of the donor animals. If they were four or more days old and contained primitive type A spermatogonia at the beginning of the culture, then an appreciable percentage of the tubules contained spermatocytes. If the donor rats were newborn or one to three days old and their testes contained only gonocytes, very few spermatocytes were formed (Steinberger and Steinberger, 1970; Steinberger et al., 1970) suggesting that the transformation of gonocytes to primitive type A spermatogonia was particularly inefficient under the conditions of culture. It is particularly interesting that there are these two 'weak points' in spermatogenesis which show up under *in vitro* conditions; pachytene is also the stage affected by many noxious agents (see Chapter 11).

Cultures of certain individual cell types from the testis can also be prepared. Peritubular cells and Leydig cells can be grown in culture reasonably well (Dufau et al., 1971; de Kretser et al., 1971; Bressler and Ross, 1973; Steinberger, 1975). Sertoli cells can also be isolated and kept alive in cultures (Welsh and Wiebe, 1975; Dorrington et al., 1974; Steinberger et al., 1975), and will under these conditions secrete androgen-binding protein (Fritz et al., 1974, 1975; Dorrington et al., 1975; Steinberger et al., 1975) and oestrogens (see Section 6.1.3.1).

## References to Chapter 7

Abbatiello, E. R., Kaminsky, M. and Weisbroth, S. (1976) The effect of prostaglandins $F_{1a}$ and $F_{2a}$ on spermatogenesis. *Int. J. Fertil.* **21**, 82–8.

Abdel-Raouf, M. (1960) The postnatal development of the reproductive organs in bulls with special reference to puberty. *Acta. Endocrinol.* Suppl. **49**, 109.

Abdel-Raouf, M. (1961) The proliferation of germ cells in the testes of bull calves and young bulls. *Acta Vet. Scand.* **2**, 22–31.

Allen, E. (1918) Studies on cell division in the albino rat. III. Spermatogenesis: the origin of the first spermatocytes and the organization of the chromosomes, including the accessory. *J. Morphol.* **31**, 133–85.

Amann, R. P. (1962a) Reproductive capacity of dairy bulls. III. The effect of ejaculation frequency, unilateral vasectomy and age on spermatogenesis. *Am. J. Anat.* **110**, 49–67.

Amann, R. P. (1962b) Reproductive capacity of dairy bulls. IV. Spermatogenesis and testicular germ cell degeneration. *Am. J. Anat.* **110**, 69–78.

Amann, R. P. (1970) Sperm production rates. In *The Testis*, eds. A. D. Johnson, W. R. Gomes and N. L. Vandemark, I, 433–82, New York: Academic Press.

Amann, R. P. and Almquist, J. O. (1962) Reproductive capacity of dairy bulls. VIII. Direct and indirect measurement of testicular sperm production. *J. Dairy Sci.* **45**, 774–81.

Amann, R. P. and Lambiase, J. T., Jr. (1969) The male rabbit. III. Determination of daily sperm production by means of testicular homogenates. *J. Anim. Sci.* **28**, 369–74.

André, J. (1962) Contribution à la connaissance du chondriome. Etude de ses modifications ultrastructurales pendant la spermatogenèse. *J. Ultrastruct. Res.* Suppl. **3**, 1–185.

Aron, M. and Luxembourger, M-M. (1968) Facteurs de l'évolution spermatogénétique et de l'évolution de la glande interstitielle avant la maturité sexuelle. Etude chez le rat par l'homotransplantation du testicule néo-natal dans le testicule de l'animal mûr. *Arch. Anat. Hist. Embryol.* (Strasbourg) **51**, 43–52.

Asdell, S. A. and Salisbury, G. W. (1941) The rate at which spermatogenesis occurs in the rabbit. *Anat. Rec.* **80**, 145–53.

Attal, J. and Courot, M. (1963) Développement testiculaire et établissement de la spermatogenèse chez le taureau. *Ann. Biol. anim. Biochim. Biophys.* **3**, 219–41.

Austin, C. R. and Sapsford, C. S. (1951) The development of the rat spermatid. *J. Roy. Microscop. Soc.* **71**, 397–406.

Barr, A. B., Moore, D. J. and Paulsen, C. A. (1971) Germinal cell loss during human spermatogenesis. *J. Reprod. Fertil.* **25**, 75–80.

Bartke, A. (1970) The yield of spermatogenesis increases in maturing laboratory mice. *J. Reprod. Fertil.* **22**, 579–81.

Beaumont, H. M. and Mandl, A. M. (1963) A quantitative study of primordial germ cells in the male rat. *J. Embryol. Exp. Morphol.* **11**, 715–40.

Bedford, J. M. and Nicander, L. (1971) Ultrastructural changes in the acrosome and sperm membranes during maturation of spermatozoa in the testis and epididymis of the rabbit and monkey. *J. Anat.* **108**, 527–43.

Benda, C. (1887) Untersuchungen uber den Bau des funktionirenden Samenkanalchens einiger Saugethiere und Folgerungen fur die Spermatogenese dieser Wirbelthierklasse. *Arch. Mikroskop. Anat. Entwicklungsmech.* **30**, 49–110.

Benda, C. (1891) Neue Mittheilungen uber die Entwicklung der Genitaldrusen und uber die Metamorphose der Samenzellen. *Arch. Anat. Physiol.* 549–52.

Berndtson, W. E. and Desjardins, C. (1974) The cycle of the seminiferous epithelium and spermatogenesis in the bovine testis. *Am. J. Anat.* **140**, 167–80.

Bessesen, A. N. and Carlson, H. A. (1923) Postnatal growth in the weight of the body and of the various organs in the guinea pig. *Am. J. Anat.* **31**, 483–521.

Bishop, M. W. H. and Walton, A. (1960) Spermatogenesis and the structure of mammalian spermatozoa. In *Marshall's Physiology of Reproduction*, ed. A. S. Parkes, vol. **1**, Part 2, 1–129. London: Longmans.

Bouin, P. (1904) Sur la durée de l'établissement de la spermatogénèse chez le cheval. *C. Rend. Soc. Biol.* **11**, 658–9.

Boyd, L. J. and Van Demark, N. L. (1957) Spermatogenic capacity of the male bovine. I. A measurement technique. *J. Dairy-Sci.* **40**, 689–97.

Bramble, F. J., Broughton, A. C., Whittam, T. R., Eccles, S. S., Murray, M. A. F. and Jacobs, H. S. (1975) Independent control of follicle-stimulating and luteinizing hormone in men: results of unilateral orchidectomy. *J. Endocrinol.* **65**, 11P.

Bressler, R. S. and Ross, M. H. (1973) On the character of the monolayer outgrowth and the fate of the peritubular myoid cells in cultured mouse testis. *Exptl. cell. Res.* **78**, 295–302.

Brökelmann, J. (1963) Fine structure of germ cells and Sertoli cells during the cycle of the seminiferous epithelium in the rat. *Z. Zellforsch. Mikroskop. Anat.* **59**, 820–50.

Brown, H. H. (1885) On spermatogenesis in the rat. *Quart. J. Microscop. Sci.* **25**, 343–69.

Burgos, M. H. and Fawcett, D. W. (1955) Studies on the fine structure of the mammalian testis. 1. Differentiation of the spermatids in the cat. *J. Biophys. Biochem. Cytol.* **1**, 287–300.

Burgos, M. H. and Vitale-Calpe, R. (1967) Mechanism of spermiation in the toad. *Am. J. Anat.* **120**, 227–52.

Burgos, M. H., Sacerdote, F. L. and Russo, J. (1973) Mechanism of sperm release. In *The Regulation of Mammalian Reproduction,* eds. S. S. Segal, R. Crozier, P. A. Corfman and P. G. Cowdliffe, 166–86. Springfield: C. C. Thomas.

Bustos-Obregon, E., Courot, M., Flechon, J. E., Hochereau-de-Reviers, M. T. and Holstein, A. F. (1975) Morphological appraisal of gametogenesis—Spermatogenic process in mammals with particular reference to man. *Andrologia* **7**, 141–63.

Carrick, F. N. and Setchell, B. P. (1977) The evolution of the scrotum. In *Reproduction and Evolution,* eds. J. N. Calaby and C. H. Tyndale-Biscoes, 165–70. Australian Academy of Science, Canberra.

Challice, C. E. (1953) Electron microscope studies of spermatogenesis in some rodents. *J. Roy. Microscop. Soc.* **73**, 115–27.

Chowdhury, A. K. and Steinberger, E. (1975) A study of germ cell morphology and duration of spermatogenic cycle in the baboon, *Papio anubis. Anat. Rec.* **185**, 155–70.

Cleland, K. W. (1951) The spermatogenic cycle of the guinea pig. *Aust. J. Sci. Res.* **4** 3, 344–69.

Clermont, Y. (1954) Cycle de l'épithélium séminal et mode de renouvellement des spermatogonies chez le hamster. *Rev. Can. Biol.* **13**, 208–45.

Clermont, Y. (1960) Cycle of the seminiferous epithelium of the guinea pig. *Fertil. Steril.* **11**, 563–73.

Clermont, Y. (1962) Quantitative analysis of spermatogenesis in the rat: a revised model for the renewal of spermatogonia. *Am. J. Anat.* **111**, 111–29.

Clermont, Y. (1963) The cycle of the seminiferous epithelium in man. *Am. J. Anat.* **112**, 35–45.

Clermont, Y. (1966) Renewal of spermatogonia in man. *Am. J. Anat.* **118**, 509–24.

Clermont, Y. (1969) Two classes of spermatogonial stem cells in the monkey (*Cercopithecus aethiops*). *Am. J. Anat.* **126**, 57–72.

Clermont, Y. (1972) Kinetics of spermatogenesis in mammals: Seminiferous epithelium cycle and spermatogonial renewal. *Physiol. Rev.* **52**, 198–236.

Clermont, Y. and Antar, M. (1973) Duration of the cycle of the seminiferous epithelium and the spermatogonial renewal in the monkey, *Macaca arctoides. Am. J. Anat.* **136**, 153–66.

Clermont, Y. and Bustos-Obregon, E. (1968) Re-examination of spermatogonial renewal in the rat by means of seminiferous tubules mounted *in toto. Am. J. Anat.* **122**, 237–48.

Clermont, Y. and Harvey, S. C. (1965) Duration of the cycle of the seminiferous epitherlium of normal, hypophysectomized and hypophysectomized-hormone treated albino rats. *Endocrinology* **76**, 80–9.

Clermont, Y. and Hermo, L. (1975) Spermatogonial stem cells in the albino rat. *Am. J. Anat.* **142**, 159–76.

Clermont, Y. and Leblond, C. P. (1953) Renewal of spermatogonia in the rat. *Am. J. Anat.* **93**, 475–502.

Clermont, Y. and Leblond, C. P. (1955) Spermiogenesis of man, monkey, ram and other mammals as shown by the 'periodic acid-Schiff' technique. *Am. J. Anat.* **96**, 229–53.

Clermont, Y. and Leblond, C. P. (1959) Differentiation and renewal of spermatogonia in the monkey, *Macacus rhesus. Am. J. Anat.* **104**, 237–73.

Clermont, Y. and Mauger, A. (1974) Existence of a spermatogonial chalone in the rat testis. *Cell. Tissue Kinet.* **7**, 165–72.

Clermont, Y. and Perey, B. (1957) Quantitative study of the cell population of the seminiferous tubules in immature rats. *Am. J. Anat.* **100**, 241–60.

Clermont, Y. and Trott, M. (1969) Duration of the cycle of the seminiferous epithelium in the mouse and hamster determined by means of $^3$H-thymidine and radioautography. *Fertil. Steril.* **20**, 805–17.

Clermont, Y., Leblond, C. P. and Messier, B. (1959) Durée du cycle de l'épithélium séminal du rat. *Arch. Anat. Microscop. Morphol. Exptl.* Suppl. **48**, 37–56.

Courot, M. (1962) Développement du testicule chez l'agneau. Etablissement de la spermatogenèse. *Ann. Biol. anim., Biochim. Biophys.* **2**, 25–41.

Courot, M. (1971) Etablissement de la spermatogénèse chez l'agneau (*Ovis aries*); étude expérimentale de son contrôle gonadotrope; importance des cellules de la ligné Sertolienne. Thése de doctorat d'etat ès-sciences naturelles. Université Paris VI.

Courot, M. and Flechon, J. E. (1966) Ultrastructure de la manchette de la spermatide chez le bélier et le taureau. *Ann. Biol. anim. Biochim. Biophys.* **6**, 479–82.

Courot, M., Hochereau-de Reviers, M-T. and Ortavant, R. (1970) Spermatogenesis. In *The Testis*, eds. A. D. Johnson, W. R. Gomes and N. L. Vandemark, **I**, 339–432, New York: Academic Press.

Curtis, G. M. (1918) The morphology of the mammalian seminiferous tubule. *Am. J. Anat.* **24**, 339–94.

Desclin, J. and Ortavant, R. (1963) Influence des hormones gonadotropes sur la durée des processus spermatogénétiques chez le rat. *Ann. Biol. anim. Biochim. Biophys.* **3**, 329–42.

Desjardins, C., Kirton, K. T. and Hafs, H. D. (1968) Sperm output of rabbits at various ejaculation frequencies and their use in the design of experiments. *J. Reprod. Fertil.* **15**, 27–32.

Dietert, S. E. (1966) Fine structure of the formation and fate of the residual bodies of mouse spermatozoa with evidence for the participation of lysosomes. *Anat. Rec.* **154**, 338–9.

Dorrington, J. H., Roller, N. F. and Fritz, I. B. (1974) The effects of FSH on cell preparations from the rat testis. *Curr. Top. molec. Endrocrinol.* **1**, 237–4.

Dorrington, J. H., Roller, N. F. and Fritz, I. B. (1975) Effects of follicle stimulating hormone on cultures of Sertoli cell preparations. *Mol. Cell. Endocrinol.* **3**, 57–70.

Duesberg, J. (1909) La spermiogénèse chez le rat. *Arch. Zellforsch.* **2**, 137–80.

Dufau, M. L., Kretser, D. M. de and Hudson, B. (1971) Steroid metabolism by isolated rat seminiferous tubules in tissue culture. *Endocrinology* **88**, 825–32.

Dym, M. and Clermont, Y. (1970) Role of spermatogonia in the repair of the seminiferous epithelium following X-irradiation of the rat testis. *Am. J. Anat.* **128**, 265–82.

Dym, M. and Fawcett, D. W. (1971) Further observations on the numbers of spermatogonia, spermatocytes and spermatids connected by intracellular bridges in the mammalian testis. *Biol. Reprod.* **4**, 195–215.

Dym, M. and Romrell, L. J. (1975) Intraepithelial lymphocytes in the male reproductive tract of rats and rhesus monkeys. *J. Reprod. Fertil.* **42**, 1–8.

Ebner, H. von (1871) Untersuchung uber den Bau der Samenkanalchen und die Entwicklung der Spermatozoiden bei den Saugetieren und beim Menschen. *Untersuch. Inst. Physiol. Histol. Graz,* **2**, 200–36.

Ebner, V. von (1888) Zur Spermatogenese bei den Saugetieren. *Arch. mikrosk. Anat.* **31**, 236–92.

Eddy, E. M. (1970) Cytochemical observations on the chromatoid body of the male germ cells. *Biol. Reprod.* **2**, 114–28.

Eddy, E. M. (1974) Fine structural observations on the form and distribution of nuage in germ cells of the rat. *Anat. Rec.* **178**, 731–58.

Edwards, J. (1940) The effect of unilateral castration on spermatogenesis. *Proc. Roy. Soc.* (London) B. **128**, 407–21.

Ericsson, R. J. (1973) Prostaglandins $E_1$ and $E_2$ and reproduction in the male rat. *Adv. Biosci* **9**, 737–42.

Fawcett, D. W. (1956) The fine structure of chromosomes in the meiotic prophase of vertebrate spermatocytes. *J. Biophys. Biochem. Cytol.* **2**, 403–6.

Fawcett, D. W. (1958) The structure of the mammalian spermatozoon. *Intern. Rev. Cytol.* **7**, 195–234.

Fawcett, D. W. (1965) The anatomy of the mammalian spermatozoon with particular reference to the guinea pig. *Z. Zellforsch. Mikroskop. Anat.* **67**, 279–96.

Fawcett, D. W. (1970) A comparative view of sperm ultrastructure. *Biol. Reprod.* Suppl. **2**, 90–127.

Fawcett, D. W. (1972) Observations on cell differentiation and organelle continuity in spermatogenesis. In *The Genetics of the Spermatozoon,* eds. R. A. Beatty and S. Gluecksohn-Waelsch, 37–67, Edinburgh.

Fawcett, D. W. (1975a) The mammalian spermatozoon. *Develop. Biol.* **44**, 394–436.

Fawcett, D. W. (1975b) Ultrastructure and function of the Sertoli cell. *Handbk. Physiol.* Section 7, *Endocrinology* **V**, 21–55.

Fawcett, D. W. and Phillips, D. M. (1969a) Observations on the release of spermatozoa and on changes in the head during passage through the epididymis. *J. Reprod. Fertil.* Suppl. **6**, 405–18.

Fawcett, D. W. and Phillips, D. M. (1969b) The fine structure and development of the neck region of mammalian spermatozoa. *Anat. Rec.* **165**, 153–84.

Fawcett, D. W., Ito, S. and Slautterback, D. (1959) The occurrence of intercellular bridges in groups of cells exhibiting synchronous differentiation. *J. Biophys. Biochem. Cytol.* **5**, 453–60.

Fawcett, D. W., Eddy, E. M. and Phillips, D. M. (1970) Observations on the fine structure and relationships of the chromatoid body in mammalian spermatogenesis. *Biol. Reprod.* **2**, 129–53.

Fawcett, D. W., Anderson, W. and Phillips, D. M. (1971) Morphogenetic factors influencing the shape of the sperm head. *Develop. Biol.* **26**, 220–51.

Ferreira, A. L., Lison, L. and Valeri, V. (1967) Caryometric study of spermatogenesis in the rat. *Z. Zellforsch. Mikroskop. Anat.* **76**, 31–55.

Firlit, C. F. and Davis, J. R. (1965) Morphogenesis of the residual body of the mouse testis. *Quart. J. Microscop. Sci.* **106**, 93–8.

Foote, R. H., Swierstra, E. E. and Hunt, W. L. (1972) Spermatogenesis in the dog. *Anat. Rec.* **173**, 341–52.

Ford, E. H. R. and Woollam, D. H. M. (1966) The fine structure of the sex vesicle and sex chromosomes association in spermatocytes of mouse, golden hamster and field vole. *J. Anat.* **100**, 787–99.

Franklin, L. E. (1971) An association of endoplasmic reticulum with the Golgi apparatus in golden hamster spermatids. *J. Reprod. Fertil.* **27**, 67–72.

Fritz, I. B., Kopec, B., Lam, K. and Vernon, R. G. (1974) Effects of FSH on levels of androgen-binding protein in the testis. *Curr. Top. Molec. Endocrinol.* **1**, 311–27.

Fritz, I. B., Louis, B. G., Tung, P. S., Griswold, M., Rommerts, F. G. and Dorrington, J. H. (1975) Biochemical responses of cultured Sertoli cell-enriched preparations to follicle stimulating hormone and dibutyryl cyclic AMP. *Curr. Top. Molec. Endocrinol.* **2**, 367–82.

Furst, C. (1887) Uber die Entwicklung den Samenkörperchen bei den Beutelthieren. *Arch. Mikroskop. Anat. Entwicklungsmech.* **30**, 336–65.

Gardner, P. J. (1966) Fine structure of the seminiferous tubule of the Swiss mouse. The spermatid. *Anat. Rec.* **155**, 235–50.

Gardner, P. J. and Holyoke, E. A. (1964) Fine structure of the seminiferous tubule of the Swiss mouse. I. The limiting membrane, Sertoli cell, spermatogonia and spermatocytes. *Anat. Rec.* **150**, 391–404.

Gatenby, J. B. and Beams, H. W. (1935) The cytoplasmic inclusions in the spermatogenesis of man. *Quart. J. Microscop. Sci.* **78**, 1–29.

Go, V. L. W., Vernon, R. G. and Fritz, I. B. (1971) Studies on spermatogenesis in rats. III. Effects of hormonal treatment on differentiation kinetics of the spermatogenic cycle in regressed hypophysectomized rats. *Can. J. Biochem.* **49**, 768–75.

Gondos, B. and Zemjanis, R. (1970) Fine structure of spermatogonia and intercellular bridges in *Macaca nemistrina*. *J. Morphol.* **131**, 431–46.

Gregoire, A. T., Bratton, R. W. and Foote, R. H. (1958) Sperm output and fertility of rabbits ejaculated either once a week or once a day for forty-three weeks. *J. Anim. Sci.* **17**, 243–8.

Grocock, C. A. and Clarke, J. R. (1976) Duration of spermatogenesis in the vole (*Microtus agrestis*) and bank vole (*Clethrionomys glareolus*). *J. Reprod. Fertil.* **47**, 133–5.

Hafs, H. D., Hoyt, R. S. and Bratton, R. W. (1959) Libido, sperm characteristics, sperm output, and fertility of mature dairy bulls ejaculated daily or weekly for thirty-two weeks. *J. Dairy Sci.* **42**, 626–36.

Hafs, H. D., Knisely, R. C. and Desjardins, C. (1962) Sperm output of dairy bulls with varying degrees of sexual preparation. *J. Dairy Sci.* **45**, 788–93.

Hall, E. R. and Hupp, E. W. (1970) Localization of irradiation damage in rat seminiferous tubules. *Nature* **225**, 85–6.

Hannah-Alava, A. (1965) The premeiotic stages of spermatogenesis. *Advan. Genet.* **13**, 157–226.

Heller, C. G. and Clermont, Y. (1963) Spermatogenesis in man: an estimate of its duration. *Science* **140**, 184–6.

Heller, C. G. and Clermont, Y. (1964) Kinetics of the germinal epithelium in man. *Rec. Progr. Horm. Res.* **20**, 545–75.

Henderson, S. A. (1966) Time of chiasma formation in relation to the time of deoxyribonucleic acid synthesis. *Nature* **211**, 1043–7.

Henricson, B. and Backström, L. (1963) Spermatocytogenesis in the boar. *Acta Anat.* **53**, 276–88.

Hermann, D. D. (1889) Die postfoetale Histiogenese des Hodens der Maus bis zur Pubertät. *Arch. Mikroskop. Anat. Entwicklungsmech.* **34**, 429–37.

Hilscher, W. (1964) Beiträge zur Orthologie und Pathologie der Spermatogoniogenese der Ratte. *Beitr. Pathol. Anat. Allgem. Pathol.* **130**, 69–132.

Hilscher, W. (1967) DNA synthesis; proliferation and regeneration of the spermatogonia in the rat. *Arch. Anat. Microscop. Morphol. Exptl.* Suppl. **56**, 75–84.

Hilscher, W. and Hilscher, B. (1976) Kinetics of the male gametogenesis. *Andrologia* **8**, 105–16.

Hilscher, W. and Makoski, H. B. (1968) Histologische und autoradiographische Untersuchungen zur Praspermatogenese und Spermatogenese de Ratte. *Z. Zellforsch. Mikroskop. Anat.* **86**, 327–50.

Hilscher, W. and Maurer, W. (1962) Autoradiographische Bestimmung der Dauer der DNS-Verdopplung und ihres zeitlichen Verlaufs bei Spermatogonien der Ratte durch Doppelmarkierung mit $^{14}C$ und $^{3}H$ Thymidin. *Naturwissenschaften* **49**, 352.

Hilscher, B., Hilscher, W. and Maurer, W. (1969) Autoradiographische Untersuchungen uber den Modus der proliferation und Regeneration des Samenepithels der Wistarratte. *Z. Zellforsch. Mikroskop. Anat.* **94**, 593–604.

Hilscher, B., Hilscher, W., Delbruck, G. and Lerouge-Benard, B. (1972) Autoradiographische Bestimmung der S-phasen-daur der Gonocyten bei der Wistarratte durch Einfach-und Doppelmarkierung. *Z. Zellforsch.* **125**, 229–51.

Hilscher, B., Hilscher, W., Bulthoff, B., Kramer, U., Birke, A., Pelzer, H. and Gauss, G. (1974) Kinetics of gametogenesis. I. Comparative histological and autoradiographic studies of oocytes and transitional prespermatogonia during oogenesis and prespermatogenesis. *Cell Tissue Res.* **154**, 443–70.

Hochereau, M-T. (1963a) Etude comparée de la vague spermatogénétique chez le taureau et chez le rat. *Ann. Biol. anim. Biochim. Biophys.* **3**, 5–20.

Hochereau, M-T. (1963b) Constance des fréquences relatives des stades du cycle de l'épithelium séminifère chez le taureau et chez le rat. *Ann. Biol. anim. Biochim. Biophys.* **3**, 93–102.

Hochereau, M-T. (1967) Synthèse de l'ADN au cours des multiplications et du renouvellement des spermatogonies chez le taureau. *Arch. Anat. microscop.* **56**, 85–96.

Hochereau-de Reviers, M-T. (1970) Etude des divisions spermatogoniales et du renouvellement de la spermatogonie souche chez le taureau. Dr Sc. Thesis-Paris-CNRS AO-3976.

Hochereau-de Reviers, M-T. (1971) Action de l'hémicastration sur l'évolution du testicule restant chez le rat. *Bull. Ass. Anat.* **151**, 364–71.

Hochereau-de Reviers, M-T. (1975) Augmentation de l'efficacité de la spermatogénèse par l'hémicastration chez le rat et le bélier. *Ann. Biol. anim. Biochim. Biophys.* **15**, 621–31.

Hochereau-de Reviers, M-T. (1976) Variation in the stock of testicular stem cells and in the yield of spermatogonial divisions in ram and bull testes. *Andrologia* **8**, 137–46.

Hochereau-de Reviers, M-T., Loir, M. and Pelletier, J. (1975) Seasonal variations in the response of the testis and LH levels to hemicastration of adult rams. *J. Reprod. Fertil.* **46**, 203–9.

Holstein, A. F. (1976) Ultrastructure observations on the differentiation of spermatids in man. *Andrologia* **8**, 157–65.

Hoof, L. van (1912) L'évolution de l'élément chromatique dans la spermatogénèse du rat. *Cellule,* **27**, 291–345.

Horstmann, E. (1961) Elektronenmikroskopische Untersuchungen zur Spermiohistogenese beim Menschen. *Z. Zellforsch. Mikroskop. Anat.* **54**, 68–89.

Howland, B. E. and Skinner, K. R. (1975) Changes in gonadotropin secretion following complete or hemicastration in the adult rat. *Horm. Res.* **6**, 71–7.

Huckins, C. (1965) Duration of spermatogenesis in pre- and postpuberal Wistar rats. *Anat. Rec.* **151**, 364.

Huckins, C. (1971a) The spermatogonial stem cell population in adult rats. I. Their morphology, proliferation and maturation. *Anat. Rec.* **169**, 533–58.

Huckins, C. (1971b) The spermatogonial stem cell population in adult rats. II. A radioautographic analysis of their cell cycle properties. *Cell Tissue Kinet.* **4**, 313–34.

Huckins, C. (1971c) The spermatogonial stem cell population in adult rats. III. Evidence for a long-cycling population. *Cell Tissue Kinet.* **4**, 335–49.

Huckins, C. (1971d) Cell cycle properties of differentiating spermatogonia in adult Sprague-Dawley rats. *Cell Tissue Kinet.* **4**, 139–54.

Huckins, C. and Clermont, Y. (1968) Evolution of gonocytes in the rat testis during late embryonic and early postnatal life. *Arch. Anat. Hist. Embryol.* (Strasbourg) **51**, 343–54.

Johnson, H. A. and Hammond, H. D. (1963) The rate of mitochondrial increase in the murine spermatocyte. *Exptl. Cell Res.* **31**, 608–10.

Kennelly, J. J. (1972) Coyote reproduction. I. The duration of the spermatogenic cycle and epididymal sperm transport. *J. Reprod.* **31**, 163–70.

Kennelly, J. J. and Foote, R. H. (1964) Sampling boar testes to study spermatogenesis quantitatively and to predict sperm production. *J. Anim. Sci.* **23**, 160–7.

Kierszenbaum, A. L. and Tres, L. L. (1975) Structural and transcriptional features of the mouse spermatid genome. *J. Cell. Biol.* **65**, 258–70.

Kingsley-Smith, B. V. and Lacy, D. (1959) Residual bodies of seminiferous tubules of the rat. *Nature* **184**, 249–51.

Knudsen, O. (1954) Cytomorphological investigations into the spermiocytogenesis of bulls with normal fertility and bulls with acquired disturbances in spermiogenesis. *Acta. Path. Microbiol. Scand.* Suppl. **101**, 1–79.

Knudsen, O. (1958) Studies on spermiocytogenesis in the bull. *Intern. J. Fertil.* **3**, 389–403.

Kramer, M. F., Lange, A. de and Visser, M. B. H. (1964) Spermatogonia in the bull. *Z. Zellforsch.* **63**, 735–58.

Kretser, D. M. de, Catt, K. J., Dufau, M. L. and Hudson, B. (1971) Studies on rat testicular cells in tissue culture. *J. Reprod. Fertil.* **24**, 311–18.

Lacy, D. (1960) Light and electron microscopy and its use in the study of factors influencing spermatogenesis in the rat. *J. Roy. Microscop. Soc.* **79**, 209–25.

Land, R. B. and Carr, W. R. (1975) Testis growth and plasma LH concentration following hemicastration and its relation with female prolificacy in sheep. *J. Reprod. Fertil.* **45**, 495–501.

La Valette St George, von (1865) Ueber die Genese der Samenkörper. I. *Arch. mikrosk. Anat.* **1**, 404–14.

La Valette St George, von (1876) Ueber die Genese der Samenkörper. IV. Die Spermatogenese bei den Amphibien. *Arch. mikrosk. Anat.* **12**, 797–825.

La Valette St George, von (1878) Ueber die Genese der Samenkörper. V. Die Spermatogenese bei den Saugethieren und dem Menschen. *Arch. mikrosk. Anat.* **15**, 261–314.

La Valette St George, von (1887) Spermatologische Beitrage. V. Ueber die Bildung der Spermatocysten bei den Lepidopteren. *Arch. mikrosk. Anat.* **30**, 426–34.

Leblond, C. P. and Clermont, Y. (1952a) Spermiogenesis of rat, mouse, hamster and guinea pig as revealed by the periodic acid-fuchsin sulfurous acid technique. *Am. J. Anat.* **90**, 167–215.

Leblond, C. P. and Clermont, Y. (1952b) Definition of the stages of the cycle of the seminiferous epithelium in the rat. *Ann. NY Acad. Sci.* **55**, 548–73.

Leblond, C. P., Steinberger, E. and Roosen-Runge, E. C. (1963) Spermatogenesis. In *Mechanisms Concerned with Conception*, ed. C. G. Hartman, 1–72, Oxford: Pergamon Press.

Lenhossek, M. von (1898) Untersuchungen uber Spermatogenese. *Arch. Mikrosk. Anat. Entwicklungsmech.* **51**, 215–318.

Liang, D. S. and Liang, M. D. (1970) Testicular hypertrophy in rats. *J. Reprod. Fertil.* **22**, 537–40.

Lino, B. F., Braden, A. W. H. and Turnbull, K. E. (1967) Fate of unejaculated spermatozoa. *Nature* **213**, 594–5.

Luxembourger, M-M. and Aron, M. (1966) Etude, par l'homotransplantation de testicule néo-natal de rat dans le testicule du rat mûr, des facteurs de l'évolution spermatogénétique avant la maturité et des facteurs de l'involution de la lignée séminale dans le testicule privé de voies excrétrices. *Compt. rend sean. Soc. Biol.* **160**, 391–4.

MacMillan, K. L. and Hafs, H. D. (1968) Gonadal and extragonadal sperm numbers during reproductive development of Holstein bulls. *J. Anim. Sci.* **27**, 697–700.

Martinet, L. and Meunier, M. (1975) Role de la durée quotidienne d'éclairement sur la response, apres hemicastration, du testicule restant chez le campagnol des champs (*Microtus arvalis*). *Ann. Biol. anim. Biochim. Biophys.* **15**, 607–9.

Mauss, J. and Hackstedt, G. (1972) The effect of unilateral orchidectomy and unilateral cryptorchidism on sperm output in the rat. *J. Reprod. Fertil.* **30**, 289–92.

Meistrich, M. L., Eng, V. W. S. and Loir, M. (1973) Temperature effects on the kinetics of spermatogenesis in the mouse. *Cell Tissue Kinet.* **6**, 379–93.

Meistrich, M. L., Reid, B. O. and Barcellona, W. J. (1976) Changes in sperm nuclei during spermiogenesis and epididymal maturation. *Exptl. Cell Res.* **99**, 72–8.

Memon, G. N. (1973) Effects of intratesticular injections of prostaglandins on the testes and accessory sex glands of rats. *Contraception* **8**, 361.

Merkle, U. (1956) Untersuchungen über den Ablauf der Spermatogenese in den Samenkanalchen bei Ratte und Meerschweinchen. *Z. Mikorskop. Anat. Forsch.* **62**, 130–52.

Merkle, U. (1957) Volumetrische Befunde an den Samenzellen und den Sertoli-Zellen der Ratte. *Z. Mikorskop. Anat. Forsch.* **63**, 252–73.

Meves, F. (1899) Uber struktur und Histogenese der Samenfaden des Meerschweinchens. *Arch. Mikroskop. Anat. Entwicklungsmech.* **54**, 329–402.

Moens, P. B. and Go, V. L. W. (1972) Intercellular bridges and division patterns of rat spermatogonia. *Z. Zellforsch.* **127**, 201–8.

Moens, P. B. and Hugenholtz, A. D. (1975) The arrangement of germ cells in the rat seminiferous tubule: an electron-microscopic study. *J. cell Sci.* **19**, 487–507.

Monesi, V. (1962) Autoradiographic study of DNA synthesis and the cell cycle in spermatognia and spermatocytes of mouse testis using tritiated thymidine. *J. cell Biol.* **14**, 1–18.

Moore, C. R. (1926) The biology of the mammalian testis and scrotum. *Quart. Rev. Biol.* **1**, 4–50.

Moree, R. (1947) The normal spermatogenetic wave-cycle in *Peromyscus. Anat. Rec.* **99**, 163–75.

Nagano, T. (1968) Fine structural relation between the Sertoli cell and the differentiating spermatid in the human testes. *Z. Zellforsch. Mikroskop. Anat.* **89**, 39–43.

Oakberg, E. F. (1956a) A description of spermiogenesis in the mouse and its use in analysis of the cycle of the seminiferous epithelium and germ cell renewal. *Am. J. Anat.* **99**, 391–413.

Oakberg, E. F. (1956b) Duration of spermatogenesis in the mouse and timing of stages of the cycle of the seminiferous epithelium. *Am. J. Anat.* **99**, 507–16.

Oakberg, E. F. (1957a) Duration of spermatogenesis in the mouse. *Nature* **180**, 1137–8.

Oakberg, E. F. (1957b) Duration of spermatogenesis in the mouse and timing of stages of the cycle of the seminiferous epithelium. *Anat. Rec.* **169**, 515–31.

Oakberg, E. F. (1971a) Spermatogonial stem-cell renewal in the mouse. *Anat. Rec.* **169**, 515–32.

Oakberg, E. F. (1971b) A new concept of spermatogonial stem-cell renewal in the mouse and its relationship to genetic effects. *Mutation Res.* **11**, 1–7.

Ortavant, R. (1954) Etude des générations spermatogoniales chez le bélier. *Compt. Rend. Soc. Biol.* **148**, 1958–61.

Ortavant, R. (1956a) Action de la durée d'éclairement sur les processus spermatogénétiques chez le bélier. *Compt. Rend. Soc. Biol.* **150**, 471–4.

Ortavant, R. (1956b) Autoradiographie des cellules germinales du testicule de bélier. Durée des phenomènes spermatogénétiques. *Arch. Anat. Microscop. Morphol. Exptl.* **45**, 1–10.

Ortavant, R. (1958) Le cycle spermatogénétique chez le bélier. D.Sc. Thesis, University of Paris, Paris.

Ortavant, R. (1959) Spermatogenesis and morphology of the spermatozoon. In *Reproduction in Domestic Animals,* eds. H. H. Cole and P. T. Cupps, 1st edition, **II**, 1–50, New York: Academic Press.

Ortavant, R., Orgebin, M. C. and Singh, G. (1961) *Etude Comparative de la Durée des Phénomènes Spermatogénétiques chez les Animaux Domestiques,* Symposium on the use of radioisotopes in animal biology and the medical sciences, Mexico City, 21 Nov–1 Dec 1961. London: Academic Press, 321–7.

Orgebin-Crist, M. C. (1968) Gonadal and epididymal sperm reserves in the rabbit: Estimation of the daily sperm production. *J. Reprod. Fertil.* **15**, 15–25.

Oud, J. L. and Rooji, D. G. de (1977) Spermatogenesis in the Chinese hamster. *Anat. Rec.* **187**, 113–24.

Palkovits, M., Czeizel, E. and Palkovich, I. (1966) Uber die karyometrische Untersuchung der Spermatogenese bei Albinoratten. *Z. Zellforsch. Mikroskop. Anat.* **74**, 449–53.

Parvinen, M. and Jokelainen, P. T. (1974) Rapid movements of the chromatoid body in living early spermatids of the rat. *Biol. Reprod.* **11**, 85–92.

Parvinen, M. and Vanha-Perttula, T. (1972) Identification and enzyme quantitation of the stages of the seminiferous-epithelial wave in the rat. *Anat. Rec.* **174**, 435–50.

Perey, B., Clermont, Y. and Leblond, C. P. (1961) The wave of the seminiferous epithelium in the rat. *Am. J. Anat.* **108**, 47–77.

Phillips, D. M. (1970) Development of spermatozoa in the Woolly Opossum with special reference to the shaping of the sperm head. *J. Ultrastruct. Res.* **33**, 369–80.

Ploen, L. (1971) A scheme of rabbit spermateliosis based upon electron microscopical observations. *Z. Zellforsch.* **115**, 553–64.

Rauh, W. (1925) Ursprung der mannlichen Keimzellen und die chromatischen Vorgange bis zur Entwicklung der Spermatocyten. *Z. Anat. Entwicklungsgeschichte* **76**, 561–77.

Regaud, C. (1901) Etudes sur la structure des tubes seminiferes et sur la spermatogénèse chez les mammifères. *Arch. Anat. Microscop. Morphol. Exptl.* **4**, 101–55 and 231–380.

Renson, G. (1882) De la spermatogénèse chez les mammifères. *Arch. Biol.* (Liege) **3**, 291–334.

Rolshoven, E. (1941) Zur Frage des 'Alterns' der generativen Elementen in den Hoden Kanalchen. *Anat. Anz.* **91**, 1–8.

Rooij, D. G. de (1968) Stem cell renewal and duration of spermatogonial cycle in the goldhamster. *Z. Zellforsch. Mikroskop. Anat.* **89**, 133–6.

Rooij, D. G. de and Kramer, M. F. (1968) Spermatogonial stem cell renewal in rats and mice. *Z. Zellforsch. Mikroskop. Anat.* **85**, 206–9.

Rooij, D. G. de and Kramer, M. F. (1968) Spermatogonial stem cell renewal in rats and mice. *Z. Zellforsch. Mikroskop. Anat.* **113**, 25–33.

Roosen-Runge, E. C. (1951) Quantitative studies on spermatogenesis in the albino rat. II. The duration of spermatogenesis and some effects of colchicine. *Am. J. Anat.* **88**, 163–73.

Roosen-Runge, E. C. (1952) Kinetics of spermatogenesis in mammals. *Ann. N.Y. Acad. Sci.* **55**, 574–84.

Roosen-Runge, E. C. (1955a) Quantitative studies on spermatogenesis in the albino rat. III. Volume changes in the cells of the seminiferous tubules. *Anat. Rec.* **123**, 385–98.

Roosen-Runge, E. C. (1955b) Untersuchungen uber die Regeneration samen bildener Zellen in der normalen Spermatogenese der Ratte. *Z. Zellforsch. Mikroskop. Anat.* **41**, 221–35.

Roosen-Runge, E. C. (1962) The process of spermatogenesis in mammals. *Biol. Rev.* **37**, 343–77.

Roosen-Runge, E. C. (1969) Comparative aspects of spermatogenesis. *Biol. Reprod.* Suppl. **1**, 24–39.

Roosen-Runge, E. C. (1974) Die Spermatogenese im Lichte der Evolution. *Verh. Anat. Ges.* **68**, 23–37.

Roosen-Runge, E. C. (1977) *The Process of Spermatogenesis in Animals*, London: Cambridge University Press.

Roosen-Runge, E. C. and Giesel, L. O. (1950) Quantitative studies on spermatogenesis in the albino rat. *Am. J. Anat.* **87**, 1–30.

Roosen-Runge, E. Ç. and Leik, J. (1968) Gonocyte degeneration in the postnatal male rat. *Am. J. Anat.* **122**, 275–99.

Rowley, M. J. and Heller, C. G. (1971) Quantitation of the cells of the seminiferous epithelium of the human testis employing the Sertoli cell as a constant. *Z. Zellforsch.* **115**, 461–72.

Ross, M. H. (1976) The Sertoli cell junctional specialization during spermiogenesis and at spermiation. *Anat. Rec.* **186**, 79–104.

Russell, L. and Clermont, Y. (1976) Anchoring device between Sertoli cells and late spermatids in rat seminiferous tubules. *Anat. Rec.* **185**, 259–78.

Sandoz, D. (1970) Evolution des ultrastructures au cours de la formation de l'acrosome du spermatozoide chez la souris. *J. Microscop.* (Paris) **9**, 535–58.

Sapsford, C. S. (1962) Changes in the cells of the sex cords and the seminiferous tubules during development of the testis of the rat and the mouse. *Aust. J. Zool.* **10**, 178–92.

Sapsford, C. S. and Rae, C. A. (1969) Ultrastructural studies on Sertoli cells and spermatids in the bandicoot and ram during the movement of mature spermatids into the lumen of the seminiferous tubule. *Aust. J. Zool.* **17**, 415–45.

Sapsford, C. S., Rae, C. A. and Cleland, K. W. (1967) Ultrastructural studies on spermatids and Sertoli cells during early spermiogenesis in the bandicoot *Perameles nasuta* Geoffroy (Marsupialia). *Aust. J. Zool.* **15**, 881–909.

Sapsford, C. S., Rae, C. A. and Cleland, K. W. (1969a) Ultrastructural studies on maturing spermatids and on Sertoli cells in the bandicoot (*Perameles nasuta*). *Aust. J. Zool.* **17**, 195–292.

Sapsford, C. S., Rae, C. A. and Cleland, K. W. (1969b) The fate of residual bodies and degenerating germ cells and the lipid cycle in Sertoli cells in the bandicoot *Perameles nasuta*. *Aust. J. Zool.* **17**, 729–53.

Sapsford, C. S., Rae, C. A. and Cleland, K. W. (1970) Ultrastructural studies on the development and form of the principal piece sheath of the bandicoot spermatozoon. *Aust. J. Zool.* **18**, 21–48.

Sawada, T. (1957) An electron microscope study of spermatid differentiation in the mouse. *Okajimas Folia Anat. Japon.* **30**, 73–80.

Schmidt, F. C. (1964) Licht-und Elektronenmikroskopische Untersuchungen am menschlichen Hoden und Nebenhoden. *Z. Zellforsch. Mikroskop Anat.* **63**, 707–27.

Schoenfeld, H. (1902) La spermatogénèse chez le taureau et chez le mammifères en general. *Arch. Biol.* (Liege) **18**, 1–72.

Schuler, H. M. and Gier, H. T. (1976) Duration of the cycle of the seminiferous epithelium in the prairie vole (*Microtus ochrogater ochrogaster*). *J. exptl. Zool.* **197**, 1–12.

Sertoli, E. (1877) Sulla struttura dei canaliculi seminifi dei testicoli studiata in rapporto allo sviluppo dei nemaspermi. *Arch. Sci. Med.* **2**, 107–46 and 267–95.

Setchell, B. P. and Carrick, F. N. (1973) Spermatogenesis in some Australian marsupials. *Aust. J. Zool.* **21**, 491–9.

Setchell, B. P. and Waites, G. M. H. (1971) The exocrine secretion of the testis and spermatogenesis. *J. Reprod. Fertil.* Suppl. **13**, 15–27.

Setchell, B. P. and Waites, G. M. H. (1972) The effects of local heating on the flow and composition of rete testis fluid in the rat, with some observations on the effects of age and unilateral castration. *J. Reprod. Fertil.* **30**, 225–33.

Setchell, B. P., Duggan, M. C. and Evans, R. W. (1973) The effects of gonadotrophins on fluid secretion and production of spermatozoa by the rat and hamster testis. *J. Endocrinol.* **56**, 27–36.

Setchell, B. P., Davies, R. V., Gladwell, R. T., Hinton, B. T., Main, S. J., Pilsworth, L. and Waites, G. M. H. (1978) The movements of fluid in the seminiferous tubules and rete testis. *Ann. Biol. anim. Biochim. Biophys.* [In press.]

Söderström, K-O. and Parvinen, M. (1976) Incorporation of [³H] uridine by the chromatoid body during rat spermatogenesis. *J. Cell Biol.* **70**, 239–46.

Sotelo, J. R. and Trujillo-Cenoz, O. (1958) Submicroscopic structure of meiotic chromosomes during prophase. *Exptl. Cell Res.* **14**, 1–8.

Spangaro, S. (1902) Uber die histologischen veranderungen des Hodens, Nebenhodens und samenleiters von Geburt an bis zum Greisenalter mit besonderer Berucksichtigung der Hodenatrophie, des elastischen Gewebes und des vorkommen von Krystallen im Hoden. *Anat. Heft. Abt. I.* **18**, 593–771.

Steinberger, A. (1967) Relationship between the yield of spermatocytes in rat testis organ culture and the age of the donor animal. *Anat. Rec.* **157**, 327.

Steinberger, A. (1975) *In vitro* techniques for the study of spermatogenesis. *Method. Enzymol.* **39**, 283–96.

Steinberger, A. and Steinberger, E. (1965) Differentiation of rat seminiferous epithelium in organ culture. *J. Reprod. Fertil.* **9**, 243–8.

Steinberger, A. and Steinberger, E. (1966a) Stimulatory effect of vitamins and glutamine on the differentiation of germ cells in rat testes organ culture grown in chemically defined media. *Exptl. Cell Res.* **44**, 429–35.

Steinberger, A. and Steinberger, E. (1966b) *In vitro* culture of rat testicular cells. *Exptl. Cell Res.* **44**, 443–52.

Steinberger, A. and Steinberger, E. (1967) Factors affecting spermatogenesis in organ cultures of mammalian testes. *J. Reprod. Fertil.* Suppl. **2**, 117–24.

Steinberger, A. and Steinberger, E. (1970) *In vitro* growth and development of mammalian testes. In *The Testis*, eds. A. D. Johnson, W. R. Gomes and N. L. Vandemark, **II**, 363–91, New York: Academic Press.

Steinberger, A., Steinberger, E. and Perloff, W. H. (1964) Mammalian testes in organ culture. *Exptl. Cell Res.* **36**, 19–27.

Steinberger, A., Elkington, J. S. H., Sanborn, B. M., Steinberger, E., Heindel, J. J. and Lindsey, J. N. (1975) Culture and FSH responses of Sertoli cells isolated from sexually mature rat testis. *Curr. Top. Molec. Endocrinol.* **2**, 399–411.

Steinberger, E. and Steinberger, A. (1975) Spermatogenic function of the testis. *Handbk. Physiol.* Section 7, **V**, 1–19.

Steinberger, E. and Tjioe, D. Y. (1968) A method for quantitative analysis of human seminiferous epithelium. *Fertil. Steril.* **19**, 960–70.

Steinberger, E., Steinberger, A. and Perloff, W. H. (1964) Initiation of spermatogenesis *in vitro*. *Endocrinology* **74**, 788–92.

Steinberger, E., Steinberger, A. and Ficher, M. (1970) Study of spermatogenesis and steroid metabolism in cultures of mammalian testes. *Rec. Progr. Horm. Res.* **26**, 547–88.

Sud, B. N. (1961a) The 'chromatoid body' in spermatogenesis. *Quart. J. Microscop. Sci.* **102**, 273–92.

Sud, B. N. (1961b) Morphological and histochemical studies of the chromatoid body and related elements in the spermatogenesis of the rat. *Quart. J. Microscop. Sci.* **102**, 495–505.

Suoranto, H. (1971) Tubular damage caused by local thermal injury or microembolization of the rat testis. *Virchows Arch. Abt. B. Zellpath.* **8**, 299–308.

Susi, F. R. and Clermont, Y. (1970) Fine structural modifications of the rat chromatoid body during spermatogenesis. *Am. J. Anat.* **129**, 177–92.

Susi, F. R., Leblond, C. P. and Clermont, Y. (1971) Changes in the Golgi apparatus during spermiogenesis in the rat. *Am. J. Anat.* **130**, 251–68.

Swierstra, E. E. (1967) Duration of spermatogenesis in the boar. *J. Anim. Sci.* **26**, 952–3.

Swierstra, E. E. (1968a) Cytology and duration of the cycle of the seminiferous epithelium of the boar, duration of spermatozoan transit through the epididymis. *Anat. Rec.* **161**, 171–86.

Swierstra, E. E. (1968b) A comparison of spermatozoa production and spermatozoa output of Yorkshire and Lacombe boars. *J. Reprod. Fertil.* **17**, 459–70.

Swierstra, E. E. (1971) Sperm production of boars as measured from epididymal sperm reserves and quantitative testicular histology. *J. Reprod. Fertil.* **27**, 91–102.

Swierstra, E. E. and Foote, R. H. (1963) Cytology and kinetics of spermatogenesis in the rabbit. *J. Reprod. Fertil.* **5**, 309–22.

Swierstra, E. E. and Foote, R. H. (1965) Duration of spermatogenesis and spermatozoan transport in the rabbit based on cytological changes, DNA synthesis and labelling with tritiated thymidine. *Am. J. Anat.* **116**, 401–11.

Swierstra, E. E., Gebauer, M. R. and Pickett, B. W. (1974) Reproductive physiology of the stallion. I. Spermatogenesis and testis composition. *J. Reprod. Fertil.* **40**, 113–23.

Thibault, C. (1969) La spermatogénèse chez les mammifères. In *Traité de Zoologie*, ed. P-P. Grassé, **16**, Fasc. VI, 716–98, Paris: Masson.

Tiba, T. (1965) Methodologisch orienterte quantitative Untersuchung der Spermatogenese beim Bullen. *Japan J. Vet. Res.* **13** Suppl. **2**, 1–127.

Tres, L. L. and Solari, A. J. (1968) The ultrastructure of the nuclei and the behaviour of the sex chromosomes of human spermatogonia. *Z. Zellforsch. Mikroskop. Anat.* **91**, 75–89.

Turner, C. D. (1938) Intra-ocular homotransplantation of prepuberal testes in the rat. *Am. J. Anat.* **63**, 101–59.

Turner, C. D. and Asakawa, H. (1963) Complete spermatogenesis in intratesticular homotransplants of fetal and neonatal testes in the rat. *Proc. Soc. Exptl. Biol. Med.* **112**, 132–5.

Voglmayr, J. K. and Mattner, P. E. (1968) Compensatory hypertrophy on the remaining testis following unilateral orchidectomy in the adult ram. *J. Reprod. Fertil.* **17**, 179–81.

Walton, J. S., Evins, J. D. and Waites, G. M. H. (1978) Feedback control of follicle-stimulating hormone in pre- and post-pubertal rams as revealed by hemicastration. *J. En-*

*docrinol.* **77,** 75–84.

Wartenberg, H. (1976) Comparative cytomorphologic aspects of the male germ cells especially of the 'gonia'. *Andrologia* **8,** 117–30.

Welsh, M. J. and Wiebe, J. P. (1975) Rat Sertoli cells: a rapid method for obtaining viable cells. *Endocrinology* **96,** 618–24.

Willett, E. L. and Ohms, J. I. (1957) Measurement of testicular size and its relation to production of spermatozoa by bulls. *J. Dairy Sci.* **40,** 1559–69.

Williams, R. G. (1950) Studies of living interstitial cells and pieces of seminiferous tubules in autogenous grafts of testis. *Am. J. Anat.* **86,** 343–69.

Woollam, D. H. M. and Ford, E. H. R. (1964) The fine structure of the mammalian chromosome in meiotic prophase with special reference to the synaptinemal complex. *J. Anat.* **98,** 163–73.

Woolley, D. M. (1970) The midpiece of the mouse spermatozoon: Its form and development as seen by surface replication. *J. Cell Sci.* **6,** 865–79.

Woolley, D. M. and Fawcett, D. W. (1973) The degeneration and disappearance of the centrioles during the development of the rat spermatozoon. *Anat. Rec.* **177,** 289–302.

Yao, T. S. and Eaton, D. N. (1954) Postnatal growth and histologic development of reproductive organs in male goats. *Am. J. Anat.* **95,** 401–31.

# 8 Fluid secretion and the entry of substances into the tubules

## 8.1 Anatomy of the excurrent ducts

In this chapter, only the adult animal is discussed. The development of the testis and the excurrent ducts have been considered in Chapter 2.

Each seminiferous tubule is usually a long convoluted tube (see Section 1.3.1), at both ends of which there is a segment lined only by Sertoli cells. These segments join the tubuli recti, which open into the rete testis (von Mihalkovics, 1873; Messing, 1877; Stieda, 1877: Spangaro, 1902; Roosen-Runge, 1961), a complex of intercommunicating channels receiving the spermatozoa and fluid from the seminiferous tubules. The rete leads in turn to the efferent ductules (see Benoit, 1926; Setchell, 1970a) through which the testicular semen passes to the epididymis, where the fluid portion is largely resorbed and the spermatozoa finally mature.

### 8.1.1 Tubuli recti

As their names suggest, tubuli recti are short straight tubules lined by simple cuboidal cells some of which in the guinea pig contain large amounts of glycogen (Fawcett and Dym, 1974). Where the tubuli recti open into the rete testis the cells lining the tubule appear to form a valve or plug (Spangaro, 1902; May, 1923; Lohmuller, 1925; Cunningham, 1928; Stieve, 1930). It has been suggested that this valve-like arrangement would prevent fluid re-entering the tubules from the rete (Roosen-Runge, 1961), but there is no evidence whether in life they are normally open or closed.

### 8.1.2 Rete testis

The rete testis in some species is located on one edge (e.g. humans) or at one pole of the testis (e.g. rats and mice). In others it extends superficially down two sides of the testis (e.g. some marsupials), while in others it has become surrounded by seminiferous tubules during development so that it lies along the long axis of the testis, extending for about three-quarters of the length (e.g. sheep, goats, cattle, pigs, guinea pigs, rabbits and carnivores) (see Benoit, 1926; van den Broek, 1933; Setchell, 1970a; Dym, 1976). In the adult rat the rete has a fairly simple sac-like shape and lies near the point at which the artery first reaches the epididymal surface of the testis (Roosen-Runge, 1961). In the human testis, it extends along most of the epididymal margin of the testis (Cooper, 1830); the surrounding connective tissue is well developed and is referred to as the mediastinum testis (Figure 8.1).

The rete is more complicated in the human than in the rat, but in both species, the seminiferous tubules normally open on only one side of the rete, the other being closely apposed to the main testicular artery in its course down the epididymal margin of the testis. In the rat the artery again approaches close to the rete after the

233

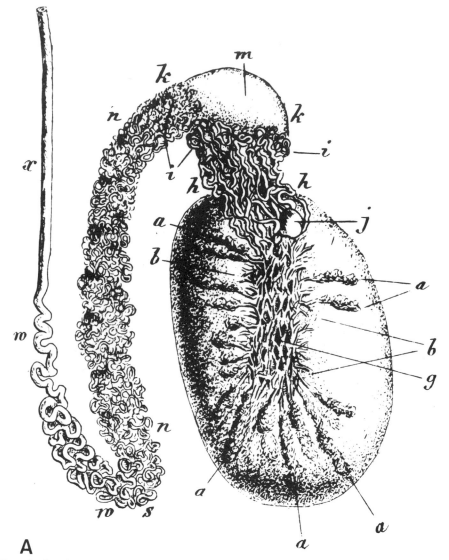

**A**

Figure 8.1 (continued opposite and overleaf) Pictures illustrating the anatomy of the rete testis in human, rat and ram.

A. Human testis with mercury injected through the ductus deferens (x), filling the convoluted proximal part of the ductus (w), most of the epididymal duct (s, n, k) except part (m) of the head, the efferent ducts (h, i) the rete testis (g), most of the tubuli recti (b) and some seminiferous tubules (a). The diagram also shows a small cyst (j) on the efferent ducts. (Reproduced from Lauth, 1830)

B. Rat testis cut in longitudinal section showing the position of the rete testis (R). The efferent ducts had been ligated 24h previously (part of the ligature is visible in this section (L)) to dilate the rete, which otherwise is very difficult to see at this magnification. (Photographed from a section prepared by S. J. Main, M. H. Ross and G. M. H. Waites)

C. Ram testis cut in longitudinal section. The rete testis is embedded in a fibrous mediastinum (M) in the centre of the testicular parenchyma (P). E, head of epididymis; T, tail of epididymis; V, vascular cone.

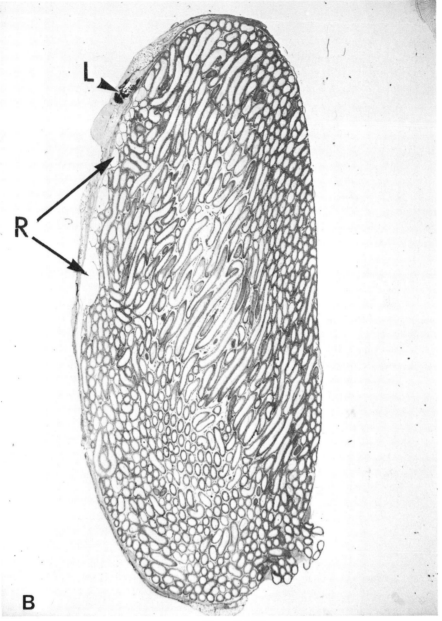

Figure 8.1 (B)

artery has undulated up the free surface of the testis and entered the parenchyma. In man, the main arterial branches run from the free surface towards the rete near which they turn to supply the parenchyma (see Section 3.3.1).

In macropod and phalangerid marsupials, the rete extends superficially down the epididymal and free surfaces as three or four separate channels lying immediately

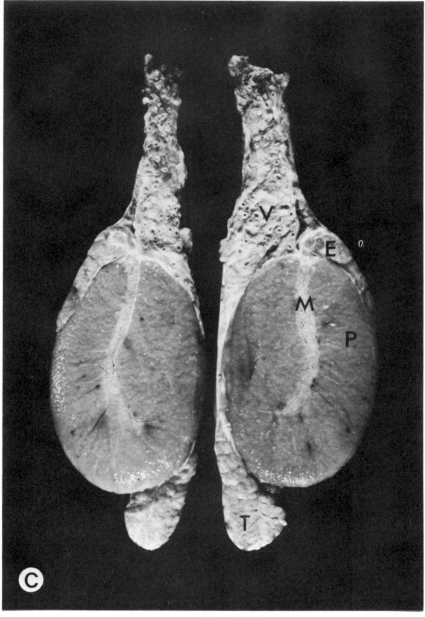

Figure 8.1 (C)

beneath the arteries and veins respectively (see Setchell, 1977). In the ram, the rete consists of about ten separate channels in any cross-section (Figure 8.2). These are enclosed in a well developed fibrous mediastinum which also contains the point of flexion of the branches of the testicular arteries which run in from the surface of the testis (see Section 3.3.1).

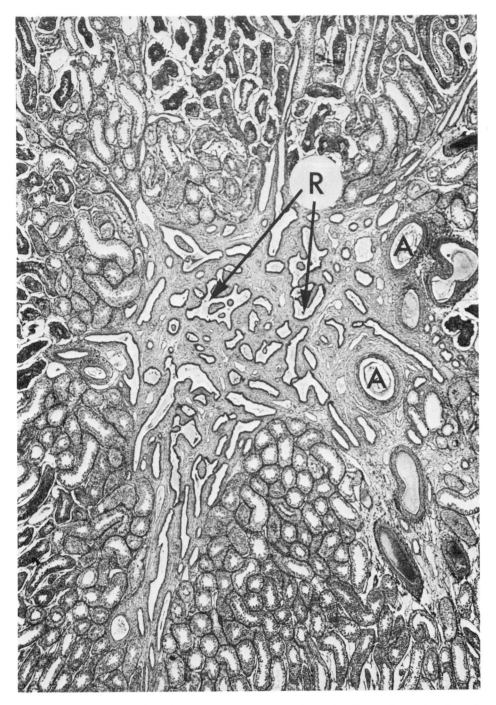

Figure 8.2  A cross-section through a ram rete testis, about half way along its length, showing the multiple channels (R) set in the fibrous mediastinum. Unfortunately some of the seminiferous tubules are poorly fixed. Note in this section the coils of the arteries (A) where they turn near the rete before beginning to branch to supply the parenchyma (see Section 3.3.1).

In all species so far studied the lining of the rete is a single layer of cuboidal or low columnar cells, joined at their luminal margins by well developed 'tight junctions' (Benoit, 1926; Ladman and Young, 1958; Leeson, 1962; Dym, 1976).

### 8.1.3    Efferent ducts

Efferent ducts are usually three to seven in number and extend from the rete testis into the first part of the head of the epididymis (see Benoit, 1926; Setchell, 1970a). Their length depends on the relationship between the testis and epididymis. In laboratory rodents the epididymis is well separated from the testis. Consequently, the ducts are comparatively long and easy to find; they run from an avascular area surrounded by small veins at the pole of the testis, about one-third of the length of the rete from its capital end. The efferent ducts run at an angle to the blood vessels to a point about 3mm from the capital end of the epididymis. In older animals, they become embedded in fat and more difficult to see.

In marsupials, the epididymis is even more clearly separated from the testis and the efferent ducts proportionally even longer than in the rat, but they lie in a bundle which also contains the arteries and veins of the testis. The efferent ducts run from the apex of the horseshoe-shaped rete and enter the caput epididymidis.

In most eutherian mammals the epididymis is closely attached to the testis and the efferent ducts are correspondingly short and difficult to find. However, in sheep, goats, cattle and pigs the head of the epididymis can be dissected off the testis and the ducts isolated. In these species which have a central rete, the ducts leave one end of the rete just in front of the attachment of the vascular cone. The small veins on the free surface of the testis run either side of the ducts and often very close to them.

The lining of the efferent ducts is essentially similar to that of the rete testis except that the columnar cells are slightly higher. Both epithelia have 'tight junctions' at their luminal border and the cells of the ducts have some cilia (Benoit, 1926; Young, 1933; Mason and Shaver, 1952; Montagna, 1952; Burgos, 1957; Ladman and Young, 1958; Lucas, 1963; Holstein, 1964; Morita, 1966; Sedar, 1966; Ladman, 1967; Martan et al., 1967).

## 8.2    Composition of fluids from the testis

### 8.2.1    Rete testis fluid (RTF)

Rete testis fluid can be collected from catheters implanted chronically in the rete testis of sheep, cattle and pigs and acutely from wallabies, rats, rabbits, hamsters and monkeys (Figure 8.3; Voglmayr et al., 1966, 1967, 1970; Setchell et al., 1969a; Burgos and Vitale-Calpe, 1969; Setchell, 1970a; Tuck et al., 1970; Waites and Einer-Jensen, 1974; Cooper et al., 1976).

Compared with ejaculated semen, the fluid contains a low concentration of spermatozoa, about $1 \times 10^8$/ml in sheep, cattle and pigs, and about $3 \times 10^7$/ml in rats, giving a spermatocrit (i.e. the volume of the cells as a percentage of the fluid volume) of about 1 per cent.

RTF is isomolar with plasma in the ram, rat and hamster (Voglmayr et al., 1967;

A. RAM

B. RABBIT

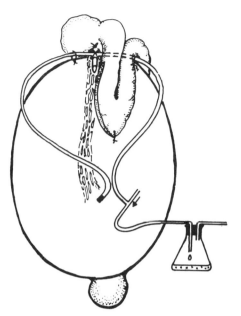

C.  RAT

D.   MONKEY

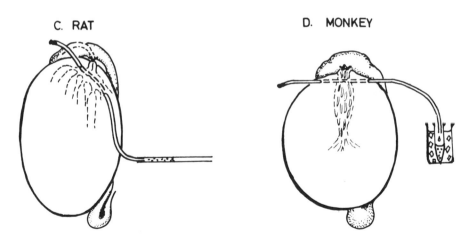

Figure 8.3  A diagram of the techniques for collection of rete testis fluid in various species. A. Ram and B. Rabbit: Direct collection by introducing a catheter into the extra-testicular rete through the ductuli efferentes.

A. the fluid flows down one arm of the 'T'-piece catheter to a 'T' junction outside the scrotum from which it is drawn by gentle suction to a container at 4°C;

C Rat and D Rhesus Monkey: Collection after prior efferent duct ligation: C, side-hole catheter in the rete testis 12–24h after ligating the ductuli efferentes;

D, side-hole catheter in mediastinum testis 20–27h after ligating the ductuli efferentes.

(Reproduced from Waites, 1977)

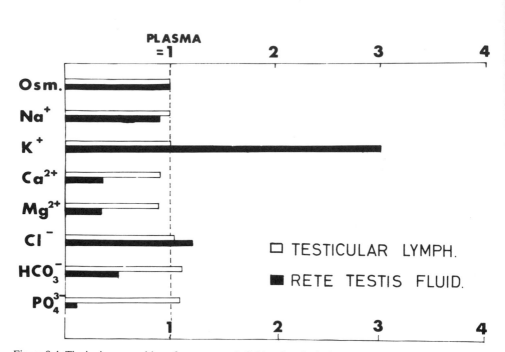

Figure 8.4 The ionic composition of ram rete testis fluid and testicular lymph, calculated as ratios to the concentrations in plasma.

Tuck *et al.*, 1970; Johnson and Howards, 1976). The ionic composition is essentially similar for fluid from all species so far studied. Compared with blood plasma and testicular lymph, there is in RTF more potassium, less sodium, magnesium and calcium, more chloride and less bicarbonate and phosphate (Figure 8.4). There is practically no glucose and no detectable fructose in RTF (see Setchell, 1970a, 1974a; Setchell and Waites, 1975a, b), but in all species except the wallabies, there are high concentrations of free myo-inositol which appears to be the only carbohydrate present (Figure 8.5; Setchell *et al.*, 1968; Middleton, 1973). Acetate, lactate and pyruvate are present in RTF in concentrations about half those in plasma. Appreciable concentrations are found in RTF of a substance which reduces 2,6-dichloroindophenol, but which does not appear to be ascorbic acid, as was originally thought.

There is about 1/50th as much protein present in RTF as in blood plasma with a greater preponderance of albumin and a relatively high concentration of $\alpha_2$-macroglobulin. The concentration of immunoglobulin in RTF is very low—about one-thousandth of that in blood plasma (Johnson and Setchell, 1968). Many of the minor proteins are not identical with any of the serum proteins and many serum proteins do not appear in RTF (Kormano *et al.*, 1971; Koskimies and Kormano, 1973).

Most of the enzymes present in blood plasma appear in RTF but in much lower

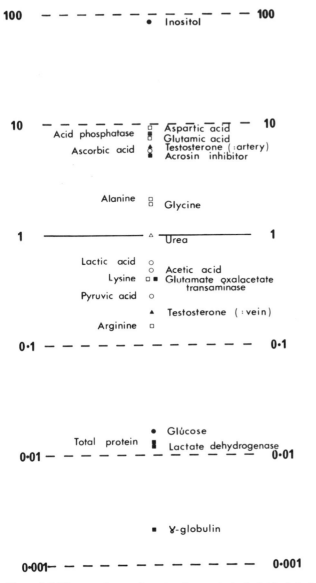

Figure 8.5 The organic constitutents of ram rete testis fluid, plotted on a logarithmic scale, as fractions of the concentrations in plasma.

concentrations. When allowance is made for the difference in protein concentration, the activity of most of the enzymes is similar in blood plasma and RTF. Several enzymes are present in higher activity in RTF, and by characterizing the isoenzymes present it would appear that some proteins are liberated from the testicular cells. However it is strange that the X-isoenzyme of lactic dehydrogenase, specific to the germinal cells of the testis (see Section 9.2), was not present in RTF. There are appreciable concentrations in RTF of a specific inhibitor of acrosomal proteinase or acrosin (Suominen and Setchell, 1972).

Rete testis fluid also contains a high concentration of an androgen-binding protein (ABP) which appears to be synthesized within the seminiferous tubules, probably by the Sertoli cells. In rats, there is no ABP in serum or lymph (French and Ritzen, 1973a, b); ABP in the rabbit is very similar to, if not identical with, the plasma androgen-binding globulin formed in the liver, but the production of the two proteins is under different control (Weddington et al., 1975). ABP production by Sertoli cells is increased in immature or hypophysectomized animals by treatment with FSH (see Hansson et al., 1975a, b) and testosterone (Tindall and Means, 1976). ABP is also formed in rats whose seminiferous tubules contain only Sertoli cells, produced by X-irradiation in utero (Tindall et al., 1974, 1975; see Section 11.1). However ABP cannot be detected in boar RTF so its general significance is still obscure. RTF contains appreciable concentrations of 'inhibin', a peptide hormone which is produced inside the tubules and which regulates pituitary secretion of FSH (see Section 6.2).

Certain free amino-acids are present in much higher concentrations in RTF than in blood plasma, but in this regard there is some variation between species. In the ram, bull, boar and wallaby, there are higher concentrations in RTF of glutamic acid, aspartic acid, glycine and alanine; in RTF of rats, glutamic acid is not higher than plasma but proline and lysine are, in addition to the other amino-acids mentioned above (Setchell et al., 1967; Tuck et al., 1970; Setchell, 1970a, b; Sexton et al., 1971).

Testosterone is present in ram, bull and monkey RTF in concentrations about a quarter of those in testicular lymph and in blood plasma from the internal spermatic vein (Voglmayr et al., 1966; Setchell et al., 1971; Cooper and Waites, 1974; Waites and Einer-Jensen, 1974; Amann and Ganjam, 1976; Ganjam and Amann, 1973, 1976); in rats and rabbits, RTF contains almost as much testosterone as testicular venous blood (Cooper and Waites, 1974; Bartke et al., 1975; Guerrero et al., 1975; Setchell and Main, 1975; Cooper et al., 1976). Rabbit and rat RTF also contain appreciable concentrations of 5α-dihydrotestosterone (Bartke et al., 1975; Guerrero et al., 1975), whereas bull RTF contains comparatively little of this steroid, but considerable amounts of dehydro-epiandrosterone and traces of other steroids (Ganjam and Amann, 1976; see Table 6.1).

In RTF of boars, there are high concentrations of oestrogens but the concentrations do not exceed those in testicular lymph or blood plasma from the internal spermatic vein (R. B. Heap and B. P. Setchell, unpublished observations).

### 8.2.2    Seminiferous tubule fluid (STF)

**8.2.2.1    Micropuncture free-flow fluid**    It has proved possible to collect fluid from the seminiferous tubules of rats and hamsters by micropuncture (using very fine glass tubes) (Figure 8.6). Only small volumes can be collected and the analyses have so far been limited to ions, proteins, inositol, and testosterone. Tubular fluid is appreciably different from rete testis fluid, and both fluids are quite unlike blood plasma or lymph (Figure 8.7). Rat tubular fluid is either isosmolar (Tuck et al., 1970) or

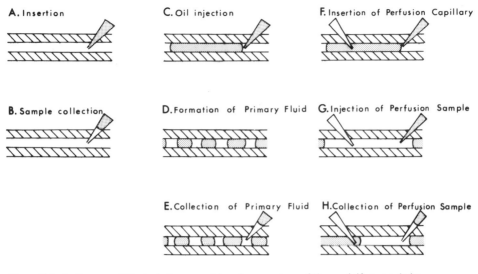

Figure 8.6 A diagram of the techniques used in micropuncture of the seminiferous tubules.
A. Insertion of oil-filled micropipette into seminiferous tubule
B. Withdrawal of sample
C. Injection of column of oil into seminiferous tubule
D. Formation of primary fluid by the cells lining the tubule, so that the oil column is broken up by droplets of aqueous fluid
E. This primary fluid can be sampled by introducing another micropipette into a droplet
F, G and H. Microperfusion of a length of seminiferous tubule; this is first filled with oil, which is then broken by a length of injected aqueous fluid of any chosen composition. At various times, portions of this fluid can be withdrawn for analysis.
(See Tuck et al., 1970; Henning and Young, 1971)

slightly hyperosmolar (Levine and Marsh, 1971) compared with plasma; hamster tubular fluid is appreciably hyperosmolar (Johnson and Howards, 1976).

Tubular fluid contains more potassium than fluid from the rete but less sodium and chloride (Tuck et al., 1970; Levine and Marsh, 1971). Presumably the anion deficit is made up by bicarbonate but no measurements have yet been made.

Tubular fluid contains more protein than RTF (Setchell et al., 1978) and there are a number of proteins there which are not present in rete testis fluid; rete testis fluid contains a higher proportion of albumin (Kormano et al., 1971) and γ-globulin (Koskimies et al., 1973) than tubular fluid. The inside of the tubule is very slightly negative electrically compared with the outside ($-7\cdot2$ mV) (Tuck et al., 1970). The concentration of inositol in free-flow tubular fluid is similar to that in rete fluid (Hinton et al., 1977). Testosterone is present in tubular fluid from rats in concentrations appreciably greater than those in RTF (Comhaire and Vermeulen, 1976).

Measurements of spermatocrit (i.e. the fraction of the fluid occupied by spermatozoa after centrifugation) in RTF and STF suggested that there were more spermatozoa in STF than RTF (Tuck et al., 1970; Levine and Marsh, 1971; Howards et al., 1976). However these measurements appear to have been affected by artefacts

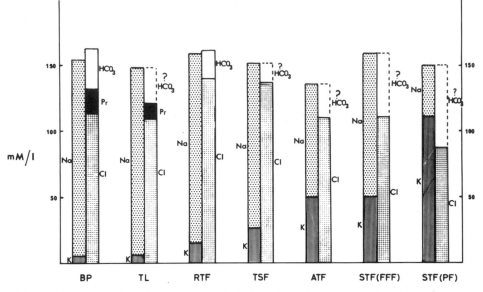

Figure 8.7 A graph of the ionic composition of rat blood plasma (BP), testicular lymph (TL) rete testis fluid (RTF), total secreted fluid (TSF—see Section 8.2.5), seminiferous tubular fluid by the difference technique (ATF—see Section 8.2.2.2), free-flow tubular fluid (STF(FFF)—see Section 8.2.2.1) and primary tubular fluid (STF(PF)—see Section 8.2.3) collected by micropuncture.
Na, sodium; K, potassium; Cl, chloride; HCO₃, bicarbonate, measured where shown as solid line or assumed to make up cation–anion equilibrium where shown with broken line and question mark); Pr, protein.

and when the spermatozoa were counted in measured volumes of the fluids, the concentrations found were almost identical (Setchell *et al.*, 1978).

**8.2.2.2    Difference technique**    If a rat testis is decapsulated, the cells dispersed and then centrifuged, a supernatant fluid can be separated (Figure 8.8). There is much more of this fluid if the efferent ducts have been ligated 24h earlier. From histological measurements, it can be calculated that about 70 per cent of the fluid from the unligated testis and about 90 per cent of the fluid from the ligated testis is fluid from the lumina of the seminiferous tubules. The rete and its enclosed fluid are removed with the capsule.

If it is assumed that both testes contain similar amounts of extratubular fluid, a calculation can be made of the composition of the 'additional' seminiferous tubular fluid, i.e.

Concentration in additional seminiferous tubular fluid =

$$\frac{\text{(Amount in fluid from ligated testis)} - \text{(Amount in fluid from unligated testis)}}{\text{(Weight of fluid from ligated testis)} - \text{(Weight of fluid from unligated testis)}}$$

The calculated ionic composition of this fluid is almost identical with free flow tubules fluid supporting the idea that this is mainly fluid from the lumina of the seminiferous tubules (Setchell *et al.*, 1976). The calculated concentration of inositol in 'additional tubular fluid' is also very similar to that in free flow tubular fluid collected by micropuncture (Hinton *et al.*, 1977).

### 8.2.3     Primary tubular fluid

During the micropuncture experiments in rats, it was noticed that if a column of oil was injected into a tubule, fluid continued to be secreted by the cells lining the tubules and broke up the oil into segments with droplets of secreted fluid (Figure 8.6). This fluid was called primary fluid and when sampled it proved to have more potassium and less sodium and chloride than free-flow fluid (Figure 8.7). Like free-flow fluid in rats, primary fluid was isosmotic with plasma and the transepithelial potential difference was also slightly negative ($-1\cdot2$mV, lumen negative). The high potassium concentration of primary fluid is most unusual being approached only by certain salivary secretions (Denton, 1950; Young and Martin, 1971), milk (Linzell and

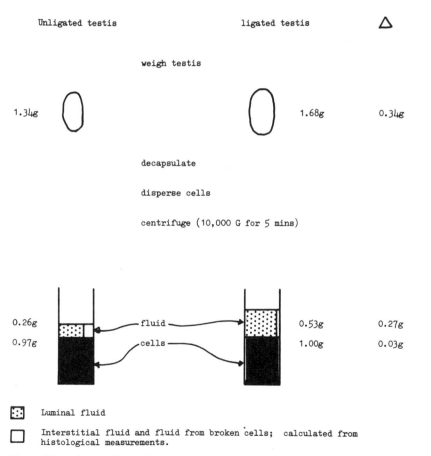

Figure 8.8  A diagram illustrating the procedure for calculating the composition of seminiferous tubular fluid by the difference technique.

Peaker, 1971) endolymph from the cochlea of the ear (Bosher and Warren, 1968) and the secretions of certain parts of the gut of insects (Maddrell, 1971).

If the remaining anion in primary fluid is bicarbonate and if the $P_{CO_2}$ is comparable to that in blood, then this fluid must be distinctly alkaline (Tuck *et al.*, 1970).

### 8.2.4   Perfusion fluid

A length of seminiferous tubule can be filled with an aqueous fluid, either isotonic mannitol, or a fluid mixed to resemble either free-flow fluid or primary fluid, and sampled after varying periods (Figure 8.6). It was found that over about two hours, the composition of all these fluids had changed to something between free-flow and primary fluid (Henning and Young, 1971).

### 8.2.5   Total secreted fluid

An estimate can be made of the composition of all the fluid retained in the testis following ligation of the efferent ducts by analysing the ligated testis and subtracting the equivalent values for the unligated testis, i.e.

Concentration in total secreted fluid =

$$\frac{(\text{Amount in ligated testis}) - (\text{Amount in unligated testis})}{(\text{Weight of ligated testis}) - (\text{Weight of unligated testis})}$$

The ionic composition of this fluid is between that of rete testis fluid and free-flow tubular fluid (Setchell, 1970c).

## 8.3   Flow rate of fluid from the testis

Fluid flows from a catheter placed in the rete testis at a rate of between 10 and $20\mu l$ fluid/g testis/h (Voglmayr *et al.*, 1967, 1970; Tuck *et al.*, 1970; Waites and Einer-Jensen, 1974). The flow is continuous, not intermittent and there appears to be no diurnal variation. Similar rates ($22\mu l/g/h$ for 45-day old rats, $14\mu l/g/h$ for adult rats) for total fluid secretion can be calculated from the rate at which the testis gains weight after efferent duct ligation (Setchell, 1970c).

Measurements of the rate of tubular secretion are much more difficult. The nature of fluid movement in the tubule can be deduced from the behaviour of a small droplet of coloured oil in the lumen. Movement can be in either direction, sometimes slow and continuous, occasionally very rapid and irregular while in some tubules a rapid movement in one direction was followed by a slow progression back in the opposite direction. However no measurements of fluid flow can be made because of the irregular nature of the movements. Similar variations in direction and velocity of movement have been observed during the infusion of aqueous solutions of dye, although most sections of tubule do seem to have a 'preferred' direction of movement.

Fluid flow in the tubules has been estimated in two ways. First by sampling as much fluid as possible from one tubule over a period of about 1h; about $1\mu l/h$ could consistently be obtained. Second, by infusing aqueous solutions of Lissamine green into one tubule at varying rates, and assessing visually the degree of dilution of the infused dye. This technique is obviously subject to considerable error, but it is clear that flow rates of $0.5\mu l/h$ tubule or more are usual (Setchell *et al.*, 1977). *In vitro* measurements on isolated tubules suggest a secretion rate of $0.44ml \, mm^{-1} \, min^{-1}$ or $48\mu l/g/h$ if the tubules comprise 90 per cent of the tissue and have an external diameter of $0.25mm$ (Cheung *et al.*, 1976).

## 8.4       Source of fluid and mechanism of production

From the data on ionic, protein and testosterone concentrations of STF and RTF, it has been concluded that two distinct secretions are involved. In the tubules there would seem to be the secretion of a comparatively small volume of a solution rich in potassium and bicarbonate, which is mixed with a larger volume of a secretion with plasma-like sodium and potassium concentration produced by either the tubuli recti or the rete testis (Tuck *et al.*, 1970). The mixing could take place in the tubules if fluid is drawn in from the rete by the peristalsis-like movements of the tubules (see Section 1.3.1). How much of the fluid flowing from the rete would originate in the tubules? If we assume that 'primary fluid' (see page 245) is representative of the fluid secreted in the tubules and that the ionic composition of the fluid secreted in the rete is similar to that of plasma, it can be calculated from the potassium values that 10 per cent of the rete testis fluid arises in the tubules. Similar calculations from the data for sodium concentrations give a value of 14 per cent. As the chloride concentration is higher and the bicarbonate is lower in rete testis fluid than in plasma or in primary fluid, similar calculations cannot be based on the concentrations of the anions. Alternatively, if we accept the two fluid theory in a simpler form, i.e. assuming that free-flow fluid represents the fluid secreted in the tubules and that the ionic composition of the rete secretion is similar to that of plasma, then it can be calculated that 22 per cent of RTF comes from the tubules if the values for sodium are used as the basis for calculation and 33 per cent if the potassium values are used.

The total fluid production by the testis of an adult rat is about $20\mu l/h$. Fluid movement along all the tubules would then be not more than $6.6\mu l/h$, and if there are about 30 tubules, as Clermont and Huckins (1961) state, each tubule would have a flow of less than $0.22\mu l/h$ if they all contribute to the same extent. If all the fluid leaving the rete originated in the tubules, then each would contribute about $0.7\mu l/h$. This latter figure is nearer the observed values, which therefore do not support the 'two-fluid theory'. The similar concentrations of inositol and spermatozoa in tubular and rete fluid also do not support this 'two-fluid' hypothesis.

It would therefore appear more likely that there is no appreciable secretion by the rete, and that the majority of the fluid leaving the rete originates in the tubules. The fall in potassium concentration could be explained by the greater permeability of the rete to potassium (see page 255). The fall in testosterone could be due to the lower concentration of steroids in the blood supplying the rete compared with that bathing the tubules, because of the peculiar anatomy of the testicular arteries (see page 61).

The difference in protein concentration is more difficult to explain but could be produced by selective absorption of protein by the cells of the tubuli recti.

The secretion of fluid in the testis is probably the result of a form of standing gradient (Diamond, 1971). These gradients are thought to be the usual way of forming secretions (Figure 8.9). Diamond suggests that epithelia can be classified as 'forward-facing' or 'backward-facing' depending on the direction of fluid flow in relation to the position of the tight junctions. The secretion of solute into the intercellular clefts leads to the establishment of an osmotic gradient; the locally high osmotic concentration leads to an inflow of water, the gradient being maintained by continuing secretion of solute.

The epithelium of the seminiferous tubules can be thought of as a forwards-facing secretory epithelium. The micropuncture evidence would suggest that the secreted

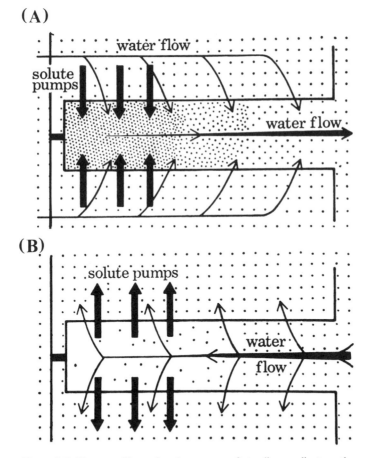

Figure 8.9 Diagrams illustrating the concept of standing gradients as they may apply to fluid secretion in the testis. A: forward-facing gradient, as assumed to be formed by the Sertoli cells, with the boundary tissue of the tubule to the left and the lumen to the right. B: backward-facing gradient as assumed to be formed by the cells lining the rete testis, with the lumen to the left. The thinner lines represent cell membranes, the thicker lines the 'tight' junctions. (Modified from Diamond, 1971)

solute is potassium bicarbonate and indeed it can be calculated that these two ions are held there against the electrochemical gradient, whereas this is not so for sodium and chloride (Tuck *et al.*, 1970). If there is secretion in the rete, the epithelium is backward-facing and the secreted solute would be probably sodium chloride. Whatever its origin, the production of fluid by the testis must be due to an active process as it is independent of blood pressure and continues despite an increase in intratesticular pressure. Secretion is dependent on a supply of glucose and oxygen and the maintenance of the correct temperature (Linzell and Setchell, 1969). Furthermore, metabolic alkalosis and inhibitors of carbonic anhydrase depress secretion (Setchell and Brown, 1972).

## 8.5    Control of fluid secretion

Almost all the information so far obtained relates to the control of the total fluid secretion by the testis and does not differentiate between effects on fluid secretion by the tubules and by the rete testis.

### 8.5.1    Fluid secretion and spermatogenesis

In young animals fluid secretion begins just before the first spermatozoa are shed by the germinal epithelium (Setchell, 1970c) and the fluid collected from these animals contains practically no spermatozoa (Figure 8.10; Setchell and Waites, 1972). Similarly, if the testes of rams and rats are heated to slightly above body temperature for between 1 and 3 hours, very few spermatozoa are released from the testis between about 20 and 40 days later. During the period of low sperm concentration, the rate of fluid secretion is unchanged or only slightly reduced.

In rats treated while still fetuses with Busulphan (Myleran) the testes are completely devoid of germ cells, the tubules contain only Sertoli cells and the testes are very much smaller than normal. These testes still secrete fluid at about the normal rate per unit weight of testis (Setchell, 1969) but the fluid in the tubular lumina contains plasma-like concentrations of sodium, potassium and chloride and may not represent normal Sertoli cell secretion (Tuck, 1969; Levine and Marsh, 1975).

The testes of rats X-irradiated *in utero* at 19 days *post coitum* also contain no germ cells. There appears to be little fluid secretion by the testes of these rats, and again the tubule fluid appears to be plasma-like in its concentrations of sodium and potassium (Setchell *et al.*, 1977).

In adult rats treated with Busulphan the difference in weight between ligated and unligated testes increased, both in absolute terms and in mg/g testis suggesting that fluid secretion was temporarily increased by this treatment; the maximal increase occurred at a time when most of the spermatocytes were absent and when the weight of the unligated testis was minimal (Laporte and Gillet, 1975).

### 8.5.2    Hormones and drugs

The beginning of fluid secretion just before the liberation of the first spermatozoa would perhaps suggest that both processes are under the control of gonadotrophic

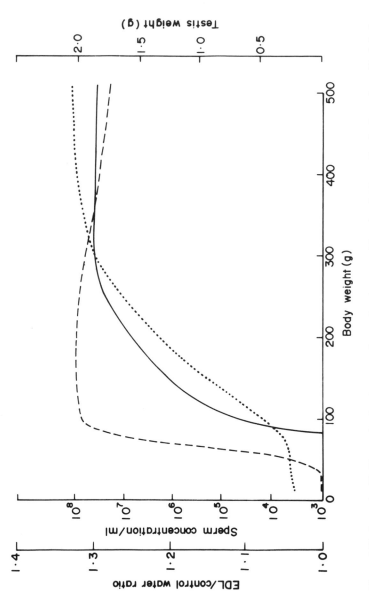

Figure 8.10 A graph of fluid secretion by the testes of rats of various ages. Note that fluid secretion begins quite suddenly before the first spermatozoa are released. The lines were drawn from data from 80 rats. Fluid secretion (– – – – –) was measured as the ratio of the water content of one testis with its efferent ducts ligated 24h previously to that of the contralateral testis of the same animal. Note that the sperm concentration (————) is plotted logarithmically, and testis weight (············) and fluid secretion linearly.

hormones. The secretion of fluid by the testes of immature rats and of adult rats treated with oestrogen implants can be stimulated slightly by several days treatment with FSH, but not by any of the other pituitary hormones (Setchell *et al.*, 1973). Burgos and Vitale-Calpe (1969) have suggested that LH increased the flow of fluid from a catheter in the rete testis of hamsters but we found no evidence in rats or hamsters in support of this suggestion, using weight gain of the testis after efferent duct ligation to measure fluid secretion (Setchell *et al.*, 1973). Long-term administration of placental gonadotrophins to rats and shorter periods of treatment in rams were also without effect on fluid secretion (Barack, 1968; Setchell *et al.*, 1973); hypophysectomy of rats did not cause any decrease in fluid secretion until the testis begin to regress and decrease in weight (Barack, 1960; Setchell, 1970c). Neither atropine nor pilocarpine affected the flow of rete testis fluid; oxytocin was also ineffective (Setchell and Linzell, 1968) except for some suggestion that the concentration of spermatozoa was increased (Voglmayr, 1975a). Thus it would appear that the flow of fluid is not hormonally controlled except in a very general way associated with the structural integrity of the testis.

Inhibitors of carbonic anhydrase cause a short-lived decrease in secretion rate (Setchell and Brown, 1972). Presumably the effects do not last longer because of the compensatory changes in the composition of the blood, which occur after the activity of carbonic anhydrase is decreased.

### 8.5.3    Temperature
During local heating, fluid secretion by the ram testis is decreased both *in vivo* (Setchell *et al.*, 1971) and in the isolated perfused testis (Linzell and Setchell, 1969). This may be an important step in the damage due to the germinal cells caused by heat (see Section 11.1). However, returning the testes of rats to their abdominal cavities has no immediate effect on fluid secretion (Setchell, 1970c).

The effect of cold has not been fully studied but preliminary observations with isolated rat seminiferous tubules (Cheung *et al.*, 1976) and in anaesthetized rats, rams and boars would suggest that decreases in testicular temperature lead to falls in the rate of fluid production.

### 8.5.4    Changes in blood composition
Fluid production by the perfused testis is dependent on the maintenance of an adequate glucose concentration in the circulating blood, and an adequate supply of oxygen (Linzell and Setchell, 1969). If the pH of the blood is raised by infusing sodium acetate or sodium bicarbonate, fluid secretion by ram and rat testes is decreased (Setchell and Brown, 1972).

## 8.6    Factors affecting composition of fluid

The chemical composition of rete testis fluid seems to remain surprisingly constant under a wide variety of conditions. Human chorionic gonadotrophin increases the testosterone concentration of ram and rat RTF (Bartke *et al.*, 1975) but does not seem to change the ionic composition of ram rete testis fluid. The concentration of spermatozoa in RTF of rams and rats decreases by more than a thousand times after the testis is heated but these changes are not accompanied by appreciable changes in ionic composition, or in the concentration of lactate, protein or testosterone. There was a slight decrease in the concentrations of glutamate and inositol in ram RTF as the concentrations of spermatozoa fell, but in rats there was no correlation of sperm concentration with the concentrations of inositol or glycine (Setchell *et al.*, 1971; Setchell and Waites, 1972).

The concentration of inositol in RTF of rats was increased by feeding or infusing galactose (Middleton, 1973), but infusions of galactose, glucose, or glucose plus insulin were without effects in rams. In rats, significant concentrations of galactitol appear after galactose administration. As already mentioned (Section 8.5), tubular fluid in rats with only Sertoli cells in their tubules is plasma-like in its ionic composition. This is so for rats treated *in utero* with Busulphan or X-rays.

## 8.7    Entry of substances into the testis: the blood–testis barrier

The peculiar composition of the fluid inside the rete testis and seminiferous tubules suggests that substances do not diffuse readily into or out of the tubules. This has been found to be the case.

### 8.7.1    Dyes
It was observed many years ago that a number of injected dyes could not be detected inside the seminiferous tubules in histological sections (Ribbert, 1904; Bouffard, 1906; Goldmann, 1909; Pari, 1910; de Bruyn *et al.*, 1950; Goldacre and Sylven, 1959, 1962). More recently, Kormano (1967a, 1968) has shown that the exclusion of dyes does not occur in young animals. Johnson (1969, 1970a) has also shown that the barrier to acriflavine is broken down by prior administration of cadmium salts or autoimmune damage.

### 8.7.2    Small molecules
The studies with substances of low molecular weight have in most cases involved intravenous infusions of the marker and the measurement of its concentration in rete testis fluid, testicular lymph or tubular fluid sampled by micropuncture, or the calculation of its concentration in total secreted or additional tubular fluid. Alternatively, it is possible to infuse a radioactive marker and measure its volume of distribution in the tissue.

Tritiated water in blood plasma readily equilibrates with water in lymph, rete testis fluid and tubular fluid; substances such as urea, ethanol, carbon dioxide and 3-O-methyl glucose (which is transported like glucose but not metabolized—glucose itself cannot be studied because it is metabolized as quickly as it enters the tubules and none appears in the fluid) enter ram and rat RTF slightly more slowly but reach equilibrium with blood plasma in less than 3 hours. Urea enters the tubules of the hamster testis more slowly. All the ions studied, $Na^+$, $K^+$, $Rb^+$, $Cl^-$, $CNS^-$ and $I^-$

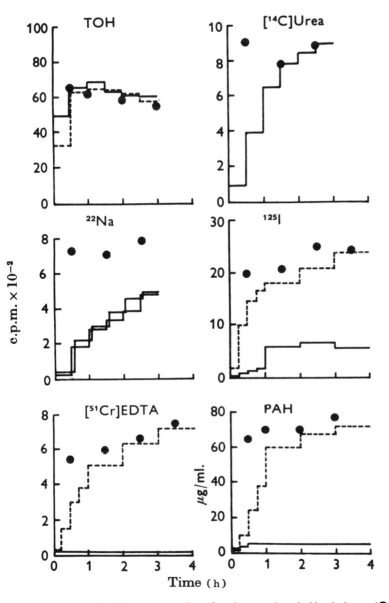

Figure 8.11 A graph of the concentration of various markers in blood plasma (●), testicular lymph (— — —) and rete testis fluid (————) of rams during intravenous infusions of the markers. The two lines for rete testis fluid in the $^{22}$Na panel refer to the two testes of the same animal. (Reproduced from Setchell et al., 1969)

pass slowly into the fluid, as do galactose and creatinine, and do not reach equilibrium within the course of the usual experiments of 3 to 6 hours. There are other substances which do not enter RTF at all such as inulin and [$^{51}$Cr]-EDTA (which are filtered by the kidney glomerulus), *p*-aminohippurate (which is actively excreted by the proximal convoluted tubules of the kidneys) glutamate and inositol (Figures 8.11, 8.12 and 8.13; Setchell, 1967; Setchell *et al.*, 1969b; Waites *et al.*, 1972; Middleton and Setchell, 1972; Howards *et al.*, 1976).

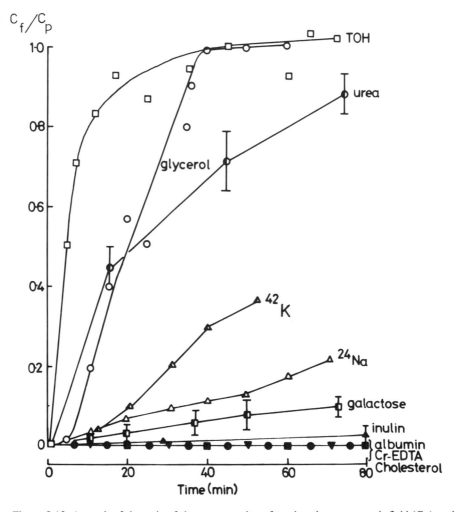

Figure 8.12 A graph of the ratio of the concentration of markers in rat rete testis fluid ($C_f$) to that in blood plasma ($C_p$). Data replotted, for tritiated water (□), $^{42}$K (▲), $^{24}$Na (△) albumin (■) and Cr-EDTA (▼) from Main and Waites (1973) and Main (1976); for glycerol (○) from Edwards *et al.* (1975); for urea (◑; m ± SD), galactose (▯; m ± SD) and inulin (▲, m ± SD) from Okumura *et al.* (1975); and for cholesterol (●) from Waites *et al.* (1973) and Cooper and Waites (1975). (Reproduced from Waites, 1977)

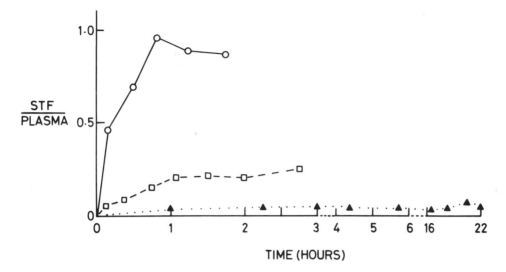

Figure 8.13 A graph of the concentration (plotted as a ratio to that in plasma) of tritiated water ( ○ ), urea ( □ ) and inulin ( ▲ ) in hamster seminiferous tubular fluid collected by micropuncture at various times after intravenous injection or the beginning of an intravenous infusion of the marker. (Data of Howards *et al.*, 1976)

A number of these markers have also been used in experiments in which their concentration has been calculated in total secreted fluid and additional tubular fluid (see p. 244) and their volumes of distribution in the testis calculated. The studies confirm that inulin and sucrose do not enter tubular fluid, and 3-O-methyl glucose, sodium, potassium and rubidium enter slowly. The rete appears to be more permeable than the tubules to potassium (Figure 8.14; Setchell *et al.*, 1977).

Studies with isolated seminiferous tubules, prepared by a technique originally described by de Graaf in 1668, are difficult to interpret because markers like inulin and sucrose, which are confined to a volume of about 10 per cent of the normal testis *in vivo*, rapidly penetrate through a much larger fraction of volume of isolated tubules, whether they are incubated as teased-out bundles or in special chambers with their cut ends occluded. This suggests that the permeability of isolated tubules is abnormal (Setchell and Singleton, 1971).

### 8.7.3    Proteins

Rete testis fluid contains much less protein than blood plasma or lymph and the composition of the proteins in the fluid is quite distinct (see Section 8.2.1). This suggests that proteins do not pass readily into the seminiferous tubules. Some non-quantitative studies with proteins tagged either with fluorescent or radioactive markers have been widely quoted as indicating that proteins, and particularly pituitary gonadotrophins, do enter the seminiferous tubules (Mancini *et al.*, 1965, 1967; Castro *et al.*, 1970, 1972), although other studies using fluorescent antibodies

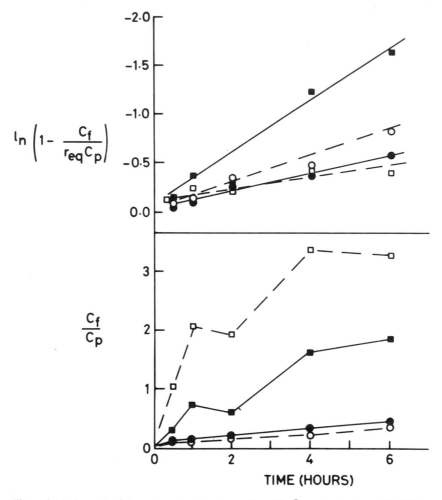

Figure 8.14 A graph of the entry of radioactive sodium (●,○) and potassium (■,□) into rete testis fluid (●,■) and seminiferous tubular fluid (○,□) of rats, and the calculation of the transfer constants ($k_{out}$) for these markers and fluids. $K_{out}$ is the slope (min$^{-1}$) of the line of ln $(1 - \frac{C_f}{r_{eq}C_p})$ plotted against time; $C_f$, concentration in fluid; $C_p$, concentration in plasma; $r_{eq}$, the ratio of the concentrations of the non-radioactive substances in the fluids to that in plasma. This latter calculation takes into account the different concentrations of non-radioactive sodium and potassium in the fluids. (Data of S. J. Main and B. P. Setchell, unpublished)

to specific proteins suggest that very little protein enters the tubules (Figure 8.15) except after damage by the injection of cadmium salts ʹGupta et al., 1967).

Furthermore, there is evidence obtained by electron microscopy that peroxidase, a protein with a molecular weight of about 30,000 Daltons is excluded from the tubules. The exact site of the block to entry has been demonstrated, partially at the myoid cells in the peritubular tissue and finally at the junctions between adjacent Sertoli cells (see Section 8.7.6).

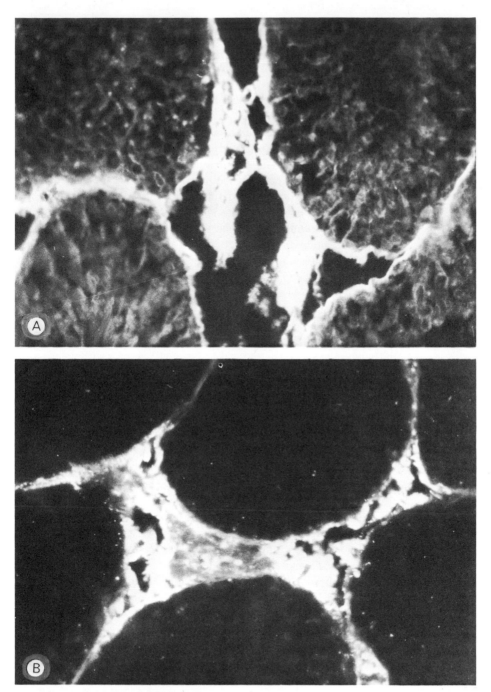

Figure 8.15 A. A section of a testis of a guinea pig after incubation with fluorescein-conjugated rabbit antiserum to guinea pig $\gamma$-globulin.
B. A section of a testis of a guinea pig which had been injected intravenously 30 minutes previously with fluorescein-labelled albumin. Note that in both sections the staining is virtually confined to the interstitial tissue. (Reproduced from Johnson, 1972)

The apparent contradiction is probably due to quantitative differences in the rates of entry implied. If the transfer of iodine-labelled albumin from blood plasma to rete testis fluid was studied, entry could be demonstrated but the specific radioactivity (i.e. cpm/mg protein) did not reach equilibrium between blood and RTF for about 4 days (Setchell and Wallace, 1972).

Insulin, which is present in rete testis fluid at about the same concentration per mg protein as in blood plasma, enters more quickly, as one might expect for a smaller molecule, but it still requires about 24 hours to reach equilibrium.

It is not possible to quantitate the entry of iodinated LH into the tubules of the rat testis because it breaks down too quickly in the blood, but entry is certainly not rapid. The entry into the tubules of iodinated FSH, prepared by three different techniques has been measured and shown to be very slow (Figure 8.16). Even with FSH a certain amount of breakdown does occur in the circulation and the free radioactive iodide so formed enters the tubules more quickly. However when correction is made for this, and the radioactivity inside the tubules separated by precipitation with trichloracetic acid or specific antisera, it can be shown that very little unchanged FSH reaches the lumina of the tubules even over 24 hours (Setchell et al., 1976).

This does not mean that FSH is not able to reach any of the cells in the tubules. It has been shown that FSH acts biochemically on the Sertoli cells (see Section 10.2.2) and the hormone would have ready access to the basal surfaces of these cells without crossing the blood-testis barrier.

### 8.7.4　Steroids

The possible production of steroids by the cells inside the seminiferous tubules is discussed elsewhere (see p. 117). Here we will consider only the evidence for the entry of preformed steroids into the tubules, although the two problems are naturally closely related. Studies have been done with isolated seminiferous tubules but these are impossible to interpret quantitatively because of the anomalous results found for the inulin and sucrose spaces of this preparation (see Section 8.7.2).

The earlier in vivo experiments involved injection of radioactively labelled steroids into the animal and then autoradiography of the testis or separation of the testis into tubules and interstitial tissue. The results of these experiments were also rather confusing, possibly because of differences in technique (see Setchell and Waites, 1975b).

The situation has now been clarified to some extent by experiments in which labelled steroids were infused intravenously and measurements were made of their entry into rete testis fluid and additional tubular fluid. Testosterone and dehydroepiandrosterone appeared to be readily transferred from blood to rete testis fluid, whereas cholesterol was excluded. Between these extremes the appearance of radioactivity in rete testis fluid suggested the following order of entry rates: progesterone > pregnenolone > $5\alpha$-reduced androgens > oestrogens > corticosteroids (Figure 8.17). However some transformation of labelled androstenedione and progesterone occurred during passage into the rete testis fluid (Cooper and Waites, 1975). The entry rates for testosterone, androstenedione, $5\alpha$-dihydrotestosterone, $5\alpha$-androstane-$3\alpha,17\beta$-diol and cholesterol into additional tubular fluid were similar to those into rete testis fluid (Figure 8.18).

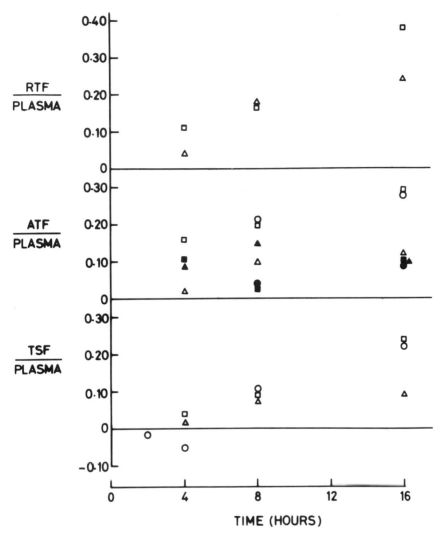

Figure 8.16 A graph of the concentrations (plotted as ratios to those in blood plasma) of total radioactivity (open symbols) and trichloracetic acid preciptable radioactivity (filled symbols) in rete testis fluid (RTF), seminiferous tubular fluid by the difference technique (ATF) and total secreted fluid (TSF) after injection of radioactive iodine-labelled rat FSH, prepared by the chloramine-T technique (○,●), using peroxidase (□,■) or succinimidyl-hydroxyiodo-phenol propionate (△,▲). (Data from Setchell et al., 1976)

However as testicular venous blood samples were collected in these latter experiments as well as arterial blood, it was noticed that [3]H-androstenedione in arterial blood was almost all transformed to [3]H-testosterone in venous blood, rete testis and tubular fluids (Setchell and Main, 1975). This emphasizes the danger of assuming that the composition of arterial blood is representative of the fluid bathing the tubules.

In summary, it would appear that, in the rat, testosterone does penetrate readily into the seminiferous tubules but that many other steroids do not. In other species, the situation may be different. In rams, although testosterone does pass readily from blood to rete testis fluid, the final concentration reached there is only about 30 per cent of that in plasma (Cooper and Waites, 1975). In boars, radioactive

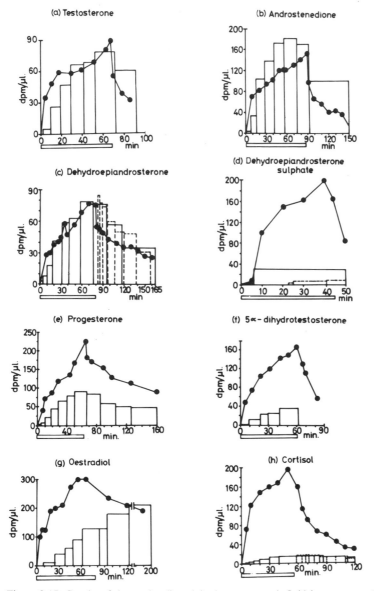

Figure 8.17 Graphs of the total radioactivity in rat rete testis fluid (_____ and _ _ _ _) and blood plasma (●) during and after intravenous infusion (bar under graph) of various tritium-labelled steroids. (Data of Cooper and Waites, 1975), (reproduced from Waites, 1977)

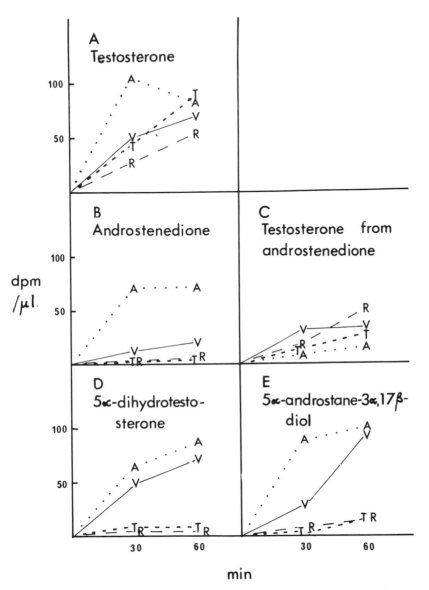

Figure 8.18 Graphs of the radioactivity, running on thin layer chromatograms with authentic infused compound (A, B, D and E) or with testosterone during infusion of androstenedione (C) in rete testis fluid (R), tubular fluid (T) testicular venous (V) and arterial blood (A) during intravenous infusion of tritium-labelled steroids. Note that during the infusion of androstenedione most of the radioactivity in testicular venous blood is not androstenedione but testosterone, and that androstenedione itself does not appear in the rete testis or tubular fluid. (Data from Setchell and Main, 1975)

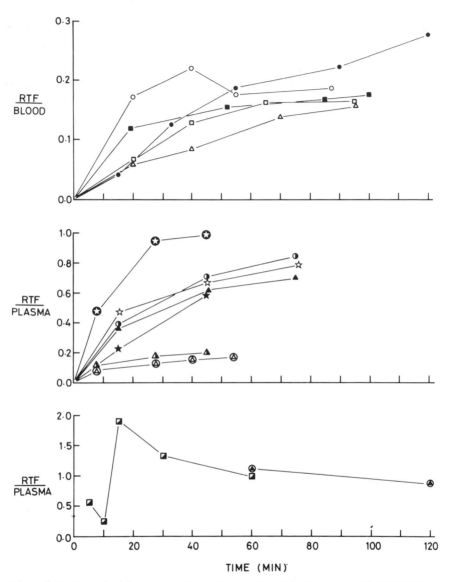

Figure 8.19 A graph of the concentrations of total radioactivity in rat rete testis fluid (plotted as a ratio to that in blood or blood plasma) after the injection of various compounds.

o  methylene dimethanesulphonate          ▲ sulphaguanidine
•  ethane dimethanesulphonate             ⊕ salicylic acid
◻  propane dimethanesulphonate            ☆ sulphamethoxypyridizine
▪  butane dimethanesulphonate             ★ sulphanilamide
△  nonane dimethanesulphonate             ✪ thiopental
▲  barbital                               ⊕ dimethynitrosamine
◑  pentobarbital                          ⬓ methyl methane sulphonate

(Data from Waites *et al.*, 1973; Okumura *et al.*, 1975; and Setchell and Main, 1977)

pregnenolone or progesterone do not pass readily from blood to rete testis fluid. Some radioactivity does appear in the fluid; it is not pregnenolone, progesterone or testosterone but it is in the form of a highly polar steroid which is also found in venous blood and lymph (R. B. Heap and B. P. Setchell, unpublished observations, see Setchell, 1974b).

### 8.7.5    Drugs

Few studies have been made of the penetration of drugs through the blood–testis barrier. A number of antifertility drugs have been shown, not surprisingly, to pass readily through the barrier (Waites et al., 1972; Edwards et al., 1975). On the other hand, the penetration of a number of barbiturates and sulphonamides was related to their lipid solubility, not to molecular size (Okumura et al., 1975). Methyl methanesulphonate and dimethylnitrosamine, both potent mutagens, appear to penetrate the blood–testis barrier readily (Figure 8.19; Setchell and Main, 1977). Actinomycin D does not appear to enter the tubules (Ro and Busch, 1965).

### 8.7.6    Anatomy of the blood–testis barrier

The myoid contractile cells in the walls of the seminiferous tubules form the first barrier to diffusion in rodents but not in primates. However, even in rodents, the effectiveness of this barrier is very variable. Inside the tubules the only cells abutting on the tunica propria are spermatogonia and Sertoli cells. All the other germinal cells, spermatocytes, spermatids and developing spermatozoa are sandwiched between pairs of Sertoli cells or embedded in their luminal surfaces. Pairs of Sertoli cells are joined above the spermatogonia but below the spermatocytes by specialized junctions. It has been suggested that these junctions are the principal component of the blood–testis barrier and that they divide the intercellular spaces into a 'basal' compartment around the spermatogonia and between them and Sertoli cells, and an 'adluminal' compartment between the Sertoli cells and the other germinal cells (Figure 8.20).

The sites of restricted permeability have been defined by examining the testes of animals injected with various electron opaque markers. In the rat and guinea pig, the larger markers such as carbon, thorium and ferritin spread throughout the interstitial tissue but do not pass the myoid contractile cells, which form tight end-to-end junctions with one another. Smaller markers, like peroxidase and lanthanum, are also stopped at the narrow junctions between pairs of myoid contractile cells at most sites. In certain areas of the tubule the markers get past these cells and then penetrate between the spermatogonia and the Sertoli cells and between adjacent Sertoli cells as far as their specialized junctions. There the markers stop (Figure 8.21). The penetration of the markers through the peritubular layer seems to be random and is not associated with any particular stage of the spermatogenic cycle (Dym and Fawcett, 1970; Fawcett et al., 1970).

The detailed structure of these Sertoli cell junctions has been examined using freeze-fracture preparations in the electron microscope, as well as thin sections. In section, the occluding junctions are characterized by: (1) the presence of an extensive series of focal, pentalaminar contacts where the opposing membranes appear to fuse;

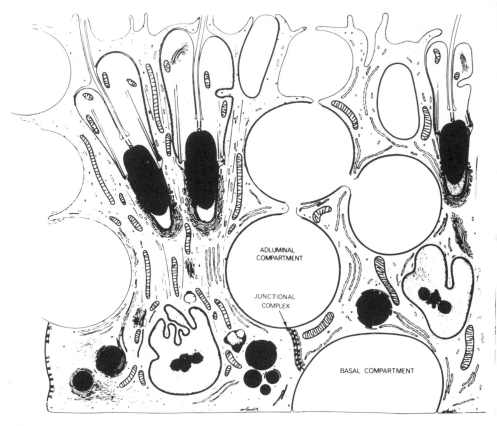

Figure 8.20 Drawing illustrating the manner in which the occluding junctional complexes between Sertoli cells divide the seminiferous epithelium into a basal compartment occupied by the spermatogonia and pre-leptotene spermatocytes and an adluminal compartment containing more advanced stages of the germ cell population. The occluding Sertoli–Sertoli junctions are the principal component of the blood–testis barrier. (Reproduced from Fawcett, 1975)

(2) occasional small septilaminar areas that resemble gap junctions, but which are more variable in length; (3) discrete bundles of filaments, usually found in the superficial cytoplasm of both cells, subjacent to the series of focal membrane contacts (the filaments run parallel to the cell surface); and (4) cisternae of the endoplasmic reticulum in both cells, orientated parallel to the cell boundary and lying deep to the layer of filaments. The cisternae are irregularly fenestrated and therefore present discontinuous profiles of varying length. They often bear ribosomes on the membrane towards the cell body, but are agranular on the side adjacent to the filament bundles (Figure 8.22). In freeze-fractured testes, the outer membrane half (B-face) of the junctions show up to 40 parallel rows of particles. The particles vary in size from 65 to 110Å; the rows are arranged parallel to the wall of the tubule (Figure 8.23). On the inner membrane half (A-face), there are occasionally elongated aggregations of particles of uniform size, about 70Å, arranged in one or more closely

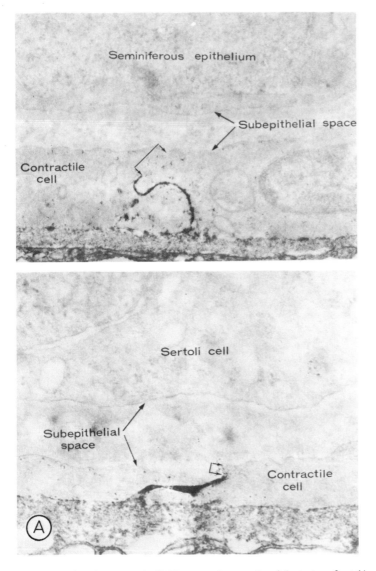

Figure 8.21 (continued overleaf) Electron micrographs of the testes of rat (A and B) and monkey (C) showing the restrictions on the penetration of lanthanum nitrate infused intra-arterially with the fixative.

(A) The marker has penetrated through the capillary walls, filled the lymphatic sinusoids and penetrated between the two myoid or contractile cells. It is stopped by the junction between these cells and has not entered the subepithelial space.

(B) In this section the marker has penetrated between the myoid cells (at the bottom of the picture) filled the subepithelial space, and penetrated between the spermatogonium and the Sertoli cells on either side of it. It has also penetrated into the clefts between pairs of Sertoli cells as far as the junctional complexes (arrows) where it stops.

(C) In this section, as in B, the marker has penetrated around the spermatogonium and between the Sertoli cells as far as the junctional complex. (Reproduced from Dym and Fawcett, 1970, and Dym, 1973)

packed rows. These are probably atypical gap junctions and are found adjacent to the linear depressions on the A-face which correspond to the particles on the B-face described above.

In immature testes, occluding junctions are absent; typical gap junctions are common, but gradually disappear. In the second week after birth, the linear arrays of particles appear on the B-face. Initially they are variable in direction but gradually adopt a consistent orientation parallel to the cell base, at the time of the establishment of the blood–testis barrier (Gilula *et al.*, 1976).

There is some suggestion that the rete may be a 'weak point' in the blood–testis barrier and that it is more permeable to dyes and protein than the tubules themselves

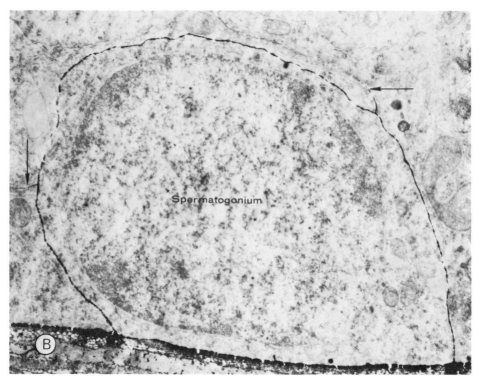

Figure 8.21 (B)

(Waksman, 1960; Kormano, 1967a; Johnson, 1970a; Kormano *et al.*, 1971; Koskimies and Kormano, 1973). Furthermore the immunological reaction which follows immunization of the animal with testis and a suitable adjuvant often begins at the rete and then spreads along the tubules (see Section 11.4).

### 8.7.7    Disturbances of the blood–testis barrier

Alterations in the function of the blood–testis barrier may be important in the development of certain forms of testicular damage. The entry of some substances into the seminiferous tubules is probably linked to the process of fluid secretion;

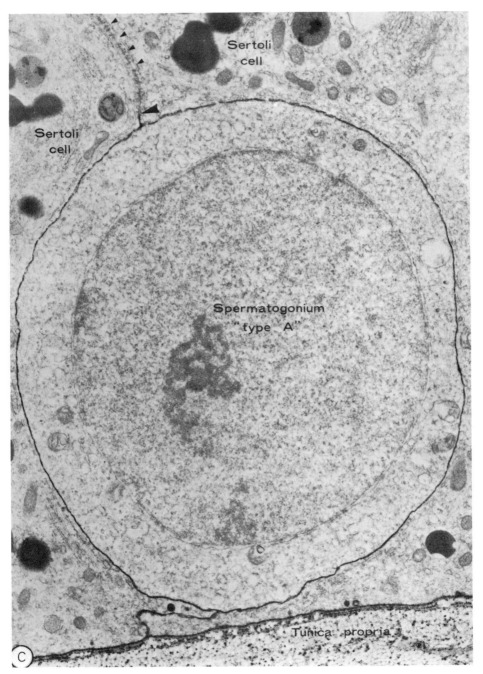

Figure 8.21 (C)

Cistern of the
  reticulum

Filaments

Cell membrane

Intercellular space

Cell membrane

Cistern of the
  reticulum

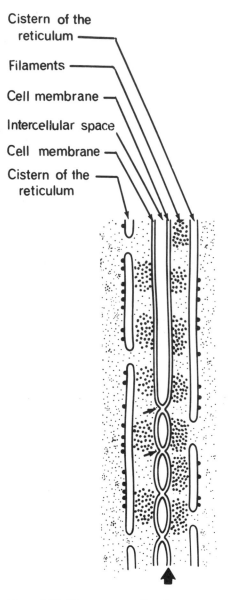

Figure 8.22 Diagram presenting the components of the junctional complex between Sertoli cells. The opposing membranes are fused at multiple levels indicated by the arrows. The subjacent cytoplasm contains bundles of filaments coursing parallel to the cell surface. Deep to these in each cell are cisternae of the endoplasmic reticulum, bearing ribosomes on the side toward the cell body. If the intercellular cleft is filled with electron-opaque material, the membrane fusions can be shown to be linear. If the Sertoli cell membrane is split by freeze-cleaving along the line indicated by the broad arrow, parallel linear arrays of intramembranous particles are seen. See Figure 8.20 for the location of the Sertoli–Sertoli junctional complex. (Reproduced from Fawcett, 1975)

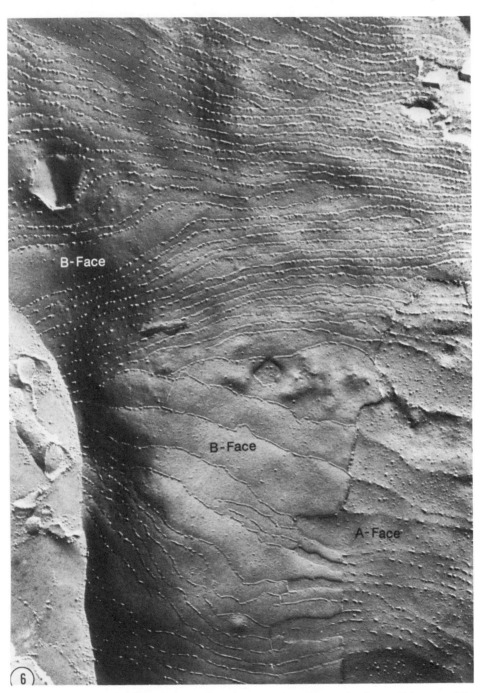

Figure 8.23 Replica of a freeze-fractured Sertoli junctional membrane showing multiple rows of intramembrane particles. The linear arrays of particles course circumferentially around the base of the cell and are roughly parallel. Unlike the juxtaluminal occluding junctions in other epithelia, the rows do not anastomose, and they are preferentially associated with the B-face (i.e. the outer half-membrane). Where the rows on the B-face appear discontinuous, the missing particles are associated with the complementary A-face, as can be seen at the lower right of the picture. (Reproduced from Gilula *et al.*, 1976)

while the testes of rams were being heated, fluid secretion was decreased although returning the testes of rats to the abdomen had no effect on fluid secretion (see Section 8.5.3). Results of direct measurements of the permeability of the blood–testis barrier with $^{86}$Rb give confusing results, depending on whether the observations were made with isolated tubules, measurement of the volume of distribution *in vivo* or direct measurements of the appearance of the marker in RTF (see Setchell and Waites, 1975b). The normally low transfer of albumin into RTF was unaffected by temperature, but the transfer of potassium, sodium and lysine was increased by heat; the normally rapid entry of testosterone was unaffected but heat increased the entry rate of more slowly penetrating steroids, dihydrotestosterone, oestradiol and cortisol (Main and Waites, 1977).

Cadmium salts also produce an increase in permeability of the blood–testis barrier (Setchell and Waites, 1970), but these changes may be more marked in the rete than in the tubules (Koskimies, 1973a, b).

Immunization of the animal with its own testis mixed with suitable adjuvants also produces some complex changes in the blood–testis barrier, but again the rete may be involved rather than the seminiferous tubules (see Section 11.4).

An indication of the stability of the Sertoli junctions is given by experiments in which hypertonic solutions were perfused through the blood vessels of the testis just before fixation. This treatment is enough to open the blood–brain barrier and to dissociate other occluding junctions in a matter of seconds. The Sertoli junctions and the cells in the adluminal compartment are unaffected by this treatment (Figure 8.24), although the spermatogonia in the basal compartment are shrunken and detached from the basement lamina, so that the basal compartment appears empty (Gilula *et al.*, 1976).

## 8.7.8      Significance of the blood–testis barrier

As the blood–testis barrier develops only at the time of puberty and the beginning of spermatogenesis, its principal function must be to create within the tubules the correct milieu for spermatogenesis and in particular for meiosis. We do not know how this is achieved but the unique composition of tubular fluid probably mirrors the conditions which must be established.

Some of the substances present in high concentration in the fluid such as glycine, glutamine and aspartic acid may be important in the synthesis of purine and pyrimidine bases (see Setchell and Waites, 1975b), but it is not even known whether the testis synthesizes its own bases or extracts them as such from the blood. Glutamic acid is one of the amino-acids found necessary for the normal development of duck and chicken gonads in tissue culture (Stengen-Haffen, 1957) and glutamine has been shown to stimulate the differentiation of testicular cells in chemically defined media (see Section 7.6.2).

In view of the relatively high concentration of $\alpha_2$-macroglobulin in RTF, it is interesting to recall that $\alpha_2$-macroglobulin will stimulate DNA synthesis by bone marrow cells after X-irradiation (Berenblum *et al.*, 1968) and that these or similar proteins are present in elevated concentrations in blood plasma during periods of rapid cell division (Heim, 1968). It has been shown that $\alpha_2$-macroglobulin binds and

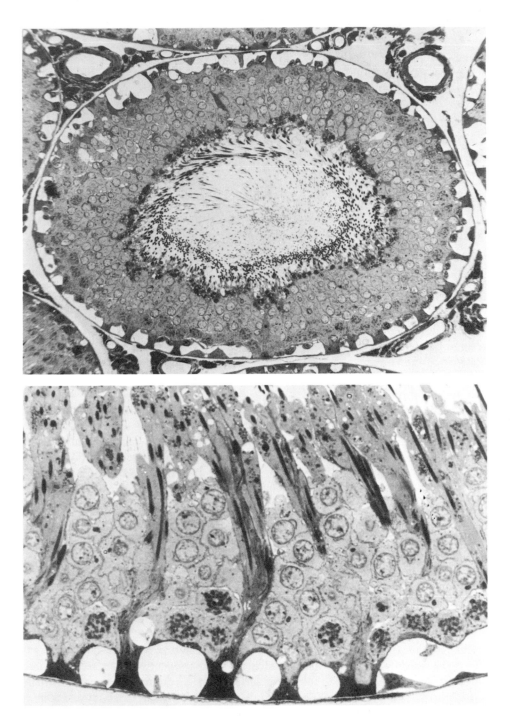

Figure 8.24 Photomicrograph of a seminiferous tubule from a rat testis perfused with hypertonic lithium chloride solution a few minutes before fixation. The enlarged lower picture shows that the basal cytoplasm of the Sertoli cells is condensed and shrunken but the cells remain attached to the basement lamina. The spermatogonia are shrunken and detached from the basement lamina so that the basal compartment appears empty. The spermatocytes and spermatids show no osmotic damage. (Reproduced from Gilula *et al.*, 1976)

inhibits a number of proteinases (Barrett and Starkey, 1973).

Another clue to a possible functional significance of the barrier may lie in the separation of the spermatogonia and the other germinal cells inside the basal and adluminal compartments respectively (see Section 8.7.6). The spermatogonia are diploid cells dividing mitotically like stem cells elsewhere in the body, the only differences being that the spermatogonial divisions are synchronized along appreciable lengths of the tubules. This synchrony may be due, at least in part, to the persistence of cellular linkages after the cells have divided (see Section 7.2) or to other factors associated with the spermatogenic cycle. The spermatocytes, on the other hand, enter the long prophase of the meiotic division at about the same time as they move from the basal to the adluminal compartment and lose contact with the tunica propria. The exact timing of this separation is difficult to establish but it may be argued that the Sertoli cells, by removing the spermatocytes from the basal compartment, commit them to enter into meiosis, whereas those cells remaining in the basal compartment continue to divide mitotically.

Probably incidental to this main function, the barrier also restricts the entry of proteins into the tubules. This means that hormones like FSH which act inside the tubules must do so via the Sertoli cells or spermatogonia, which are effectively outside the barrier (see Section 10.2.2). The virtual exclusion of proteins also has immunological consequences. The body does not recognize the proteins of its own haploid germ cells as self because its immunological system normally does not come in contact with them.

Similarly, circulating antibodies which might react with the haploid cells are normally excluded from the tubule. The differential permeability to various steroids means that a non-permeant steroid formed inside the tubule from a permeant one will accumulate there without the expenditure of energy.

Another important consequence of the barrier is that it enables a high concentration to be maintained in the fluid of a peptide which selectively inhibits the acrosomal proteinase or 'acrosin' which enables the spermatozoa to penetrate the ovum at the moment of fertilization. Acrosin also allows spermatozoa to penetrate other cells and it is therefore not surprising that this inhibitory peptide is present when the spermatozoa leave the germinal epithelium (Suominen and Setchell, 1972, 1976) and is not just added to the semen at the moment of ejaculation as was believed previously. Its action is probably reinforced by the action of $\alpha_2$-macroglobulin which also combines with, and inhibits, proteinases. This latter inhibition is more pronounced for large substrate molecules, and proteinases bound to $\alpha_2$-macroglobulins can be further inhibited by small inhibitor molecules. The $\alpha_2$-macroglobulin and the peptide inhibitor together may thus be important in dealing with any proteinases which leak out of the acrosome. It will also be important to see whether the macroglobulin molecules attach themselves to the sperm surface. If they do, their removal may be a part of capacitation. The small peptide inhibitor from RTF can block fertilization *in vitro* (Suominen *et al.*, 1973), has a molecular weight of about 6000 (Suominen and Setchell, 1976) and would probably be lost from the fluid into the bloodstream if the barrier did not exist.

One final point to remember concerns the action of drugs on the seminiferous tubules. A drug could be excluded or, on the other hand, could be transformed into a metabolite which is retained or even concentrated inside the tubules, with possible mutagenic dangers (Setchell and Main, 1977). The blood–testis barrier should be kept in mind with all drugs, not just those which we want to get inside the tubules.


## 8.8    Fate of rete testis fluid (RTF)

Almost the entire output of fluid from the testis is reabsorbed in the head of the epididymis, leaving the spermatozoa behind as a dense suspension (Levine and Marsh, 1971). It has not been determined whether the secretory epithelium of the rete is adjacent to the resorptive epithelium of the efferent ducts and head of the epididymis without an intermediate stage. The secretory-resorptive borderline may relate to the different embryological origins of the two parts of the excurrent duct system (see Section 2.3.1).

The resorption of fluid is probably achieved by means of another standing osmotic gradient, in this case a forward-facing one because the tight junctions are along the luminal margins of the cells. In other words, the structure is similar to that of the rete epithelium but the direction of fluid movement is opposite.


## 8.9    Functional significance of fluid secretion

### 8.9.10    Liberation and transport of spermatozoa
The current of rete testis fluid is the principal means of transport for the spermatozoa out of the seminiferous tubules and from the testis to the epididymis (Setchell and Waites, 1974). As they leave the testis the spermatozoa are almost completely immotile, and the rete testis and efferent ducts are poorly endowed with smooth muscle. There is a much greater density of smooth muscle fibres in the distal parts of the epididymis, beyond the region of fluid reabsorption.

The role of fluid secretion in liberating the spermatozoa from the cytoplasm of the Sertoli cells is less certain. The formation of fluid-filled vacuoles in the Sertoli cell cytoplasm may be important in some amphibia (Burgos and Vitale-Calpe, 1967), but the evidence for an analogous mechanism in mammals (Vitale-Calpe and Burgos, 1970) is tenuous.


### 8.9.2    Nutrition of spermatozoa
Clearly once the spermatozoa are shed from the Sertoli cells, they must rely on endogenous reserves or on what reaches them through the fluid. No decision has yet been reached on the relative importance of these two sources. Under some circumstances, the spermatozoa can certainly utilize some of their reserves (Setchell et al., 1969), and no natural constituent of RTF has been shown to be an important substrate for the sperm. However, glucose probably enters the fluid readily, and the sperm may metabolize all that reaches them. RTF does stimulate the oxygen uptake

of testicular spermatozoa, when this is compared with oxygen uptake in a saline mixture of similar composition (Evans and Setchell, 1978).

### 8.9.3   Effects of RTF on the epididymis
The cells lining the epididymal duct are exposed to the fluid contained in the lumen and thus the function of the epididymis may be locally regulated by the ipsilateral testis (see Section 6.1.5.3).

## 8.10   Testicular spermatozoa

When the spermatozoa leave the testis, they are virtually immotile and incapable of fertilizing ova (Voglmayr *et al.*, 1967; Setchell *et al.*, 1969a; Lambiase and Amann, 1973). The metabolism of ram testicular spermatozoa has been the subject of some detailed studies. While this topic is perhaps outside the scope of this book, it is worth noting that the metabolism of these sperm is quite different from that of the testis as a whole on the one hand and from ejaculated or epididymal spermatozoa on the other (see Voglmayr, 1975b). The latter difference is a manifestation of the need for 'maturation' of the spermatozoa after they leave the testis to achieve the capacity for fertility and mobility. The former reflects the variety in cell anatomy and function within the testis and should warn us against arguing from the tubules as a whole to any individual cell type.

### References to Chapter 8

Amann, R. P. and Ganjam, V. K. (1976) Steroid production by the bovine testis and steroid transfer across the pampiniform plexus. *Biol. Reprod.* **15**, 695–703.

Aoki, A. and Fawcett, D. W. (1975) Impermeability of Sertoli cell junctions to prolonged exposure to peroxidase. *Andrologia* **7**, 63–76.

Aoki, A., Einstein, J. and Fawcett, D. W. (1976) Tannic acid as an electron-dense probe in the testis. *Am. J. Anat.* **146**, 449–53.

Baillie, A. H. (1962) Observations comparing the effects of epididymal obstruction at various levels on the mouse testis with those of ischaemia. *J. Anat.* **96**, 335–54.

Barack, B. M. (1968) Transport of spermatozoa from seminiferous tubules to epididymis in the mouse: a histological and quantitative study. *J. Reprod. Fertil.* **16**, 35–48.

Barrett, A. J. and Starkey, P. M. (1973) The interaction of $\alpha_2$-macroglobulin with proteinases; characteristics and specificity of the reaction and a hypothesis concerning its molecular mechanism. *Biochem. J.* **133**, 709–24.

Bartke, A., Harris, M. E. and Voglmayr, J. K. (1975) Regulation of testosterone and dihydrotestosterone levels in rete testis fluid. Evidence for androgen biosynthesis in seminiferous tubules *in vivo*. *Curr. Top. Molec. Endocrinol.* **2**, 197–212.

Benoit, J. (1926) Recherches anatomiques, cytologiques et histophysiologiques sur les voies excrétrices du testicule, chez les mammifères. *Arch. Anat.* (Strasbourg) **5**, 173–412.

Berenblum, I., Burger, M. and Knyszynsli, A. (1968) Regeneration of bone marrow cells and thymus induced by 19S $\alpha_2$-globulin in irradiated mice. *Nature* (London) **217**, 857–8.

Bosher, S. K. and Warren, R. L. (1968) Observations on the electrochemistry of the cochlear endolymph of the rat: a quantitative study of its electrical potential and ionic composition as determined by means of flame spectrophotometry. *Proc. Roy. Soc.* (London) B. **171**, 227–47.

Bouffard, G. (1906) Injection des couleurs de benzidine aux animaux normaux. *Ann. Inst. Pasteur* **20**, 539–46.

Broek, A. J. P. van den (1933) Gonaden und Ausführungsgänge. In *Handbuch der vergleichenden Anatomie der Wirbeltier*, eds. L. Bolk *et al.*, **6**, 1–54, Berlin: Urban & Schwarzenberg.

Bruyn, P. P. H. de, Robertson, R. C. and Farr, R. S. (1950) *In vivo* affinity of diaminoacridines for nuclei. *Anat. Rec.* **108**, 279–307.

Burgos, M. H. (1957) Fine structure of the efferent ducts of the hamster testis. *Anat. Rec.* **127**, 401.

Burgos, M. H. and Vitale-Calpe, R. (1967) The mechanism of spermiation in the toad. *Am. J. Anat.* **120**, 227–52.

Burgos, M. H. and Vitale-Calpe, R. (1969) Gonadotrophic control of spermiation. In *Progress in Endocrinology*, ed. C. Gual, 1030–7, Amsterdam: Excerpta Medica.

Castro, A. E., Seiguer, A. C. and Mancini, R. E. (1970) Electron microscopic study of the localization of labelled gonadotrophins in the Sertoli and Leydig cells of the rat testis. *Proc. Soc. Exptl. Biol. Med.* **133**, 582–6.

Castro, A. E., Alonso, A. and Mancini, R. E. (1972) Localization of follicle-stimulating and luteinizing hormones in the rat testis using immunohistological tests. *J. Endocrinol.* **52**, 129–36.

Cavicchia, J. C. and Burgos, M. H. (1977) Tridimensional reconstruction and histology of the intratesticular seminal pathway in the hamster. *Anat. Rec.* **187**, 1–10.

Cheung, Y. M., Hwang, J. C. and Wong, P. Y. D. (1976) *In vitro* measurement of the secretory rate in isolated seminiferous tubules of rats. *J. Physiol.* **254**, 17P–19P.

Clermont, Y. (1958) Contractile elements in the limiting membrane of the seminiferous tubules of the rat. *Exptl. Cell Res.* **15**, 438–40.

Comhaire, F. H. and Vermeulen, A. (1976) Testosterone concentration in the fluids of seminiferous tubules, the interstitium and the rete testis of the rat. *J. Endocrinol.* **70**, 229–35.

Connell, C. J. (1977) The effect of hCG on pinocytosis within the canine inter-Sertoli cell tight junction. A preliminary report. *Am. J. Anat.* **148**, 149–53.

Cooper, A. (1830) *Observations on the Structure and Diseases of the Testis*, London: Longman, Rees, Orme, Brown and Green.

Cooper, T. G. and Waites, G. M. H. (1974) Testosterone in rete testis fluid and blood of rams and rats. *J. Endocrinol.* **62**, 619–29.

Cooper, T. G. and Waites, G. M. H. (1975) Steroid entry into rete testis fluid and the blood–testis barrier. *J. Endocrinol.* **65**, 195–205.

Cooper, T. G., Danzo, B. J., Dipietro, D-L., McKenna, T. J. and Orgebin-Crist, M-C. (1976) Some characteristics of rete testis fluid from rabbits. *Andrologia* **8**, 87–94.

Cunningham, J. T. (1928) On ligature of the vas deferens in the cat, and researches on the efferent ducts of the testis in cat, rat and mouse. *Brit. J. exptl. Biol.* **6**, 12–25.

Denton, D. A. (1956) The effect of $Na^+$ depletion on the $Na^+:K^+$ ratio of the parotid saliva of the sheep. *J. Physiol.* (London) **131**, 516–25.

Diamond, J. M. (1971) Water-solute coupling and ion selectivity in epithelia. *Phil. Trans. Roy. Soc.* (London) B. **262**, 141–51.

Doorn, L. G. van, Bruijn, H. W. A. de, Galjaard, H. and Molen, H. J. van der (1974) Intercellular transport of steroids in the infused rabbit testis. *Biol. Reprod.* **10**, 47–53.

Dym, M. (1973) The fine structure of the monkey (*Macaca*) Sertoli cell and its role in maintaining the blood–testis barrier. *Anat. Rec.* **175**, 639–56.

Dym, M. (1974) The fine structure of monkey Sertoli cells in the transitional zone at the junction of the seminiferous tubules with the tubuli recti. *Am. J. Anat.* **140**, 1–26.

Dym, M. (1976) The mammalian rete testis—a morphological examination. *Anat. Rec.* **186**, 493–524.

Dym, M. and Fawcett, D. W. (1970) The blood–testis barrier in the rat and the physiological compartmentation of the seminiferous epithelium. *Biol. Reprod.* **3**, 308–26.

Edwards, E. M., Jones, A. R. and Waites, G. M. H. (1975) The entry of $\alpha$-chlorohydrin into body fluids of male rats and its effect upon the incorporation of glycerol into lipids. *J. Reprod. Fertil.* **43**, 225–32.

Ericsson, R. J. (1970) Male antifertility compounds: U-5897 as a rat chemosterilant. *J. Reprod. Fertil.* **22**, 213–22.

Evans, R. W. and Setchell, B. P. (1978) The effect of rete testis fluid on the metabolism of testicular spermatozoa. *J. Reprod. Fertil.* **52**, 15–20.

Fawcett, D. W. (1975) Ultrastructure and function of the Sertoli cell. *Handb. Physiol.* Section 7, *Endocrinology*, **V**, 21–55.

Fawcett, D. W. and Dym, M. (1974) A glycogen-rich segment of the tubuli recti and proximal portion of the rete testis in the guinea pig. *J. Reprod. Fertil.* **38**, 401–10.

Fawcett, D. W., Leak, L. V. and Heidger, P. M. (1970) Electron microscopic observations on the structural components of the blood–testis barrier. *J. Reprod. Fertil.* Suppl. **10**, 105–22.

Flickinger, C. J. (1972) Ultrastructure of the rat testis after vasectomy. *Anat. Rec.* **174**, 477–94.

Flickinger, C. J. and Fawcett, D. W. (1967) Junctional specialization of Sertoli cells in the seminiferous epithelium. *Anat. Rec.* **158**, 207–22.

Frederik, P. M., Klepper, D., Vusse, G. J. van der and Molen, H. J. van der (1976) Dynamics of steroid uptake in rat testis studied by quantitative autoradiography. *Mol. Cell. Endocr.* **5**, 123–36.

French, F. S. and Ritzen, E. M. (1973a) Androgen-binding protein in efferent duct fluid of rat testis. *J. Reprod. Fertil.* **32**, 479–83.

French, F. S. and Ritzen, E. M. (1973b) A high-affinity androgen-binding protein (ABP) in rat testis: Evidence for secretion into efferent duct fluid and absorption by epididymis. *Endocrinology* **93**, 88–95.

Ganjam, V. K. and Amann, R. P. (1973) Testosterone and dihydrotestosterone concentrations in the fluid milieu of spermatozoa in the reproductive tract of the bull. *Acta Endocrinol.* **74**, 186–200.

Ganjam, V. K. and Amann, R. P. (1976) Steroids in fluids and sperm entering and leaving the bovine epididymis, epididymal tissue and accessory gland secretions. *Endocrinology* **99**, 1618–30.

Gilula, N. B., Fawcett, D. W. and Aoki, A. (1976) The Sertoli cell occluding junctions and gap junctions in mature and developing mammalian testes. *Develop. Biol.* **50**, 142–68.

Goldacre, R. J. and Sylven, B. (1959) A rapid method for studying tumour blood supply using systemic dyes. *Nature* **184**, 63–4.

Goldacre, R. J. and Sylven, B. (1962) On the access of blood-borne dyes to various tumour regions. *Br. J. Cancer* **16**, 306–22.

Goldmann, E. E. (1909) Die aussere und innere Sekretion des gesunden und kranken Organismus im Lichte der 'vitalen Farbung'. *Bruns' Beitr. Klin. Chir.* **64**, 192–265.

Graaf, R. de (1668) Tractatus de Virorum Organis Generationi Inservientibus. Translated by H. D. Jocelyn and B. P. Setchell in Regnier de Graaf on the Human Reproductive Organs. *J. Reprod. Fert.* Suppl. **17**, 1–76.

Gravis, C. J., Chen, I. and Yates, R. D. (1977) Stability of the infra-epithelial component of the blood–testis barrier in epinephrine-induced testicular degeneration in Syrian hamsters. *Am. J. Anat.* **148**, 19–32.

Guerrero, R., Ritzen, E. M., Purvis, K., Hansson, V. and French, F. S. (1975) Concentration of steroid hormones and androgen binding protein (ABP) in rabbit efferent duct fluid. *Curr. Top. Molec. Endocrinol.* **2**, 213–21.

Gupta, R. J., Barnes, G. W. and Skelton, F. R. (1967) Light microscopic and immunopathological observations on cadmium chloride-induced injury in mature rat testis. *Am. J. Pathol.* **51**, 191–205.

Hammarström, L. (1966) Autoradiographic studies on the distribution of $C^{14}$-labelled ascorbic acid and dehydroascorbic acid. *Acta Physiol. Scand.* **70** Suppl. 289, 1–84.

Hansson, V., Ritzen, E. M., French, F. S. and Nayfeh, S. N. (1975a) Androgen transport and receptor mechanisms in testis and epididymis. *Handb. Physiol.* Sect. 7, **V**, 173–201.

Hansson, V., Weddington, S. C., Naess, O., Attramadal, A., French, F. S., Kotite, N., Nayfeh, S. N., Ritzen, E. M. and Hagenas, L. (1975b) Testicular androgen binding protein (ABP)—a parameter of Sertoli cell secretory function. *Curr. Top. Molec. Endocrinol.* **2**, 323 36.

Harris, M. E. and Bartke, A. (1974) Concentration of testosterone in testis fluid of the rat. *Endocrinology* **95**, 701–6.

Harris, M. E. and Bartke, A. (1975) Maintenance of rete testis fluid testosterone and dihydrotestosterone levels by pregnenolone and other $C_{21}$ steriods in hypophysectomized rats. *Endocrinology* **96**, 1396–402.

Harrison, R. G. (1953) The effect of ligation of the vasa efferentia on the rat testis. *Proc. Soc. Study Fertil.* **5**, 97–100.

Heim, W. G. (1968) Relation between rat slow $\alpha_2$-globulin and $\alpha_2$-macroglobulin of other mammals. *Nature* (London) **217**, 1057–8.

Henning, R. D. and Young, J. A. (1971) Electrolyte transport in the seminiferous tubules of the rat studied by the stopped-flow microperfusion technique. *Experientia* **27**, 1037–9.

Hinton, B. T., Setchell, B. P. and White, R. W. (1977) The determination of myo-inositol in micropuncture samples from the testis and epididymis of the rat. *J. Physiol.* (London) **265**, 14P–15P.

Holstein, A. F. (1964) Elektronenmikroskopische Untersuchungen an den Ductuli Efferentes des Nebenhodens normaler und kastrierter Kaninchen. *Z. Zeliforsch. Mikroskop. Anat.* **64**, 767–77.

Howards, S. S., Jessee, S. J. and Johnson, A. L. (1976) Micropuncture studies of the blood-seminiferous tubule barrier. *Biol. Reprod.* **14**, 264–9.

Johnson, A. L. and Howards, S. S. (1976) Hyperosmolality in intraluminal fluids from hamster testis and epididymis: a micropuncture study. *Science* **195**, 492–3.

Johnson, M. H. (1969) The effect of cadmium chloride on blood testis barrier of the guinea pig. *J. Reprod. Fertil.* **19**, 551–3.

Johnson, M. H. (1970a) An immunological barrier in the guinea pig testis. *J. Pathol.* **101**, 129–39.

Johnson, M. H. (1970b) Changes in the blood testis barrier of the guinea pig in relation to histological damage following iso-immunization with testis. *J. Reprod. Fertil.* **22**, 119–27.

Johnson, M. H. (1972) The distribution of immunoglobulin and spermatozoal autoantigen in the genital tract of the male guinea pig. *Fertil. Steril.* **23**, 383–92.

Johnson, M. H. and Setchell, B. P. (1968) Protein and immunoglobulin content of rete testis fluid of rams. *J. Reprod. Fertil.* **17**, 403–6.

Kormano, M. (1967a) Dye permeability and alkaline phosphatase activity of testicular capillaries in the postnatal rat. *Histochemie* **9**, 327–38.

Kormano, M. (1967b) Distribution of injected L-3, 4-dihydroxyphenylalanine in the adult rat testis and epididymis. *Acta Physiol. Scand.* **71**, 125–6.

Kormano, M. (1968) Penetration of intravenous trypan blue into the rat testis and epididymis. *Acta histochem.* **30**, 133–6.

Kormano, M. and Niemi, M. (1964) The use of tetracycline fluorescence in experimental testicular pathology. *Med. exp.* **11**, 383–8.

Kormano, M. and Penttila, A. (1968) Distribution of endogenous and administered 5-hydroxytryptamine in the rat testis and epididymis. *Ann. Med. Exptl. Biol. Fenn.* **46**, 468–73.

Kormano, M., Koskimies, A. I. and Hunter, R. L. (1971) The presence of specific proteins, in the absence of many serum proteins, in the rat seminiferous tubule fluid. *Experientia* **27**, 1461–3.

Koskimies, A. I. (1973a) Effect of cadmium on protein composition of fluids in rat rete testis and seminiferous tubules. *Ann. Med. Exptl. Biol. Fenniae.* (Helsinki) **51**, 74–81.

Koskimies, A. I. (1973b) Rapid appearance of excess serum protein in the rete testis in response to Cd-induced increase of vascular permeability. In *Immunology of Reproduction*, 126, Sofia: Bulgarian Academy of Sciences Press.

Koskimies, A. I. (1973c) Electrophoretic studies on proteins in testicular fluids. *Acta Inst. Anat. Universitatis Helsinkiensis* Suppl. **6**, 1–37.

Koskimies, A. I. and Kormano, M. (1973) The proteins in fluids from the seminiferous tubules and rete testis of the rat. *J. Reprod. Fertil.* **34**, 433–44.

Koskimies, A. I. and Reijonen, K. (1976) Early effects of vasectomy on the protein composition of fluid from the rete testis and epididymis of the rat. *J. Reprod. Fertil.* **47**, 141–3.

Koskimies, A. I., Kormano, M. and Alfthan, O. (1973) Proteins of the seminiferous tubule fluid in man—Evidence for a blood–testis barrier. *J. Reprod. Fertil.* **32**, 79–88.

Ladman, A. J. (1967) Fine structure of the ductuli efferentes of the opossum. *Anat. Rec.* **157**, 559–76.

Ladman, A. J. and Young, W. C. (1958) Electron microscopic study of the ductuli efferentes and rete testis of the guinea pig. *J. Biophys. Biochem. Cyt.* **4**, 219–26.

Lambiase, J. T. and Amann, R. P. (1973) Infertility of rabbit testicular spermatozoa collected in their native fluid environment. *Fertil. Steril.* **24**, 65–7.

Laporte, P. and Gillet, J. (1975) Influence de la spermatogénèse sur la secretion du fluide testiculaire chez le rat adulte. *C. rend. Acad. Sci. Paris.* **281**, 1397–400.

Lauth, E. A. (1830) Mémoire sur le testicule humaine. *Mem. Soc. Hist. Nat.* (Strasbourg) **1**, 1–42.

Leeson, T. S. (1962) Electron microscopy of the rete testis of the rat. *Anat. Rec.* **144**, 57–67.

Levine, N. and Marsh, D. J. (1971) Micropuncture studies of the electrochemical aspects of fluid and electrolyte transport in individual seminiferous tubules, the epididymis and the vas deferens in rats. *J. Physiol.* (London) **213**, 557–70.

Levine, N. and Marsh, D. J. (1975) Micropuncture study of the fluid composition of 'Sertoli cell-only' seminiferous tubules in rats. *J. Reprod. Fertil.* **43**, 547–50.

Linzell, J. L. and Peaker, M. (1971) Mechanism of milk secretion. *Physiol. Rev.* **51**, 564–97.

Linzell, J. L. and Setchell, B. P. (1969) Metabolism, sperm and fluid production of the isolated perfused testis of the sheep and goat. *J. Physiol.* **201**, 129–43.

Lohmüller, W. (1925) Die Ubergangsstellen der gewundenen in die 'geraden' Hodenkanalchen beim Menschen. *Z. Mikroskop. Anat. Forsch.* **3**, 147–78.

Lucas, A. M. (1963) Ciliated epithelium. In *Special Cytology*, ed. E. V. Cowdry, 2nd edition **I**, 407–74,

Maddrell, S. H. P. (1971) Fluid secretion by the Malpighian tubules of insects. *Phil. Trans. Roy. Soc.* (London) B. **262**, 197–207.

Main, S. J. (1975) The blood–testis barrier and temperature. PhD. dissertation, University of Reading.

Main, S. J. and Waites, G. M. H. (1977) The blood–testis barrier and testicular damage to the testis of the rat. *J. Reprod. Fertil.* **51**, 439–50.

Mancini, R. E., Vilar, O., Alvarez, B. and Seiguer, A. C. (1965) Extravascular and intratubular diffusion of labeled serum proteins in the rat testis. *J. Histochem. Cytochem.* **13**, 376–85.

Mancini, R. E., Castro, A. and Seiguer, A. C. (1967) Histologic localization of follicle-stimulating and luteinizing hormones in the rat testis. *J. Histochem. Cytochem.* **15**, 516–25.

Marin-Padilla, M. (1964) The mesonephric-testicular connection in man and some mammals. *Anat. Rec.* **148**, 1–14.

Martan, J., Hruban, Z. and Slesers, A. (1967) Cytological studies of ductuli efferentes of the opposum. *J. Morphol.* **121**, 81–101.

Mason, K. A. and Shaver, S. L. (1952) Some functions of the caput epididymis. *Ann. N.Y. Acad. Sci.* **55**, 585–93.

May, F. (1923) Kurze Mitteilung uber den anatomischen Aufbau der Ubergangsstellen der Tubuli contorti in die Tubuli recti in menschlichen Hoden. *Arch. Pathol. Anat. Physiol.* **243**, 474–7.

Messing, W. (1877) Anatomische Untersuchungen uber die Testikel der Saugetiere unter besonderer Berucksichtigung des Corpus Highmori. Inaugural Dissertation, Dorpat University.

Middleton, A. (1973) Glucose metabolism in rat seminiferous tubules. PhD. dissertation, University of Cambridge.

Middleton, A. and Setchell, B. P. (1972) The origin of inositol in rete testis fluid. *J. Reprod. Fertil.* **30**, 473–5.

Mihalkovics, V. von (1873) Beitrage zur Anatomie und Histologie des Hodens. *Ber. Verhandl. K. Sachs. Ges. Wiss.* **24**, 217–56.

Montagna, W. (1952) Some cytochemical observations on human testes and epididymides. *Ann. N. Y. Acad. Sci.* **55**, 629–42.

Moore, C. R. (1931) Supplementary observations on mammalian testis activity. I. Vasa efferentia ligation. II. Atypical scrota. *Anat. Rec.* **48**, 105–19.

Morita, I. (1966) Some observations on the fine structure of the human ductuli efferentes testis. *Arch. Histol.* (Okayama) **26**, 341–65.

Neaves, W. R. (1973) Permeability of Sertoli cell tight junctions to lanthanum after ligation of ductus deferens and ductuli efferentes. *J. Cell Biol.* **59**, 559–72.

Nicander, L. (1967) An electron microscopical study of cell contacts in the seminiferous tubules of some mammals. *Z. Zellforsch. Mikroskop. Anat.* **83**, 375–97.

Odeblad, E. and Bostrom, H. (1952) Uptake of radioactive sulphate in the genito-urinary system of the male rabbit. *Acta path. Microbiol.* Scand. **32**, 448–52.

Okumura, K., Lee, I. P. and Dixon, R. I. (1975) Permeability of selected drugs and chemicals across the blood–testis barrier of the rat. *J. Pharmacol. Exptl. Therapeut.* **194**, 89–95.

Oslund, R. M. (1926) Ligation of the vasa efferentia in rats. *Am. J. Physiol.* **77**, 83–90.

Pari, G. A. (1910) Uber die Verwendbarkeit vitaler Karmineinspritzungen fur die pathologische Anatomie. *Frankfurter Z. Pathol.* **4**, 1–28.

Parvinen, M. and Niemi, M. (1971) Distribution and conversion of exogenous cholesterol and sex steroids in the seminiferous tubules and interstitial tissue of the rat testis. *Steroidologia* **2**, 129–37.

Ribbert, H. (1904) Die Abscheidung intravenos injizierten gelosten Karmins in den Geweben. *Z. Allgem. Physiol.* **4**, 201–14.

Ro, T. S. and Busch, H. (1965) Concentration of [$^{14}$C] actinomycin D in various tissues following intravenous injection. *Biochim. Biophys. Acta* **108**, 317–8.

Rolshoven, E. (1936) Ursachen und Bedeutung der intratubularen Sekretstromung im Saugerhoden. *Z. Anat. Entwicklungsgeschichte* **105**, 374–408.

Roosen-Runge, E. C. (1951) Motions of the seminiferous tubules of rat and dog. *Anat. Rec.* **109**, 413.

Roosen-Runge, E. C. (1961) The rete testis in the albino rat: Its structure, development and morphological significance. *Acta Anat.* **45**, 1–30.

Ross, M. H. (1970) The Sertoli cell and the blood-testicular barrier: an electronmicroscopic study. *Advan. Andrology* **1**, 83–6.

Ross, M. H. (1974) The organization of the seminiferous epithelium in the mouse testis following ligation of the efferent ductules. A light microscopic study. *Anat. Rec.* **180**, 565–80.

Ross, M. H. (1977) Sertoli-Sertoli junctions and Sertoli-spermatid junctions after efferent duct ligation and lanthanum treatment. *Am. J. Anat.* **148**, 49–56.

Ross, M. H. and Dobler, J. (1975) The Sertoli cell junctional specializations and their relationship to the germinal epithelium as observed after efferent ductule ligation. *Anat. Rec.* **183**, 267–92.

Russell, L. (1977) Movement of spermatocytes from the basal to the adluminal compartment of the rat testis. *Am. J. Anat.* **148**, 313–28.

Schinz, H. R. and Slotopolsky, B. (1924) Beitrag zur experimentellen Pathologie des Hodens. *Denkschr. Schweiz. Naturforsch. Ges.* **61**, 27–137.

Sedar, A. W. (1966) Transport of exogenous peroxidase across the epithelium of the ductuli efferentes. *J. cell. Biol.* **31**, 102A.

Seiguer, A. C. and Mancini, R. E. (1971) The permeability of the 'blood–testis' barrier to ferrocyanide. *J. Reprod. Fertil.* **27**, 269–72.

Setchell, B. P. (1967) The blood–testicular fluid barrier in sheep. *J. Physiol.* (London) **189**, 63–65P.

Setchell, B. P. (1969) Do the Sertoli cells secrete the rete testis fluid? *J. Reprod. Fertil.* **19**, 391 2.

Setchell, B. P. (1970a) Testicular blood supply, lymphatic drainage and secretion of fluid. In *The Testis*, eds. A. D. Johnson, W. R. Gomes and N. L. VanDemark, **I**, 101–239, New York: Academic Press.

Setchell, B. P. (1970b) Fluid secretion by the testes of an Australian marsupial *Macropus eugenii. Comp. Biochem. Physiol.* **36**, 411–14.

Setchell, B. P. (1970c) The secretion of fluid by the testes of rats, rams and goats with some observations on the effect of age, cryptorchidism and hypophysectomy. *J. Reprod. Fertil.* **23**, 79–85.

Setchell, B. P. (1974a) Secretions of the testis and epididymis. *J. Reprod. Fertil.* **37**, 165–77.

Setchell, B. P. (1974b) The entry of substances into the seminiferous tubules. In *Male Fertility and Sterility*, ed. R. E. Mancini and L. Martini, 37–57, New York: Academic Press.

Setchell, B. P. (1977) Reproduction in male marsupials. In *Biology of Marsupials*, eds. D. P. Gilmore and B. Stonehouse, 411–57. London: Macmillan.

Setchell, B. P. and Brown, B. W. (1972) The effect of metabolic alkalosis, hypotension and inhibitors of carbonic anhydrase on fluid secretion by the rat testis. *J. Reprod. Fertil.* **28**, 235–40.

Setchell, B. P. and Linzell, J. L. (1968) Effects of some drugs, hormones and physiological factors on the flow of rete testis fluid in the ram. *J. Reprod. Fertil.* **16**, 320–1.

Setchell, B. P. and Main, S. J. (1975) The blood–testis barrier and steroids. *Curr. Top. Molec. Endocrinol.* **2**, 223–33.

Setchell, B. P. and Main, S. J. (1978) Drugs and the blood–testis barrier. *Environmental Health Perspectives.* [In press.]

Setchell, B. P. and Singleton, H. M. (1971) The penetration of rubidium, sucrose and inulin into rat seminiferous tubules *in vivo* and *in vitro. J. Physiol.* **217**, 15P–16P.

Setchell, B. P. and Waites, G. M. H. (1970) Changes in the permeability of the testicular capillaries and of the 'blood–testis barrier' after injection of cadmium chloride in the rat. *J. Endocrinol.* **47**, 81–6.

Setchell, B. P. and Waites, G. M. H. (1971) The exocrine secretion of the testis and spermatozoa. *J. Reprod. Fertil.* Suppl. **13**, 15–27.

Setchell, B. P. and Waites, G. M. H. (1972) The effects of local heating of the testis on the flow and composition of rete testis fluid in the rat, with some observations on the effects of age and unilateral castration. *J. Reprod. Fertil.* **30**, 225–33.

Setchell, B. P. and Waites, G. M. H. (1974) Fluid secretion by the testis in the transport and survival of spermatozoa. In *Transport, Survie, et Pouvoir Fecondant des Spermatozoides*, eds. E. S. E. Hafez and C. G. Thibault, 11–34, Paris: Inserm.

Setchell, B. P. and Waites, G. M. H. (1975a) Fluid secretion by the testis in the transport and survival of spermatozoa. In *The Biology of Spermatozoa*, eds. E. S. E. Hafez and C. Thibault, 2–9, Basel: Karger.

Setchell, B. P. and Waites, G. M. H. (1975b) The blood–testis barrier. In *Handbook of Physiology*, Section 7, *Endocrinology* **V**. Male Reproductive System, 143–72.

Setchell, B. P. and Wallace, A. L. C. (1972) The penetration of iodine-labelled follicle-stimulating hormone and albumin into the seminiferous tubules of sheep and rats. *J. Endocrinol.* **54**, 67–77.

Setchell, B. P., Hinks, N. T., Voglmayr, J. K. and Scott, T. W. (1967) Amino-acids in the ram testicular fluid and semen and their metabolism by spermatozoa. *Biochem. J.* **105**, 1061–5.

Setchell, B. P., Dawson, R. M. C. and White, R. W. (1968) The high concentration of free myo-inositol in rete testis fluid from rams. *J. Reprod. Fertil.* **17**, 219–20.

Setchell, B. P., Scott, T. W., Voglmayr, J. K. and Waites, G. M. H. (1969a) Characteristics of testicular spermatozoa and the fluid which transports them into the epididymis. *Biol. Reprod.* Suppl. **1**, 40–66.

Setchell, B. P., Voglmayr, J. K. and Waites, G. M. H. (1969b) A blood–testis barrier restricting passage from blood into rete testis fluid but not into lymph. *J. Physiol.* (London) **200**, 73–85.

Setchell, B. P., Voglmayr, J. K. and Hinks, N. T. (1971) The effect of local heating on the flow and composition of rete testis fluid in the conscious ram. *J. Reprod. Fertil.* **24**, 81–9.

Setchell, B. P., Duggan, M. C. and Evans, R. W. (1973) The effect of gonadotrophins on fluid secretion and sperm production by the rat and hamster testis. *J. Endocrinol.* **56**, 27–36.

Setchell, B. P., Hinton, B. T., Jacks, F. and Davies, R. V. (1976) Restricted penetration of iodinated follicle-stimulating and luteinizing hormone into the seminiferous tubules of the rat testis. *Molec. Cell. Endocrinol.* **6**, 59–69.

Setchell, B. P., Davies, R. V., Gladwell, R. T., Hinton, B. T., Main, S. J., Pilsworth, L. and Waites, G. M. H. (1978) The movement of fluid in the seminiferous tubules and rete testis. *Ann. Biol. Animal. Biochem. Biophys.* [In press.]

Sexton, T. J., Amann, R. P. and Flipse, R. J. (1971) Free amino acids and protein in rete testis fluid, vas deferens plasma, accessory sex gland fluid and seminal plasma of the conscious bull. *J. Dairy Sci.* **54**, 412–16.

Slotopolsky, B. and Schinz, H. R. (1925) Histologische zur Steinach Unterbindung. *Z. Mikroskop. Anat. Forsch.* **2**, 225–53.

Smith, G. (1962) The effects of ligation of the vasa efferentia and vasectomy on testicular function in the adult rat. *J. Endocrinol.* **23**, 385–99.

Spangaro, S. (1902) Uber die histologischen Veranderungen des Hodens, Nebenhodens and Samenleiters von Geburt an bis zum Greisenalter. *Anat. Heft. Abt I* **18**, 593–771.

Stenger-Haffen, K. (1957) Etudes des besoins nutritifs des gonades embryonnaires d'oiseau cultivées en milieux synthétiques. *Arch. Anat. microscop. Morphol. exper.* **46**, 521—607.

Stieda, L. (1877) Ueber den Bau des Menschenhoden. *Arch. Mikrosk. Anat.* **14**, 17–50.

Stieve, H. (1930) Männliche Genitalorgane. In *Handbuch der mikroskopischen Anatomie des Menschen*, ed. W. von Mollendorff, vol. 7, part 2, 1–394, Berlin: Springer.

Suominen, J. and Setchell, B. P. (1972) Enzymes and trypsin inhibitor in the rete testis fluid of rams and boars. *J. Reprod. Fertil.* **30**, 235–45.

Suominen, J. J. O. and Setchell, B. P. (1976) Proteinase inhibitors in testicular and epididymal fluid. *Protides of Biol. Fluids* **23**, 171–5.

Suominen, J., Kaufman, M. H. and Setchell, B. P. (1973) Prevention of fertilization *in vitro* by an acrosin inhibitor from rete testis fluid of the ram. *J. Reproduct. Fertil.* **34**, 385–8.

Tindall, D. J. and Means, A. R. (1976) Concerning the hormonal regulation of androgen-binding protein in rat testis. *Endocrinology* **99**, 809–18.

Tindall, D. J., Schrader, W. T. and Means, A. R. (1974) The production of androgen-binding protein by Sertoli cells. *Curr. Top. Molec. Endocrinol.* **1**, 167–75.

Tindall, D. J., Vitale, R. and Means, A. R. (1975) Androgen-binding protein as a biochemical marker of formation of the blood-testis barrier. *Endocrinology* **97**, 636–48.

Tuck, R. R. (1969) An investigation of the fluid secreted by the seminiferous tubules and the rete testis of the rat. BSc (Med) thesis, University of Sydney.

Tuck, R. R., Setchell, B. P., Waites, G. M. H. and Young, J. A. (1970) The composition of fluid collected by micropuncture and catheterization from the seminiferous tubules and rete testis of rats. *European J. Physiol.* (Pflügers, Arch.) **318**, 225–43.

Vitale-Calpe, R. and Aoki, A. (1969) Fine structure of the intratesticular excretory pathway in the guinea pig. *Z. Anat. Entwickl. Gesch.* **129**, 135–53.

Vitale-Calpe, R. and Burgos, M. H. (1970) The mechanism of spermiation in the hamster. I. Ultrastructure of spontaneous spermiation. *J. Ultrastruct. Res.* **31**, 381–93.

Vitale-Calpe, R., Fawcett, D. W. and Dym, M. (1973) The normal development of the blood–testis barrier and the effects of clomiphene and estrogen treatment. *Anat. Rec.* **176**, 333–44.

Voglmayr, J. K. (1975a) Output of spermatozoa and fluid by the testis of the ram and response to oxytocin. *J. Reprod. Fertil.* **43**, 119–22.

Voglmayr, J. K. (1975b) Metabolic changes in spermatozoa during epididymal transit. *Handbk. Physiol.* Section 7, *Endocrinology* **V**, 437–51.

Voglmayr, J. K., Waites, G. M. H. and Setchell, B. P. (1966) Studies on spermatozoa and fluid collected directly from the testis of the conscious ram. *Nature* **210**, 861–3.

Voglmayr, J. K., Scott, T. W., Setchell, B. P. and Waites, G. M. H. (1967) Metabolism of testicular spermatozoa and characteristics of testicular fluid collected from the conscious ram. *J. Reprod. Fertil.* **14**, 87–99.

Voglmayr, J. K., Larsen, L. H. and White, I. G. (1970) Metabolism of spermatozoa and composition of fluid collected from the rete testis of living bulls. *J. Reprod. Fertil.* **21**, 449–60.

Voglmayr, J. K., Kavanaugh, J. F., Griel, L. C., Jr and Amann, R. P. (1972) A modified technique for cannulating the rete testis of the bull. *J. Reprod. Fertil.* **31**, 291–4.

Wagenen, G. van (1925) Changes in the testis of the rat following ligation of the ductuli efferenti. *Anat. Rec.* **29**, 339.

Wagenen, G. van (1926) Degeneration and regeneration of the seminiferous tubules after ligation of the ductuli efferentes in the rat. *Anat. Rec.* **32**, 225.

Waites, G. M. H. (1977) Fluid secretion. In *The Testis*, eds. A. D. Johnson and W. R. Gomes, **IV**, 91–123. New York: Academic Press.

Waites, G. M. H. and Einer-Jensen, N. (1974) Collection and analysis of rete testis fluid from macaque monkeys. *J. Reprod. Fertil.* **41**, 505–8.

Waites, G. M. H., Jones, A. R., Main, S. J. and Cooper, T. G. (1972) The entry of antifertility and other drugs into the testis. *Adv. Biosci.* **10**, 101–16.

Waksman, B. H. (1960) The distribution of experimental auto-allergic lesions. *Am. J. Pathol.* **37**, 673–85.

Weddington, S. C., Brandtzaeg, P., Sletten, K., Christensen, T., Hansson, V., French, F. S., Petrusz, P., Nayfeh, S. N. and Ritzen, E. M. (1975) Purification and characterization of rabbit testicular and androgen-binding protein (ABP). *Curr. Top. Molec. Endocrinol.* **2**, 433–51.

Willson, J. T., Jones, N. A., Katsch, S. and Smith, S. W. (1973) Penetration of the testicular-tubular barrier by horse radish peroxidase induced by adjuvant. *Anat. Rec.* **176**, 85–100.

Young, J. A. and Martin, C. J. (1971) The effect of a sympatho- and a parasympathomimetic drug on the electrolyte concentrations of primary and final saliva of the rat submaxillary gland. *Pflugers Arch.* **327**, 285–302.

Young, W. C. (1933) Die Resorption in den Ductuli efferentes der Maus und ihre Bedeutung fur das Problem der Unterbindung im Hoden-Nebenhodensystem. *Z. Zellforsch. Mikrosk. Anat.* **17**, 729–57.

# 9 Metabolism in the testis

## 9.1 Composition of the testis

### 9.1.1 Water and ions

The testis contains more water than most other organs, e.g. muscle, liver, kidney and brain. It has about 90 per cent or 6·5g water/g dry weight (Setchell, 1970a), compared with 75–80 per cent or 3 to 4g/g for the other tissues.

The sodium and potassium content of the dry tissue is comparable with that of other tissues, and if allowance is made for an extracellular volume of 12 per cent wet tissue containing plasma-like concentrations, then the concentration of these ions in testicular cells is comparable to the concentrations in other cells of the body.

### 9.1.2 Carbohydrates

Despite, or probably because of, its dependence on glucose for a large fraction of its $CO_2$ production, the testis contains very little free glucose; in fact, if extracellular fluid is assumed to contain as much glucose as plasma, then the cells must contain a negative amount of glucose; that is to say that the extravascular extracellular fluid must contain less glucose than blood plasma and the cells probably contain none or very little (Middleton, 1973).

Glycogen has been demonstrated histochemically, in the testis, particularly in the Sertoli cells (Montagna, 1952; Cavazos and Mellampy, 1954; Nicander, 1957), but quantitatively, the testis contains very much less glycogen (0·2 to 5mg/g) than muscle or liver (see Free, 1970). The testes of fetal and newborn rats contain up to 20 times as much glycogen as those of adults (Leidermann and Mancini, 1969; Gunaga et al., 1972). Inositol is present in high concentrations in testes (Hauser, 1969) and rete testis fluid (Setchell et al., 1968).

The rat testis contains appreciable concentrations of ascorbic acid, which decrease after puberty (Coste et al., 1953). The concentration of ascorbic acid in the adrenal gland is decreased by injections of its trophic hormone, but this does not happen in the testis (Collizzi and Tusini, 1953; Noach and van Rees, 1958; Llaurado and Eik-Nes, 1961), probably because most of the ascorbic acid in the testis can be demonstrated histochemically inside the tubules, not associated with the steroid-producing cells. The intratubular localization is supported by the observation that the concentration and total amount of ascorbic acid in the testis is reduced when the testis is made cryptorchid (Pestis and Raoul, 1969).

### 9.1.3 Lipids

The testis is comparatively rich in lipid. In most species there is about 20mg/g wet weight, i.e. 20 per cent of the total dry weight. Three-quarters of this lipid is phospholipid, in the rat mostly as phosphatidyl ethanolamine or phosphatidyl choline

(Oshima and Carpenter, 1968). In the ram, phosphatidyl choline was the predominant phospholipid (Scott and Setchell, 1968). The neutral lipids present are mainly triglyceride and cholesterol (Oshima and Carpenter, 1968).

Cholesterol is present in the testis both as free cholesterol and cholesterol esters. The ratio of free to esterified cholesterol in the interstitial tissue is between 10 and 20 (Bartke, 1971; Bartke and Shire, 1972; van der Molen *et al.*, 1973); a similar value is found in the whole testes (Perlman, 1950a; Renston *et al.*, 1975) in contrast to the adrenal gland where this value lies between 0·05 and 0·10 (Long, 1947; Garren *et al.*, 1971). The testicular concentration of free cholesterol remains reasonably constant, while the esterified cholesterol falls during maturation in rabbits (Renston *et al.*, 1975).

The fatty acid composition of the various lipid fractions reveals some interesting peculiarities. In all species so far examined palmitic (16:0) acid (the '16' indicates the length of the carbon chain, the '0' the number of double bonds) is the most abundant but there is a high proportion of either docosapentaenoic or docosahexanoic acid (22:5, or 22:6) as well as high proportions of arachidonic (20:4) and oleic (18:1) acids. The 22:5 acid appears in rat testes only at maturity (Davis *et al.*, 1966; Oshima and Carpenter, 1968). The rat contains high concentrations of 22:5 in its testes, the sheep testis has 22:6 and the pig testis approximately equal amounts of the two fatty acids.

The lipid composition of the testis is changed in a number of conditions. For example, hypophysectomy in the rat leads to an increase in total lipid, (in mg/g testis) mainly in cholesterol esters and triglyceride fractions, with a decrease in phospholipids (Gambal and Ackerman, 1967; Nakamura *et al.*, 1968; Jensen and Privett, 1969; Nakamura and Privett, 1973). Similar changes are produced by cryptorchidism in the rate (Davis and Coniglio, 1967), but in the rabbit, triglyceride and cholesterol concentrations rise without much change in the phospholipids (Fleeger *et al.*, 1968a, b). The rise in esterified and total cholesterol was apparent one day after hypophysectomy, before the testis weight had begun to fall (Perlman, 1950a, b; Gambal and Ackerman, 1967; Nakamura *et al.*, 1968; Hafiez and Bartke, 1972). Changes in the lipid composition of the testes of hypophysectomized rats were partially reversed by treatment with FSH or LH, although normal values were not reached (Nakamura and Privett, 1973).

The fatty acid composition of the testes of rats is also changed by making them cryptorchid or treating them with cadmium. Reductions are then seen in the percentage of 16:0 and 22:5 with increases in the percentages 18:2, 20:4 and 22:4 (Davis and Coniglio, 1967). Rather surprisingly, after hypophysectomy in mature normal animals, there is a dramatic increase in the percentage of 22:5 in the cholesterol esters and glycerol ether diesters, but no change or a slight decrease in 22:5 in the fatty acids of the polar lipids (Nakamura *et al.*, 1968). These changes were reversed by treatment with LH and FSH (Nakamura and Privett, 1973). Hypophysectomy does not have such an effect on immature rats (Jensen and Privett, 1969), nor in essential fatty acid deficient rats, even when these are supplemented with methyl linoleate (Jensen *et al.*, 1968). In another study (Goswami and William, 1967), although there is no direct comparison with intact rats, the testicular lipids of mature hypophysectomized rats contained much less than normal amounts of $22:5$ and this was increased by injections of FSH, LH or testosterone propionate.

Fatty acid deficiency has been reported to cause a decrease in phospholipids and in the fatty acids $22:5$, $20:4$, $20:3\omega6$ and $18:2$, but an increase in $20:3\omega9$ and to a lesser extent in $18:2$ and $16:1$ (Ahluwalia et al., 1967; Bieri et al., 1969).

Vitamin E deficiency in rats does not cause any striking changes in the amount of the various classes of phospholipids, but the amounts of $22:5$ in lecithin, cephalin and sphingomyelin are very much less with either no change or an increase in $20:4$ and $22:4$ depending on how the results are presented. In the neutral lipids, the proportion of $22:5$ and $16:0$ decreases, with no change in the proportion of $20:4$. The change occurs first in the phospholipids (Bieri and Andrews, 1964; Bieri and Prival, 1966; Carpenter, 1971).

The testes contains appreciable concentrations of prostaglandins although the levels are much lower than in other parts of the male tract; in mouse and rat testes there is more PGF than PGE, whereas in pig testis the ratio is reversed. The concentration of both PGF and PGE in mouse testis falls with age between 2 and 8 weeks (Carpenter and Wiseman, 1970; Michael, 1973; Bartke and Koerner, 1974; Carpenter, 1974; Badr, 1975, 1976; Badr et al., 1975). These studies do not appear to have distinguished between capsule and parenchyma, and it has recently been shown that the capsule contains $PGE_2$ and $PGF_{2a}$ in concentrations about 100 times higher than those in decapsulated testes (Gerozissis and Dray, 1977).

The testes of several species contain appreciable concentrations of a sulpho-glycero-galactolipid, 1-O-hexadecyl-2-O-hexadecanoyl-glycerylmonogalacto-side sulphate (Ishizuka et al., 1973; Suzuki et al., 1973; Kornblatt et al., 1974). Its concentration in rat testis rises sharply at about 20 days of age, at the time when the spermatocytes appear, but this lipid is also found in later cell types. The concentration of the enzyme responsible for its formation rises and then falls, suggesting that this lipid, once formed, is stable during spermatogenesis (Kornblatt et al., 1974).

### 9.1.4    Nitrogenous compounds

**9.1.4.1    Proteins** Spermatozoa contain a specialized contractile protein, not present in precursor cells (Shelanski and Taylor, 1968; Mohri, 1968), and also unique acidic (Kadohama and Turkington, 1974) and basic proteins, rich in arginine serine and cysteine. A different basic protein can be isolated from the testis when the animals are old enough for spermatids to be present. This protein disappears when the testis is made cryptorchid or when spermatogenesis is abnormal (Lam and Bruce, 1971; Kistler et al., 1973, 1975a; Kistler and Williams-Ashman, 1975). Its amino-acid sequence has been established (Kistler et al., 1974, 1975b); it is rich in arginine and lysine but contains no cysteine. There is also a high concentration in the testes of a lysine-rich histone, which is present in lower concentrations elsewhere in the body (Kistler and Geroch, 1975; Kistler and Williams-Ashman, 1975).

The most striking feature of the nuclear proteins of the testis is the substitution at about or soon after the time of meiosis of an arginine-rich histone for the somatic type histone rich in lysine (Monesi, 1964a, 1965a, 1967; Gledhill et al., 1966; Vaughn, 1965, 1966; Bloch, 1969). Earlier reports (Lison, 1955; Alfert, 1956) suggested that in the rat and guinea pig, the process continued in the maturing sperm with the replacement of the histone by protamine as occurs in invertebrates, but this observation does not appear to have been substantiated by later work. The significance of the change in the histone is not entirely clear but may be associated with repression of the genetic activity of the spermatids and spermatozoa (Allfrey et al., 1963). Testis-specific histones have recently been reported from mature testes of several species (Shires et al., 1975; Branson et al., 1975; Grimes et al., 1975; Mills et al., 1977).

During the later stages of spermiogenesis, there is an accumulation of basic protein in the cytoplasm of the spermatids in the form of a large safranophilic mass, the sphere chromatophile. This body consists of ribonucleoprotein and glycogen (Daoust and Clermont, 1955; Firlit and Davis, 1965) and arises from RNA-containing granules in the cytoplasm known as von Ebner granules (Regaud, 1901). The protein in these granules may be the lysine-rich nuclear protein which is lost from the nucleus at about this time (Vaughn, 1965, 1966).

**9.1.4.2   Amino-acids**   The testis has a high concentration of certain amino-acids, particularly glutamic acid, aspartic acid and glycine (Mellampy et al., 1955; Hopwood and Gassner, 1962; Härkönen et al., 1970). Their function is unknown.

**9.1.4.3   Polyamines**   The testis contains appreciable concentrations of the polyamines spermidine and spermine (Tabor and Tabor, 1964), which were originally isolated from semen as their names imply. Crystals of spermine phosphate were described by Leuwenhoeck in 1678. Spermidine is formed from ornithine and S-adenosyl-L-methionine by the following steps:

$$L\text{-ornithine} \rightarrow putrescine + CO_2$$

$$S\text{-adenosyl-L-methionine} \rightarrow 5'deoxy\text{-}5'S\text{-}(3 \text{ methylthiopropylamine}) \text{ sulphonium adenosine (decarboxylated 'SAM')} + CO_2$$

$$Putrescine + decarboxylated \text{ } SAM \rightarrow spermidine + methylthioadenosine$$

The enzymes involved are discussed in Section 9.2.3.3. Spermine is formed from spermidine by a reaction similar to the last-mentioned.

$$Spermidine + decarboxylated \text{ } SAM \rightarrow spermine + methylthioadenosine$$

(Tabor and Tabor, 1976).

The concentration of spermidine in the testis rises rapidly after birth in the rat to reach a peak at about 14 days after birth. Thereafter it falls to a minimum at about 35 days, rises to a plateau between 40 and 60 days and then falls again (Figure 9.1). Hypophysectomy was followed by a reduction in spermidine concentration (Macindoe and Turkington, 1973).

Spermine is present in adult rat testes in about the same concentrations as sper-

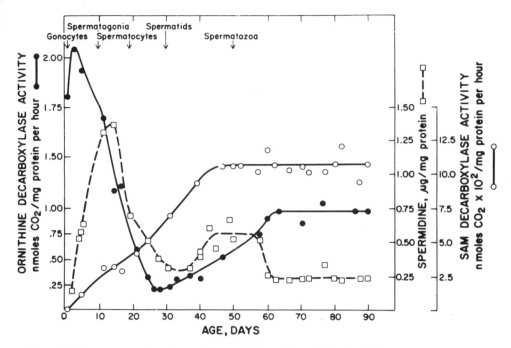

Figure 9.1 The concentration of spermidine and the specific activity of ornithine decarboxylase and S-adenosyl-methionine (SAM) decarboxylase in the testes of rats of various ages. (Reproduced from Macindoe and Turkington, 1973)

midine (Tabor and Tabor, 1964) but variations with age have apparently not been investigated.

The significance of these compounds is not known but it is interesting that both spermidine and spermine will cause compaction of bacteriophage DNA (Gosule and Schallman, 1976) and therefore may be involved in condensation of the spermatid nucleus.

**9.1.4.4    Nucleotides**    The testis contains adenosine triphosphate (ATP) in concentrations similar to those in several tissues, but lower than in skeletal muscle (Means and Hall, 1968b).

Cyclic adenosine monophosphate (AMP) and cyclic guanosine monophosphate (GMP) concentrations have been measured in the testes of rats of various ages. Cyclic AMP decreases from 5 to 40 days then increases again, whereas cyclic GMP decreases throughout the period of observation (5 to 90 days) (Figure 9.2). In postnatal rats, cyclic AMP and GMP are distributed in cells throughout the tubule. As the rat matures, the cyclic AMP bound to receptor-proteins was localized, by immunocytochemical techniques, in the cytoplasm of cells lining the tubular wall, while cyclic GMP was found on the membranes of the primary spermatocytes and associated with the chromosomes of cells undergoing meiosis (Ong *et al.*, 1975; Spruill and Steiner, 1976; Macindoe *et al.*, 1977).

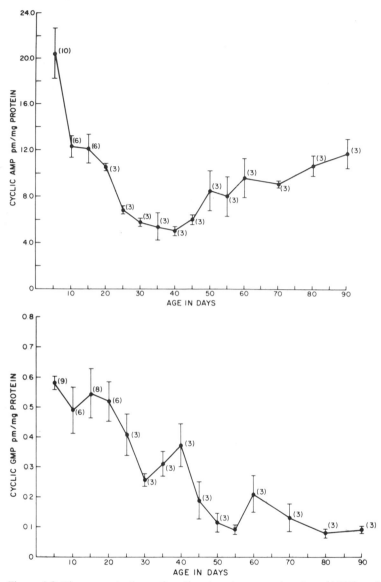

Figure 9.2 The concentrations of cyclic adenosine monophosphate (AMP) and of cyclic guanosine monophosphate (GMP) in the testes of rats of various ages. (Reproduced from Spruill and Steiner, 1976)

**9.1.4.5    Biogenic amines**    The concentrations of norepinephrine, dopamine, 5-hydroxytryptamine and histamine were highest in the testes of neonatal rats, and decreased progressively during development. Using histofluorescent techniques, these amines could be located in the walls of the seminiferous tubules and the interstitial cells (Kormano and Penttila, 1968; Assaykeen and Thomas, 1965; Zieher *et al.*, 1971).

## 9.2    Testicular enzymes and cofactors

As one would expect in a tissue with an active oxidative metabolism, the testis has a full complement of enzymes catalysing the various steps in the metabolic pathways. Here comment will be made only on those enzymes which have been found in the testis in unusual forms or in particularly high concentrations (see Blackshaw, 1970; Fritz, 1974; Gomes and VanDemark, 1974).

### 9.2.1    Enzymes concerned with carbohydrate metabolism

**9.2.1.1    Lactic dehydrogenase (LDH)**    Lactic dehydrogenase is a widely distributed enzyme which is normally found as four different isoenzymes. However in mature testes a fifth isoenzyme is found, called LDH-X. LDH-X contains subunits (C type) different from those of skeletal muscle (A type) or heart muscle (B type) (Zinkham *et al.*, 1964; Zinkham, 1968) although in guinea pig, a second testis specific isoenzyme is found, a hybrid of A and C type subunits (Batellino and Blanco, 1970). LDH-X is unusual in that it catalyses the oxidation of DL-$\alpha$-hydroxybutyrate and DL-$\alpha$-hydroxyvalerate as well as lactate (Allen, 1961). It is located in the tubules and appears at the time that the pachytene spermatocytes are first formed (Blackshaw and Elkington, 1970a, b; Shen and Lee, 1976). The highest concentrations are found in the 'heavy mitochondrial' fraction of the testis, presumably derived from spermatids and the midpieces of spermatozoa (Clausen, 1969; de Domenech *et al.*, 1972). The concentration of LDH-X in the testis falls after treatments which remove germinal cells (Blackshaw and Elkington, 1970b; Ewing and Schanbacher, 1970; Sherins and Hodgen, 1976) but the functions of the different isoenzymes are obscure (see Hawtrey *et al.*, 1975). No LDH-X could be demonstrated in ram rete testis fluid (see Section 8.2.1).

**9.2.1.2    Hexokinase**    Testis contains a specific isoenzyme of hexokinase. This was first demonstrated in *Drosophila* but has now been confirmed in mammals. It is low in immature animals and reaches a maximum before the first spermatozoa are shed. The concentration of this enzyme is reduced by cryptorchidism (Pilkis, 1970; Cheng, 1971; Sosa *et al.*, 1972).

**9.2.1.3    Sorbitol dehydrogenase (SDH)**    This enzyme is not specific to the testis, nor does it have isoenzymes which are. However it has been widely used as a marker for spermatogenesis because it is present in the mature testis of every species of animal so far examined and its concentration increases sharply when spermatogenesis begins, at about the time of the appearance of the primary spermatocytes. The concentrations of this enzyme in the testis fall in cryptorchidism, immunological aspermatogenesis, avitaminosis A or after hypophysectomy (Bishop, 1967, 1968a; Katsh and Aguirre, 1968; Mills and Means, 1972; Shen and Lee, 1976; Sherins and Hodgen, 1976).

**9.2.1.4     Phosphofructokinase**   Two isoenzymes are present in the testis (Tsai and Kemp, 1973) and the concentration of this enzyme falls in cryptorchidism (Ewing and Schanbacher, 1970).

**9.2.1.5     Glucose-6-phosphate dehydrogenase**   The activity of this enzyme is highest in the testes of young mice and decreases after puberty (Shen and Lee, 1976).

**9.2.1.6     α-Glycerophosphate dehydrogenase**   At birth in rats and mice, this enzyme is in very low concentrations, but its activity rises steadily until puberty and then remains reasonably constant (Schenkman et al., 1965; Shen and Lee, 1976). Activity is decreased by oestrogen treatment in rats. There are slightly lower concentrations of this enzyme in the testes of cat and bull than in rat testis, and much lower levels in the testes of human, dog and horse (Schenkman et al., 1965).

**9.2.1.7     Pyruvate dehydrogenase**   The activity of this enzyme in rat testis decreased within hours and then stabilized at a lower level when the testis was returned to the abdomen (Free and Payvar, 1974).

**9.2.1.8     Malate dehydrogenase (MDH)**   MDH can be histochemically localized in the interstitial tissue of rat testis (Ambadkar and George, 1964) and its activity in the testes of mice is greatest before puberty (Shen and Lee, 1976). Two isoenzymes are present in ram testis and rete testis fluid, a cytoplasmic and a mitochondrial isoenzyme; only the cytoplasmic isoenzyme was present in blood plasma (see Section 8.2.1).

**9.2.1.9     N-acetyl-β-glucosaminidase and β-galactosidase**   Two isoenzymes of each of these lysozomal enzymes have been discovered in the testes (Majumder and Turkington, 1974). The specific activity of β-galactosidase isoenzyme I is highest in testes from 4-day old rats, and falls as the testis develops. Isoenzyme II was undetectable in the testes of young rats and appeared only with the development of acrosome-containing spermatids. In the mature testis, this latter isoenzyme represents about 15 per cent of total activity and appears to be identical to a tissue-specific isoenzyme which represents about 90 per cent of the activity in epididymal spermatozoa. These results suggest that isoenzyme I is associated with Sertoli cells and isoenzyme II with post-meiotic cells. β-galactosidase II disappears from the testis after hypophysectomy, and can be induced with LH plus FSH, or testosterone. The isoenzymes of N-acetyl-β-glucosamidase both increase with the formation of spermatocytes, and then decrease with the appearance of spermatids and spermatozoa (Figure 9.3). Isoenzyme II represents more than 90 per cent of the activity at all stages, and a third sperm-specific isoenzyme represents less than 5 per cent of the total testis activity (Majumder et al., 1975). Histochemically, N-acetyl-β-glucosaminidase activity is almost all confined to the contents of the seminiferous tubules, with no particular association with any of the cell types (Pugh and Walker, 1961).

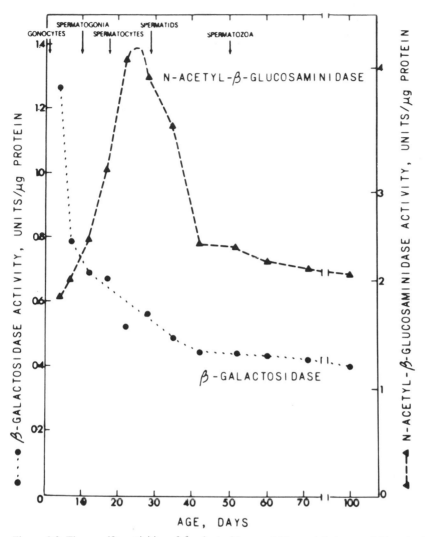

Figure 9.3 The specific activities of β-galactosidase and N-acetyl-β-glucosaminidase in the testes of rats of various ages. (Reproduced from Majumder *et al.*, 1975)

### 9.2.2    Enzymes involved in lipid metabolism

**9.2.2.1    Carnitine acetyl transferase (CAT)**    The concentration of this enzyme in the testis rises sharply at the time of puberty, and falls in cryptorchidism and after hypophysectomy (Figure 9.4). Its rise coincides with the appearance of late pachytene and diplotene spermatocytes (Marquis and Fritz, 1965; Schanbacher *et al.*, 1974). The high level of activity in diplotene spermatocytes was confirmed by separating the cells on a staput gradient (Vernon *et al.*, 1971).

Its function in the germinal cells is unknown; CAT is usually involved with fatty

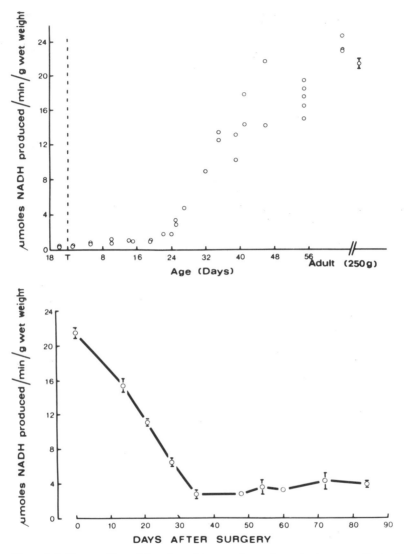

Figure 9.4 The specific activity of carnitine acetyl transferase in the testes of rats of various ages (T, birth; the 18 before the T indicates 18 days *post coitum*) and after hypophysectomy. (Reproduced from Vernon *et al.*, 1971)

acid metabolism, but spermatogonia have a high rate of palmitate oxidation but a low level of CAT (Lin and Fritz, 1972).

### 9.2.3    Enzymes concerned with amino-acid and protein metabolism

### 9.2.3.1    Aminopeptidases   Four different aminopeptidases have been demonstrated in the testis although none are specific for this tissue. Two are in-

terstitial, two are tubular. The two interstitial and one tubular enzyme decrease at puberty and increase after cryptorchidism; the other tubular enzyme increases at puberty and decreases after cryptorchism. Four dipeptidyl aminopeptidases have also been described in the testis; two of these are lysosomal enzymes, one of which is found only in the interstitium while the other is present in the tubules, but occurs mainly in the interstitium. These enzymes did not change with age (Vanha-Perttula, 1973a, b, c).

**9.2.3.2    γ-Glutamyl transpeptidase (GGT)**    This enzyme may be a marker of Sertoli cell function. Its specific activity increased markedly at the time of the cessation of mitoses in the Sertoli cells, just before the formation of the first primary spermatocytes, and then remained constant throughout puberty (Hodgen and Sherins, 1973). In vitamin A deficient rats, testicular levels of GGT were high despite virtually complete loss of germ cells and large falls in lactic dehydrogenase and sorbitol dehydrogenase. In human clinical cases, the enzyme was in high concentration even when germ cells were few, except when there was 'profound immaturity of seminiferous tubules and Sertoli cells' (Sherins and Hodgen, 1976).

**9.2.3.3    L-ornithine decarboxylase and the S-adenosyl-L-methionine decarboxylase complex (SAM decarboxylase)**    These enzymes are involved in the formation of the polyamines spermidine and spermine in the testis (see Section 9.1.4.4). The SAM decarboxylase seems to be distributed throughout the germinal epithelium and its specific activity (enzyme activity per mg per total protein) rises steadily from birth to puberty in rats. The L-ornithine decarboxylase, on the other hand seems to be confined to the Sertoli cells. Its specific activity is highest at about 2 days after birth when the Sertoli cell is the predominant cell type, falls during the early stages of spermatogenesis, up to about 30 days, and then rises again to reach a plateau at about one-half its maximal values (Figure 9.1).

Both enzymes are reduced by hypophysectomy at 28 days of age. The specific activity of SAM-decarboxylase can be maintained at normal levels by daily injections of FSH + LH or testosterone, but these treatments were much less effective in maintaining the specific activity of the ornithine decarboxylase, probably because of its short half-life (Macindoe and Turkington, 1973).

These authors suggested that the distribution of the enzymes indicates that the ornithine decarboxylase in the Sertoli cells produces the putrescine which is then transferred to the spermatids for the other reactions. The mechanism of transfer is unknown.

**9.2.4    Enzymes concerned with nucleic acid and nucleotide metabolism**

**9.2.4.1    Uridine diphosphatase (UDPase)**    This enzyme is a component of smooth endoplasmic reticulum in a number of cell types and also appears to be membrane bound in the testis. It increases in the rat testis from birth to 26 days of age and then decreases sharply with the formation of the spermatids. The enzyme could be detected histochemically in spermatogonia and spermatocytes but not in spermatids. Hypophysectomy of adult rats led to an increase in the specific activity of UDPase as the spermatids disappeared (Xuma and Turkington, 1972).

**9.2.4.2    5′-nucleotidase**   This enzyme is a marker of plasma membranes in a variety of cells. In the testis it is also membrane-bound and in the rat its specific activity rises from birth until about 26 days of age, when it reaches a plateau. Hypophysectomy had no effect on this enzyme (Xuma and Turkington, 1972).

**9.2.4.3    Phosphodiesterase and adenyl cyclase**   Two isoenzymes (*c* and *f*) of phosphodiesterase can be demonstrated in the rat testis by starch gel electrophoresis; the *f* isoenzyme is confined to the testis. It appears first in rats at the time of the elongation of the spermatids (40 days) (Monn *et al.*, 1972) and its development has been suggested as an explanation for the lack of effect of FSH on adult rat testes (see Section 10.2.1).

However, age has no effect on the concentration of phosphodiesterase in the testes of rats irradiated *in utero* to eliminate the germ cells (Sertoli cell only rats) (see Section 11.1). The activity of phosphodiesterase in the testis is reduced by FSH treatment in young rats, whereas no effect is seen in adult animals.

Three isoenzymes of phosphodiesterase have been demonstrated in the testis by ion-exchange chromatography. One is calcium dependent and hydrolyses cyclic GMP as well as cyclic AMP; the second is also calcium dependent but hydrolyses only cyclic GMP; the third does not require calcium and is specific for cyclic AMP. The second is found only in mature animals, the first only in immature animals. There is also a heat-stable calcium-binding protein which activates phosphodiesterase, and the concentration of this protein increases between 5 and 47 days of age in normal rats but not in Sertoli cell only rats (Means *et al.*, 1976, 1977).

The amount of adenyl cyclase in the testis increases progressively from 25 to 60 days of age (Hollinger, 1970). Two forms of this enzyme are present in rat testis. There is a particulate enzyme which is stimulated by fluoride and gonadotrophins (see Sections 10.1.2.2 and 10.2.2.2), but is independent of $Mg^{2+}$ or $Mn^{2+}$; there is also a soluble enzyme which is stimulated by $Mn^{2+}$, but not by fluoride or gonadotrophins. The particulate enzyme is associated with non-germinal cells, whereas the soluble enzyme is present only in post-meiotic germ cells. There is also a soluble guanylate cyclase in rat testis. This enzyme is also $Mn^{2+}$ dependent, and is located in the tubules, but not in germ cells (Braun and Dods, 1975; Braun *et al.*, 1977).

**9.2.5    Enzymes concerned with steroid metabolism**

There have been many studies on the various enzymes involved in the formation of androgens from cholesterol and there is a great deal known about their distribution, sub-cellular location, cofactor requirement and substrate specificity. Some of this information is given in Table 9.1. Although some of the enzymes occur inside the seminiferous tubules, it is possible to find a complete set of enzymes only in the interstitial tissue, where most of the testosterone is produced. The significance of the enzymes inside the tubules is not known, but they are presumably concerned with

**Table 9.1**
**Enzymes related to androgen formation from cholesterol**

| Name of enzyme | E.C. No. | Required cofactor | Substrate | Product |
|---|---|---|---|---|
| 20α-Hydroxylase | 1.41.1.9 | NADPH, $O_2$ | Cholesterol<br>22R-Hydroxycholesterol | 20α-Hydroxycholesterol<br>20α-22R-Dihydroxycholesterol |
| 22R-Hydroxylase | | NADPH, $O_2$ | Cholesterol<br>20α-Hydroxycholesterol | 22R-Hydroxycholesterol<br>20α-22R-Dihydroxycholesterol |
| $C_{20}$-$C_{22}$ Lyase | | NADPH, $O_2$ | 20α-22R-Dihydroxycholesterol | Pregnenolone |
| Δ⁵-3β-Hydroxysteroid dehydrogenase with isomerase | 1.1.1.51<br>5.3.3.1 | NAD and NADP | Pregnenolone<br>17α-Hydroxypregnenolone<br>Dehydroepiandrosterone<br>Δ⁵-Androstene-3β, 17β-diol | Progesterone<br>17α-Hydroxyprogesterone<br>Δ⁴-Androstenedione<br>Testosterone |
| 17α-Hydroxylase | 1.14.1.7 | NADPH, $O_2$ | Progesterone<br>Pregnenolone | 17α-Hydroxyprogesterone<br>17α-Hydroxypregnenolone |
| $C_{17}$-$C_{20}$ Lyase | | NADPH, $O_2$ | 17α-Hydroxyprogesterone<br>17α-Hydroxypregnenolone | Androstenedione<br>Dehydroepiandrosterone |
| 17β-Hydroxysteroid dehydrogenase | 1.1.1.64 | NADPH | Androstenedione<br><br>Dehydroepiandrosterone | Testosterone<br><br>Δ⁵-Androstene-3β, 17β-diol |

transformations of steroids formed in the interstitium. The enzymes concerned with the formation of pregnenolone from cholesterol are mitochondrial, whereas those involved in the formation of testosterone from pregnenolone form part of the endoplasmic reticulum. Most of the enzymes require NADPH as a cofactor, not the more common NADH, the exception being $\Delta^5$-3$\beta$-hydroxysteroid dehydrogenase which needs both NAD and NADP in the oxidized forms. This is also the only enzyme not requiring oxygen as an electron-receptor. Most of the enzymes act on a number of substrates, and therefore which reaction they catalyse depends on availability of substrate (Baillie, 1961, 1974; Baillie et al., 1966; Murato et al., 1965; Inano et al., 1967a, b; Tamaoki et al., 1969, 1975). Spermatocytes contain appreciable amounts of the enzyme 5$\alpha$-reductase which forms 5$\alpha$-dihydrotestosterone (DHT) from testosterone; Sertoli cells also contain this enzyme and a high level of 3$\alpha$-hydroxysteroid dehydrogenase which converts DHT to 5$\alpha$-androstan-3$\alpha$, 17$\beta$-diol (Dorrington and Fritz, 1975).

In the testis, there are also a large number of other steroid-metabolizing enzymes, not directly concerned with the formation of androgens. Some are concerned with the formation of other physiologically important steroids such as oestrogens, but the function of many of them is quite obscure (Tamaoki et al., 1969).

### 9.2.6   Miscellaneous enzymes

**9.2.6.1   Hyaluronidase**   The testis is the usual source for the commercial preparation of this enzyme. It is acrosomal and therefore confined to the spermatids and spermatozoa (see Fritz, 1974), appearing at about 33 days of age in rats and 20 days in mice, and reaching a peak at 60 days of age in rats (Males and Turkington, 1970) and 40 days in mice (Shen and Lee, 1976). Hypophysectomy of rats at 26 days of age prevents its appearance, and in adults causes a fall (Steinberger and Nelson, 1955).

**9.2.6.2   Esterases**   The testis is rich in non-specific esterase activity which increases sharply in the rat from the 27th day onwards (Huggins and Moulton, 1948; Holmes and Masters, 1967). Histochemically, it appears to be located mainly in the Leydig and Sertoli cells (Niemi et al., 1962, 1966; Niemi and Kormano, 1965a). The increase is associated with the appearance of specific electrophoretic bands, which can also be induced in immature rats by hCG (Meyer et al., 1974). The esterase isoenzyme pattern in rete testis fluid is quite different from that in the testis (see Section 8.2.1).

**9.2.6.3   Acid phosphatase**   Four isoenzymes are present in the testes. Isoenzymes I and II are lysosomal and interstitial. Isoenzyme III is tubular, but its activity is unchanged during puberty or after cryptorchidism, and it is therefore probably located in the Sertoli cells or spermatogonia. Isoenzyme IV is testis specific; it increases rapidly at the time of puberty and falls following cryptorchidism (Vanha-Perttula, 1971a, b; Vanha-Perttula and Nikkanen, 1973). It also showed higher concentrations in sections of tubules containing large number of late spermatids, and low concentrations in sections of tubules where the spermatozoa had just been shed (Parvinen and Vanha-Perttula, 1972). This evidence was taken to indicate that the isoenzyme may be produced by the relatively less mature germ cells, but is concentrated in post-meiotic cells.

**9.2.6.4    Carbonic anhydrase**    Five isoenzymes of this enzyme were found in the rat testes, and one of these was not present in erythrocytes or kidneys (Hodgen *et al.*, 1971). A non-tubular site was suggested, and histochemical studies support this in the rat but not in the ram (Niemi and Hyyppa, 1970; Cohen *et al.*, 1976).

**9.2.6.5    Monoamine oxidase and catalase**    Monoamine oxidase has been observed in the interstitial cells and walls of the seminiferous tubules of rat testes. Its activity is related to androgen synthesis (Ellis *et al.*, 1972). The reactions it catalyses give rise to hydrogen peroxide, which is broken down by catalase. The activity of this latter enzyme was reduced by hypophysectomy and increased by irradiation (Buhrley and Ellis, 1973), but catalase activity in rabbit testis drops sharply as the first spermatids appear (Ihrig *et al.*, 1974).

**9.2.6.6    Adenosine triphosphatase**    Adenosine triphosphatase (ATPase) has been demonstrated histochemically in the testis (Tice and Barnett, 1963). There appears to be no detectable Na–K stimulated (ouabain sensitive) activity in the testis (Setchell *et al.*, 1972b, but cf. Gupta, 1975), but a bicarbonate-stimulated enzyme, associated with mitochondria, has been demonstrated in testis homogenates from rats, gerbils and hamsters (Hollinger, 1971; Setchell *et al.*, 1972b).

Delhumeau-Ongay *et al.* (1973a) have reported that an ATPase stimulated by calcium and magnesium, but not by Na or K was present in rat testis, reaching a peak of activity at 25 days of age. Cryptorchidism, hypophysectomy or treatment with fluoroacetamide caused increases in ($Ca^{24}$-$Mg^{24}$)–ATPase activity of the testis (Delhumeau-Ongay *et al.*, 1973b).

The ATPase has been localised in the Sertoli cells near the spermatids and spermatocytes (Tice and Barnett, 1963; Chakraborty and Nelson, 1974; Chang *et al.*, 1974; Barham, 1976), but other workers have described enzyme in association with cytoplasmic filaments contained within the Sertoli cell processes, and in the lymphatic endothelium (Gravis *et al.*, 1976).

**9.2.7    Cofactors**

**9.2.7.1    Cytochromes**    Several of the enzymes concerned with androgen production, 17α-hydroxylase and $C_{17}$-$C_{20}$ lyase have for their active sites the microsomal CO-binding haemoprotein, cytochrome P-450 (Machino *et al.*, 1969; Inano *et al.*, 1970). This cytochrome is located with the enzymes in the interstitial tissue and not in the tubules (Menard *et al.*, 1975). A testis-specific cytochrome c has recently been found in the cells of the germinal epithelium (Hennig, 1975; Goldberg *et al.*, 1977).

## 9.3      Oxidative metabolism

### 9.3.1      Oxygen uptake

**9.3.1.1      *In vivo***   The testis has a reasonable oxygen uptake *in vivo* compared
with other organs in the body, but because of the comparatively slow blood flow
(Chapter 3), about one half of the arterial oxygen is removed during one passage of
the blood through the testis (Setchell and Waites, 1964). This is a higher fraction
than most other organs and, because most animals do not seem able to increase
testicular blood flow, the testis is always on the verge of hypoxia. This is particularly
so because the seminiferous tubules are avascular and therefore their centres are at
least $100\mu m$ from the nearest blood vessels. No active transport of oxygen has ever
been convincingly demonstrated, and therefore we must assume that the tubules rely
on oxygen diffusing in through their walls. This is a process which is defined by the
equation (Setchell, 1970b), modified from Krogh

$$T_0 - T_i = \frac{10^4 p}{d}\left(\frac{R^2 - r^2}{4} - 1 \cdot 15 r^2 \log \frac{R}{r}\right)$$

where      $T_0 =$ oxygen tension outside the tubules.
$T_i =$ oxygen tension in the tubular lumen.
$p =$ oxygen uptake by the tubules, about $0 \cdot 0059 ml/g/min$ in the ram
testis (Setchell and Waites, 1964); it is also necessary to assume that
the uptake is independent of oxygen tension over the range involved.
$d =$ diffusion coefficient of oxygen in the testis; $0 \cdot 14 ml/cm^2\text{-min-atm-}\mu m$
for most tissues.
$R =$ outside radius of the seminiferous tubules, approximately $100\mu m$.
$r =$ radius of the tubular lumen, approximately $20\mu m$.

Using these values, the outside–inside difference should be about 7mm Hg (Setchell,
1970b); the important value to know is the true tension outside the tubules. Free and
VanDemark (1968) found that the oxygen tension in blood from the internal spermatic
vein was about 23mm Hg. However Cross and Silver (1962) measured the oxygen
tension in the testis itself by inserting a platinum electrode and obtained values of
about 11mm Hg. The latter value may be affected by the damage caused by inserting
the electrode; the former may be an overestimate because it includes some blood from
tissues with higher blood flow (for example the head of the epididymis) and because the
venous blood may gain oxygen from arterial blood during its passage through the spermatic cord. This latter effect has not been directly demonstrated but, in rams breathing
pure oxygen, the oxygen tension of arterial blood falls as the blood passes through the
spermatic cord and certain radioactive tracers can cross from artery to vein or vice versa
in the spermatic cord (see Section 3.2.2). However it seems reasonably certain that the
oxygen tension at the centres of the tubules is between 4 and 16mm Hg, a reasonably low
value, and direct measurements support this conclusion (Free *et al.*, 1976).

**9.3.1.2**    *In vitro*    The most remarkable feature of oxygen uptake by teased-out and sliced preparations of rat testis *in vitro* is its pronounced dependence on the presence of glucose (Dickens and Greville, 1933; Ewing and VanDemark, 1963a, b) or other suitable substrates. These include lactate (Dickens and Simer, 1930; Elliott *et al.*, 1973), pyruvate (Elliott *et al.*, 1937; Featherstone *et al.*, 1955; Serfaty and Boyer, 1956) and succinate (Elliott *et al.*, 1937; Serfaty and Boyer, 1956). Both lactate and pyruvate are present in blood plasma, testicular lymph and rete testis fluid and therefore are normally available for metabolism. Other tricarboxylic acid cycle intermediates are less effective substrates, but do cause some stimulation of oxygen uptake. Other substances which are not normally found in the testis, such as fructose, also stimulate oxygen uptake (Dickens and Greville, 1933; Serfaty and Boyer, 1956). Acetate has no effect.

The effect of glucose in stimulating oxygen uptake was much less obvious with dog and human testes (Kato, 1958) and is abolished when the testis is made cryptorchid (Tepperman *et al.*, 1949) or when spermatogenesis is arrested by incorporating nitrofurans in the diet (Paul *et al.*, 1953). These findings suggest that the spermatids are the cells responsible for the effect of glucose.

A different picture again emerged using homogenates of ram testes; glucose, citrate and glutamate had no effect on oxygen uptake but acetate, $\beta$-hydroxybutyrate and octanoate stimulated it. With testicular mitochondria, citrate, glutamate, acetate and $\beta$-hydroxybutyrate stimulated but glucose and octanoate had no effect (Setchell *et al.*, 1965). These results emphasize the importance of permeability (see p. 252) and substrate transport in the testis (see p. 302).

It is interesting that the testis is one of the few tissues whose oxygen uptake does not increase in thyroidectomized rats after treatment with thyroxine (Barker and Klitgaard, 1952). This could be explained if thyroxine does not penetrate through the blood-testis barrier (Section 8.7).

**9.3.2**    **Carbon dioxide production and respiratory quotient (RQ)**
The production of carbon dioxide by the testis has been measured both *in vivo* and *in vitro*. The RQ (moles of $CO_2$ produced/moles of $O_2$ taken up) is about 0·9 *in vivo* (Himwich and Nahum, 1929; Annison *et al.*, 1963; Setchell and Waites, 1964; Waites and Setchell, 1964; Setchell and Hinks, 1967) and *in vitro*, with glucose added to the medium (Dickens and Simer, 1930; Tepperman *et al.*, 1949). When no exogenous substrate was added the RQ fell to about 0·76 (Dickens and Simer, 1930). Predictions from RQ about substrates utilized are fraught with great uncertainty but values about 0·9 certainly point toward substantial carbohydrate oxidation, and the fall with no added substrates indicates a switch to lipid metabolism. Beyond this, it is probably unwise to say anything from these data.

**9.4**    **Carbohydrate metabolism**

**9.4.1**    **Uptake of glucose**
*In vitro*, glucose is rapidly removed from the media by teased out seminiferous

tubules from rats (Tepperman and Tepperman, 1950; Paul *et al.*, 1953) and cotton rats (Ewing *et al.*, 1965) and testis slices from rabbits (Ewing and VanDemark, 1963a,b; Ewing *et al.*, 1964; Ewing, 1967).

*In vivo*, the ram testis normally removes about 15 per cent of the arterial glucose, enough to account for about 75 per cent of the oxygen if all the glucose were oxidized (Setchell and Waites, 1964; Setchell and Hinks, 1967; Setchell, 1970b; Middleton, 1973).

This uptake is much higher than that seen elsewhere in the body except in the brain (see Pappenheimer and Setchell, 1973; Lindsay and Setchell, 1976) and pregnant uterus (Setchell *et al.*, 1972a), and appears to be independent of glucose concentration in the blood, except at very low concentrations. This suggests that glucose may be transported into the tubules by a process known as facilitated diffusion. This term is used for selective transport down a concentration gradient and should be distinguished from active transport, which implies movement against the concentration gradient. Much of the evidence for this facilitated diffusion depends on the use of an analogue of glucose, 3-0-methyl glucose (3-OMG), which in other tissues is transported by the same mechanism as glucose but is not metabolized. The main evidence for the existence of a transport mechanism *in vivo* is that the amount of 3-OMG entering the testis decreases as the plasma concentration of glucose increases. The same applies *in vitro* with teased-out seminiferous tubules (Figure 9.5) and it has been shown that mannose and to some extent galactose compete for the same carrier, but that fructose, sucrose or inositol do not. If the 3-OMG is already in the tubules, it can be displaced by raising the external glucose concentration, a phenomenon known as counter flow, which is characteristic of facilitated diffusion. The entry of 3-OMG is unaffected by Na concentration or by the presence of DNP, phloridzin or ouabain so the carrier mechanism appears to be more similar to that in brain, liver and erythrocytes rather than that in muscle (Setchell and Middleton, 1971; Middleton, 1973).

As conditions can be controlled more precisely *in vitro*, it is possible to show that the transport mechanism conforms to Michaelis saturation kinetics with a $K_m$ of about 11mM. An estimate of the $K_m$ *in vivo* has also been made in rats from the uptake of 3-OMG at different plasma glucose concentrations, and the value of 8mM is reasonably similar to the value obtained *in vitro*. This would suggest that glucose uptake should depend on plasma glucose concentration. Measurements in the ram suggest that this is not so, but the characteristics of the carrier may be different in that species (Middleton, 1973).

### 9.4.2    Endproducts of glucose metabolism

As we have already mentioned (p. 301), the carbon dioxide production by the testis *in vivo* and *in vitro* is approximately equivalent to the oxygen uptake, suggesting a dependence on glucose. Using [$^{14}$C] labelled glucose, it has been shown *in vivo* that, in the ram testis, a substantial proportion of the $CO_2$ is derived from blood glucose and that a large proportion of the glucose taken up is oxidized (Setchell and Hinks, 1967). A somewhat smaller proportion of the glucose is converted to $CO_2$ *in vitro* by rat seminiferous tubules (Middleton, 1973) or testis slices (Hollinger and Davis,

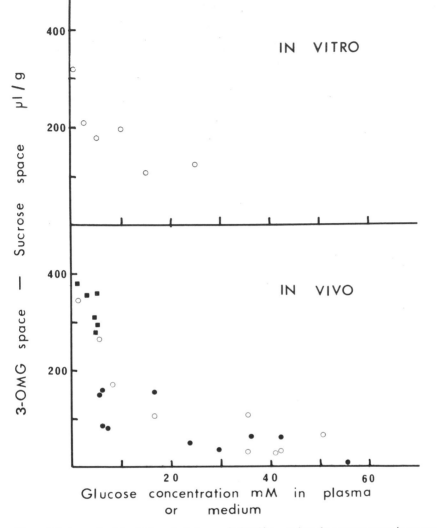

Figure 9.5  A graph of the 3-O-methyl glucose (3-0MG) space less the sucrose space in rat seminiferous tubules *in vitro* in media with varying concentrations of glucose and in the testes of rats *in vivo* with varying concentrations of glucose in the plasma:O, normal rats;●, streptozoticin-diabetic rats;■, alloxan diabetic rats. Some rats had been injected with insulin or infused with glucose. (Unpublished data of A. Middleton and B. P. Setchell)

1968) and considerable amounts appear as lactate, inositol and some amino-acids, in particular glutamate, alanine and aspartic acid (Middleton, 1973).

The conversion of glucose to amino-acids (Setchell *et al.*, 1967) and inositol (Middleton and Setchell, 1972) by the ram testis *in vivo* has also been reported but in neither of these studies was it possible to quantitate the importance of these processes because of the slowness with which the specific radioactivity of these endproducts rose in the rete testis fluid.

### 9.4.3    Pathways of glucose metabolism

Using glucose specifically labelled in the 1 and 6 positions, it has been shown that only a small proportion of the glucose is metabolized through the pentose cycle (Setchell and Hinks, 1967; Free and VanDemark, 1969).

## 9.5    Lipid metabolism

### 9.5.1    Phospholipids

The metabolism of testicular phospholipids has been studied mainly *in vivo* by injecting the animal with radioactive inorganic phosphate ($[^{32}P]P_i$), removing the testes after various times and measuring the incorporation of radioactivity into the various individual phospholipids. In the ram, infusions of $[^{32}P] P_i$ into the testicular artery led to the incorporation of label into all the phospholipids. However the majority of the radioactivity appeared in phosphatidylinositol (68 per cent after 3h and 39 per cent after 5h), a phospholipid which is present in comparatively small amounts (about 6 per cent of total phospholipid). The most abundant phospholipid, phosphatidylcholine (40 per cent of total phospholipid), accounted for only 16 per cent of the radioactivity after 3h rising to 33 per cent after 5h. The next most abundant phospholipid, ethanolamine plasmalogen (11 per cent of total), was even less active metabolically, and the highest specific radioactivity (counts/min/mg P) was observed in phosphatidic acid, which comprizes only about 1 per cent of the total phospholipid (Scott and Setchell, 1968). Yokoe and Hall (1970a) have shown that $[^{32}P] P_i$ is incorporated into phosphatidyl ethanolamine, phosphatidyl choline and sphingomyelin in rat testes. The highest specific radioactivity was found in phosphatidyl ethanolamine and the lowest in sphingomyelin. It is hard to understand why these authors did not study the compcunds which Scott and Setchell (1968) found to be most actively labelled, monophosphatidyl inositol and phosphatidic acid. However Yokoe and Hall (1970a) did show in addition that $[^3H]$ choline is incorporated into phosphatidyl choline and sphingomyelin and $[^3H]$ ethanolamine into phosphatidyl ethanolamine. They also showed that all these incorporations can be stimulated in hypophysectomized rats by injections of FSH which acts mainly on the tubular synthesis of phospholipids. LH also stimulates incorporation into phosphatidyl choline. Antisera to ICSH decreased the incorporation of $[^{32}P] P_i$ into phospholipids, and the anti-FSH sera seemed to have a slight inhibitory effect (Gambal, 1967).

### 9.5.2    Neutral lipids and cholesterol

It was shown some years ago that the testis, like the liver and other tissues can synthesize cholesterol from acetate (Srere *et al.*, 1950). In fact, the testis seems to depend more on endogenous synthesis than dietary sources, unlike most other tissues of the body (Morris and Chaikoff, 1959). When radioactive glucose was infused in the testicular artery of the ram, some label was incorporated into cholesterol and into mono-, di- and triglycerides, but most (61 per cent) appeared in triglycerides (Scott and Setchell, 1968).

### 9.5.3    Fatty acids

Radioactive acetate injected directly into the testes of rats is converted into palmitate (16:0), stearate (18:0), oleate (18:1), arachidonate (20:4), docosatetraenoate (22:4) and docosapentaenoate (22:5) (Davis and Coniglio, 1966; Evans et al., 1971). 16:0 is largely synthesized de novo in the testis whereas 18:0 is formed partially by chain elongation (Coniglio et al., 1971). Incorporation of label from acetate into palmitate, palmitoleate, stearate and oleate has also been shown for slices of rabbit testis in vitro (Hall et al., 1963). The pattern of incorporation is quite different in the two types of experiment (Evans et al., 1971).

Radioactive linoleate (18:2) injected into the testis is incorporated into various un-saturated fatty acids via the pathway:

$$18:2 \rightarrow 20:3 \rightarrow 20:4 \rightarrow 22:4 \rightarrow 22:5 \rightarrow 24:5$$
$$\searrow$$
$$24:4$$

(Davis and Coniglio, 1966; Nakamura and Privett, 1969; Bridges and Coniglio, 1970a, b). Furthermore there is good evidence that injected 22:5 is converted into 20:4 and 16:0 (Bridges and Coniglio, 1970c). When the acids are injected, radio-activity appeared as $CO_2$, and appreciable transport to other organs occurred (Bridges and Coniglio, 1970a). However, much higher proportions remain in the testicular lipids when the labelled fatty acid was injected as an emulsion, and incor-poration was demonstrated into higher fatty acids of all lipid classes. Slightly more activity was incorporated into the phospholipids and other polar lipids than into the neutral lipids (Nakamura and Privett, 1969).

About 40 per cent of a dose of [1-$^{14}$C-] palmitic acid injected into the testes of rats appeared in the expired air within 4h while about 40 per cent remained in the testis at that time; after 2 weeks 6 per cent of the injected label was still in the testis. Initially the label was largely in phospholipids, but at later times, triglycerides contained proportionally more. There was comparatively little conversion of the injected palmitate to other fatty acids (Coniglio et al., 1972). Radioactive palmitic acid in-jected directly into the ram testis appeared in the ejaculated sperm between 17 and 63 days later. The [$^{14}$C] appeared mainly in phospholipids, with phosphatidyl inositol showing the highest specific radioactivity. Some radioactivity appeared in 1,2-diglycerides but these were the only neutral lipids labelled (Neill and Masters, 1974).

Hypophysectomy of mature rats reduced the incorporation of [$^{14}$C] linoleate (18:2) into arachidonate (20:4) and both FSH and LH partially reversed this effect (Goswami and Williams, 1967; Nakamura and Privett, 1973); LH also stimulated incorporation of 18:2 into 22:5 in the former study, in which effects of LH could be mimicked with testosterone propionate. The conversion of linoleate to 20:4 and 22:5 was very much less in alloxan-diabetic rats showing pronounced atrophy of the testes (Peluffo et al., 1970). However as the results were expressed as the percentage of the radioactivity in each fatty acid, it is not possible to say whether metabolism was impaired or that there was simply less of these acids present. Fatty acids (20:4

and 20:3) are converted to prostaglandins by rat and rabbit testis respectively (Carpenter and Wiseman, 1970; Christ and Van Dorp, 1972).

## 9.6    Protein and amino-acid metabolism

### 9.6.1    Incorporation of amino-acids into protein

A number of amino-acids have been shown to be incorporated into proteins of the early stages of spermatogenesis. Those studied include lysine, arginine, leucine, tyrosine, histidine, phenylalanine and tryptophane. The incorporation of lysine in the rat testis reaches two peaks, one during the second peak of spermatogonial mitosis in stage XII and a second, larger peak in the pre-leptotene spermatocytes (Figure 9.6; Davis and Firlit, 1965). In the mouse, Monesi (1965a) has shown that amino-acid incorporation into protein by spermatogonia is almost equal during interphase and prophase but declines during metaphase and anaphase, i.e. at the stage at which RNA synthesis ceases. The rate of synthesis of protein is greatest in A type sper-

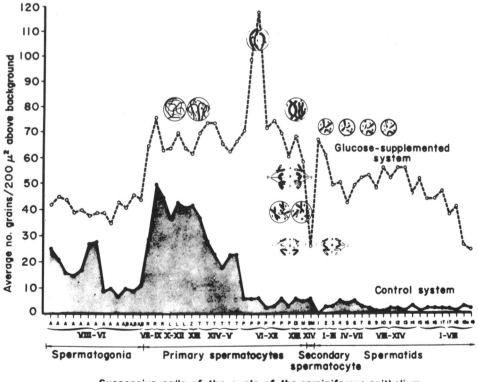

Figure 9.6  The incorporation of ³H-lysine into protein in germinal cells of various stages of development in the testes of rats and the effects of glucose. (Reproduced from Davis and Langford, 1970)

matogonia than in B type. In the primary spermatocytes of the mouse, the incorporation of amino-acids into protein begins before the synthesis of RNA and reaches a peak during pachytene in stages II to V (Monesi, 1965a), i.e. apparently later than in the rat.

In maturing spermatids, as might be expected from the appearance of the arginine-rich histone, arginine is the only amino-acid to be incorporated into nuclear proteins and this only occurs between stages XI and IV with a peak at stage XIII (Monesi, 1965a; Loir, 1972b).

A number of amino-acids are incorporated into cytoplasmic proteins of stage I to XI spermatids but later spermatids incorporate less, and mature spermatids none (Figure 9.7; Monesi, 1965a). The lysine-rich proteins of the chromatophile sphere

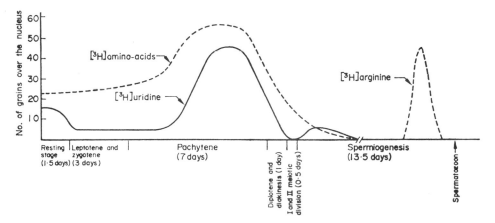

Figure 9.7 The pattern of ³H-uridine and ³H-amino-acid incorporation into RNA and nuclear protein respectively, during spermatogenesis in the mouse. (Reproduced from Monesi, 1971)

found in the cytoplasm of late spermatids and residual bodies appear to be synthesized when the cells are pre-leptotene spermatocytes (Vaughn, 1966).

A preparation of polyribosomes from mature rat testes will incorporate [¹⁴C]-valine into protein *in vitro* under appropriate conditions (Means *et al.*, 1969).

### 9.6.2    Effect of temperature and glucose on incorporation of lysine into protein

The incorporation of radioactive lysine into the protein of rat testicular slices is greater at 32 than at 26 or 38°C. This contrasts with a greater incorporation at 38°C into slices of other rat tissues such as liver, kidney or spleen and production of labelled glutamate and $CO_2$ from glucose by testis slices. Slices of mouse and hamster testis also incorporate more lysine into protein at 32 than at 38°C, but with slices of rabbit, guinea pig and dog testes (Buyer and Davis, 1966) and of cryptorchid and immature rat testes, the situation is reversed (Davis *et al.*, 1964).

Further studies by Davis and Morris (1963) have shown that maximal incorporation of lysine occurs at 32°C in the absence of added glucose, and when glucose (9mM) is added to the medium, the level of incorporation is increased several-fold, and the temperature at which maximal incorporation occurs is increased to 36°C. As glucose is normally available to the tubular cells, this latter situation is probably near to the physiological situation, but literal interpretation of *in vitro* data is always dangerous. However it is interesting that the stimulating effect of glucose is greater for the testis than for slices of many other tissues (Davis and Morris, 1963). It is also curious that while the incorporation of lysine into protein in the absence of glucose is seen in spermatogonia and leptotene and zygotene primary spermatocytes, when glucose is added, the greatest incorporation is into the protein of pachytene spermatocytes and substantial incorporation is seen into the proteins of spermatids (Figure 9.6; Davis and Firlit, 1965). However, it was not made clear whether the incorporation was nuclear or cytoplasmic, and rather surprisingly, in view of these authors' own earlier results on the effects of temperature on incorporation, the studies were done at 37.5°C. Also, it is difficult to reconcile this incorporation of lysine into spermatid protein with the observations by Monesi (1965a) quoted earlier.

The stimulatory effect of glucose does not seem to occur with testicular slices from immature rats (Davis and Firlit, 1965; Means and Hall, 1968a, b) and is lost after hypophysectomy (Means and Hall, 1968a).

The effect of glucose may be mediated by maintenance of the concentration of ATP during incubation (Means and Hall, 1968b), by stimulation of amino-acid transport, and by a direct effect on protein synthesis (Means and Hall, 1969). Protein synthesis, particularly that by spermatids is inhibited by 5-thio-D-glucose (Nakamura and Hall, 1976, 1977; see also Section 11.2.3).

### 9.6.3    Effect of hormones

Follicle stimulating hormone injected into young (21-day old) rats, young (9-day old) mice or hypophysectomized adult rats causes a pronounced stimulation of the incorporation of [³H]-lysine into protein in the testes. No effect is seen in intact adult rats, and no effect is seen when FSH is added *in vitro*. This effect is not just due to stimulated transport of amino-acids, but appears to be due to a specific stimulation of peptide synthesis in the ribosomes (see Section 10.2.2).

## 9.7    Nucleic acid metabolism

Almost all attention directed to nucleic acid metabolism in the testis has been concerned with the incorporation of purine or pyrimidine bases and nucleosides into nucleic acids. No information appears to be available which would establish whether the testis synthesizes its own precursors or whether these are absorbed as such from the blood.

### 9.7.1    DNA

The synthesis of DNA in the adult testis is confined to the spermatogonia and the pre-leptotene spermatocytes, as judged by the incorporation of [³H] thymidine immediately after injection (Messier and Leblond, 1960; Courot *et al.*, 1970) (Figure 9.8). There are as many as 5 synchronized mitotic divisions of spermatogonia, each

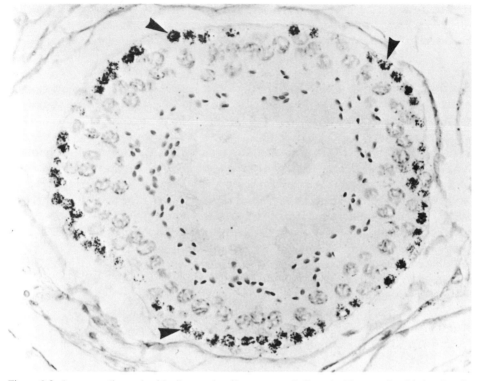

Figure 9.8  An autoradiograph with silver grains (3 groups are indicated with arrow heads) showing the incorporation of ³H thymidine into DNA in pre-leptotene spermatocytes of a Bennett's wallaby, *Macropus rufogriseus*, 1h after injection of the label into a testicular artery. The section is at stage 2 of the spermatogenic cycle; incorporation also occurs into spermatogonia at stages 6, 7 and 8 but not into more developed cells.

associated with DNA synthesis, as well as the slow, uncoordinated divisions of stem ($A_0$) spermatogonia (see p. 211).

The synthesis of DNA by the pre-leptotene spermatocytes in the earliest stages of the meiotic prophase, is more obvious because of the larger number of cells involved. Once this synthesis has occurred, there is no further detectable synthesis of DNA by the germinal cells (Monesi, 1962, 1971, but cf. Lima-de-Faria *et al.*, 1968). Presumably because the Sertoli cell population does not divide in the seminiferous tubules of adult testes, no synthesis of DNA by these cells has been observed, but they do incorporate ³H thymidine into DNA in rats less than 14-days old (Steinberger and Steinberger, 1971; Nagy, 1972).

However, it is curious that the incorporation of thymidine is maximal between 1 and 3h after injection (Courot *et al.*, 1970) whereas [¹⁴C] adenine is incorporated into DNA with a maximal incorporation between 1 and 8 days after injection, much later than in other tissues (Pelc and Howard, 1957). It has been suggested that this is because the adenine is also incorporated into RNA and that the adenine in RNA later becomes available for DNA synthesis (Pelc, 1958). However it seems also possible that adenine itself is not incorporated into DNA in the testis, but must be metabolized elsewhere in the body before it can enter the tubules or be converted to DNA.

### 9.7.2   RNA

Spermatogonia incorporate [³H]-uridine into nuclear RNA, type A cells being more active than type B. This RNA is initially nuclear, but is released slowly into the cytoplasm during interphase, or discharged during metaphase (Monesi, 1964b, 1965a; Kierzenbaum and Tres, 1974a).

The synthesis of RNA by spermatocytes is now well-established but much of the RNA formed is not ribosomal RNA (Utakoji, 1966; Muramatsu *et al.*, 1968). Most of the RNA which is synthesized by the spermatocytes is mainly in the form of DNA-like heterogeneous molecules with a rapid turnover (Muramatsu *et al.*, 1968) and it appears to be more stable than bacterial messenger RNA (Monesi, 1965). The sex chromosomes however are completely inactive and synthesis only occurs in the autosomal chromosomes (Monesi, 1965b). There is also a lack of [³H]-uridine labelling of the nucleolus associated with the sex chromosomes (Galdieri and Monesi, 1973), but nucleoli are also organized by autosomal bivalents in proximity to a chromosomal region known as the terminal or basal knob, and [³H]-uridine is incorporated into these autosomal nucleoli (Kierzenbaum and Tres, 1974a; Tres and Kierzenbaum, 1975). RNA synthesis can also be demonstrated in association with loosely extended chromosome loops along autosomal bivalents (Kierzenbaum and Tres, 1974b).

RNA synthesis, measured by [³H]-uridine incorporation is at a fairly low level during the pre-leptotene synthesis of DNA, falls to negligible values during leptotene and zygotene and then rises to a fairly sharp peak during pachytene, coinciding with the peak in amino-acid incorporation (Figure 9.7). The rate then declines during diplotene and diakinesis and synthesis practically ceases during metaphase and anaphase (Monesi, 1964b, 1965a; Utakoji, 1966; Loir, 1972a) although some post-meiotic transcription can be demonstrated by uridine incorporation (Monesi, 1965a) and assay of RNA polymerase (Moore, 1971). RNA synthesis during pachytene is stimulated by FSH + testosterone (Parvinen and Söderström, 1976). Much of the spermatocyte RNA synthesized is not transferred immediately to the cytoplasm as is usual, but remains in the nucleolus associated with the chromosomes. However the RNA does pass rapidly into the cytoplasm during diakinesis and metaphase and can be detected there throughout spermiogenesis until it is finally eliminated in the residual body (Monesi, 1964b, 1965a, 1971). Presumably this RNA carries the genetic information which controls the transformation of spermatids into spermatozoa.

The Sertoli cells are also active in incorporating uridine or cytidine into nuclear RNA. Sometimes the labelling is concentrated around the nucleolus but is usually absent from the karyosomes. This labelling eventually disappears from the nucleus but it has not been possible to determine whether it remains in Sertoli cell cytoplasm or passes into the cytoplasm of the germinal cells where label appears at about the same time (Monesi, 1964c, 1965a; Kierzenbaum, 1974). Studies have also been made of the incorporation of $^{32}PO_4$ into 18 and 28S ribosomal RNA in testes from which various cell types were missing after X-irradiation. The results suggest that either the spermatocytes may synthesize some ribosomal RNA but at a slower rate than the Sertoli cells, or that ribosomal RNA is synthesized in the Sertoli cells and then transferred to the spermatocytes (Monesi, 1974).

## References to Chapter 9

Ahluwalia, B. and Holman, R. T. (1965) Variations in lipid composition of the bovine testes during prepartum and postpartum life. *J. Dairy Sci.* **48**, 823.

Ahluwalia, B., Rahm, J. J., Mohrhaurer, H. and Holman, R. T. (1965) Influence of dietary polyunsaturated fatty acids upon fatty acid composition of rat testes. *J. Dairy Sci.* **48**, 819.

Ahluwalia, B., Pincus, G. and Holman, R. T. (1967) Essential fatty acid deficiency and its effects upon reproductive organs of male rabbits. *J. Nutr.* **92**, 205–14.

Alfert, M. (1956) Chemical differentiation of nuclear proteins during spermatogenesis in the salmon. *J. Biophys. Biochem. Cytol.* **2**, 109–14.

Allen, J. M. (1961) Multiple forms of lactic dehydrogenase in tissues of the mouse: Their specificity, cellular localization, and response to altered physiological conditions. *Ann. N.Y. Acadm. Sci.* **94**, 937–51.

Allfrey, V. G., Littau, V. C. and Mirsky, A. E. (1963) On the role of histones in regulating ribonucleic acid synthesis in the cell nucleus. *Proc. Nat. Acad. Sci.* **49**, 414–21.

Ambadkar, P. M. and George, J. C. (1964) Histochemical localization of certain oxidative enzymes in the rat testis. *J. Histochem. Cytochem.* **12**, 587–90.

Annison, E. F., Scott, T. W. and Waites, G. M. H. (1963) The role of glucose and acetate in the oxidative metabolism of the testis and epididymis of the ram. *Biochem. J.* **88**, 482–8.

Assaykeen, T. and Thomas, J. A. (1965) Endogenous histamine in male organs of reproduction. *Endocrinology* **76**, 839–43.

Ayala, S., Gaspar, G., Brenner, R. R., Peluffo, R. O. and Kunau, W. (1973) Fate of linoleic, arachidonic, and docosa-7,10,13,16-tetraenoic acids in rat testicles. *J. Lipid Res.* **14**, 296–305.

Badr, F. M. (1975) Prostaglandin levels in tissues of the male reproductive system in six strains of mice. *Endocrinology* **96**, 540–3.

Badr, F. M. (1976) Effect of sexual maturation and androgens on prostaglandin levels in tissues of the male reproductive system in mice. *Endocrinology* **98**, 1523–7.

Badr, F. M., Barcikowsli, B. and Bartke, A. (1975) Effect of castration, testosterone treatment and hereditary sterility on prostaglandin concentration in the male reproductive system of mice. *Prostaglandins* **9**, 289–97.

Baillie, A. H. (1961) Observations on the growth and histochemistry of the Leydig tissue in the postnatal prepubertal mouse testis. *J. Anat.* **95**, 357–70.

Baillie, A. H. (1964) Further observations on the growth and histochemistry of the Leydig tissue in the postnatal prepubertal mouse testis. *J. Anat.* **98**, 403–19.

Baillie, A. H., Ferguson, M. M. and Hart, D. McK. (1966) *Developments in Steroid Histochemistry*, New York: Academic Press.

Baldwin, J. and Temple-Smith, P. (1973) Distribution of LDHX in mammals: Presence in marsupials and absence in the monotremes *Platypus* and *Echidna*. *Comp. Biochem. Physiol.* **46B**, 805–11.

Baldwin, J., Temple-Smith, P. and Tidemann, C. (1974) Changes in testis specific lactate dehydrogenase isoenzymes during the seasonal spermatogenic cycle of the marsupial *Schoinobates volans* (Petauridae). *Biol. Reprod.* **11**, 377–86.

Barham, S. S., Berlin, J. D. and Brackeen, R. D. (1976) The fine structural localization of testicular phosphatases in man: the control testis. *Cell Tiss. Res.* **166**, 497–510.

Barker, S. B. and Klitgaard, H. M. (1952) Metabolism of tissues excised from thyroidectomized rats. *Am. J. Physiol.* **170**, 81–6.

Bartke, A. (1971) Concentration of free and esterified cholesterol in the testes of immature and adult mice. *J. Reprod. Fert.* **25**, 153–6.

Bartke, A. and Koerner, S. (1974) Androgenic regulation of the concentration of prostaglandin F in the male reproductive system of rats and mice. *Endocrinology* **95**, 1739–43.

Bartke, A. and Shire, G. J. M. (1972) Differences between mouse strains in testicular cholesterol levels and androgen target organs. *J. Endocrinol.* **55**, 173–84.

Battellino, L. J. and Blanco, A. (1970) Testicular lactate dehydrogenase isozyme nature of multiple forms in guinea pig. *Biochim. Biophys. Acta.* **212**, 205–12.

Battellino, L. J., Jaime, F. R. and Blanco, A. (1968) Kinetic properties of rabbit testicular lactate dehydrogenase isozyme. *J. Biol. Chem.* **243**, 5185–92.

Baust, P., Goslar, H. G. and Tonutti, E. (1966) Das Verhalten der Esterasen im Hoden von Ratte, Maus and Meerschweinchen nach Oestrogenbehandlung. *Z. Zellforsch. Mikroskop. Anat.* **69**, 686–98.

Bieri, J. G. and Andrews, E. L. (1964) Fatty acids in rat testes as affected by vitamin E. *Biochem. Biophys. Res. Comm.* **17**, 115–19.

Bieri, J. G. and Prival, E. L. (1965) Lipid composition of testes from various species. *Comp. Biochem. Physiol.* **15**, 275–82.

Bieri, J. G. and Prival, E. L. (1966) Effect of deficiencies of $\alpha$-tocopherol, retinol and zinc on the lipid composition of the rat testis. *J. Nutr.* **89**, 55–61.

Bieri, J. G., Mason, K. E. and Prival, E. L. (1969) Essential fatty acid deficiency and the testis: Lipid composition and the effect of preweaning diet. *J. Nutr.* **97**, 163–72.

Bishop, D. W. (1967) Significance of testicular sorbitol dehydrogenase (SDH). *J. Gen. Physiol.* **50**, 2504.

Bishop, D. W. (1968a) Sorbitol dehydrogenase in relation to spermatogenesis and fertility. *J. Reprod. Fertil.* **17**, 410–11.

Bishop, D. W. (1968b) Testicular enzymes as fingerprints in the study of spermatogenesis. In *Reproduction and Sexual Behaviour*, ed. M. Diamond, 261–86, Bloomington Ind.: Indiana University Press.

Bishop, D. W., Schrank, W. W., Musselman, A. D. and Meucke, E. C. (1967) Testis sorbitol dehydrogenase and activity changes induced aspermato-genesis and cryptorchidism. *Federation Proc.* **26**, 646.

Blackshaw, A. W. (1970) Histochemical localization of testicular enzymes. In *The Testis*, eds. A. D. Johnson, W. R. Gomes and N. L. Vandemark, **II**, 73–123, New York: Academic Press.

Blackshaw, A. W. and Elkington, J. S. H. (1970a) Developmental changes in lactate dehydrogenase isoenzymes in the testis of the immature rat. *J. Reprod. Fertil.* **22**, 69–75.

Blackshaw, A. W. and Elkington, J. S. H. (1970b) The effect of age and hypophysectomy on growth and the isoenzymes of lactate dehydrogenase in the mouse testis. *Biol. Reprod.* **2**, 268–74.

Blackshaw, A. W. and Samisoni, J. I. (1966a) The effects of cryptorchism in the guinea pig on the isoenzymes of testicular lactate dehydrogenase. *Aust. J. Biol. Sci.* **19**, 841–48.

Blackshaw, A. W., Miller, O. C. and Graves, C. N. (1968) Lactate dehydrogenase isoenzymes of calf and adult bull testes. *J. Dairy Sci.* **51**, 950.

Blank, M. L., Wykle, R. L. and Snyder, F. (1973) The retention of arachidonic acid in ethanolamine plasmalogens in rat testes during essential fatty acid deficiency. *Biochim. Biophys. Acta* **316**, 28–34.

Blanco, A. and Zinkham, W. H. (1963) Lactate dehydrogenases in human testis. *Science* **139**, 601–2.

Bloch, D. P. (1969) A catalog of sperm histones. *Genetics Supplement* **61**, 93–111.

Branson, R. E., Grimes, S. R., Yonuschot, G. and Irvin, J. L. (1975) The histones of the rat testis. *Arch. Biochem. Biophys.* **168**, 403–12.

Braun, T. and Dods, F. R. (1975) Development of a $Mn^{2+}$-sensitive, 'soluble' adenylate cyclase in rat testis. *Proc. Nat. Acad. Sci. U.S.* **72**, 1097–101.

Braun, T., Frank, H., Dods, R. and Sepsenwol, S. (1977) $Mn^{2+}$-sensitive, soluble adenylate cyclase in rat testis. Differentiation from other testicular nucleotide cyclases. *Biochim. Biophys. Acta* **481**, 227–35.

Bridges, R. B. and Coniglio, J. G. (1970a) The biosynthesis 9,12,15,18-tetracosatetraenoic and of 6,9,12,15,18-tetracosapentaenoic acid by rat testes. *J. Biol. Chem.* **245**, 46–9.

Bridges, R. B. and Coniglio, J. G. (1970b) The metabolism of linoleic and arachidonic acids in rat testis. *Lipids* **5**, 628–35.

Bridges, R. B. and Coniglio, J. G. (1970c) The metabolism of 4,7,10,13,16-[5-$^{14}$C] docosapentaenoic acid in the testis of the rat. *Biochim. Biophys. Acta* **218**, 29–35.

Buhrley, I. F. and Ellis, L. C. (1973) Catalase activity in rat testicular preparations: Distribution and changes in activity induced by hypophysectomy and X-irradiation. *Fertil. Steril.* **24**, 956–61.

Butler, W. R., Johnson, A. D., Gomes, W. R. and VanDemark, N. L. (1967) Testis lipids after stimulation or inhibition of testicular elements. *J. Dairy Sci.* **50**, 1005–6.

Buyer, R. and Davis, J. R. (1966) Species variation in the effect of temperature on the incorporation of lysine-U-$^{14}$C into protein of testis slices. *Comp. Biochem. Physiol.* **17**, 151–5.

Caldwell, K. (1976) Sperm diaphorase: Genetic polymorphism of a sperm-specific enzyme in man. *Science* **191**, 1185–7.

Calvin, H. I., Kosto, B. and Williams-Ashmann, H. G. S. (1967) Enzymic incorporation of deoxyribonucleotides into deoxyribonucleic acid by mammalian testis. *Arch. Biochem. Biophys.* **118**, 670–81.

Carpenter, M. P. (1971) The lipid composition of maturing rat testis. The effect of $\alpha$-tocopherol. *Biochim. Biophys. Acta* **231**, 397–406.

Carpenter, M. P. (1974) Prostaglandins of rat testis. *Lipids* **9**, 397–406.

Carpenter, M. P. and Wiseman, B. (1970) Prostaglandins of rat testis. *Fed. Proc.* **29**, 248.

Casillas, E. R. (1972) The distribution of carnitine in male reproductive tissues and its effect on palmitate oxidation by spermatozoal particles. *Biochim. Biophys. Acta* **280**, 545–51.

Cavazos, L. F. and Mellampy, R. M. (1954) A comparative study of the periodic acid reactive carbohydrate in vertebrate testes. *Am. J. Anat.* **95**, 467–95.

Chakraborty, J. and Nelson, L. (1974) Organization and redistribution of adenosinetriphosphatase during spermiogenesis in the mouse. *Biol. Reprod.* **10**, 85–97.

Chang, J. P., Yokohama, M., Brinkley, B. R. and Mayahara, H. (1974) Electron microscopic cytochemical study of phosphatases during spermiogenesis in Chinese hamster. *Biol. Reprod.* **10**, 601–10.

Cheng, H. C. (1971) Isoenzymes of hexokinase in rat testes at various developmental and endocrine states. PhD. dissertation, Ohio State University, Columbus. Quoted by Gomes and VanDemark, 1974.

Christ, E. J. and Van Dorp, D. A. (1972) Comparative aspects of prostaglandin biosynthesis in animal tissues. *Biochim. Biophys. Acta* **270**, 537–45.

Clausen, J. (1969) Lactate dehydrogenase isoenzymes of sperm cells and testes. *Biochem. J.* **111**, 207–18.

Cohen, J. P., Hoffer, A. P. and Rosen, S. (1976) Carbonic anhydrase localization in the epididymis and testis of the rat: Histochemical and biochemical analysis. *Biol. Reprod.* **14**, 339–46.

Colizzi, C. and Tusini, G. (1953) Sulle variazioni in acido ascorbico nel testicolo sotto l'influsso degli ormoni gonadotropi. *Atti. Soc. Lombarda Sci. Med. Biol.* **8**, 265–70.

Collins, F. D. and Shotlander, V. L. (1961) Studies on phospholipids. 7. The distribution of complex phospholipids in various species and tissues. *Biochem. J.* **79**, 316–20.

Coniglio, J. G., Zseltvay, R. R. and Wharton, R. A. (1971) Biosynthesis of stearic acid in rat testis. *Biochim. Biophys. Acta* **239**, 374–5.

Coniglio, J. G., Bridges, R. B., Aguilar, H. and Zseltvay, R. R. (1972) The metabolism of 1-$^{14}$C-palmitic acid in the testis of the rat. *Lipids* **7**, 368–71.

Coniglio, J. G., Grogan, W. M. and Rhamy, R. K. (1975) Lipids and fatty acid composition of human testes removed at autopsy. *Biol. Reprod.* **12**, 255–9.

Coste, F., Delbarre, F. and Lacronique, F. (1953) Variations du taux de l'acide ascorbique des endocrines sous l'influence des stimulines hypophysaires. Taux de l'acide ascorbique des testicules du rat en fonction de l'age. *C. rend. Soc. Biol.* **147**, 608–11.

Cross, B. A. and Silver, I. A. (1962) Neurovascular control of oxygen tension in the testis and epididymis. *J. Reprod. Fertil.* **3**, 377–95.

Daoust, R. and Clermont, Y. (1955) Distribution of nucleic acids in germ cells during the cycle of the seminiferous epithelium in the rat. *Am. J. Anat.* **96**, 255–84.

Darzynkiewicz, Z., Gledhill, B. L. and Ringertz, N. R. (1969) Changes in deoxyribonucleoprotein during spermiogenesis in the bull. $^3$H-actinomycin D binding capacity. *Exptl. Cell Res.* **58**, 435–38.

Davis, J. R. and Firlit, C. F. (1965) Effect of glucose on uptake of L-lysine-H$^3$ in cells of the seminiferous epithelium. *Am. J. Physiol.* **209**, 425–32.

Davis, J. R. and Langford, G. A. (1970) Testicular proteins. In *The Testis*, eds. A. D. Johnson, W. R. Gomes and N. L. Vandemark, II, 259–306, New York: Academic Press.

Davis, J. R. and Morris, R. N. (1963) Effect of glucose on incorporation of L-lysine-U-C[14] into testicular proteins. *Am. J. Physiol.* **205**, 833–6.

Davis, J. R., Firlit, C. F. and Hollinger, M. A. (1963) Effect of temperature on incorporation of L-lysine-U-C[14] into testicular proteins. *Am. J. Physiol.* **204**, 696–8.

Davis, J. R., Morris, R. N. and Hollinger, M. A. (1964) Incorporation of L-lysine-U-C[14] into proteins of cryptorchid testis slices. *Am. J. Physiol.* **207**, 50–4.

Davis, J. R., Morris, R. N. and Hollinger, M. A. (1965) Comparison of testicular protein labelling in cryptorchidism induced in prepubertal and adult rats. *J. Reprod. Fertil.* **10**, 149–52.

Davis, J. T. and Coniglio, J. G. (1966) The biosynthesis of docosapentaenoic and other fatty acids by rat testes. *J. Biol. Chem.* **241**, 610–12.

Davis, J. T. and Coniglio, J. G. (1967) The effect of cryptorchidism, cadmium and anti-spermatogenic drugs on fatty acid composition of rat testis. *J. Reprod. Fertil.* **14**, 407–13.

Davis, J. T., Bridges, R. B. and Coniglio, J. G. (1966) Changes in lipid composition of the maturing rat testis. *Biochem. J.* **98**, 342–6.

Delhumeau-Ongay, G., Trejo-Bayona, R. and Lara-Vivas, L. (1973a) Changes of ($Ca^{2+}$-$Mg^{2+}$)-adenosine-triphosphatase activity in rat testis throughout maturation. *J. Reprod. Fertil.* **33**, 513–17.

Delhumeau-Ongay, G., Trejo-Bayona, R., Alvarez-Buyilla, R. and Lara-Vivas, L. (1973b) Increase of ($Ca^{2+}$-$Mg^{2+}$)-adenosine-triphosphatase activity in rat testis undergoing regression of the germinal epithelium. *J. Reprod. Fertil.* **34**, 149–53.

Denduchis, B., Gonzalez, N. and Mancini, R. E. (1972) Concentration of hydroxyproline in testes of hypophysectomized patients before and after treatment with gonadotrophins and testosterone. *J. Reprod. Fertil.* **31**, 111–14.

Dickens, F. and Greville, G. D. (1933) Metabolism of normal and tumour tissue. VIII. Respiration in fructose and in sugar-free media. *Biochem. J.* **27**, 832–41.

Dickens, F. and Simer, F. (1930) The metabolism of normal and tumour tissue. II. The respiratory quotient, and the relationship of respiration to glycolysis. *Biochem. J.* **24**, 1301–26.

Dickens, F. and Simer, F. (1931) The metabolism of normal and tumour tissue. IV. The respiratory quotient in bicarbonate-media. *Biochem. J.* **25**, 985–93.

Domenech, E. M. de, Domenech, C. E., Aoki, A. and Blanco, A. (1972) Association of the testicular lactate dehydrogenase isozyme with a special type of mitochondria. *Biol. Reprod.* **6**, 136–47.

Dorrington, J. H. and Fritz, I. B. (1975) Cellular localization of 5α-reductase and 3α-hydroxysteroid dehydrogenase in the seminiferous tubule of the rat testis. *Endocrinology* **96**, 879–89.

Eisenberg, F., Jr. (1967) D-myoinositol 1-phosphate as product of cyclization of glucose-6-phosphate and substrate for a specific phosphatase in rat testis. *J. Biol. Chem.* **242**, 1375–82.

Eisenberg, F. and Bolden, A. H. (1963) Biosynthesis of inositol in rat testis homogenate *Biochem. Biophys. Res. Commun.* **12**, 72–7.

Elliott, K. A. C., Grieg, M. E. and Benoy, M. P. (1937) The metabolism of lactic and pyruvic acids in normal and tumour tissues. III. Rat liver, brain and testis. *Biochem. J.* **31**, 1003–20.

Ellis, L. C., Jaussi, A. W., Baptista, M. H. and Urry, R. L. (1972) Correlation of age changes in monoamine oxidase activity and androgen synthesis by rat testicular minced and teased-tubular preparations *in vitro*. *Endocrinology* **90**, 1610–8.

Evans, O. B., Zseltvay, R., Wharton, R. and Coniglio, J. G. (1971) Fatty acid synthesis in rat testes injected intratesticularly or incubated with 1-$^{14}$C acetate. *Lipids* **6**, 706–11.

Evrev, T. I. (1975) Autoantigenicity of LDH-X Isozymes. In *Isozymes II. Physiological Function*, ed. C. L. Markert, New York: Academic Press.

Ewing, L. L. (1967) The effect of aging on testicular metabolism in the rabbit. *Am. J. Physiol.* **212**, 1261–7.

Ewing, L. L. and Schanbacher, L. M. (1970) Early effects of experimental cryptorchidism on the activity of selected enzymes in the rat testis. *Endocrinology* **87**, 129–34.

Ewing, L. L. and VanDemark, N. L. (1963a) Factors affecting testicular metabolism and function. II. Effect of temperature elevation *in vivo* on subsequent metabolic activity of rabbit testicular tissue *in vitro*. *J. Reprod. Fertil.* **6**, 9–16.

Ewing, L. L. and VanDemark, N. L. (1963b) Factors affecting testicular metabolism and function. III. Effect of *in vitro* temperature elevation on tissue slices and perfused testes in the rabbit. *J. Reprod. Fertil.* **6**, 17–24.

Ewing, L. L., Noble, D. J. and Ebner, K. E. (1964) The effect of epinephrine and competition between males on the *in vitro* metabolism of rabbit testes. *Can. J. Physiol. Pharmacol.* **42**, 527–32.

Ewing, L. L., Green, P. M. and Stebler, A. M. (1965) Metabolic and biochemical changes in testis of cotton rat (*Sigmodon hispidus*) during breeding cycle. *Proc. Soc. Exptl. Biol. N.Y.* **118**, 911–13.

Ewing, L. L., Means, A. R., Beames, C. G. and Montgomery, R. D. (1966) Biochemical changes in rat testis during postnatal maturation. *J. Reprod. Fertil.* **12**, 295–308.

Featherstone, R. M., Nelson, W. O., Welden, F., Marburger, E., Boccabella, A. and Boccabella, R. (1955) Pyruvate oxidation in testicular tissues during furadoxyl-induced spermatogenic arrest. *Endocrinology* **56**, 727–36.

Field, J. B., Pastan, I., Herring, B. and Johnson, P. (1960) Studies of pathways of glucose metabolism of endocrine tissues. *Endocrinology* **67**, 801–6.

Firlit, C. F. and Davis, J. R. (1965) Morphogenesis of the residual body of the mouse testis. *Quart. J. Microsc. Sci.* **106**, 93–8.

Firlit, C. F. and Davis, J. R. (1966) Radio-autographic incorporation of L-lysine-$^{3}$H into protein of cells of the germinal epithelium in cryptorchidism. *J. Reprod. Fertil.* **11**, 125–31.

Fleeger, J. L., Bishop, J. P., Gomes, W. R. and VanDemark, N. L. (1968a) Testicular lipids. I. Effect of unilateral cryptorchidism on lipid classes. *J. Reprod. Fertil.* **15**, 1–7.

Fleeger, J. L., Bishop, J. P., Gomes, W. R. and VanDemark, N. L. (1968b) Testicular lipids. II. Effect of unilateral cryptorchidism on fatty acids of testicular phospholipids and triglycerides. *J. Reprod. Fertil.* **15**, 9–14.

Foote, R. H. and Koefoed-Johnson, H. H. (1959) The use of adenine-8-C$^{14}$ for studying spermatogenesis in the rabbit. *J. Anim. Sci.* **18**, 1553.

Free, M. J. (1970) Carbohydrate metabolism in the testis. In *The Testis*, eds. A. D. Johnson, W. R. Gomes and N. L. Vandemark, **II**, 125–92, New York: Academic Press.

Free, M. J. and Payvar, F. (1974) Pyruvate dehydrogenase in the normal and cryptorchid rat testis. *Biol. Reprod.* **10**, 69–73.

Free, M. J. and VanDemark, N. L. (1968) Gas tensions in spermatic and peripheral blood of rams with normal and heated testes. *Am. J. Physiol.* **214**, 863–5.

Free, M. J. and VanDemark, N. L. (1969) Radiorespirometric studies on glucose metabolism in testis tissue from rat, rabbit and chicken. *Comp. Biochem. Physiol.* **30**, 323–33.

Free, M. J., Vera Cruz, N. C., Johnson, A. D. and Gomes, W. R. (1968) Metabolism of glucose-1-,$^{14}$C and glucose-6-,$^{14}$C by testis tissue from cryptorchid and testosterone propionate treated rabbits. *Endocrinology* **82**, 183–7.

Free, M. J., Massie, E. D. and VanDemark, N. L. (1969) Glucose metabolism by the cryptorchid rat testis. *Biol. Reprod.* **1**, 354–66.

Free, M. J., Schluntz, G. A. and Jaffe, R. B. (1976) Respiratory gas tensions in tissues and fluids of the male rat reproductive tract. *Biol. Reprod.* **14**, 481–8.

Fritz, I. B. (1974) Selected topics on the biochemistry of spermatogenesis. *Cur. Top. Cell. Regul.* **7**, 129–74.

Galdieri, M. and Monesi, V. (1973) Ribosomal RNA synthesis in spermatogonia and Sertoli cells of the mouse testis. *Exptl. Cell Res.* **80**, 120–6.

Galdieri, M. and Monesi, V. (1974) Ribosomal RNA in mouse spermatocytes. *Exptl. Cell Res.* **85**, 287–95.

Gambal, D. (1967) Antigonadotropic hormones and lipogenesis in the testis and seminal vesicles of the rat. *Arch. Biochem. Biophys.* **118**, 709–15.

Gambal, D. and Ackerman, R. J. (1967) Hormonal control of rat testicular phospholipids. *Endocrinology* **80**, 231–9.

Garren, L. D., Gill, G. N., Masui, H. and Walton, G. M. (1971) On the mechanism of action of ACTH. *Rec. Progr. Horm. Res.* **27**, 433–74.

Geer, B. W., Gabliani, G. I. and Schramm, P. (1977) Changes in enzyme levels in the testis and liver of the 13-lined ground squirrel (*Spermophilus tridecemlineatus*) at the time of arousal from hibernation. *Arch. Intern. Physiol. Biochem.* **85**, 221–32.

George, J. C. and Ambadkar, P. M. (1963) Histochemical demonstration of lipids and lipase activity in rat testis. *J. Histochem. Cytochem.* **11**, 420–5.

Geremia, R., Goldberg, R. B. and Bruce, W. R. (1976) Kinetics of histone and protamine synthesis during meiosis and spermiogenesis in the mouse. *Andrologia* **8**, 147–56.

Gerozissis, K. and Dray, F. (1977) Prostaglandins in the isolated testicular capsule of immature and young adult rats. *Prostaglandins*, **13**, 777–83.

Gilchrist, R. D. and Hoffmann, R. A. (1961) Changes in ascorbic acid and cholesterol in the cryptorchid rat testis. *Am. Zoologist* **1**, 356.

Gledhill, B. L., Gledhill, M. P., Rigler, R. and Ringertz, N. R. (1966) Changes in deoxyribonucleoprotein during spermiogenesis in the bull. *Exptl. Cell Res.* **41**, 652–65.

Goldberg, E. and Hawtrey, C. (1968) Effect of experimental cryptorchism on the isozymes of lactate dehydrogenase in mouse testes. *J. Exptl. Zool.* **167**, 411–7.

Goldberg, E., Sberna, D., Wheat, T. E., Urbanski, G. J. and Margoliash, E. (1977) Cytochrome c: immunofluorescent localization of the testis-specific form. *Science* **196**, 1010–1.

Gomes, W. R. (1971) Oxygen consumption by seminiferous tubules and interstitial tissue of normal and cryptorchid rat testes. *J. Reprod. Fertil.* **26**, 427–9.

Gomes, W. R. and VanDemark, N. L. (1974) The male reproductive system. *Ann. Rev. Physiol.* **34**, 307–29.

Gosule, L. C. and Schellman, J. A. (1976) Compact form of DNA induced by spermidine. *Nature* **259**, 333–5.

Goswami, A. and Coniglio, J. G. (1966) Effect of pyridoxine deficiency on the metabolism of linoleic acid in the rat. *J. Nutr.* **89**, 210–16.

Goswami, A. and Williams, W. L. (1967) Effect of hypophysectomy and replacement therapy on fatty acid metabolism in the rat testis. *Biochem. J.* **105**, 537–43.

Goswami, A., Skipper, J. K. and Williams, W. L. (1968) Stimulation of fatty acid synthesis *in vitro* by gonadotrophin-induced testicular ribonucleic acid. *Biochem. J.* **108**, 147–52.

Gravis, C. J., Yates, R. D. and Chen, I-L. (1976) Light and electron microscopic localization of ATPase in normal and degenerating testes of Syrian hamsters. *Am. J. Anat.* **147**, 419–32.

Grimes, S. R., Chae, C.-B and Irving, J. L. (1975) Effect of age and hypophysectomy upon relative proportions of various histones in rat testis. *Biochem. Biophys. Res. Comm.* **64**, 911–17.

Grootegoed, J. A., Grolle-Hey, A. H., Rommerts, F. F. G. and Molen, H. J. van der (1977) Ribonucleic acid synthesis *in vitro* in primary spermatocytes isolated from rat testis *Biochem. J.* **168**, 23–31.

Gunaga, K. P., Chitra Rao, M., Sheth, A. R. and Rao, S. S. (1972) The role of glycogen during the development of the rat testis and prostate. *J. Reprod. Fertil.* **29**, 157–62.

Gupta, G. S. (1975) Radiation effects on testes: Evidences for ($Na^+ + K^+$) stimulated ATPase in germinal and non-germinal cells of testis of rat. *Indian J. Biochem. Biophys.* **12**, 189–91.

Hafiez, A. A. and Bartke, A. (1972) Effect of hypophysectomy on cholesterol metabolism in the testes of rats and mice. *J. Endocrinol.* **52**, 321–6.

Hall, P. F., Nishizawa, E. E. and Eik-Nes, K. B. (1963) Synthesis of fatty acids by testicular tissue *in vitro*. *Can. J. Biochem. Physiol.* **41**, 1267–74.

Handa, S., Yamato, K., Ishizuka, I., Suzuki, A. and Yamakawa, T. (1974) Biosynthesis of seminolipid: sulfation *in vivo* and *in vitro*. *J. Biochem.* (Japan) **75**, 77–83.

Härkönen, M. and Kormano, M. (1970) Acute cadmium-induced changes in the energy metabolism of the rat testis. *J. Reprod. Fertil.* **21**, 221–6.

Harkonen, M. and Kormano, M. (1971) Energy metabolism of the normal and cryptorchid rat testis. *J. Reprod. Fertil.* **25**, 29–39.

Harkonen, M., Suvanto, O. and Kormano, M. (1970) Glutamate in postnatal rat testis. *J. Reprod. Fertil.* **21**, 533–6.

Hauser, G. (1969) Myo-inositol transport in slices of rat kidney cortex. III. Hormonal, metabolic and efflux studies. *Ann. N.Y. Acad. Sci.* **165**, 630–45.

Hawtrey, C. O., Naidu, A., Nelson, C. and Wolfe, D. (1975) The physiological significance of LDH-X. In *Isozymes II. Physiological Function*, ed. C. L. Markert, New York: Academic Press.

Henderson, S. A. (1964) RNA synthesis during male meiosis and spermiogenesis. *Chromosoma* **15**, 345–66.

Hennig, B. (1975) Change of cytochrome c structure during development of the mouse. *Eur. J. Biochem.* **55**, 167–83.

Himwich, H. E. and Nahum, L. H. (1929) The respiratory quotient of testicle. *Am. J. Physiol.* **88**, 680–5.

Hitzeman, J. W. (1962) Development of enzyme activity in the Leydig cells of the mouse testis. *Anat. Rec.* **143**, 351–62.

Hitzeman, J. W. (1965) Production and utilization of reduced nicotinamide adenine dinucleotide phosphate in the mouse testis. *J. Exptl. Zool.* **160**, 107–16.

Hodgen, G. D. and Sherins, R. J. (1973) Enzymes as markers of testicular growth and development in the rat. *Endocrinology* **93**, 985–9.

Hodgen, G. D., Gomes, W. R. and VanDemark, N. L. (1971) A testicular isoenzyme of carbonic anhydrase. *Biol. Reprod.* **4**, 224–8.

Hollinger, M. A. (1970) Studies on adenyl cyclase in rat testis. *Life Sci.* **9**, 533–40.

Hollinger, M. A. (1971) Metabolism of ATP by testis mitochondria of 25-day-old rats. *J. Reprod. Fertil.* **25**, 443–5.

Hollinger, M. A. and Davis, J. R. (1968) Aerobic metabolism of uniformly labelled ($^{14}$C) glucose in tissue slices of rat testis. *J. Reprod. Fertil.* **17**, 343–55.

Hollinger, M. A. and Hwang, F. (1972) Effect of glucose on the synthesis of testicular proteins separated by disc electrophoresis. *Biochim. Biophys. Acta* **281**, 652–7.

Holman, R. T. and Hofstetter, H. H. (1965) The fatty acid composition of the lipids from bovine and porcine reproductive tissues. *J. Am. Oil Chemists' Soc.* **42**, 540–4.

Holmes, R. S. and Masters, C. J. (1967) The developmental multiplicity of isoenzyme status of rat esterases. *Biochim. Biophys. Acta* **146**, 138–50.

Hopwood, M. L. and Gassner, F. X. (1962) The free amino-acids of bovine semen. *Fertil. Steril.* **13**, 290–303.

Howland, B. E. and Zebrowski, E. J. (1972) Hyposecretion of gonadotrophins in alloxan-treated male rats. *J. Reprod. Fertil.* **31**, 115–18.

Huggins, C. and Moulton, S. H. (1948) Esterases of testis and other tissues. *J. Exptl. Med.* **88**, 169–79.

Hunt, E. L. and Bailey, D. Q. (1961) The effects of alloxan diabetes on the reproductive system of young male rats. *Acta Endocrinol.* **38**, 432–40.

Hupka, S. and Dumont, J. E. (1963) *In vitro* effect of adrenaline and other amines on glucose metabolism in sheep thyroid, heart, liver, kidney and testicular slices. *Biochem. Pharmacol.* **12**, 1023–35.

Ihrig, T. J., Renston, R. H., Renston, J. P. and Gondos, B. (1974) Catalase activity in the developing rabbit testis. *J. Reprod. Fertil.* **39**, 105–8.

Inano, H., Nakano, H., Shikita, M. and Tamaoki, B-I. (1967a) The influence of various factors upon testicular enzymes related to steroidogenesis. *Biochim. Biophys. Acta* **137**, 540–8.

Inano, H., Egusa, M. and Tamaoki, B-I. (1967b) Studies on the enzymes related to steroidogenesis in testicular tissue of guinea pig. *Biochim. Biophys. Acta* **144**, 165–7.

Inano, H., Inano, A. and Tamaoki, B-I. (1970) Studies on enzyme reaction related to steroid biosynthesis. II. Submicrosomal distribution of the enzymes related to androgen production from pregnenolone and of the cytochrome P-450 in testicular gland of rat. *J. Steroid*

*Biochem.* **1**, 83–91.

Ishizuka, I. and Yamakawa, T. (1974) Absence of seminolipid in seminoma tissue with concomitant increase in spingoglycolipids. *J. Biochem.* (Japan) **76**, 221–3.

Ishizuka, I., Suzuki, M. and Yamakawa, T. (1973) Isolation and characterization of a novel sulfoglycolipid, 'seminolipid', from boar testis and spermatozoa. *J. Biochem.* (Tokyo) **73**, 77–87.

Jensen, B. and Privett, O. S. (1969) Effect of hypophysectomy on lipid composition in the immature rat. *J. Nutr.* **99**, 210–16.

Jensen, B., Nakamura, M. and Privett, O. S. (1968) Effect of hypophysectomy on the metabolism of essential fatty acids in rat testes and liver. *J. Nutr.* **95**, 406–12.

Johnson, A. D., Gomes, W. R., Free, M. J. and VanDemark, N. L. (1968) Testicular lipids. III. Effect of surgery and unilateral or bilateral cryptorchidism. *J. Reprod. Fertil.* **16**, 409–14.

Johnson, A. D., Gomes, W. R. and VanDemark, N. L. (1969) Effect of elevated ambient temperature on lipid levels and cholesterol metabolism in the ram testis. *J. Anim. Sci.* **29**, 469–75.

Johnson, A. D., Gomes, W. R. and VanDemark, N. L. (1971) Testicular lipids. IV. Effect of unilateral and bilateral cryptorchidism on the fatty acids of esterified cholesterol in the rat and rabbit. *J. Reprod. Fertil.* **25**, 425–30.

Joshi, M. and Macleod, J. (1961) Metabolism of testicular tissue in immature and mature rats. *J. Reprod. Fertil.* **2**, 198–9.

Kadohama, N. and Turkington, R. W. (1974) Changes in acidic chromatin proteins during the hormone-dependent development of rat testis and epididymis. *J. biol. Chem.* **249**, 6225–33.

Kato, S. (1958) Studies on the TCA cycle metabolism of the male reproductive organs and their accessory glands. *Acta. Urol.* (Japan) **49**, 659.

Katsh, S. and Aguirre, A. (1968) Biochemical responses in organs of guinea pigs immunized with aspermatogenic antigen. *Int. Arch. Allergy Appl. Immunol.* **33**, 141–50.

Kerr, J. B. and De Kretser, D. M. (1975) Cyclic variations in Sertoli cell lipid content throughout the spermatogenic cycle in the rat. *J. Reprod. Fertil.* **43**, 1–8.

Kierszenbaum, A. L. (1974) RNA synthetic activities of Sertoli cells in the mouse testis. *Biol. Reprod.* **11**, 365–76.

Kierszenbaum, A. L. and Tres, L. L. (1974a) Nucleolar and perichromosomal RNA synthesis during meiotic prophase in the mouse testis. *J. Cell Biol.* **60**, 39–53.

Kierszenbaum, A. L. and Tres, L. L. (1974b) Transcription sites in spread meiotic prophase chromosomes from mouse spermatocytes. *J. Cell Biol.* **63**, 923–5.

Kistler, W. S. and Geroch, M. E. (1975) An unusual pattern of lysine rich histone components is associated with spermatogenesis in rat testis. *Biochem. Biophys. Res. Comm.* **63**, 378–84.

Kistler, W. S. and Williams-Ashman, H. G. (1975) On three varieties of specific basic proteins associated with mammalian spermatogenesis. *Curr. Top. Molec. Endocr.* **2**, 423–32.

Kistler, W. S., Geroch, M. E. and Williams-Ashman, H. G. (1973) Specific basic proteins from mammalian testes. Isolation and properties of small basic proteins from rat testes and epididymal spermatozoa. *J. Biol. Chem.* **248**, 4532–43.

Kistler, W. S., Noyes, C. and Heinrikson, R. L. (1974) Partial structural analysis of a highly basic low molecular weight protein from rat testis. *Biochem. Biophys. Res. Comm.* **57**, 341–7.

Kistler, W. S., Geroch, M. E. and Williams-Ashman, H. G. (1975a) A highly basic small protein associated with spermatogenesis in the human testis. *Invest. Urol.* **12**, 346–50.

Kistler, W. S., Noyes, C., Hsu, R. and Heinrikson, R. L. (1975b) The amino acid sequence of a testis-specific basic protein that is associated with spermatogenesis. *J. Biol. Chem.* **250**, 1847–53.

Kormano, M. and Penttila, A. (1968) Distribution of endogenous and administered 5-hydroxytryptamine in the rat testis and epididymis. *Ann. Med. exp. Fenn.* **46**, 468–73.

Kormano, M., Harkonen, M. and Kontinen, E. (1964) Effect of experimental cryptorchidism on the histochemically demonstrable dehydrogenases of the rat testis. *Endocrinology* **74**, 44–51.

Kornblatt, M. J., Knapp, A., Levine, M., Schachter, H. and Murray, R. K. (1974) Studies on the structure and formation during spermatogenesis of the sulfoglycerogalactolipid of rat testis. *Can. J. Biochem.* **52**, 689–97.

Kosto, B., Calvin, H. I. and Williams-Ashman, H. G. (1967) Note on enzymatic synthesis of deoxyribonucleic acid by human testis. *Invest. Urol.* **4**, 389–95.

Lam, D. M. K. and Bruce, W. R. (1971) The biosynthesis of protamine during spermatogenesis of the mouse: extraction, partial characterization and site of synthesis. *J. cell Physiol.* **78**, 13–24.

Leuwenhoeck, A. van (1678) Observationes D. Antonii Leuwenhoeck, de natis e semine genitali animalculis. *Phil. Trans. R. Soc. Lond.* **12**, 1040–3.

Leiderman, B. and Mancini, R. E. (1968) Aerobic and anaerobic lactate production in the prepuberal and adult rat testis. *Proc. Soc. Exptl. Biol. Med.* **128**, 818–21.

Leiderman, B. and Mancini, R. E. (1969) Glycogen content in the rat testis from postnatal to adult ages. *Endocrinology* **85**, 607–9.

LeVier, R. R. and Spaziani, E. (1968) Influence of temperature on the incorporation of palmitic acid-U-$^{14}$C into rat testicular lipids. *J. Reprod. Fertil.* **15**, 365–72.

Lewin, L. M., Yannai, Y., Sulimovici, S. and Kraicer, P. (1976) Studies on the metabolic role of myo-inositol. Distribution of radioactive myo-inositol in the male rat. *Biochem. J.* **156**, 375–80.

Lima-de-Faria, A., German, J., Ghatnekar, M., McGovern, J. and Anderson, L. (1968) *In vitro* labelling of human meiotic chromosomes with $^{3}$H thymidine. *Hereditas* **60**, 249–61.

Lin, C. H. and Fritz, I. B. (1972) Studies on spermatogenesis in rats. IV. Rates of oxidation of palmitate and pyruvate by various cell populations. *Can. J. Biochem.* **50**, 963–8.

Lindsay, D. B. and Setchell, B. P. (1976) The oxidation of glucose, ketone bodies and acetate by the brain of normal and ketonaemic sheep. *J. Physiol.* (London) **259**, 801–23.

Lison, L. (1955) Variation de la basophile pendant la maturation du spermatozoïde chez le rat et sa signification histochimique. *Acta Histochem.* **2**, 47–67.

Llaurado, J. G. and Eik-Nes, K. B. (1961) Immobility of ascorbic acid in the stimulated testis of rat and dog. *Gen. Comp. Endocrinol.* **1**, 154–60.

Lofts, B. (1960) Cyclical changes in the distribution of the testis lipids of a seasonal mammal (*Talpa europaea*). *Quart. J. Microscop. Sci.* **101**, 199–205.

Loir, M. (1972a) Métabolism de l'acide ribonucléique et des protéines dans les spermatocytes et les spermatides de bélier (*Ovis aries*). I. Incorporation et devenir de la $^{3}$H-uridine. *Ann. Biol. anim. Bioch. Biophys.* **12**, 203–19.

Loir, M. (1972b) Métabolisme de l'acide ribonucléique et des protéines dans les spermatocytes et les spermatides du bélier (*Ovis aries*). II. Variation de l'incorporation et devenir de la ³H-lysine, de la ³H-arginine et de la ³⁵S-cystine. *Ann. biol. Anim. Biochim. Biophys.* **12**, 411–29.

Loir, M. and Hochereau-De Reviers, M. T. (1972) Deoxyribonucleoprotein changes in ram and bull spermatids. *J. Reprod. Fertil.* **31**, 127–30.

Loisel, G. (1903) Les graisses du testicule chez quelques mammifères. *Compt. Rend. Soc. Biol.* **55**, 1009–12.

Long, C. H. N. (1947) The relation of cholesterol and ascorbic acid to the secretion of the adrenal. *Rec. Progr. Horm. Res.* **1**, 99–117.

Lynch, K. M. and Scott, W. W. (1951) Lipid distribution in Sertoli cell and Leydig cell of the rat testes as related to experimental alterations of the pituitary–gonad system. *Endocrinology* **49**, 8–14.

McEnery, W. B. and Nelson, W. O. (1950) Cytochemical studies on testicular lipids. *Anat. Rec.* **106**, 221–2.

McEnery, W. B. and Nelson, W. O. (1953a) Organic phosphorus compounds in the testes of the rat at various ages. *Endocrinology* **52**, 93–103.

McEnery, W. B. and Nelson, W. O. (1953b) Organic phosphorus compounds in the testis of the rat under experimental conditions which impair spermatogenesis. *Endocrinology* **52**, 104–14.

Machino, A., Inano, H. and Tamaoki, B-I. (1969) Studies on enzyme reactions related to steroid biosynthesis. 1. Presence of the cytochrome P-450 in testicular tissue and its role in the biogenesis of androgens. *J. Steroid Biochem.* **1**, 9–16.

Macindoe, J. H. and Turkington, R. W. (1973) Hormonal regulation of spermidine formation during spermatogenesis in the rat. *Endocrinology* **92**, 595–605.

Macindoe, J. H., Sullivan, W. and Wray, H. L. (1977) Immunofluorescent localization of cyclic AMP in developing rat testis. *Endocrinology* **101**, 568–76.

Majumder, G. C. and Turkington, R. W. (1974) Acrosomal and lysosomal isoenzymes of β-galactosidase and N-acetyl-β-glucosaminidase in rat testis. *Biochem.* **13**, 2857–64.

Majumder, G. C., Lessin, S. and Turkington, R. W. (1975) Hormonal regulation of isoenzymes of N-acetyl-β-glucosaminidase and β-galactosidase during spermatogenesis in the rat. *Endocrinology* **96**, 890–7.

Males, J. L. and Turkington, R. W. (1970) Hormonal regulation of hyaluronidase during spermatogenesis in the rat. *J. Biol. Chem.* **245**, 6329–34.

Males, J. L. and Turkington, R. W. (1971) Hormonal control of lysozomal enzymes during spermatogenesis in the rat. *Endocrinology* **88**, 579–88.

Mancine, R. E., Penhos, J. C., Izquierdo, I. A. and Heinrich, J. J. (1960) Effects of acute hypoglycemia on rat testis. *Proc. Soc. Exptl. Biol. Med.* **104**, 699–703.

Mancini, R. E., Alonso, A., Barquet, J., Alvarez, B. and Nemirovsky, M. (1964) Histoimmunological localization of hyaluronidase in the bull testis. *J. Reprod. Fertil.* **8**, 325–30.

Marquis, N. R. and Fritz, I. B. (1965) The distribution of carnitine, acetylcarnitine and carnitine acetyltransferase in rat tissue. *J. Biol. Chem.* **240**, 2193–6.

Massie, E. D., Gomes, W. R. and VanDemark, N. L. (1969) Oxygen tension in testes of normal and cryptorchid rats. *J. Reprod. Fertil.* **19**, 559–61.

Mazzanti, L., Lopez, M. and Berti, M. G. (1965) Atrofia del testicolo prodotta dal monofluoroacetato sodico nel ratto albino. *Experientia* **21**, 446–7.

Means, A. R. and Hall, P. F. (1968a) Protein biosynthesis in the testis: I. Comparison between stimulation by FSH and glucose. *Endocrinology* **82**, 597–602.

Means, A. R. and Hall, P. F. (1968b) Protein biosynthesis in the testis: II. Role of adenosine triphosphate (ATP) in stimulation by glucose. *Endocrinology* **83**, 86–96.

Means, A. R. and Hall, P. F. (1969) Protein biosynthesis in the testis: III. Dual effect of glucose. *Endocrinology* **84**, 285–97.

Means, A. R., Hall, P. F., Nicol, L. W., Sawyer, W. H. and Baker, C. A. (1969) Protein biosynthesis in the testis. IV. Isolation and properties of polyribosomes. *Biochemistry* **8**, 1488–95.

Means, A. R., Fakunding, J. L. and Tindall, D. J. (1976) Follicle stimulating hormone regulation of protein kinase activity and protein synthesis in testis. *Biol. Reprod.* **14**, 54–63.

Means, A. R., Dedman, J. R., Fakunding, J. L. and Tindall, D. J. (1978) Mechanism of action of FSH in the male rat. In *Hormone Receptors and Mechanism of Action*, eds. B. W. O'Malley and L. Birnbaumer, New York: Academic Press. [In press.]

Mellampy, R. M., Cavazos, L. F. and Duncan, G. W. (1955) Composition and histochemistry of bull, ram, boar and rooster testis. Michigan State Univ. Symp. 34–44.

Menard, R. H., Latif, S. A. and Purvis, J. L. (1975) The intratesticular localization of cytochrome P-450 and cytochrome P-450-dependent enzymes in the rat testis. *Endocrinology* **97**, 1587–92.

Messier, B. and Leblond, C. P. (1960) Cell proliferation and migration as revealed by radioautography after injection of thymidine-$H^3$ into male rats and mice. *Am. J. Anat.* **106**, 247–85.

Meyer, E. H. H., Forsgren, K. Deimling, O. von and Engel, W. (1974) Induction of non-specific carboxyl esterase in the immature rat testis by human chorionic gonadotrophin. *Endocrinology* **95**, 1737–9.

Meyer, G. F. (1969) Experimental studies on spermiogenesis in drosophila. *Genet. Suppl.* **61**, 79–92.

Michaël, C. M. (1973) Prostaglandins in swine testes. *Lipids* **8**, 92 3.

Middleton, A. (1973) Glucose metabolism in rat seminiferous tubules. Ph.D. dissertation, University of Cambridge.

Middleton, A. and Setchell, B. P. (1972) The origin of inositol in the rete testis fluid of the ram. *J. Reprod. Fertil.* **30**, 473–5.

Mills, N. C. and Means, A. R. (1972) Sorbitol dehydrogenase of rat testis: changes of activity during development, after hypophysectomy and following gonadotrophic hormone administration. *Endocrinology* **91**, 147–56.

Mills, N. C., Nguyen, T. van, and Means, A. R. (1977) Histones of rat testis chromatin during early postnatal development and their interactions with DNA. *Biol. Reprod.* **17**, 760–8.

Mohri, H. (1968) Amino-acid composition of 'Tubulin' constituting microtubules of sperm flagella. *Nature* **217**, 1053–4.

Molen, H. J. van der, Bijleveld, M. J., Vusse, G. J. van der and Cooke, B. A. (1973) Effects of gonadotrophins on cholesterol and cholesterol esters as precursors of steroid production in the testis. *J. Endocrinol.* **57**, vi–vii.

Monesi, V. (1962) Autoradiographic study of DNA synthesis and cell cycle in spermatogonia and spermatocytes of mouse testis, using tritiated thymidine. *J. Cell Biol.* **14**, 1–18.

Monesi, V. (1964a) Autoradiographic evidence of a nuclear histone synthesis during mouse spermiogenesis in the absence of detectable quantities of nuclear ribonucleic acid. *Exptl. Cell Res.* **36**, 683–8.

Monesi, V. (1964b) Ribonucleic acid synthesis during mitosis and meiosis in the mouse testis. *J. Cell Biol.* **22**, 521–32.

Monesi, V. (1964c) Il metabolismo dell'acido ribonucleico durante lo sviluppo delle cellule germinale maschili e nelle cellule del Sertoli, nel topo adulto. *Arch. Ital. Anat. Embriol.* **69**, 89–118.

Monesi, V. (1965a) Synthetic activities during spermatogenesis in the mouse. *Exptl. Cell Res.* **39**, 197–224.

Monesi, V. (1965b) Differential rate of ribonucleic acid synthesis in the autosomes and sex chromosomes during male meiosis in the mouse. *Chromosa* **17**, 11–21.

Monesi, V. (1967) Ribonucleic acid and protein synthesis during differentiation of male germ cells in the mouse. *Arch. Anat. Microscop.* **56** Suppl. 3–4, 61–74.

Monesi, V. (1971) Chromosome activities during meiosis and spermiogenesis. *J. Reprod. Fertil.* Suppl. **13**, 1–14.

Monesi, V. (1974) Nucleoprotein synthesis in spermatogenesis. In *Male Fertility and Sterility*, eds. R. E. Mancini and L. Martini, 59–87, New York:Academic Press.

Monn, E., Desautel, M. and Christiansen, R. O. (1972) Highly specific testicular adenosine 3′,5′-monophosphate phosphodiesterase associated with sexual maturation. *Endocrinology* **91**, 716–20.

Montagna, W. (1952) The distribution of lipids, glycogen and phosphatases in the human testis. *Fertil. Steril.* **3**, 27–32.

Montagna, W. and Hamilton, J. B. (1951) Histological studies of human testes. I. The distribution of lipids. *Anat. Rec.* **109**, 635–59.

Moonsammy, G. I. and Stewart, M. A. (1971) Hexose phosphate levels in testes of galactose-fed rats. *J. Reprod. Fertil.* **27**, 113–14.

Moore, G. P. M. (1971) DNA-dependent RNA synthesis in fixed cells during spermatogenesis in mouse. *Exptl. Cell Res.* **68**, 462–5.

Morin, R. J. (1967) *In vitro* incorporation of acetate-1-$^{14}$C into sphingomyelin, phosphatidyl choline and phosphatidyl ethanolamine of rabbit testes. *Proc. Soc. Exptl. Biol. Med.* **126**, 229–32.

Morris, M. D. and Chaikoff, I. L. (1959) The origin of cholesterol in liver, small intestine, adrenal gland, and testis of the rat: dietary versus endogenous contributions. *J. Biol. Chem.* **234**, 1095–7.

Morris, R. N. and Collins, A. C. (1971) Biosynthesis of myo-inositol by rat testis following surgically induced cryptorchidism or treatment with triethylenemelamine. *J. Reprod. Fertil.* **27**, 201–10.

Morris, R. N. and Davis, J. R. (1966) Effect of testosterone on the incorporation of L-lysine-U-$^{14}$C into protein of rat testis slices. *Arch. Intern. Pharmacodyn.* **162**, 432–6.

Mukherjee, A. B. and Cohen, M. M. (1968) DNA synthesis during meiotic prophase in male mice. *Nature* **219**, 489–90.

Muramatsu, M., Utakoji, T. and Sugano, H. (1968) Rapidly-labeled nuclear RNA in Chinese hamster testis. *Exptl. Cell Res.* **53**, 278–83.

Murota, S., Shikita, M. and Tamaoki, B-I. (1965) Intracellular distribution of the enzymes related to androgen formation in mouse testes. *Steroids* **5**, 409–13.

Nagy, F. (1972) Cell division kinetics and DNA synthesis in the immature Sertoli cells of the rat testis. *J. Reprod. Fertil.* **28**, 389–96.

Nakamura, M. and Hall, P. F. (1976) Inhibition by 5-thio-D-glucopyranose of protein biosynthesis *in vitro* in spermatids from rat testis. *Biochim. Biophys. Acta* **447**, 474–83.

Nakamura, M. and Hall, P. F. (1977) Effect of 5-thio-D-glucose on protein synthesis *in vitro* by various types of cells from rat testes. *J. Reprod. Fertil.* **49**, 395–7.

Nakamura, M. and Privett, O. S. (1969) Metabolism of lipids in rat testes: Interconversions and incorporation of linoleic acid into lipid classes. *Lipids* **4**, 41–9.

Nakamura, M. and Privett, O. S. (1973) Studies on metabolism of linoleic-1-$^{14}$C acid in testes of hypophysectomized rats. *Lipids* **8**, 224–31.

Nakamura, M., Jensen, B. and Privett, O. S. (1968) Effect of hypophysectomy on the fatty acids and lipid classes of rat testes. *Endocrinology* **82**, 137–42.

Neill, A. R. and Masters, C. J. (1974) The distribution of $^{14}$C-label in the lipids of ram semen following the intra-testicular injection of [1-$^{14}$C] palmitic acid. *J. Reprod. Fertil.* **38**, 311–23.

Nicander, L. (1957) A histochemical study on glycogen in the testes of domestic and laboratory animals, with special reference to variations during the spermatogenetic cycle. *Acta Morphol. Neerl. Scand.* **1**, 233–40.

Niemi, M. and Hyyppa, M. (1970) Quoted by Setchell, 1970b.

Niemi, M. and Ikonen, M. (1963) Histochemistry of the Leydig cells in the postnatal prepuberal testis of the rat. *Endocrinology* **72**, 443–8.

Niemi, M. and Kormano, M. (1965a) Histochemical demonstration of a C-esterase activity in the seminiferous tubules of the rat testis. *J. Reprod. Fertil.* **10**, 49–54.

Niemi, M. and Kormano, M. (1965b) Cyclical changes in significance of lipids and acid phosphatase activity in the seminiferous tubules of the rat testis. *Anat. Rec.* **151**, 159–70.

Niemi, M., Harkonen, M. and Kokko, A. (1962) Localization and identification of testicular esterases in the rat. *J. Histochem. Cytochem.* **10**, 186–93.

Niemi, M., Harkonen, M. and Ikonen, M. (1966) A chemical and histochemical study on the significance of the non-specific esterase activity in the adult rat testis. *Endocrinology* **79**, 294–300.

Noach, E. L. and Rees, G. P. van (1958) Ascorbic acid in the gonads of rats. *Acta Endocrinol.* **27**, 502–8.

Ong, S.-H., Whitley, T. H., Stowe, N. W. and Steiner, A. L. (1975) Immunohistochemical localization of 3′:5′-cyclic AMP and 3′:5′-cyclic GMP in rat liver, intestine and testis. *Proc. Nat. Acad. Sci. USA* **72**, 2022–6.

Oshima, H., Sarada, T., Ochi-Ai, K. and Tamaoki, B.-I. (1967) Intracellular distribution and properties of steroid 16α-hydroxylase in human testes. *J. Clin. Endocrinol. Metab.* **27**, 1249–54.

Oshima, M. and Carpenter, M. P. (1968) The lipid composition of the prepuberal and adult rat testis. *Biochim. Biophys. Acta* **152**, 479–97.

Pappenheimer, J. R. and Setchell, B. P. (1973) Cerebral glucose transport and oxygen consumption in sheep and rabbits. *J. Physiol.* (London) **233**, 529–51.

Parez, M., Petel, J. P. and Vendrely, C. (1960) Sur la teneur en acide desoxyribonucleique des spermatozoïdes de taureaux presentant differents degres de fecondite. *Compt. Rend. Sean. Acad. Sci. Paris* **251**, 2581–3.

Parvinen, M. and Söderström, K-O. (1976) Effects of FSH and testosterone on the RNA synthesis in different stages of rat spermatogenesis. *J. Steroid Biochem.* **7**, 1021–3.

Parvinen, M. and Vanha-Perttula, T. (1972) Identification and enzyme quantitation of the stages of the seminiferous epithelial wave in the rat. *Anat. Rec.* **174**, 435–49.

Paul, H. E., Paul, M. F., Kopko, F., Bender, R. C. and Everett, G. (1953) Carbohydrate metabolism studies on the testis of rats fed certain nitrofurans. *Endocrinology* **53**, 585–92.

Pelc, S. R. (1957) On the connection between the synthesis of RNA and DNA in the testis of the mouse. *Exptl. Cell Res.* **12**, 320–4.

Pelc, S. R. and Howard, A. (1956) A difference between spermatogonia and somatic tissues of mice in the incorporation of [8-$^{14}$C]-adenine into deoxyribonucleic acid. *Exptl. Cell Res.* **11**, 128–34.

Peluffo, R. O., Ayala, S. and Brenner, R. R. (1970) Metabolism of fatty acids of the linoleic acid series in testicles of diabetic rats. *Am. J. Physiol.* **218**, 669–73.

Perlman, P. L. (1950a) The functional significance of testis cholesterol in the rat: Effects of hypophysectomy and cryptorchidism. *Endocrinology* **46**, 341–6.

Perlman, P. L. (1950b) The functional significance of testis cholesterol in the rat: Histochemical observations on testes following hypophysectomy and experimental cryptorchidism. *Endocrinology* **46**, 347–52.

Pestis, J. and Raoul, Y. (1969) Localisation de l'acide ascorbique dans le testicule du rat. Influence des hormones gonadotropes et de la cryptorchidie sur le taux et la répartition de cette substance. *Compt. rend. sean. Soc. Biol.* **163**, 1027–33.

Pilkis, S. J. (1970) Hormonal control of hexokinase activity in animal tissues. *Biochim. Biophys. Acta* **215**, 461–76.

Posalaki, Z., Szabo, D., Bacsi, E. and Okros, I. (1968) Hydrolytic enzymes during spermatogenesis in rat. An electron microscopic and histochemical study. *J. Histochem. Cytochem.* **16**, 249–62.

Pugh, D. and Walker, P. G. (1961) The localization of N-acetyl-$\beta$-glucosaminidase in tissues. *J. Histochem. Cytochem.* **9**, 242–50.

Regaud, C. (1901) Etudes sur la structure des tubes séminifères et sur la spermatogénèse chez les mammifères. *Arch. Anat. micr.* **4**, 101–56 and 231–80.

Renston, R. H., Ihrig, T. J., Renston, J. P. and Gondos, B. (1975) Concentration of free and esterified cholesterol in the testes of maturing rabbits. *J. Reprod. Fertil.* **43**, 91–6.

Ressler, N., Olivero, E. and Josephe, R. R. (1965) Lactic dehydrogenase isoenzymes in human testis. *Nature* **206**, 829–30.

Ringertz, N. R., Gledhill, B. L. and Darzynkiewicz, Z. (1970) Changes in deoxyribonucleoprotein during spermiogenesis in the bull. Sensitivity of DNA to heat denaturation. *Exptl. Cell Res.* **62**, 204–18.

Rubin, A. (1958) Studies in human reproduction. II. The influence of diabetes mellitus in men upon reproduction. *Am. J. Obstet. Gynecol.* **76**, 25–9.

Rüsfeldt, O. (1949) Origin of hyaluronidase in the rat testis. *Nature* **163**, 874–5.

Russo, J. (1970) Glycogen content during the postnatal differentiation of the Leydig cell in the mouse testis. *Z. Zellforsch. Mikroskop. Anat.* **104**, 14–18.

Schanbacher, B. D., Gomes, W. R. and VanDemark, N. L. (1974) Testicular carnitine acetyltransferase activity and serum testosterone levels in developing and cryptorchid rats. *J. Reprod. Fertil.* **41**, 435–40.

Schenkman, J. B., Richert, D. A. and Werterfeld, W. W. (1965) α-Glycerophosphate dehydrogenase activity in rat spermatozoa. *Endocrinology* **76**, 1055–61.

Schoffling, K., Federlin, K. and Pfeiffer, E. F. (1963) Disorders of sexual function in male diabetics. *Diabetes* **12**, 519–27.

Schor, N. A., Cara, J. and Perez, A. (1963) Hormonal dependence of oxidative enzymes in the testis of the rat. *Nature* **198**, 1310.

Scott, T. W. and Setchell, B. P. (1968) Lipid metabolism in the testis of the ram. *Biochem. J.* **107**, 273–7.

Scott, T. W., Voglmayr, J. K. and Setchell, B. P. (1967) Lipid composition and metabolism in testicular and ejaculated ram spermatozoa. *Biochem. J.* **102**, 456–61.

Serfaty, A. and Boyer, J. (1956) L'influence de divers métabolites glucidiques sur l'intensité respiratoire du testicule de rat blanc. *Experientia* **12**, 386–7.

Setchell, B. P. (1970a) The secretion of fluid by the testes of rats, rams and goats with some observations on the effect of age, cryptorchidism and hypophysectomy. *J. Reprod. Fertil.* **23**, 79–85.

Setchell, B. P. (1970b) Testicular blood supply, lymphatic drainage and secretion of fluid. In *The Testis*, eds. A. D. Johnson, W. R. Gomes and N. L. VanDemark, **I**, 101–239, New York: Academic Press.

Setchell, B. P. and Hinks, N. T. (1967) The importance of glucose in the oxidative metabolism of the testis of the conscious ram and the role of the pentose cycle. *Biochem. J.* **102**, 623–31.

Setchell, B. P. and Middleton, A. (1971) Facilitated diffusion of glucose into rat seminiferous tubules. *Proc. Int. Union Physiol. Sci.* **9**, 508.

Setchell, B. P. and Waites, G. M. H. (1964) Blood flow and the uptake of glucose and oxygen in the testis and epididymis of the ram. *J. Physiol.* **171**, 411–25.

Setchell, B. P., Waites, G. M. H. and Lindner, H. R. (1965) Effect of undernutrition on testicular blood flow and metabolism and the output of testosterone in the ram. *J. Reprod. Fertil.* **9**, 149–62.

Setchell, B. P., Hinks, N. T., Voglmayr, J. K. and Scott, T. W. (1967) Amino acids in ram testicular fluid and semen and their metabolism by spermatozoa. *Biochem. J.* **105**, 1061–5.

Setchel, B. P., Dawson, R. M. C. and White, R. W. (1968) The high concentration of free myo-inositol in rete testis fluid from rams. *J. Reprod. Fertil.* **17**, 219–20.

Setchell, B. P., Bassett, J. M., Hinks, N. T. and Graham, N. McC. (1972a) The importance of glucose in the oxidative metabolism of the pregnant uterus and its contents in conscious sheep, with some preliminary observations on the oxidation of fructose and glucose by fetal sheep. *Quart. J. exp. Physiol.* **57**, 257–66.

Setchell, B. P., Smith, M. W. and Munn, E. A. (1972b) The stimulation by bicarbonate of adenosine triphosphatase activity in the seminiferous tubules of rodents and the lack of effect of ouabain. *J. Reprod. Fertil.* **28**, 413–18.

Sharma, C. and Weinhouse, S. (1962). Glucose-6-phosphatase as the product of glucose phosphorylation in testes. *Proc. Soc. Exptl. Biol. Med.* **110**, 522–4.

Shelanski, M. L. and Taylor, E. W. (1968) Properties of the protein subunit of central-pair and outer-doublet microtubules of sea urchin flagella. *J. cell. Biol.* **38**, 304–15.

Shen, R-S. and Lee, I. P. (1976) Developmental patterns of enzymes in mouse testis. *J. Reprod. Fertil.* **48**, 301–5.

Sherins, R. J. and Hodgen, G. D. (1976) Testicular gamma glutamyl-transpeptidase: an index of Sertoli cell function in man. *J. Reprod. Fertil.* **48**, 191–3.

Shires, A., Carpenter, M. P. and Chalkley, R. (1975) New histones found in mature mammalian testes. *Proc. Nat. Acad. Sci. USA.* **72**, 2714–8.

Söderström, K.-O. (1976) Characterization of RNA synthesis in mid-pachytene spermatocytes of the rat. *Exptl. Cell. Res.* **102**, 237–45.

Söderström, R. O. and Parvinen, M. (1976) RNA synthesis in different stages of rat seminiferous epithelial cycle. *Molec. Cell. Endocrinol.* **5**, 181–99.

Solari, A. J. and Tres, L. (1967) The localization of nucleic acids and the argentaffin substance in the sex vesicle of mouse spermatocytes. *Exptl. Cell Res.* **47**, 86–96.

Sosa, A., Altamirano, E., Hernandez, P. and Rosado, A. (1972) Developmental patterns of rat testis hexokinase. *Life Sci.* **11**, 499–510.

Spruill, A. and Steiner, A. (1976) Immunohistochemical localization of cyclic nucleotides during testicular development. *J. Cyclic Nucleotide Res.* **3**, 225–39.

Srere, P. A., Chaikoff, I. L., Treitman, S. S. and Burstein, L. S. (1950) The extrahepatic synthesis of cholesterol. *J. Biol. Chem.* **182**, 629–34.

Stallcup, O. T. and Roussel, J. D. (1965) Development of the LDH enzyme system in the testis and epididymis of young dairy bulls. *J. Dairy Sci.* **48**, 1511–16.

Stambaugh, R. and Buckley, J. (1967) The enzymic and molecular nature of the lactic dehydrogenase subbands and $X_4$ isozyme. *J. Biol. Chem.* **242**, 4053–9.

Stefanini, M., Martino, C. De, D'Agostino, A., Agrestini, A. and Monesi, V. (1974) Nucleolar activity of rat primary spermatocytes. *Exptl. Cell Res.* **86**, 166–70.

Steinberger, E. and Nelson, W. O. (1955) The effect of hypophysectomy, cryptorchidism, estrogen and androgen upon the level of hyaluronidase in the rat testes. *Endocrinology* **56**, 429–44.

Steinberger, A. and Steinberger, E. (1971) Replication pattern of Sertoli cells in maturing rat testis *in vivo* and in organ culture. *Biol. Reprod.* **4**, 84–7.

Steinberger, E. and Wagner, C. (1961) Observations on the endogenous respiration of rat testicular tissue. *Endocrinology* **69**, 305–11.

Suzuki, A., Ishizuka, I., Ueta, N. and Yamakawa, T. (1973) Isolation and characterization of seminolipid (1-0-alkyl-2-0-acyl-3-[β 3'-sulfogalactosyl] glycerol) from guinea pig testis and incorporation of $^{35}$S-sulfate into seminolipid in sliced testis. *Jap. J. exp. Med.* **43**, 435–42.

Suzuki, A., Ishizuka, I. and Yamakawa, T. (1975) Isolation and characterization of a ganglioside containing fucose from boar testis. *J. Biochem.* (Japan) **78**, 947–54.

Svasti, J. and Viriyachai, S. (1975) The properties of purified LDH-$C_4$ from human testis. In *Isozymes II Physiological Function*, ed. C. L. Markert, New York: Academic Press.

Tabor, C. W. and Tabor, H. (1976) 1,4-diaminobutane (putrescine), spermidine and spermine. *Ann. Rev. Biochem.* **45**, 285–306.

Tabor, H. and Tabor, C. W. (1964) Spermidine, spermine and related amines. *Pharmacol. Rev.* **16**, 245–300.

Tamaoki, B-I., Inano, H. and Nakano, H. (1969) *In vitro* synthesis and conversion of androgens in testicular tissue. In *The Gonads*, eds. K. W. McKerns, 547–613, New York: Appleton Century Crofts.

Tamaoki, B-I., Inano, H. and Suzuki, K. (1975) Testicular enzymes related to steroid metabolism. *Curr. Top. Molec. Endocrinol.* **2**, 123–32.

Tepperman, H. M. and Tepperman, J. (1950) Glucose utilization *in vitro* by normal adult, immature and cryptorchid testis. *Endocrinology* **47**, 459–61.

Tepperman, J., Tepperman, H. M. and Dick, H. J. (1949) A study of the metabolism of rat testis *in vitro*. *Endocrinology* **45**, 491–503.

Tice, L. W. and Barnett, R. J. (1963) The fine structural localization of some testicular phosphatases. *Anat. Rec.* **147**, 43–63.

Tres, L. L. (1975) Nucleolar RNA synthesis of meiotic prophase spermatocytes in the human testis. *Chromosoma* **53**, 141–51.

Tres, L. L. and Kierszenbaum, A. L. (1975) Transcription during mammalian spermatogenesis with special reference to the Sertoli cells. *Curr. Top. Molec. Endocrinol.* **2**, 455–78.

Tsai, M. Y. and Kemp, R. G. (1973) Isoenzymes of rabbit phosphofructokinase. *J. biol. Chem.* **248**, 785–92.

Utakoji, T. (1966) Chronology of nucleic acid synthesis in meiosis the male chinese hamster. *Exptl. Cell Res.* **42**, 585–96.

Valenta, M., Hyldgaard-Jensen, J. and Moustgaard, J. (1967) Three lactic dehydrogenase isozyme systems in pig spermatozoa and the polymorphism of sub-units controlled by a third locus C. *Nature* **216**, 506–7.

VanDemark, N. L., Zogg, C. A. and Hays, R. L. (1968) Effect of hyper- and hypoglycemia accompanying cryptorchidism on testis function. *Am. J. Physiol.* **215**, 977–84.

Vanha-Perttula, T. (1971a) A new type of acid phosphatase from rat testis. *Experientia* **27**, 42–4.

Vanha-Perttula, T. (1971b) Chromatographic fractionation and characterization of rat testicular acid phosphatases. *Biochim. Biophys. Acta* **227**, 390–401.

Vanha-Perttula, T. (1973a) Aminopeptidases of rat testis. I. Fractionation and characterization. *J. Reprod. Fertil.* **32**, 33–44.

Vanha-Perttula, T. (1973b) Aminopeptidases of rat testis. II. Effects of puberty, cryptorchidism and cadmium chloride treatment. *J. Reprod. Fertil.* **32**, 45–54.

Vanha-Perttula, T. (1973c) Aminopeptidases of rat testis. III. Activity of dipeptidyl aminopeptidases I and II in normal and experimental conditions. *J. Reprod. Fertil.* **32**, 55–64.

Vanha-Perttula, T. and Nikkanen, V. (1973) Acid phosphatases of the rat testis in experimental conditions. *Acta Endocrinol.* **72**, 376–90.

Vaughn, J. C. (1965) Histone metabolism: The significance of the 'sphere chromatophile' in rat spermatogenesis. *Am. Zoologist* **5**, 231.

Vaughn, J. C. (1966) The relationship of the 'sphere chromatophile' to the fate of displaced histones following histone transition in rat spermiogenesis. *J. Cell Biol.* **31**, 257–78.

Vendrely, C. (1952) L'acide desoxyribonucleique du noyau des cellules animales. Son rôle possible dans la biochimie de l'hérédité. *Bull. Biol. France Belgique* **86**, 1–87.

Vendrely, C., Knobloch, A. and Vendrely, R. (1956) Contribution a l'étude biochimique comparée de diverses desoxyribonucleoproteines d'origine animale. *Biochim. Biophys. Acta* **19**, 472–9.

Vendrely, R., Knobloch, A. and Vendrely, C. (1958) Les desoxyribonucleoproteines du noyau cellulaire et le mécanisme de transformation de la nucleohistone en nucleoprotamine. *Compt. Rend. Sean. Acad. Sci. Paris* **246**, 3128–30.

Vernon, R. G., Go, V. L. W. and Fritz, I. B. (1971) Studies on spermatogenesis in rats. II. Evidence that carnitine acetyltransferase is a marker enzyme for the investigation of germ cell differentiation. *Can. J. Biochem.* **49**, 761–7.

Waites, G. M. H. and Setchell, B. P. (1964) Effect of local heating on blood flow and metabolism in the testis of the conscious ram. *J. Reprod. Fertil.* **8**, 339–49.

Wolf, R. C. and Leathem, J. H. (1953) Hormonal and nutritional influences on the biochemical composition of the rat testis. *Endocrinology* **57**, 286–90.

Wröbel, K. H. and Kühnel, W. (1967a) Die Histotopik einiger Oxydoreduktasen in Hoden von Hund und Katze. *Histochemie* **10**, 208–15.

Wröbel, K. H. and Kühnel, W. (1967b) Zur Histochemie von Hydrolasen im Hoden des Hundes und der Katze. *Histochemie* **10**, 329–35.

Wröbel, K. H. and Kühnel, W. (1968) Enzymhistochemie am Hoden der Haussäugetiere. l. Oxydoreduktasen in Hoden von Ziege und Schwein. *Berlin. Münch. Tierärtztl. Wochschr.* **81**, 86–90.

Xuma, M. and Turkington, R. W. (1972) Hormonal regulation of uridine diphosphatase during spermatogenesis in the rat. *Endocrinology* **91**, 415–22.

Yokoe, Y. and Hall, P. F. (1970a) Testicular phospholipids: I. Action of follicle-stimulating hormone (FSH) upon the biosynthesis of phospholipids by rat testis. *Endocrinology* **86**, 18–28.

Yokoe, Y. and Hall, P. F. (1970b) Testicular phospholipids: II. Action of interstitial cell-stimulating hormone (ICSH) upon testicular phospholipids in hypophysectomized rats. *Endocrinology* **86**, 1257–63.

Yokoe, Y. and Hall, P. F. (1971) Testicular phospholipids: IV. The influence of ICSH and testosterone upon phospholipids in rat testes with intact germinal epithelium. *Endocrinology* **88**, 1092–4.

Yokoe, Y., Irby, D. C. and Hall, P. F. (1971) Testicular phospholipids. III: Site of action of ICSH in testis following regression of the germinal epithelium. *Endocrinology* **88**, 195–205.

Zieher, L. M., Debeljuk, L., Iturriza, F. and Mancini, R. E. (1971) Biogenic amine concentration in testes of rats at different ages. *Endocrinology* **88**, 351–4.

Zinkham, W. H. (1968) Lactate dehydrogenase isoenzymes of testis and sperm: biological and biochemical properties and genetic control. *Ann. N.Y. Acad. Sci.* **151**, 589–610.

Zinkham, W. H., Blanco, A. and Kupchyk, L. (1963) Lactate dehydrogenase in testis: Dissociation and recombination of subunits. *Science* **142**, 1303–4.

Zinkham, W. H., Blanco, A. and Clowry, L. J. (1964) An unusual isozyme of lactate dehydrogenase in mature testes: Localization, ontogeny, and kinetic properties. *Ann. N.Y. Acad. Sci.* **121**, 571–88.

Zinkham, W. H., Holtzman, N. A. and Isensee, H. (1968) The molecular size of lactate dehydrogenase isozymes in mature testes. *Biochim. Biophys. Acta* **160**, 172–7.

Zogg, C. A., Hays, R. L., VanDemark, N. L. and Johnson, A. D. (1968) Effect of duration of experimental cryptorchidism on testis composition and metabolic activity. *Am. J. Physiol.* **215**, 985–90.

# 10 Endocrinological control of the testis

The regression of the testes of rats following hypophysectomy, first reported by P. E. Smith in 1927 indicated clearly that the testis is controlled by the pituitary. It was soon shown that different effects were produced on the testis by different extracts of the pituitary. One extract stimulated the interstitial tissue to produce more androgens; this hormone was therefore named 'interstitial cell stimulating hormone' (ICSH) and was later shown to be identical with 'luteinizing hormone' (LH) so-called from its effects in the female. (Here, this hormone will be called LH.) Follicle stimulating hormone (FSH) was also named for its effects in the female, but was soon shown to stimulate the seminiferous tubules of the testis and with LH, completely to restore spermatogenesis in hypophysectomized rats (Greep et al., 1936; Greep and Fevold, 1937). After hypophysectomy in adults, spermatogenesis is 'arrested' at the stage of the primary spermatocytes (rat: Smith, 1930; Clermont and Morgenthaler, 1955; Clermont and Harvey, 1967; Steinberger, 1971; ground squirrel: Wells and Gomez, 1937; monkey: Smith, 1942), but there is also a reduction in the numbers of B spermatogonia formed. The major degeneration occurs between preleptotene and pachytene spermatocytes, but there is also increased degeneration of spermatids (Russell and Clermont, 1977). The development of the blood–testis barrier is delayed, but does eventually occur in rats whose gonadotrophin output was suppressed with oestrogen or clomiphene (Vitale et al., 1973). Effects of the pituitary on the peritubular myoid cells and on the maturation of the Sertoli cells have also been reported (Bressler and Ross, 1972; Bressler, 1976).

Some doubt was cast on the role of FSH in the male by the demonstration that spermatogenesis could be maintained immediately after hypophysectomy in rats with LH alone (Randolph et al., 1959; Simpson et al., 1944) or even testosterone in large doses. These treatments would not reinitiate spermatogenesis in hypophysectomized rats in which the testis had been allowed to regress or initiate it in young animals hypophysectomized before puberty. In ground squirrels and rhesus monkeys, testosterone can maintain spermatogenesis in hypophysectomized animals without any gonadotrophins and also appears to be capable of reinitiating spermatogenesis after the testis has regressed (see Section 6.1.6.2). In man, the experiments with testosterone cannot be conducted because of the unethically large dose required, but the testes of hypophysectomized men regress, and can be restored with human chorionic gonadotrophin (hCG) and human menopausal gonadotrophin (hMG) (Gemzell and Kjessler, 1964; Macleod et al., 1966; Mancini et al., 1968; Johnsen and Christiansen, 1968; MacLeod, 1970). In sheep, regression of the testis after hypophysectomy in prepubertal animals can be prevented with LH + FSH (Courot, 1967, 1971; Courot and Ortavant, 1972). In other species, either the effects of both hypophysectomy and testosterone therapy are slow to develop, as in the guinea pig (Allanson et al., 1935; Cutuly, 1941) or the effects of testosterone have not been tried (ferret: Hill and Parkes,

1933; rabbit: White, 1933; hedgehog: McPhail, 1933; cat: McPhail, 1935; mouse: Randolph *et al.*, 1959; wallaby: Hearn, 1976).

## 10.1    Luteinizing hormone (LH) or interstitial cell stimulating hormone (ICSH)

### 10.1.1    Site of action

There seems little doubt that the primary site of action of LH is on the Leydig cells. With the high protein concentration of testicular lymph (see Section 3.3.3) and the ready passage of proteins across the walls of the testicular capillaries (see Section 8.7) there is no difficulty in the hormone gaining access to the Leydig cells. LH stimulates androgen production from the Leydig cells both *in vivo* and *in vitro* (see Section 6.1.4.1) and structural changes are produced in these cells when the animals are treated with LH or hCG (de Kretser, 1967; Russo and Sacerdote, 1971; Heller and Leach, 1971; Chemes *et al.*, 1976a). Stimulation of androgen production by isolated Leydig cells can be achieved with either LH or hCG (Moyle and Ramachandran, 1973; van Damme *et al.*, 1974; Qazi *et al.*, 1974). It is curious that a response in testosterone production to LH or hCG can be demonstrated with decapsulated adult testes of rats and mice (Dufau *et al.*, 1971; Rommerts *et al.*, 1972; van Damme *et al.*, 1973), whereas teased-out preparations are unresponsive unless they have been preincubated for an hour and then fresh medium and trophic hormone added (Cooke *et al.*, 1972; Rommerts *et al.*, 1972, 1973, 1974).

Most authors agree that LH has very little action on isolated seminiferous tubules, in stimulating levels of cyclic AMP, or in activating adenyl cyclate or protein kinase, or in increasing synthesis of androgen-binding protein (see Section 10.2.2.1). An action of LH in increasing fluid flow from the testis and causing liberation of spermatozoa from the germinal epithelium has been reported by Burgos and Vitale-Calpe, but subsequent experiments in my laboratory and by others have not confirmed this finding (see Sections 7.3 and 8.5.2).

### 10.1.2    Mechanism of action of LH

#### 10.1.2.1    Binding of LH to Leydig cells    With most protein hormones, the first stage in their action is to bind the surface of the cells which they stimulate. Binding of LH and hCG to Leydig cells can be demonstrated both *in vivo* in rats by injection of labelled hormone (de Kretser *et al.*, 1969, 1971) or *in vitro* with testis homogenates both from rat and pig (Catt *et al.*, 1972a, 1974) and with dispersions of the interstitial cell fraction (Catt and Dufau, 1973; Leidenberger and Reichert, 1972). No binding of LH occurs with isolated tubules. The binding is temperature dependent, with a higher initial rate at higher temperatures, although at 34°C there is a lower equilibrium level (Catt *et al.*, 1972a; Charreau *et al.*, 1974; Dufau *et al.*, 1974), presumably due to quicker degradation of receptor and labelled hormone. Apart from hCG, no other hormones bind to these sites; apparent binding at high concentrations can usually be explained by contamination with LH. Bound labelled hor-

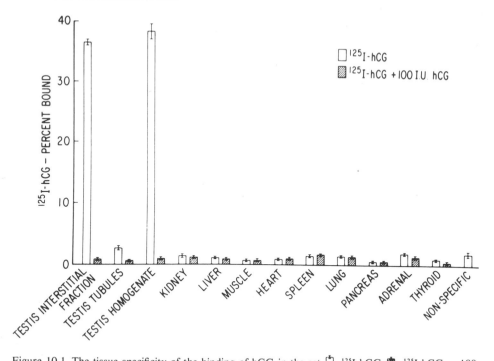

Figure 10.1 The tissue specificity of the binding of hCG in the rat. ⬜, $^{125}$I-hCG; ▨, $^{125}$I-hCG + 100 I.U. hCG. (Reproduced from Catt *et al.*, 1974)

mone can be displaced with an excess of unlabelled hormone and no comparable binding can be demonstrated in liver, muscle, kidney, lung or spleen (Figure 10.1; Catt *et al.*, 1972a, b, 1974). Binding probably occurs on the cell membrane not inside the cell as binding is maximal with a fraction of the cells which can be sedimented from saline suspensions at 1500g (Catt *et al.*, 1974). Furthermore, bound hormone can be dissociated with dilute acid in a form which retains biological activity (Dufau *et al.*, 1972a). The location of the binding at the cell membrane is supported by the observation that LH bound to sepharose so that it could not possibly enter cells, still stimulated testosterone production (Dufau *et al.*, 1971).

Much more hormone can be bound than is needed for maximal stimulation of testosterone synthesis; maximum binding of hCG by rat testis is reached at a concentration of 200ng/ml whereas maximum steroidogenesis is produced with 0·5ng/ml (Catt *et al.*, 1974) but no significance can yet be ascribed to these 'spare' receptors (Figure 10.2). The LH receptor in the testis have been solubilized (Dufau and Catt, 1973; Dufau *et al.*, 1973a, 1974). Their equilibrium association constant is about $1 \times 10^{10}$M$^{-1}$, which is concordant with a plasma LH level of about $10^{-10}$M. The binding sites are destroyed by trypsin and phospholipase A but not by neuraminidase or phospholipase C (Catt *et al.*, 1974; Dufau *et al.*, 1974). This binding has been made the basis of a radio-ligand assay for LH (Catt *et al.*, 1972b). The amount of binding for LH/hCG is decreased by pretreatment of rats with hCG (Sharpe,

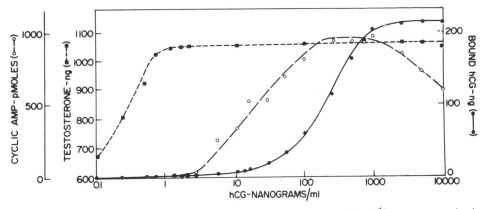

Figure 10.2 The concentrations of hCG needed to produce maximal binding (●), to cause maximal stimulation of cyclic AMP formation (O) and testosterone formation (■) in rat testis *in vitro*. (Reproduced from Catt *et al.*, 1974)

1976) or with bromocriptine, which reduces serum prolactin concentration (Aragona *et al.*, 1977).

**10.1.2.2    Activation of adenyl cyclase and protein kinases by LH**    In the testis, as in several other steroid-producing tissues, e.g. the adrenal and the corpus luteum, stimulation by the trophic hormones enhances adenyl cyclase activity and the formation of cyclic AMP (Murad *et al.*, 1969; Kuehl *et al.*, 1970; Dorrington *et al.*, 1972). Steroidogenesis can be increased by cyclic AMP and particularly by its dibutyryl derivative, which enters cells much more readily (Sandler and Hall, 1966a, b; Connell and Eik-Nes, 1968; Shin, 1967). However it is rather puzzling that the minimum doses of LH necessary to increase cyclic AMP synthesis are about 15 times larger than the minimum doses needed to stimulate testosterone secretion (Figure 10.2), but less than the amount needed to saturate the receptors (Catt and Dufau, 1973; Dufau and Catt, 1973; Catt *et al.*, 1974). This could either be due to an alternative pathway of stimulation or the existence of small, localized intracellular 'pools' of cyclic AMP which can be increased by hCG without producing detectable changes in the cyclic AMP in the whole tissue. This is so even with hCG from which sialic acid or sialic acid and galactose have been removed, thereby reducing their activity *in vivo* , or in the presence of inhibitors of phosphodiesterase inhibitors such as theophylline or 3-isobutyl-1-methyl-xanthine. However, these inhibitors do decrease the dose of hCG necessary for stimulation of both testosterone and cyclic AMP synthesis (Catt *et al.*, 1973) an observation which does support the role of cyclic AMP in the action of LH.

The response of testosterone synthesis always takes at least 20 minutes regardless of the dose of hCG whereas at higher doses, a response in cyclic AMP synthesis can be detected in less than 1 minute (Rommerts *et al.*, 1972, 1973; Cooke *et al.*, 1972, 1974; Catt *et al.*, 1974). This suggests that a complex series of metabolic events are interposed between the formation of cyclic AMP and the release of testosterone. The effect of hCG on testosterone synthesis is abolished in the presence of actinomycin D or cycloheximide, without affecting the response in cyclic AMP (Catt *et al.*, 1974; Cooke *et al.*, 1975), suggesting that a protein kinase is involved. The activity of protein kinase in total homogenate and in subcellular fractions of interstitial tissue from rat testes was increased by the addition of LH and 3-isobutyl-1-methyl-xanthine (an

## scheme for control of steroidogenesis

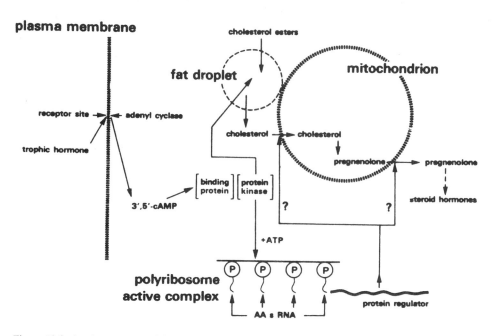

Figure 10.3  A scheme summarizing the mechanism of action of LH on the interstitial cells of the testis. (Reproduced from Rommerts *et al.*, 1974)

inhibitor of phosphodiesterase) during incubation. Cyclic AMP also increased protein kinase activity in the absence of LH (Cooke *et al.*, 1976). The details of the intermediate steps which probably also involve an apparently obligatory effect upon RNA transcription, are still under investigation (see Figure 10.3; Rommerts *et al.*, 1974). The protein kinase presumably catalyses the formation of a messenger protein which then activates the conversion of cholesterol to pregnenolone (see Section 6.1.1.3). It is not known how this is achieved, but it is fairly certain that LH acts on a specific step in the transformation of cholesterol to pregnenolone and does not increase the conversion of pregnenolone to testosterone (Hall, 1966, 1970; Sandler and Hall, 1966a, b).

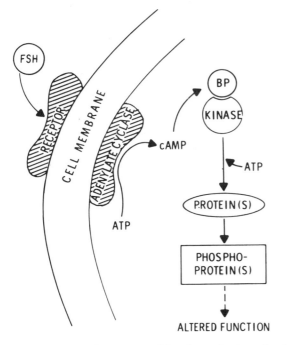

Figure 10.4 A scheme summarizing the mechanism of action of FSH on the Sertoli cells. BP, binding protein. (Reproduced from Means *et al.*, 1976)

## 10.2    Follicle stimulating hormone (FSH)

### 10.2.1    Site of action

Apart from a few reports that FSH potentiates the effect of LH on steroid production by the testis (see Section 6.1.5.1), the effects of FSH have been shown to occur in the tubules. However there is still the question of which cells are involved. In experiments concerned with the restoration of spermatogenesis in testes which have regressed following hypophysectomy or oestrogen treatment (which suppresses the pituitary by negative feedback—see Section 6.1.6.5) it was found that cell development proceeded as far as the young spermatids in animals treated with testosterone, but that FSH was necessary for the completion of spermiogenesis (Steinberger, 1971). However this is not the same as saying that FSH has direct action on the developing spermatids.

Indeed it would appear most unlikely that FSH can penetrate through the walls of the seminiferous tubule (see Section 8.7.3) and therefore it would appear that FSH must act on Sertoli cells and/or spermatogonia. It is hard to see how an action on the spermatogonia could produce the effects on spermatid maturation, so a direct effect on the Sertoli cells seems the most likely, and was suggested by Murphy (1965a) on histological grounds. An additional effect on spermatogonia is still possible as are indirect effects on the other germinal cells via the Sertoli cells.

The final convincing proof of action on the Sertoli cells came with the demonstration of effects of FSH on the tubules of rats with no germ cells but only Sertoli cells (see Means *et al.*, 1976b), and on Sertoli cells in culture (Fritz *et al.*, 1975a, b).

However the curious fact remains that an action of FSH on the tubules can be demonstrated only in very young animals (15- to 24-day old rats) or in adult animals about 10 days after hypophysectomy although FSH does bind to the tubules of older rats (Means *et al.*, 1977). No convincing explanation has yet been advanced for this anomaly. A partial explanation can be derived from the development at about 30 days of age of a specific testicular isoenzyme of phosphodiesterase (Section 9.2.4.3) which blocks the *in vitro* actions of FSH by breaking down cyclic AMP before it can act. FSH will produce effects in the testis of older rats *in vitro* in the presence of inhibitors of this enzyme. However, if cyclic AMP is an obligatory step in the function of FSH, the existence of this enzyme does not change the fact that FSH does not act in older animals *in vivo*; it only explains why. The possibility remains that FSH has an action only at puberty or at the beginning of the breeding season.

## 10.2.2    Mechanism of action of FSH

**10.2.2.1    Binding of FSH to tubular cells**    Several studies in rats using ovine FSH labelled with radioactive iodine and other markers suggested that FSH became specifically associated with the Sertoli cells (Mancini *et al.*, 1967; Castro *et al.*, 1970, 1972). These studies can be criticized on several grounds, not least of which is the enormous amounts of FSH injected, but these earlier workers may have been right for the wrong reason. When tritiated human FSH of high biological activity became available, Means and Vaitukaitis (1972) showed that in rats, this hormone bound specifically to cells of the seminiferous tubules but not to isolated Leydig cells or other tissues. Later studies showed that $^{125}$I-FSH also bound to the receptors (Desjardins *et al.*, 1974; Steinberger *et al.*, 1974; Rabin, 1974; Bhalla and Reichert, 1974a, c; Cheng, 1975) and this binding has been used as the basis for a radioreceptor assay for FSH (Reichert and Bhalla, 1974). The testicular tubular receptors for FSH were saturable, of high affinity ($K_d \simeq 10^{-10}$) but low capacity, temperature dependent and hormone specific. Addition of large amounts of FSH displaced the label, and it could also be released in a biologically active form by mild acid treatment, suggesting that it was bound to the membranes, not the interior of the cells. This suggestion is supported by the observation that isolated plasma membranes of tubular cells bound the hormone, with hormone specificity, and similar fractions from other tissues were ineffective. Removal of between 5 and 100 per cent of sialic acid from the FSH does not affect the binding (Means, 1973).

The amount of labelled FSH bound per testis in rats increases with age until about 15 days and then remains constant (Means and Huckins, 1974; Means, 1975; Means *et al.*, 1976b; Fakunding *et al.*, 1976). This time schedule corresponds with the numbers of Sertoli cells in the testis, and furthermore, the binding of FSH is unaffected by hypophysectomy (Steinberger *et al.*, 1974) which reduces the number of germ cells but not the Sertoli cells. Therefore it has been suggested that the Sertoli cells are responsible for the binding of the FSH, and this has been confirmed with isolated Sertoli cells (Heindel *et al.*, 1975). The tubules of rats X-irradiated *in utero* also bind FSH. In these rats, the tubules contain only Sertoli cells, and although the testes are smaller, they bind as much FSH as normal testes (Means and Huckins, 1974; Means, 1975; Means et al., 1976b, 1977; Fakunding et al., 1976).

The ability of the plasma membrane to bind FSH is destroyed by incubation with trypsin or pronase, so the active component would appear to be protein; if the trypsin treated cells are then exposed to an inhibitor of trypsin prepared from soybeans, the ability to bind FSH returns after an incubation of between 1 and 2 hours. It would therefore appear that the cells can resynthesize the binding sites in vitro, particularly as the recovery of binding sites can be blocked with cycloheximide, an inhibitor of protein synthesis. Treatment with phospholipase A also produces a loss of binding ability, but it is not clear whether this is because of general damage to the cell membrane or specific damage to the binding sites (Means, 1973, 1975). FSH receptors have been solubilized but there is still some doubt about their specificity when in solution (Bhalla and Reichert, 1974b, c).

**10.2.2.2    Metabolic effects of FSH**    Once the FSH is bound to the membrane of the Sertoli cell, the next step seems to be the stimulation of adenyl cyclase and the formation of cyclic AMP. This effect can be produced by FSH in various testis preparations (Murad et al., 1969; Kuehl et al., 1970; Dorrington et al., 1972, 1974; Means, 1973, 1975; Braun and Sepsenwol, 1974, 1976; Dorrington and Fritz, 1974; Means et al., 1974, 1976b, 1977; Means and Huckins, 1974; Fakunding et al., 1976) and, most significantly with isolated Sertoli cells (Dorrington et al., 1974, 1975; Heindel et al., 1975). As with LH, the concentration needed to produce maximal effects on adenyl cyclase is much less than that needed to saturate the binding sites (Means and Huckins, 1974). The next step is to stimulate protein kinase activity in the tubules, resulting in the phosphorylation of one or more species of proteins (Figure 10.4). This phosphorylation of proteins can result in altered function and if the proteins concerned have important structural or enzymic roles, this may be the basis of the biological actions of the hormone concerned. Again a stimulation of protein kinase could be demonstrated in vitro by adding FSH to incubated isolated tubules, and this is specific to FSH (Means, 1973; Means et al., 1974, 1976a, b, 1977). Maximal stimulation is seen at a concentration of FSH even lower than the concentration producing a maximal effect on adenyl cyclase, which in turn is greater than the concentration giving maximum binding (Figure 10.5; Means and Huckins, 1975). Inhibitors of protein synthesis have no effect on the stimulation of protein kinase (Means et al., 1974). Two soluble forms of protein kinase can be demonstrated in the Sertoli cells, and the ratio of their activity varies with age of the rat. However, both forms appear to respond to FSH (Means et al., 1976a, b, 1977; Lee et al., 1976).

The effects so far described can all be demonstrated in vitro. The next events in the series can only be seen in tissues of rats which had been injected with FSH, and cannot be produced by FSH added in vitro. The first of these is the stimulation of the production of rapidly labelled nuclear RNA which begins within 15 minutes of injection, is maximal by 30 minutes and has returned to control values by 60 minutes. The activity of polymerase II (the enzyme which catalyses the synthesis of adenine

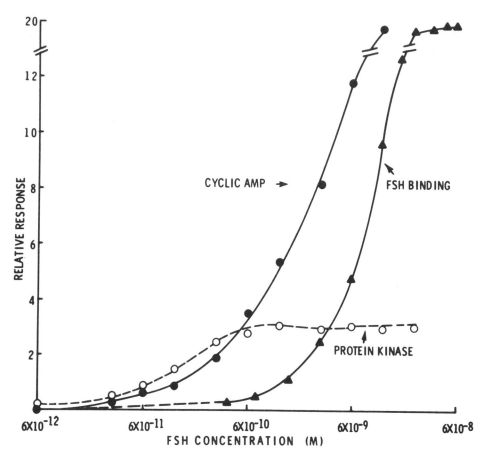

Figure 10.5 The concentrations of FSH needed to stimulate protein kinase activity (○), to increase the production of cyclic AMP (●) and to produce maximal binding (▲). (Reproduced from Means and Huckins, 1974)

and uridine-rich heterodisperse nuclear RNA) is also stimulated within 15 to 30 minutes after injection but this reaches a maximum 2h after injection and is still apparent after 6h (Figure 10.6). These events were also coincident with the stimulation of the synthesis of androgen-binding protein (see Section 10.2.2.3) and with an increase in the synthesis of translatable messenger RNA by the Sertoli cells. Then, 30 minutes after injection of FSH, chromatin template activity is increased. Again, this is a transitory effect, and control values are reached 2h after injection. The next change which appears involves polymerase I, the nucleolar enzyme which catalyses the synthesis of ribosomal RNA; the activity of the enzyme begins to increase about 1h after injection, reaches a peak after 4h and is back to control levels at 6h (Figure 10.7). This as one might expect is followed by an increase in ribosomal RNA synthesis, as determined by the incorporation of tritiated uridine into ribosomal RNA.

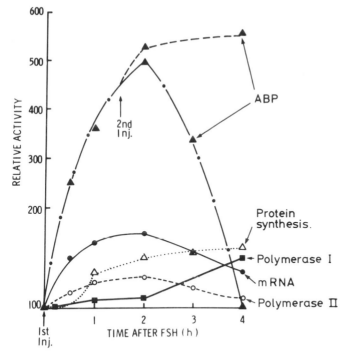

Figure 10.6     The effect of FSH on androgen binding protein (ABP), RNA and protein synthesis and polymerase I and II activities in the testes of Sertoli-cell only rats. FSH was injected intravenously at various times before removing the tissue. (Reproduced from Means and Tindall, 1975)

This is maximal at 4h and has returned to control values by 8h (Means, 1970, 1971, 1975; Means and Tindall, 1975; Means *et al.*, 1976a, 1977).

The rise in polymerase I is closely related to an increase in protein synthesis measured by incubating teased out testis with [$^{14}$C] lysine (Means and Hall, 1967, 1968, 1971). Means (1975) has suggested that this is not associated with a stimulation in the transport of amino-acids, but recently Irusta and Wasserman (1976) have obtained some contrary results. Means and Hall (1967) also suggest that an increase in amino-acid activation is not involved because FSH does not increase the incorporation of labelled amino-acids into aminoacyl transfer RNA. The increase in protein synthesis persists for at least 12 hours, and like all the other effects can only be demonstrated in the testes of rats between 15 and 24 days, or in hypophysectomized adults. FSH also increases the incorporation of $^{14}$C-valine into all polyribosome fractions, while affecting neither the proportion of testicular ribosomes appearing as polyribosomes nor the relative proportion of the various polyribosomal species. FSH has no effect on the electrophoretic pattern of labelled proteins formed by the tubules *in vivo* or by isolated polyribosomes from the tubules incubated with $^{14}$C-lysine, suggesting that FSH has a general effect on protein synthesis rather than a specific one (Means and Hall, 1969, 1971). All these effects of FSH can be demonstrated in the testes of rats with only Sertoli cells in the tubules, produced by X-irradiation *in utero* (usually referred to as Sertoli-cell only or SCO rats) (Means, 1975; see Section 11.1.1).

Human FSH injected into 9-day old mice caused an increase in [³H] lysine incorporation into protein, measured *in vivo* by injecting the lysine into the animal (Davies *et al.*, 1975). The time course of the *in vivo* incorporation is slower than that *in vitro*.

**10.2.2.3   FSH and androgen-binding protein**   The testis has been shown to produce a specific androgen-binding protein (ABP) and to secrete it into the fluid in the lumina of the seminiferous tubules. This protein is chemically indistinguishable from the testosterone-binding globulin found in the serum of many species but not rat (see Section 8.2.1). The production of this protein can be stimulated by FSH in (1) the testes of immature rats; (2) the testes of hypophysectomized adult rats; (3) the

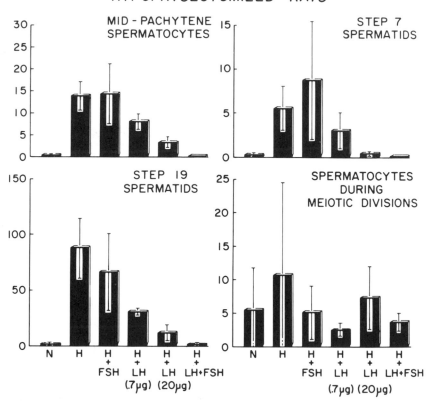

DEGENERATION OF GERMINAL CELLS IN NORMAL (N), HYPOPHYSECTOMIZED (H) AND HORMONE TREATED HYPOPHYSECTOMIZED RATS

Figure 10.7 The result of hypophysectomy and the injection of FSH and/or LH on degeneration of germ cells at the particular stages shown. Values shown are means with standard deviations. (Reproduced from Russell and Clermont, 1976)

testes of Sertoli-cell only rats; and (4) rat Sertoli cells in culture (Hansson *et al.*, 1973, 1974, 1975a, b, c; French *et al.*, 1974; Fritz *et al.*, 1974, 1975; Tindall *et al.*, 1974; Sanborn *et al.*, 1974, 1975; Means and Tindall, 1975; Ritzen, 1975; Steinberger *et al.*, 1975). ABP production can also be stimulated in Sertoli cell only rats with 8-bromocyclic AMP, a derivative of cyclic AMP which enters cells readily but is not broken down by the enzyme phosphodiesterase. The timing of the rise in ABP is closely coincident with the rise in polymerase II (see above), i.e. rising to a peak 2h after injection and returning to control levels by 4 hours (Figure 10.7). The response can be prolonged by the injection of a second dose of FSH 1·5h after the first and is prevented by inhibitors of protein synthesis such as cycloheximide or inhibitors of messenger RNA synthesis such as actinomycin D.

The timing of the ABP response is thus quite different from that of the protein synthesis response which is still proceeding at maximal rates 4h after a single injection of FSH (Means and Tindall, 1975).

The most recent studies showed that highly purified human FSH had no effect on ABP synthesis in 10- to 14-day old rats, but that testosterone or hormones which stimulate the production of testosterone produced a specific stimulation. The authors suggested that the stimulation by FSH seen previously was due to impurities in the FSH preparations used, and that intratesticular concentration of testosterone regulates the production of ABP in the testis (Tindall and Means, 1976).

**10.2.2.4    FSH and cellular events in the testis**    While FSH has never been shown to increase sperm production by the mature testis, certain cellular events have been demonstrated in the testes of immature rats and mice treated with FSH. Wet weight of the testis of 20-day old rats is increased between 2 and 4h after an injection of FSH, dry weight is increased between 3 and 6h (Means, 1975), and fluid secretion by the testis is slightly increased by 3 days treatment (shorter periods were not examined in this last experiment; Setchell *et al.*, 1973). Wet weight and protein content of the testes of 9-day old mice are increased 22h after an injection of FSH (Davies *et al.*, 1975). Greep *et al.* (1936) reported an increase in mitotic activity and an increase in primary spermatocytes after six daily injections of FSH in 35-day old rats, hypophysectomized 7 days before the beginning of treatment. Subsequent investigations by Mills and Means (1972) have shown an increase in the numbers of primary spermatocytes in intact immature rats after 5 days of FSH treatment beginning when the rats were 15-days old. There also appeared to be an increase in mitotic activity, due to a decrease in the extent of degeneration of the type A spermatogonia which normally reaches about 50 per cent in rats of this age, mainly between $A_4$ and intermediate spermatogonia. Nine hours after a single injection of FSH, degeneration of A spermatogonia had fallen to less than 10 per cent (Means, 1975). This observation was not confirmed by Russell and Clermont (1977) who found that FSH + LH reduced to normal levels the increased degeneration after hypophysectomy in adult rats of mid-pachytene spermatocytes, and stages 7 and 19 spermatids; LH alone had some effect but FSH had none. The degeneration of type A spermatogonia, like the degeneration of spermatocytes during the meiotic divisions, was apparent in normal animals but was not affected by hypophysectomy or hormone treatment (Figure 10.7). FSH treatment increased the numbers of spermatogonia and Sertoli cells precursors in 9-day old mice (Davies, 1971) and increased the mass of Sertoli cell cytoplasm and the number of polysomes per Sertoli cell; FSH + LH increased the size of the mitochondria in these cells (Davies, 1976). The incorporation of $^3$H-thymidine into spermatogonia and spermatocytes is increased by FSH in prepubertal rats, but not in adult rats, in which LH is effective (Ortavant et al., 1972).

However it must be emphasized that long-term treatment with rat FSH, human chorionic gonadotrophin or pregnant mare serum gonadotrophin did not change sperm production by rats (see Section 7.6.1.2).

## 10.3    Other pituitary hormones

Prolactin has been shown to increase $\beta$-glucuronidase activity in the testis (Evans, 1962) and to stimulate spermatogenesis in hereditary dwarf mice and in hypophysectomized mice treated with LH (Bartke and Lloyd, 1970; Bartke, 1971). It also potentiates the effect of LH on testosterone secretion by the testis of rats (Hafiez et al., 1972; Bartke and Dalterio, 1976; Bartke, 1976) and increases the concentration of esterified cholesterol in the testis (Bartke, 1969, 1971b). Treatment with prolactin also reduces the atrophy of the testes of hamsters exposed to short day-length (Bartke et al., 1975). Inhibitors of prolactin release reduce the concentration of esterified cholesterol in the testis and of testosterone in the plasma (Boys et al., 1970; Bartke, 1976). It was originally reported that prolactin did not bind to the testes (Turkington and Frantz, 1970) but more recent studies suggest that there was low, but definite specific binding in the rat testis which increased up to 70 days of age and then declined slightly. The amount of binding was much less than in the prostate (Aragano and Friesen, 1975).

In normal adult men, the concentrations of prolactin and testosterone in the peripheral circulation are correlated, and prolactin would therefore seem to have an effect on steroidogenesis (Rubin et al., 1975). On the other hand, very high levels of prolactin in serum can be associated with hypogonadism and impotence (Boyar et al., 1974; Thorner et al., 1974) and transplantation of prolactin-producing tumours into male rats leads to testicular atrophy (Fang et al., 1974). High levels of prolactin, produced by grafting several anterior pituitaries under the kidney capsule of rats and mice, caused a fall in serum LH and FSH but had no effect on the concentrations of testosterone or on testis weight (Bartke et al., 1977).

Growth hormone has been reported to act synergistically with LH in increasing androgen secretion (Lostroh et al., 1958; Woods and Simpson, 1961), with testosterone in stimulating spermatogonial divisions and testicular weight (Boccabella, 1963) and with FSH and testosterone in repopulating the seminiferous tubules after hypophysectomy (Lostroh, 1969).

## 10.4    Effect on the testis of immunization against gonadotrophins

Testicular function can be suppressed by either active or passive immunization against LH. In rats, testis size is reduced, spermatogenesis is reduced and deranged, the Leydig cells appear atrophic, the motility of epididymal spermatozoa is reduced and the accessory glands atrophy (Moudgal and Li, 1961; Hayashida, 1963; Wakabayashi and Tamaoki, 1966; Talaat and Laurence, 1971). In rabbits, the testes of actively immunized animals return to the abdominal cavity (Quadri et al., 1966; Pineda et al., 1967). If immature animals were immunized against LH, maturation of the gonads was prevented, with spermatogenesis arrested at the spermatogonia (Monastirsky et al., 1971a). Electron microscopic studies of passively immunized rats showed a reduction in the size of Leydig cells, with bizarre-shaped nuclei and a reduction in smooth endoplasmic reticulum, before there was any change in testis weight. There were also changes in the Sertoli cells, which retained spermatids longer than normal, and contained abnormally high amounts of lipid and vesicular mitochondial cristae, with a reduction in smooth endoplasmic reticulum. Mid-pachytene spermatocytes degenerated and the spermatid nuclei were not properly shaped during elongation (Madwha Raj and Dym, 1976).

Passive immunization of adult rats against FSH had no effects on testis weight, although the animals were less fertile (Turner and Johnson, 1971). When young rabbits were actively immunized against sheep FSH, their testes developed normally (Monastirsky et al., 1971a), but if newborn rats were treated with rabbit anti-ovine FSH antiserum for 35 days, their testes were smaller than normal, although spermatogenesis was 'normal' (Monastirsky et al., 1971b). Young (20-day old) rats treated for 14 days with rabbit antisera to rat FSH had testes only half normal size, with reduced numbers of spermatocytes and spermatids; testosterone was normal (Madwha Raj and Dym, 1976).

The effects of immunization can be confused by reaction of the body's homeostatic mechanism to the changes produced. For example, when anti-LH antiserum is given to an animal, testosterone concentration in blood falls and this stimulates the animals own pituitary to secrete more LH. Unless more antiserum is given, the increase in serum LH concentration can then stimulate the Leydig cells again. Also the hormone–antibody complex may have a prolonged biological half-life, but still retain biological activity. If this happens, immunization would actually intensify the action of the hormone (see Setchell and Edwards, 1975).

## 10.5    Miscellaneous endocrinological effects on the testis

The testis influences the thymus (see Section 6.1.5.6) and conversely, the thymus has been reported to have an influence on the testis. In genetically athymic ('nude') mice, spermatogenesis is deranged, testis weight is decreased (in absolute terms, but not in proportion to body weight) and androgen production (as reflected by seminal vesicle weight) is decreased (Flanagan, 1966; Shire and Pantalouris, 1974). Removal of the thymus of newborn rats leads initially to an increase and then to a decrease in plasma testosterone (Binimbi-Massengo et al., 1976) and extracts of thymus decrease the synthesis of DNA and RNA in testis fragments in vitro (Deschaux et al., 1974).

Pinealectomy leads to precocious puberty in rats (Thieblot and Blaise, 1965), and to increased secretion of testosterone (Kinson and Peat, 1971); it also abolishes testicular atrophy after blinding in hamsters (see Section 1.4). Pellets of melatonin or serotonin implanted in immature and adult rats lead to a decrease in plasma testosterone followed by a return to normal values or an increase; serotonin in adult rats (but not in equivalent doses in immature rats) reduced testis weight and interfered with spermatogenesis (Kinson and Liu, 1973; Liu and Kinson, 1973). The doses used in these experiments were mu‹ h smaller than those causing ischaemia (see Section 11.2.1) and both serotonin and melatonin can affect steroidogenesis by testis preparations *in vitro* (Ellis, 1972; Peat and Kinson, 1971). However, it is not possible to exclude an action *in vivo* via the pituitary gonadotrophins, although melatonin given to hypophysectomized rats treated with hCG produces a small reduction in testis weight (Debeljuk *et al.*, 1971). In contrast to these results, it was found that implants of melatonin or 5-methoxytryptophol prevented the regression of the testis of hamsters changed from a light: dark environment of 14:10 to 1:23h. (Reiter *et al.*, 1974, 1975.) A stimulatory effect of melatonin on spermatogenesis in young rats has also been suggested (Thieblot *et al.*, 1966).

## References to Chapter 10

Allanson, M., Hill, R. T. and McPhail, M. K. (1935) The effect of hypophysectomy on the reproductive organs of the male guinea pig. *J. Exptl. Biol.* **12**, 348–54.

Aragona, C. and Friesen, H. G. (1975) Specific prolactin binding sites in the prostate and testes in rats. *Endocrinology,* **97**, 677–84.

Aragona, C., Bohnet, H. G. and Friesen, H. G. (1977) Localization of prolactin binding in prostate and testis: the role of serum prolactin concentration on the testicular LH receptor. *Acta Endocrin.* **84**, 402–9.

Bartke, A. (1969) Prolactin changes cholesterol stores in the mouse testis. *Nature* **224**, 700–1.

Bartke, A. (1971a) Effects of prolactin on spermatogenesis in hypophysectomized mice. *J. Endocrinol.* **49**, 311–16.

Bartke, A. (1971b) Effects of prolactin and luteinizing hormone on the cholesterol stores in the mouse testis. *J. Endocrinol.* **49**, 317–24.

Bartke, A. (1976) Pituitary–testis relationship. Role of prolactin in the regulation of testicular function. *Prog. Reprod. Biol.* **1**, 136–52.

Bartke, A. and Dalterio, S. (1976) Effects of prolactin on the sensitivity of the testis to LH. *Biol. Reprod.* **15**, 90–3.

Bartke, A. and Lloyd, C. W. (1970) Influence of prolactin and pituitary isografts on spermatogenesis in dwarf mice and hypophysectomized rats. *J. Endocrinol.* **46**, 321–9.

Bartke, A., Croft, B. T. and Dalterio, S. (1975) Prolactin restores testosterone levels and stimulates testicular growth in hamsters exposed to short day-length. *Endocrinology* **97**, 1601–4.

Bartke, A., Smith, M. S., Michael, S. D., Peron, F. G. and Dalterio, S. (1977) Effects of experimentally-induced chronic hyperprolactinemia on testosterone and gonadotrophin levels in male rats and mice. *Endocrinology* **100**, 182–6.

Berswordt-Wallrabe, R. von, Steinbeck, H. and Neumann, F. (1968) Effect of FSH on the testicular structure of rats. *Endocrinologia* **53**, 35–42.

Bhalla, V. K. and Reichert, L. E. (1974a) Properties of follicle-stimulating hormone-receptor interactions. Specific binding of human follicle-stimulating hormone to rat testes. *J. Biol. Chem.* **249**, 43–51.

Bhalla, V. K. and Reichert, L. E. (1974b) Gonadotrophin receptors in rat testis. Interaction of an ethanol-soluble testicular factor with human follicle-stimulating hormone and luteinizing hormone. *J. Biol. Chem.* **249**, 7996–8004.

Bhalla, V. K. and Reichert, L. E. (1974c) FSH receptors in rat testes: chemical properties and solubilization studies. *Curr. Top. Molec. Endocrinol.* **1**, 201–20.

Binimbi-Massengo, Deschaux, P. and Fontanges, R. (1976) Contribution a l'etude de l'interaction thymus-testicules. *J. Physiol.* (Paris) **72**, 35A.

Boccabella, A. V. (1963) Reinitiation and restoration of spermatogenesis with testosterone propionate and other hormones after a long-term post-hypophysectomy regression period. *Endocrinology* **72**, 787–98.

Boyar, R. M., Kapen, S., Finkelstein, J. W., Perlow, M., Sassin, J. F., Fukushima, D. K., Weitzman, E. D. and Hellman, L. (1974) Hypothalamic-pituitary function in diverse hyper-prolactinemic states. *J. Clin. Invest.* **53**, 1588–98.

Boyns, A. R., Cole, E. N., Golder, M. P., Danutra, V., Harper, M. F., Brownsey, B., Cowley, T., Jones, G. E. and Griffiths, K. (1970) Prolactin studies with the prostate. In *Prolactin and Carconogenesis*, eds. A. R. Boyns and K. Griffiths, 207–16, Cardiff: Alpha Omega Alpha.

Braun, T. and Sepsenwol, S. (1974) Stimulation of $^{14}$C-cyclic AMP accumulation by FSH and LH on testis from mature and immature rats. *Endocrinology* **94**, 1028–33.

Braun, T. and Sepsenwol, S. (1976) LH and FSH stimulation of adenyl cyclase in seminiferous tubules from young rats: functional FSH and LH receptors unmasked by homogenization. *Mol. Cell Endocrinol.* **4**, 183–94.

Bressler, R. S. (1976) Dependence of Sertoli cell maturation on the pituitary gland in the mouse. *Am. J. Anat.* **147**, 447–56.

Bressler, R. S. and Ross, M. H. (1972) Differentiation of peritubular myoid cells of the testis: Effects of intratesticular implantation of newborn mouse testes into normal and hypophysec-tomized adults. *Biol. Reprod.* **6**, 141–59.

Castro, A. E., Seiguer, A. C. and Mancini, R. E. (1970) Electron microscopic study of the localization of labelled gonadotrophins in the Sertoli and Leydig cells of the rat testis. *Proc. Soc. Exptl. Biol. Med.* **133**, 582–6.

Castro, A. E., Alonso, A. and Mancini, R. E. (1972) Localization of follicle-stimulating and luteinizing hormones in the rat testis using immunohistological tests. *J. Endocrinol.* **52**, 129–36.

Catt, K. J. and Dufau, M. L. (1973) Interactions of LH and hCG with testicular gonadotrophin receptors. *Adv. Exptl. Med. Biol.* **36**, 379–418.

Catt, K. J., Tsuruhara, T. and Dufau, M. L. (1972a) Gonadotrophin binding sites of the rat testis. *Biochim. Biophys. Acta* **279**, 194–201.

Catt, K. J., Dufau, M. L. and Tsuruhara, T. (1972b) Radioligand-receptor assay of luteinizing hormone and chorionic gonadotrophin. *J. Clin. Endocrinol. Metab.* **34**, 123–32.

Catt, K. J., Tsuruhara, T., Mendelson, C., Ketelslegers, J-M. and Dufau, M. L. (1974)

Gonadotrophin binding and activation of the interstitial cells of the testis. *Curr. Top. Molec. Endocrinol.* **1**, 1–30.

Charreau, E. H., Dufau, M. L. and Catt, K. J. (1974) Multiple forms of solubilized gonadotrophin receptors from the rat testis. *J. Biol. Chem.* **249**, 4189–95.

Chemes, H. E., Rivarola, M. A. and Bergada, C. (1976a) Effect of HCG on the interstitial cells and androgen production in the immature rat testis. *J. Reprod. Fertil.* **46**, 279–82.

Chemes, H. E., Rivarola, M. A. and Bergada, C. (1976b) Effect of gonadotrophins and testosterone on the seminiferous tubules of the immature rat. *J. Reprod. Fertil.* **46**, 283–8.

Cheng, K-W. (1975) Properties of follicle-stimulating-hormone receptor in cell membranes of bovine testis. *Biochem. J.* **149**, 123–32.

Clermont, Y. and Harvey, S. C. (1965) Duration of the cycle of the seminiferous epithelium of normal hypophysectomized and hypophysectomized hormone treated albino rats. *Endocrinology* **76**, 80–9.

Clermont, Y. and Harvey, S. C. (1967) Effects of hormones on spermatogenesis in the rat. *Ciba Found. Colloq. Endocrinol.* **16**, 173–89.

Clermont, Y. and Morgentaler, H. (1955) Quantitative study of spermatogenesis in the hypophysectomized rat. *Endocrinology* **57**, 369–82.

Connell, G. M. and Eik-Nes, K. B. (1968) Testosterone production by rabbit testis slices. *Steroids* **12**, 507–16.

Cooke, B. A. and Kemp, A. J. W. C. M. van der (1976) Protein kinase activity in rat testis interstitial tissue. *Biochem. J.* **154**, 371–8.

Cooke, B. A., Beurden, W. M. O. van, Rommerts, F. F. G. and Molen, H. J. van der (1972) Effect of trophic hormones on 3',5'-cyclic AMP levels in rat testis interstitial tissue and seminiferous tubules. *FEBS letters* **25**, 83–6.

Cooke, B. A., Rommerts, F. F. G., Kemp, J. W. C. M. van der and Molen, H. J. van der (1974) Effects of luteinizing hormone, follicle stimulating hormone, prostaglandin $E_1$ and other hormones on adenosine-3',5'-cyclic monophosphate and testosterone production in rat testis tissues. *Mol. Cell Endocrinol.* **1**, 99–111.

Cooke, B. A., Janszen, F. H. A., Clotscher, W. F. and Molen, H. J. van der (1975) Effect of protein-synthesis inhibitors on testosterone production in rat testis interstitial tissue and Leydig-cell preparations. *Biochem. J.* **150**, 413–18.

Courot, M. (1962) Action des hormones gonadotropes sur le testicule de l'agneau impubère. Réponse particulière de la lignée sertolienne. *Ann. Biol. anim. Biochim. Biophys.* **2**, 157–62.

Courot, M. (1965) Action des hormones gonadotropes sur le testicule de l'agneau. *Ann. Biol. anim. Biochim. Biophys.* **5**, 145–9.

Courot, M. (1967) Endocrine control of the supporting and germ cells of the impuberal testis. *J. Reprod. Fertil.* Suppl. **2**, 89–100.

Courot, M. (1971) Etablissement de la spermatogénèse chez l'agneau (*Ovis aries*): étude expérimentale de son contrôle gonadotrope; importance des cellules de la lignée Sertolienne. Thèse de doctorat d'état ès-sciences naturelles, Universitè Paris VI.

Courot, M. (1976) Hormonal regulation of male reproduction (with reference to infertility in man) *Andrologia* **8**, 187–93.

Courot, M. and Ortavant, R. (1972) Contrôle gonadotrope de la spermatogénèse chez les mammifères. *Colloque de la Societe Nationale pour L'Etude de la Sterilite et de la Fecondite,* 1–18.

Cutuly, E. (1941) Androgen and spermatogenesis in the hypophysectomized guinea pig. *Proc. Soc. Exptl. Biol. N.Y.* **47**, 290–2.

Damme, M-P. Van, Robertson, D. M., Romani, P. and Diczfalusy, E. (1973) A sensitive *in vitro* bioassay method for luteinizing hormone (LH) activity. *Acta Endocrinol.* **74**, 642–58.

Damme, M-P. Van, Robertson, D. M. and Diczfalusy, E. (1974) An improved *in vitro* bioassay method for measuring luteinizing hormone (LH) activity using mouse Leydig cell preparations. *Acta Endocrinol.* **77**, 655–71.

Davies, A. G. (1971) Histological changes in the seminiferous tubules of immature mice following administration of gonadotrophins. *J. Reprod. Fertil.* **25**, 21–8.

Davies, A. G. (1976) Gonadotrophin-induced changes in the Sertoli cells of the immature mouse testis. *J. Reprod. Fertil.* **47**, 83–5.

Davies, A. G., Davies, W. E. and Sumner, C. (1975) Stimulation of protein synthesis *in vivo* in immature mouse testis by FSH. *J. Reprod. Fertil.* **42**, 415–22.

Debeljuk, L., Vilchez, J. A., Schnitman, M. A., Paulucci, O. A. and Feder, V. M. (1971) Further evidence for a peripheral action of melatonin. *Endocrinology,* **89**, 1117–9.

Debeljuk, L., Arimura, A., Shiino, M., Rennels, E. G. and Schally, A. V. (1973) Effects of chronic treatment with LH/FSH-RH in hypophysectomized pituitary-grafted male rats. *Endocrinology* **92**, 921–30.

Deschaux, P., Flores, J. L., Binimbi-Massengo and Fontanges, R. (1974) Etude *in vitro* de l'interaction thymus-testicules. *J. Physiol.* (Paris) **69**, 193A.

Desclin, J. and Ortavant, R. (1963) Influences des hormones gonadotropes sur la durée des processus spermatogénétiques chez le rat. *Ann. Biol. anim. Biochim. Biophys.* **3**, 329–42.

Desjardins, C., Zelesnik, A. J., Midgley, A. R. and Reichert, L. E. (1974) *In vitro* binding and autoradiographic localization of human chorionic gonadotrophin and follicle stimulating hormone in rat testes during development. *Curr. Top. Molec. Endocrinol.* **1**, 221–35.

Dorrington, J. H. and Fritz, I. B. (1974) Effects of gonadotropins on cyclic AMP production by isolated seminiferous tubule and interstitial cell preparations. *Endocrinology* **94**, 395–403.

Dorrington, J. H., Vernon, R. G. and Fritz, I. B. (1972) The effect of gonadotrophins on the 3′,5′-AMP levels of seminiferous tubules. *Biochem. Biophys. Res. Comm.* **46**, 1523–8.

Dorrington, J. H., Roller, N. F. and Fritz, I. B. (1974) The effects of FSH on cell preparations from the rat testis. *Curr. Top. Molec. Endocrinol.* **1**, 237–41.

Dufau, M. L. and Catt, K. J. (1973) Extraction of soluble gonadotrophin receptors from rat testis. *Nature New Biol.* **242**, 246–8.

Dufau, M. L., Catt, K. J. and Tsuruhara, T. (1971) Gonadotrophin stimulation of testosterone production by the rat testis *in vitro. Biochim. Biophys. Acta* **252**, 574–9.

Dufau, M. L., Catt, K. J. and Tsuruhara, T. (1972a) Biological activity of human chorionic gonadotrophin released from testis binding-sites. *Proc. Nat. Acad. Sci. US.* **69**, 2414–16.

Dufau, M. L., Catt, K. J. and Tsuruhara, T. (1972b) A sensitive gonadotrophin responsive system: Radioimmunoassay of testosterone production by the rat testis *in vitro. Endocrinology* **90**, 1032–40.

Dufau, M. L., Charreau, E. H. and Catt, K. J. (1973a) Characteristics of a soluble gonadotrophin receptor from the rat testis. *J. Biol. Chem.* **248**, 6973–82.

Dufau, M. L., Watanabe, K. and Catt, K. J. (1973b) Stimulation of cyclic AMP production by the rat testis during incubation with hCG *in vitro*. *Endocrinology* **92**, 6–11.

Dufau, M. L. Charreau, E. Ryan, D. and Catt, K. J. (1974) Characteristics of soluble gonadotropin receptors for LH and HCG. *Curr. Top. Molec. Endocrinol.* **1**, 47–77.

Ellis, L. C. (1972) Inhibition of rat testicular androgen synthesis *in vitro* by melatonin and serotonin. *Endocrinology* **90**, 17–28.

Evans, A. J. (1962) The *in vitro* effect of prolactin on $\beta$-glucuronidase in the testis of the rat. *J. Endocrinol.* **24**, 233–44.

Evans, H. M., Simpson, M. E. and Pencharz, R. I. (1937) An anterior pituitary gonadotrophic fraction (ICSH) specifically stimulating the interstitial tissue of testis and ovary. *Cold Spring Harbor Symp. Quant. Biol.* **5**, 229–38.

Fakunding, J. L., Tindall, D. J., Dedman, J. R., Mena, C. R. and Means, A. R. (1976) Biochemical actions of follicle-stimulating hormone in the Sertoli cell of the rat testis. *Endocrinology* **98**, 392–402.

Fang, V. S., Refetoff, S. and Rosenfield, R. L. (1974) Hypogonadism induced by a transplantable, prolactin-producing tumor in male rats: hormonal and morphological studies. *Endocrinology* **65**, 991–8.

Flanagan, S. P. (1966) 'Nude', a new hairless gene with pleiotropic effects in the mouse. *Genet. Res.* **8**, 295–309.

French, F. S., McLean, W. S., Smith, A. A., Tindall, D. J., Weddington, S. C., Petrusz, P., Sar, M., Stumpf, W. E. and Nayfeh, S. N. (1974) Androgen transport and receptor mechanisms in testis and epididymis. *Curr. Top. molec. Endocrinol.* **1**, 265–85.

Fritz, I. B. (1978) Sites of action of androgens and follicle stimulating hormone on cells of the seminiferous tubule. In *Biochemical Action of Hormones*, **5**, ed. G. Litwack, New York: Academic Press. [In press.]

Fritz, I. B., Kopec, B., Lam, K. and Vernon, R. G. (1974) Effects of FSH on levels of androgen binding protein in the testis. *Curr. Top. molec. Endocrinol.* **1**, 311–27.

Fritz, I. B., Rommerts, F. G., Louis, B. G. and Dorrington, J. H. (1975a) Regulation by FSH and dibutyryl cyclic AMP of the formation of androgen-binding protein in Sertoli cell-enriched cultures. *J. Reprod. Fertil.* **46**, 17–24.

Fritz, I. B., Louis, G. B., Tung, P. S., Griswold, M., Rommerts, F. G. and Dorrington, J. H. (1975b) Biochemical responses of cultured Sertoli cell-enriched preparations to follicle stimulating hormone and dibutyryl cyclic AMP. *Curr. Top. Molec. Endocrinol.* **2**, 367–82.

Gemzell, C. A. and Kjessler, B. (1964) Treatment of infertility after partial hypophysectomy with human pituitary gonadotrophins. *Lancet* **I**, 644.

Greep, R. O. and Fevold, H. L. (1937) The spermatogenic and secretory function of the gonads of hypophysectomized adult rats treated with pituitary FSH and LH. *Endocrinology* **21**, 611–18.

Greep, R. O., Fevold, H. L. and Hisaw, F. L. (1936) Effect of two hypophyseal gonadotropic hormones on the reproductive system of the male rat. *Anat. Rec.* **65**, 261–71.

Greep, R. O., Dyke, H. B. van and Chow, B. F. (1942) Gonadotropins of the swine pituitary. I. Various biological effects of purified thylokentrin (FSH) and pure metakentrin (ICSH). *Endocrinology* **30**, 635–49.

Griswold, M. D., Mably, E. R. and Fritz, I. B. (1976) FSH stimulation of DNA synthesis in

Sertoli cells in culture. *Mol. Cell Endocrinol.* **4**, 139–50.

Hafiez, A. A., Bartke, A. and Lloyd, C. W. (1972) The role of prolactin in the regulation of testis function: the synergistic effects of prolactin and luteinizing hormone on the incorporation of [1-$^{14}$C]-acetate into testosterone and cholesterol by testes from hypophysectomized rats *in vitro. J. Endocrinol.* **53**, 223–30.

Hall, P. F. (1963) The effect of interstitial cell-stimulating hormone on the bio-synthesis of testicular cholesterol from acetate-1-C$^{14}$. *Biochem.* **2**, 1232–7.

Hall, P. F. (1966) On the stimulation of testicular steroidogenesis in the rabbit by interstitial cell-stimulating hormone. *Endocrinology* **78**, 690–8.

Hall, P. F. and Eik-Nes, K. B. (1963) The influence of gonadotrophins *in vivo* upon the biosynthesis of androgens by homogenate of rat testis. *Biochim. Biophys. Acta* **71**, 438–47.

Hansson, V., Reusch, E., Trygstad, O., Torgersen, O., French, F. S. and Ritzen, E. M. (1973) FSH stimulation of testicular androgen binding protein (ABP). *Nature New Biol.* **246**, 56–9.

Hansson, V., Trygstad, O., French, F. S., McLean, W. S., Smith, A. A., Tindall, D. J., Weddington, S. C., Petrusz, P., Nayfeh, S. N. and Ritzen, E. M. (1974a) Androgen transport and receptor mechanisms in testis and epididymis. *Nature* **250**, 387–91.

Hansson, V., French, F. S., Weddington, S. C., Nayfeh, S. N. and Ritzen, E. M. (1974b) FSH stimulation of testicular androgen binding protein (ABP). *Curr. Top. molec. Endocrinol.* **1**, 287–90.

Hansson, V., Weddington, S. C., McLean, W. S., Smith, A. A., Nayfeh, S. N., French, F. S. and Ritzen, E. M. (1975a) Regulation of seminiferous tubular function by FSH and androgen. *J. Reprod. Fertil.* **44**, 363–75.

Hansson, V., Ritzen, E. M., French, F. S. and Nayfeh, S. N. (1975b) Androgen transport and receptor mechanisms in testis and epididymis. *Handbk. Physiol.* Sect. 7, **V**, 173–201.

Hansson, V., Weddington, S. C., Naess, O., Attremadal, A., French, F. S., Kotite, N., Nayfeh, S. N., Ritzen, E. M. and Hagenas, L. (1975c) Testicular androgen binding protein (ABP)—a parameter of Sertoli cell secretory function. *Curr. Top. molec. Endocrinol.* **2**, 323–36.

Hansson, V., Djoseland, O., Torgersen, O., Ritzen, E. M., French, F. S. and Nayfeh, S. N. (1976) Hormones and hormonal target cells in the testis. *Andrologia* **8**, 195–202.

Hartman, C. G., Millman, N. and Stavorski, J. (1960) Vasodilatation of the rat testis in response to human chorionic gonadotropin. *Fertil. Steril.* **1**, 443–53.

Hayashida, T. (1963) Inhibition of spermiogenesis, prostate and seminal vesicle development in normal animals with anti-gonadotrophic hormone serum. *J. Endocrinol.* **26**, 75–83.

Hearn, J. P. (1975) The role of the pituitary in the reproduction of the male tammar wallaby. *J. Reprod. Fertil.* **42**, 399–402.

Heindel, J. J., Rothenberg, R., Robinson, G. A. and Steinberger, A. (1975) LH and FSH stimulation of cyclic AMP in specific cell types isolated from the testis. *J. Cyclic Nucleotide Res.* **1**, 69–78.

Heller, C. G. and Leach, D. R. (1971) Quantification of Leydig cells and measurement of Leydig-cell size following administration of human chorionic gonadotrophin to normal men. *J. Reprod. Fertil.* **25**, 185–92.

Heller, C. G., Dalli, M. F., Pearson, J. E. and Leach, D. R. (1971) A method for the quantification of Leydig cells in man. *J. Reprod. Fertil.* **25**, 177–84.

Hill, M. and Parkes, A. S. (1933) Studies on the hypophysectomised ferret. II. Spermatogenesis. *Proc. Roy. Soc.* (London) B. **112**, 146–52.

Irusta, O. and Wassermann, G. F. (1974) Factors influencing the uptake of |α-$^{14}$C| aminoisobutyric acid by rat testes. *J. Endocrinol.* **60**, 463–71.

Johnson, B. H. and Ewing, L. L. (1971) Follicle-stimulating hormone and the regulation of testosterone secretion in rabbit testes. *Science* **173**, 635–7.

Johnsen, S. G. and Christiansen, P. (1968) Spermatogenesis and conception during HMG treatment of hypogonadotrophic hypogonadism. In *Gonadotrophins*, ed. E. Rosemberg, 515–25, Los Altos: Geron X.

Kinson, G. A. and Peat, F. (1971) The influences of illumination, melatonin and pinealectomy on testicular function in the rat. *Life Sciences,* **10**, 259–69.

Kinson, G. S. and Liu, C.-C. (1973) Testicular responses to melatonin and serotonin implanted peripherally in immature rats. *Life Sciences* **12**, 173–84.

Kretser, D. M. de (1967) Changes in the fine structure of human testicular interstitial cells after treatment with human gonadotrophins. *Z. Zellforsch. Mikroskop. Anat.* **83**, 344–58.

Kretser, D. M. de, Catt, K. J., Burger, H. and Smith, G. C. (1969) Radioautographic studies on the localization of $^{125}$I-labelled human luteinizing and growth hormone in immature male rats. *J. Endocrinol.* **43**, 105–11.

Kretser, D. M. de, Catt, K. J. and Paulsen, C. A. (1971) Studies on the *in vitro* testicular binding of iodinated luteinizing hormone in rats. *Endocrinology* **88**, 332–7.

Kuehl, F., Patanelli, D. J., Tarnoff, J. and Humes, J. L. (1970) Testicular adenyl cyclase: stimulation by the pituitary gonadotropins. *Biol. Reprod.* **2**, 154–63.

Lacy, D. and Lofts, B. (1965) Studies on the structure and function of the mammalian testis. I. Cytological and histochemical observations after continuous treatment with oestrogenic hormone and the effects of FSH and LH. *Proc. Roy. Soc.* (London) B. **162**, 188–97.

Leblond, C. P. and Nelson, W. O. (1937) Modifications histologiques des organes de la souris après hypophysectomie. *Compt. Rend. Sean. Soc. Biol. Paris* **124**, 9–11.

Lee, P. C., Radloff, D., Schweppe, J. S. and Jungman, R. A. (1976) Testicular protein kinases. Characterization of multiple forms and ontogeny. *J. Biol. Chem.* **251**, 914–21.

Leidenberger, F. and Reichert, L. E. (1972) Studies on the uptake of human chorionic gonadotropin and its subunits by rat testicular homogenates and interstitial tissue. *Endocrinology* **91**, 135–43.

Leidl, W., Bentley, M. I. and Gass, G. H. (1976) Longitudinal growth of the seminiferous tubules in LH and FSH treated rats. *Andrologia* **8**, 131–6.

Liu, C.-C. and Kinson, G. A. (1973) Testicular gametogenic and endocrine responses to melatonin and serotonin peripherally administered to mature rats. *Contraception* **7**, 153–63.

Lostroh, A. J. (1963) Effect of follicle-stimulating hormone on spermatogenesis in Long-Evans rats hypophysectomized for six months. *Acta Endocrinol.* **43**, 592–600.

Lostroh, A. J. (1969) Regulation by FSH and ICSH (LH) of reproductive function in the immature male rat. *Endocrinology* **85**, 438–45.

Lostroh, A. J. and Li, C. H. (1957) Stimulation of the sex accessories of hypophysectomized male rats by non-gonadotrophic hormones of the pituitary gland. *Acta Endocrinol.* **25**, 1–16.

Lostroh, A. J., Squire, P. G. and Li, C. H. (1958) Bioassay of interstitial cell-stimulating hormone in the hypophysectomized male rat by the ventral prostate test. *Endocrinology* **62**, 833–42.

Lostroh, A. J., Johnson, R. and Jordan, C. W., Jr. (1963) Effect of ovine gonadotrophins and antiserum to interstitial cell-stimulating hormone on the testis of the hypophysectomized rat. *Acta Endocrinol.* **44**, 536–44.

MacLeod, J. (1970) The effects of urinary gonadotrophins following hypophysectomy and in hypogonadotrophic eunuchoidism. *Adv. Exptl. Med. Biol.* **10**, 577–86.

MacLeod, J., Pazianos, A. and Ray, B. (1966) The restoration of human spermatogenesis and of the reproductive tract with urinary gonadotrophins following hypophysectomy. *Fertil. Steril.* **17**, 7–23.

McPhail, M. K. (1933) The adaptation of parapharyngeal hypophysectomy to the guinea-pig and hedgehog. *Proc. Roy. Soc.* (London) B. **114**, 10–20.

McPhail, M. K. (1935) Hypophysectomy of the cat. *Proc. Roy. Soc.* (London) B. **117**, 45–63.

Maddock, W. O. and Nelson, W. O. (1952) The effects of chorionic gonadotrophin in adult men: Increased estrogen and 17-ketosteroid excretion, gynaecomastia, Leydig cell stimulating and seminiferous tubule damage. *J. Clin. Endocrinol. Metab.* **12**, 985–1013.

Madhwa Raj, H. G. and Dym, M. (1976) The effects of selective withdrawal of FSH or LH on spermatogenesis in the immature rat. *Biol. Reprod.* **14**, 489–94.

Mancini, R. E., Castro, A. and Seiguer, A. C. (1967) Histologic localization of follicle-stimulating and luteinizing hormones in the rat testis. *J. Histochem. Cytochem.* **15**, 516–25.

Mancini, R. E., Seiguer, A. C. and Perez Lloret, A. (1968) The effect of gonadotropins on the testis of hypophysectomized patients. In *Gonadotropins*, ed. E. Rosemberg, 505–12, Los Altos: Geron X.

Means, A. R. (1970) Early effects of FSH upon testicular metabolism. *Adv. Exptl. Med. Biol.* **10**, 301–9.

Means, A. R. (1971) Concerning the mechanism of FSH action: rapid stimulation of testicular synthesis of nuclear RNA. *Endocrinology* **89**, 981–9.

Means, A. R. (1973) Specific interaction of $^3$H-FSH with rat testis binding sites. *Adv. Exptl. Med. Biol.* **36**, 431–48.

Means, A. R. (1975) Biochemical effects of follicle stimulating hormone on the testis. *Handbook Physiol.* Sect. 7, Vol. V, 203–18.

Means, A. R. and Hall, P. F. (1967) Effect of FSH on protein biosynthesis in the testes of the immature rat. *Endocrinology* **81**, 1151–60.

Means, A. R. and Hall, P. F. (1968) Protein biosynthesis in the testis: I. Comparison between stimulation by FSH and glucose. *Endocrinology* **82**, 597–602.

Means, A. R. and Hall, P. F. (1969) Protein biosynthesis in the testis. V. Concerning the mechanism of stimulation by follicle-stimulating hormone. *Biochem.* **8**, 4293–8.

Means, A. R. and Hall, P. F. (1971) Protein biosynthesis in the testis. VI. Action of follicle-stimulating hormone on polyribosomes in immature rats. *Cytobios* **3**, 17–24.

Means, A. R. and Huckins, C. (1974) Coupled events in the early biochemical actions of FSH on the Sertoli cells of the testis. *Curr. Top. Molec. Endocrinol.* **1**, 145–65.

Means, A. R. and Tindall, D. J. (1975) FSH-induction of androgen binding protein in testes of Sertoli-cell only rats. *Curr. Top. Molec. Endocrinol.* **2**, 383–98.

Means, A. R. and Vaitukaitis, J. (1972) Peptide hormone receptors; specific binding of ³H-FSH to testis. *Endocrinology* **90**, 39–46.

Means, A. R., MacDougall, E., Soderling, T. R. and Corbin, J. D. (1974) Testicular adenosine 3′,5′-monophosphate-dependent protein kinases: regulation by follicle stimulating hormone. *J. Biol. Chem.* **249**, 1231–8.

Means, A. R., Fakunding, J. L. and Tindall, D. J. (1976a) Follicle stimulating hormone regulation of protein kinase activity and protein synthesis in testis. *Biol. Reprod.* **14,** 54–63.

Means, A. R., Fakunding, J. L., Huckins, C., Tindall, D. J. and Vitale, R. (1976b) Follicle-stimulating hormone, the Sertoli cell and spermatogenesis. *Rec. Progr. Horm. Res.* **32**, 477–522.

Means, A. R., Dedman, J. R., Fakunding, J. L. and Tindall, D. J. (1978) Mechanism of action of FSH in the male rat. In *Hormone Receptors and Mechanism of Action*, eds. B. W. O'Malley and L. Birnbaumer, New York: Academic Press. [In press.]

Mills, N. C. and Means, A. R. (1972) Sorbitol dehydrogenase of rat testis: changes of activity during development, after hypophysectomy and following gonadotrophic hormone administration. *Endocrinology* **91**, 147–56.

Monastirsky, R., Laurence, K. A. and Tovar, E. (1971a) The effects of gonadotrophin immunization of prepubertal rabbits on gonadal development. *Fertil. Steril.* **22**, 318–25.

Monastirsky, R., Laurence, K. A. and Tovar, E. (1971b) Quoted by Laurence, K. A. *et al.* (1972) Application of immunologic techniques to the study of gonadotropin action. In *Gonadotropins*, eds. B. B. Saxena, C. G. Belling and H. M. Gandy, 329–34, New York: Wiley.

Moudgal, N. R. and Li, C. H. (1961) An immunochemical study of sheep pituitary interstitial cell-stimulating hormone. *Arch. Biochem. Biophys.* **95**, 93–8.

Moyle, W. R. and Ramachandran, J. (1973) Effect of LH on steroidogenesis and cyclic AMP accumulation in rat Leydig cell preparations and mouse tumour Leydig cells. *Endocrinology* **93**, 127–34.

Murad, F., Strauch, B. S. and Vaughan, M. (1969) The effect of gonadotropins on testicular adenyl cyclase. *Biochim. Biophys. Acta* **177**, 591–8.

Murphy, H. D. (1965a) Sertoli cell stimulation following intratesticular injections of FSH in the hypophysectomized rat. *Proc. Soc. Exptl. Biol. Med.* **118**, 1202–5.

Murphy, H. D. (1965b) Intratesticular assay of follicle-stimulating hormone in hypophysectomized rats. *Proc. Soc. Exptl. Biol. Med.* **120**, 671–5.

Musto, N., Hafiez, A. A. and Bartke, A. (1972) Prolactin increases 17β-hydroxysteroid dehydrogenase activity in the testis. *Endocrinology* **91**, 1106–8.

Ortavant, R. and Courot, M. (1964) Problèmes concernant l'action des hormones gametocinetiques sur la spermatogénèse des mammifères. *Arch. Biol.* (Liege) **75**, 625–67.

Ortavant, R. and Courot, M. (1967) Action des hormones gonadotropes sur la lignée germinale mâle adulte. *Arch. Anat. Microscop. Morphol. Exptl.* **56**, 111–24.

Ortavant, R., Courot, M., de Reviers, M-T. (1968) Activités spécifiques des differentes FSH et LH sur le testicule des mammifères. In *La specificité zoologique des hormones hypophysaires et de leurs activités*, Coll. Int. CNRS, **177**, 369–79.

Ortavant, R., Courot, M. and Hochereau-de Reviers, M. T. (1972) Gonadotrophic control of tritiated thymidine incorporation in the germinal cells of the rat testis. *J. Reprod. Fertil.* **31**,

451–3.

Peat, F. and Kinson, G. A. (1971) Testicular steroidogenesis *in vitro* in the rat in response to blinding pinealectomy and to the addition of melatonin. *Steroids* **17**, 251–64.

Pineda, M. H., Lueker, D. C., Faulkner, L. C. and Hopwood, M. L. (1967) Atrophy of rabbit testes associated with production of antiserum to bovine luteinizing hormone. *Proc. Soc. Exptl. Biol. N.Y.* **125**, 665–8.

Podesta, E. J., Dufau, M. and Catt, K. J. (1976) Characterization of two forms of cyclic 3′,5′-adenosine monophosphate dependent protein kinase in rat testicular interstitial cells. *Mol. cell Endocrinol.* **5**, 109–22.

Pokel, D., Moyle, W. R. and Greep, R. O. (1972) Depletion of esterified cholesterol in mouse testes and Leydig cell tumors by luteinizing hormone. *Endocrinology* **91**, 323–5.

Qazi, M. H., Romani, P. and Diczfalusy, E. (1974) Discrepancies in plasma LH activities as measured by radioimmunoassay and an *in vitro* bioassay. *Acta Endocrinol.* **77**, 672–85.

Quadri, S. K., Harbers, L. H. and Spies, H. G. (1966) Inhibition of spermatogenesis and ovulation in rabbits with antiovine LH rabbit serum. *Proc. Soc. Exptl. Biol. Med. N.Y.* **123**, 809–14.

Rabin, D. (1974) Binding of human FSH and its subunits to rat testis. *Curr. Top. Molec. Endocrinol.* **1**, 193–200.

Randolph, R. W., Lostroh, A. J., Gratarola, R., Squire, P. G. and Li, C. H. (1959) Effect of ovine interstitial cell-stimulating hormone on spermatogenesis in the hypophysectomized mouse. *Endocrinology* **65**, 433–41.

Reddi, A. H., Ewing, L. L. and Williams-Ashman, H. G. (1971) Protein phosphokinase reactions in mammalian testis: stimulatory effects of adenosine 3′,5′-cycle monophosphate on the phosphorylation of basic proteins. *Biochem. J.* **122**, 333–45.

Reichert, L. E. and Bhalla, V. K. (1974) Development of a radioligand tissue receptor assay for human follicle-stimulating hormone. *Endocrinology* **94**, 483–91.

Reiter, R. J., Vaughan, M. K., Blask, D. E. and Johnson, L. Y. (1974) Melatonin: its inhibition of pineal antigonadotrophic activity in male hamsters. *Science* (New York) **185**, 1169–71.

Reiter, R. J., Vaughan, M. K., Blask, D. E. and Johnson, L. Y. (1975) Pineal methoxyindoles: new evidence concerning their function in the control of pineal-mediated changes in the reproductive physiology of male golden hamsters. *Endocrinology* **96**, 205–13.

Ritzen, E. M., Hagenas, L., Hansson, V. and French, F. S. (1975) *In vitro* synthesis of testicular androgen binding protein (ABP): stimulation by FSH and androgen. *Curr. Top. Molec. Endocrinol.* **2**, 353–66.

Rommerts, F. F. G., Cooke, B. A., Kemp, J. W. C. M. van der and Molen, H. J. van der (1972) Stimulation of 3′,5′-cyclic AMP and testosterone production in rat testis *in vitro*. *FEBS letters* **24**, 251–4.

Rommerts, F. F. G., Cooke, B. A., Kemp, J. W. C. M. van der and Molen, H. J. van der (1973) Effect of luteinizing hormone on 3′,5′-cyclic AMP and testosterone production in isolated interstitial tissue of rat testis. *FEBS letters* **33**, 114–18.

Rommerts, F. F. G., Cooke, B. A. and Molen, H. J. van der (1974) The role of cyclic AMP in the regulation of steroid biosynthesis in testis tissue. *J. Steroid Biochem.* **5**, 279–85.

Rubin, R. T., Gouin, P. R., Lubin, A., Poland, R. E. and Pirke, K. M. (1975) Nocturnal increase of plasma testosterone in men: relation to gonadotropins and prolactin. *J. Clin. En-*

*docrinol. Metab.* **40**, 1027–33.

Russell, L. D. and Clermont, Y. (1977) Degeneration of germ cells in normal, hypophysectomized and hormone treated hypophysectomized rats. *Anat. Rec.* **187**, 347–66.

Russo, J. and Sacerdote, F. L. (1971) Ultrastructural changes induced by HCG in the Leydig cell of the adult mouse testis. *Z. Zellforsch. Mikroskop. Anat.* **112**, 363–70.

Sanborn, B. M., Elkington, J. S. H. and Steinberger, E. (1974) Properties of rat testicular androgen binding proteins. *Curr. Top. Molec. Endocrinol.* **1**, 291–310.

Sanborn, B. M., Elkington, J. S. H., Chowdhury, M., Tcholakian, R. K. and Steinberger, E. (1975) Hormonal influences on the level of testicular androgen-binding activity: effect of FSH following hypophysectomy. *Endocrinology* **96**, 304–12.

Sandler, R. and Hall, P. F. (1966a) Stimulation *in vitro* by adenosine-3',5'-cyclic monophosphate of steroidogenesis in rat testis. *Endocrinology* **79**, 647–9.

Sandler, R. and Hall, P. F. (1966b) The response of the rat testis to interstitial cell-stimulating hormone *in vitro*. *Comp. Biochem. Physiol.* **19**, 833–43.

Scowen, E. F. (1937–38) The effects of androsterone and testosterone on the testes of hypophysectomized guinea pigs. *Anat. Rec.* **70**, Suppl. 3, 71–2.

Setchell, B. P. and Edwards, R. G. (1975) The effect of immunization against gonadotrophins on the testis and male reproductive tract. In *Physiological Effects of Immunity against Reproductive Hormones*, eds. R. G. Edwards and M. H. Johnson, 167–80, London: Cambridge University Press.

Setchell, B. P. and Hinton, B. T. (1973) Bibliography with review on action on gonadotrophins on the testis in mammals. *Biolphy. Reprod.* **21**, 817–26 and 959–67.

Setchell, B. P., Duggan, M. C. and Evans, R. W. (1973) The effect of gonadotrophins on fluid secretion and sperm production by the rat and hamster testis. *J. Endocrinol.* **56**, 27–36.

Sharpe, R. M. (1976) hCG-induced decrease in availability of rat testis receptors. *Nature* **264**, 644–6.

Shikita, M. and Hall, P. F. (1967a) The action of human chorionic gonadotrophin *in vivo* upon microsomal enzymes of immature rat testis. *Biochim. Biophys. Act* **136**, 484–97.

Shikita, M. and Hall, P. F. (1967b) Action of human chorionic gonadotrophin *in vivo* upon microsomal enzymes in testes of hypophysectomized rats. *Biochim. Biophys. Acta* **141**, 433–5.

Shin, S. (1967) Studies on interstitial cells in tissue culture: Steroid biosynthesis in monolayers of mouse testicular interstitial cells. *Endocrinology* **81**, 440–8.

Shire, J. G. M. and Pantelouris, E. M. (1974) Comparison of endocrine function in normal and genetically athymic mice. *Comp. Biochem. Physiol.* **47A**, 93–100.

Simpson, M. E., Li, C. H. and Evans, H. M. (1944) Sensitivity of reproductive system of hypophysectomized 40 day male rats to gonadotrophic substances. *Endocrinology* **35**, 96–104.

Smith, P. E. (1927) The disabilities caused by hypophysectomy and their repair. *J. Am. med. Assn.* **88**, 158–61.

Smith, P. E. (1930) Hypophysectomy and replacement therapy in the rat. *Am. J. Anat.* **45**, 205–56.

Smith, P. E. (1942) Effect of equine gonadotropin on testes of hypophysectomized monkeys. *Endocrinology* **31**, 1–12.

Steinberger, A., Thanki, K. H. and Siegal, B. (1974) FSH binding in rat testes during maturation and following hypophysectomy. Cellular localization of FSH receptors. *Curr. Top. Molec. Endocrinol.* **1**, 177–91.

Steinberger, A., Heindel, J. J., Lindsey, J. N., Elkington, J. S. H., Sanborn, B. M. and Steinberger, E. (1975a) Isolation and culture of FSH responsive Sertoli cells. *Endocr. Res. Comm.* **2**, 261–72.

Steinberger, A., Elkington, J. S. H., Sanborn, B. M., Steinberger, E., Heindel, J. J. and Lindsey, J. N. (1975b) Culture and FSH responses of Sertoli cells isolated from sexually mature rat testis. *Curr. Top. Molec. Endocrinol.* **2**, 399–411.

Steinberger, E. (1971) Hormonal control of mammalian spermatogenesis. *Physiol. Rev.* **51**, 1–22.

Talaat, M. and Laurence, K. A. (1971) Impairment of spermatogenesis and libido through antibodies to luteinizing hormone. *Fertil. Steril.* **22**, 113–18.

Thieblot, L. and Blaise, S. (1965) Influence de la glande pinéale sur la sphère génitale. *Progr. Brain Res.* **10**, 577–84.

Thieblot, L. Berthelet, J. and Blaise, J. (1966) Effects de la mélatonine chez le rat mâle et femelle. 1. Action au niveau des gonade et des annexes. *Ann. Endocr.* **27**, 65–8.

Thorner, M. O., McNeilly, A. S., Hagan, C. and Besser, G. M. (1974) Long-term treatment of galactorrhoea and hypogonadism with bromocriptine. *Br. Med. J.* **II**, 419–22.

Tindall, D. J. and Means, A. R. (1976) Concerning the hormonal regulation of androgen binding protein in rat testis. *Endocrinology* **99**, 809–18.

Tindall, D. J., Schrader, W. T. and Means, A. R. (1974) The production of androgen binding protein by Sertoli cells. *Curr. Top. Molec. Endocrinol.* **1**, 167–75.

Turner, P. C. and Johnson, A. D. (1971) The effect of anti-FSH serum on the reproductive organs of the male rat. *Int. J. Fertil.* **16**, 169–76.

Turkington, R. W. and Frantz, W. L. (1970) The biochemical effects of prolactin. In *Prolactin and Carcinogenesis*, eds. A. R. Boyns and K. Griffiths, 39–47, Cardiff: Alpha Omega Alpha.

Vernon, R. G., Go, V. L. W. and Fritz, I. B. (1975) Hormonal requirements of the different cycles of the seminiferous epithelium during reinitiation of spermatogenesis in long-term hypophysectomized rats. *J. Reprod. Fertil.* **42**, 77–94.

Vitale, R., Fawcett, D. W. and Dym, M. (1973) The normal development of the blood–testis barrier and the effects of clomiphene and estrogen treatment. *Anat. Rec.* **176**, 333–44.

Wakabayashi, K. and Tamaoki, B-I, (1966) Influence of immunization with luteinizing hormone upon the anterior pituitary–gonad system of rats and rabbits, with special reference to histological changes and biosynthesis of luteinizing hormone and steroids. *Endocrinology* **79**, 477–85.

Wells, L. J. and Gomez, E. T. (1937) Hypophysectomy and its effect on male reproductive organs in a wild mammal with annual rut (Citellus). *Anat. Rec.* **69**, 213–27.

Wells, L. J. and Overholser, M. D. (1938) Sperm formation and growth of accessory reproductive organs in hypophysectomized ground squirrels in response to substances from blood and human urine. *Anat. Rec.* **72**, 231–47.

White, W. E. (1933–4) The effect of hypophysectomy of the rabbit. *Proc. Roy. Soc.* (London) B. **114**, 64–79.

Woods, M. C. and Simpson, M. E. (1961) Pituitary control of the testis of the hypophysec-tomized rat. *Endocrinology* **69**, 91–125.

# 11 Naturally occurring and induced dysfunctions of the testis

## 11.1 Radiation and temperature

### 11.1.1 X-irradiation

The extreme sensitivity of the testis to the damaging effects of X-rays has been known since the reports of Albers-Schönberg in 1903 (see also Bergonie and Tribondeau, 1904, 1905, 1906). Both whole body and localized irradiation produce damage to the testis. While some effects have been reported on androgen production, the most obvious effects are on the germinal cells although the mechanism of action is not known.

The A spermatogonia are the most sensitive cells and with a suitable dose (300–500R) in rats, the majority of these except the $A_s$ or $A_0$ spermatogonia are eliminated (see Regaud and Blanc, 1906; Regaud and Dubreuil, 1907; Gunsel, 1949b; Shaver, 1953a; Oakberg, 1955a, b, 1957a, b, 1959, 1975; Ellis, 1970; Dym and Clermont, 1970). As with all localized cellular lesions in the testis, the timing of the cellular development is not changed by the absence of some of the cells so a 'maturation depletion' moves through the epithelium and successive cellular stages are absent at increasing times after irradiation (Figure 11.1). The intermediate and B spermatogonia are less sensitive than the A spermatogonia but are often wrongly claimed to be the most sensitive cells because they are the cells missing 3 days after irradiation. Eventually the epithelium is repopulated from the stem-cell $A_s$ or $A_0$ sper-

| Days after x-irradiation | Types of germ cells | | | | | | | |
|---|---|---|---|---|---|---|---|---|
| | A | In B | Pl | L, Z | P (stages I–XIV) | Sptd 1–7 | Sptd 8–19 |
| 0 | + + + | + + + | + + + | + + + | + + + | + + | + + + |
| 4 | + | − | + | + + | + + + | + + | + + + |
| 8 | − | − | − | + | + + | + + | + + + |
| 13 | + | − | − | − | + | + + | + + |
| 19 | + | − | − | − | − | + + | + + |
| 26 | + + | + | + | − | − | + | + + |
| 39 | + + + | + + + | + + + | + + | − | − | + |
| 52 | + + + | + + + | + + + | + + + | + + + | − | − |
| 80 | + + + | + + + | + + + | + + + | + + + | + + + | + + + |

Figure 11.1 The cell types found in sections of rat testes at various times after local x-irradiation (300R). A, In B, various types of spermatogonia; Pl, L. Z and P, Pre-leptotene, leptotene, zygotene and pachytene spermatocytes; Sptd, spermatids. + + +, normal numbers of cells; + +, approximately 50 to 90% of normal numbers of cells; +, approximately 25 to 50% of normal numbers of cells; −, fewer than 25% of normal numbers of cells.
Note that soon after irradiation $A_1$ spermatogonia are almost absent and then a lesion moves through the germinal epithelium by 'maturation depletion'. The $A_1$ spermatogonia are replenished after about 40 days and by 84 days the tubules are normal.
(Reproduced from Dym and Clermont, 1970)

359

matogonia (Hertwig, 1938; Nebel and Murphy, 1960; Dym and Clermont, 1970).

Greater doses of X-irradiation will kill the other germinal cells, in the order spermatids, spermatocytes and spermatozoa, as the dose is increased (von Wattenwyl and Joel, 1941a, b, 1942; Oakberg and di Minno, 1955, 1960). Higher doses of X-irradiation also produce chromosome damage in spermatocytes and spermatids. Mature sperm can still fertilize eggs after 65,000R but induced genetic abnormalities prevent the embryos from developing (Chang et al., 1957). Furthermore, X-irradiation has been shown to cause reduced sperm numbers in semen and infertility (Pace et al., 1959, 1962; Freund and Murphee, 1960; Maqsood and Ashikowa, 1961; Willham and Cox, 1961; Cox and Willham, 1961; Freund and Borrelli, 1964).

The Sertoli cells are usually thought to be unaffected, but recent quantitative assessments have shown that their numbers may be halved by irradiation in immature animals (Courot, 1963; Kochar and Bateman, 1969; Erickson and Blend, 1976). The contractions of the seminiferous tubules and the peritubular myoid cells which produce them seem to be unaffected by irradiation (Kormano and Hovatta 1972; Hovatta and Kormano, 1974).

Testosterone production is not affected by moderate doses of radiation, but serum FSH and LH are elevated when the spermatids are missing from the germinal epithelium (see Section 6.2); larger doses of radiation do cause changes in steroidogenesis (Abbott, 1959; Schoen, 1964; Ellis and Berliner, 1963, 1964, 1967; Berliner et al., 1964; Berliner and Ellis, 1965; Simpson and Ellis, 1967; Ellis and van Kampen, 1970; Ellis, 1970).

If pregnant rats are irradiated (whole body 130R) on the 19th or 20th day of pregnancy, the germ cells (known as gonocytes at this stage) are killed although they appear not to be actively dividing at that time; the male offspring grow up with only Sertoli cells in their tubules. There is also a period of high sensitivity of the testis to X-rays associated with mitosis at 13 to 17 days of pregnancy. After birth, the testes become less sensitive to X-rays (Heald et al., 1939; Rugh and Jackson, 1958; Shaver, 1953b; Ershoff, 1959; Beaumont, 1960; Harding, 1961; Starkie, 1961; Hughes, 1962; Erickson and Martin, 1973; Erickson, 1976), although there is a suggestion of another peak of sensitivity at 17 days of age (Leonard et al., 1964).

The testes of newborn rats seem to be more sensitive than those of prepubertal boars and bulls, although in these species also, the gonocytes were noticeably more sensitive than the spermatogonia (Erickson, 1963, 1964; Erickson et al., 1972).

## 11.1.2    Heat and cryptorchidism

The fact that many but not all mammals carry their testes in a scrotum has been known for many years. Many authors also described individuals of these species in which one or both testes did not descend, but it was not recognized that cryptorchid testes are almost invariably sterile. For example de Graaf in 1668 wrote: 'Animals who have their testes hidden in the cavity of the abdomen are said to be more salacious, to copulate more often and to sire more offspring.' His first two observations may or may not be valid, but we now know that the third is not. By the end of the nineteenth century, it was generally recognized that the cryptorchid testis was aspermatogenic (Hunter, 1756; Goubaux and Follin, 1855; Godard, 1856; Monod and Arthaud, 1887; Felizet and Branca, 1898, 1902; Regaud and Policard, 1901; see also Crew, 1922; Moore, 1926), but many of these authors thought that the testis failed to descend because it was already abnormal, rather than the abnormality in the testis being due to its retention in the abdomen.

It was shown in 1891 by Piana and Savarese that when the scrotal testes of rats were pushed into the abdomen and retained there surgically they atrophied. Many other authors have since confirmed this observation on other species (Griffiths, 1893; Sand, 1921; Moore, 1924a, c; Moore, and Oslund, 1924; Bascom, 1925; Cunningham, 1927; Asdell and Salisbury, 1941; Nelson, 1951; Payne, 1956; Takewaki, 1960; Clegg, 1963a, b; Niemi and Kormano, 1965; Davis and Firlit, 1966; Zogg et al., 1968; Igboeli and Foote, 1969). Histological changes appear within 2 days; after 2 weeks only Sertoli cells and spermatogonia remain. The testes become normal again if returned to the scrotum, provided the period spent in the abdomen was not too long (Nelson, 1951; Takewaki, 1960; Hagen, 1971). Piana and Savarese (1891) suggested that the effects of cryptorchidism were due to the higher temperature in the abdomen, and Fukui (1923) and Moore (1923) independently showed that similar changes could be produced by locally heating the testis. It has been found that temperatures above 37°C are effective in most animals, and that the higher the temperature, the shorter the exposure needed (Moore and Chase, 1923; Young, 1927; Cunningham and Osborn, 1929; Guieysse-Pellissier, 1937; Williams and Cunningham, 1940; Imig et al., 1948; Gunsel, 1949a; Elfving, 1950; Steinberger and Dixon, 1959; Watanabe, 1959; Venkatachalam and Ramanathan, 1962; Waites and Setchell, 1964; Chowdhury and Steinberger, 1964, 1970; Fowler, 1968; Waites and Ortavant, 1968; Collins and Lacy, 1969; Setchell et al., 1971; Setchell and Waites, 1972; Fahim et al., 1975). Repeated applications of heat produce progressive damage (Bowler, 1967, 1972).

It was also found that grafts of testis survive normally only when they are transplanted to a part of the body which is at a lower temperature than the body core (see Section 7.6.1.5).

Damage to the testis can also be produced by insulating the scrotum, so as to prevent normal heat loss (Moore and Oslund, 1924; Phillips and McKenzie, 1934; Lagerlof, 1934; Gunn, 1936; McKenzie and Berliner, 1937; Glover, 1955a, b; Rock and Robinson, 1965; Robinson and Rock, 1967) or by heating the whole animal (Hart, 1922; Stieve, 1923; MacLeod and Hotchkiss, 1941; Gunn et al., 1942; Maqsood and Reincke, 1950; Hardberger, 1950; Oloufa et al., 1951; Casady et al., 1953, 1955, 1956; El Sheikh and Casida, 1955; Dutt and Hamm, 1957; Dutt and Sharpe, 1957; Soliman and Abd-El-Malek, 1959; Moule and Waites, 1963; Okauchi, 1963; Tokoyama, 1963; Procopé, 1965; DeAlba and Riera, 1966; Fowler and Dun, 1966; Skinner and Louw, 1966; Rathore and Yeates, 1967; Mazzari et al., 1968, 1970; Lindsay, 1969; McNitt and First, 1970; Wettemann et al., 1976) when indirect effects can be mediated through nervous, metabolic or hormonal pathways as well as direct effects on the testis itself.

**11.1.2.1    Cell types in the testis sensitive to heat**    Heating the testis does not damage all the cells there to the same extent. The most sensitive cells seem to be the

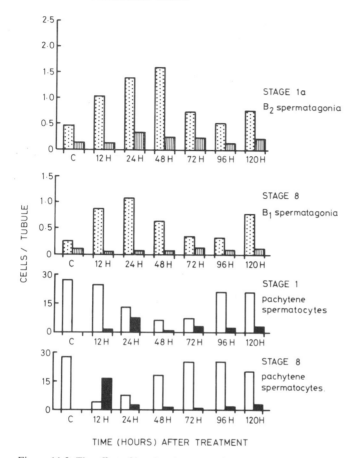

Figure 11.2  The effect of heating the testes of rams to about 40°C for 150 min on the number of $B_1$ and $B_2$ spermatogonia in prophase (dotted columns) and metaphase (lined columns) of mitosis, and on the number of normal (open columns) and degenerating (black columns) pachytene spermatocytes. C: control animals. (Data of Waites and Ortavant, 1968)

primary spermatocytes during pachytene (Figure 11.2) and the young spermatids. The later spermatids are apparently resistant and the sensitivity of the spermatogonia depends on the species. In the rat, they seem to be resistant (Chowdhury and Steinberger, 1964); in the ram, exposure of the testes to 41°C for 3h stimulates the B spermatogonia to divide and then die (Figure 11.2; Waites and Ortavant, 1968); losses of spermatogonia also occur in bulls during general heating (Skinner and Louw, 1966). From the effects of exposing rat testes to 43°C for varying times, Collins and Lacy (1969), suggest that there are 'critical periods' that the developing cells cannot pass after they have been heated. The first 'critical period' is between leptotene and pachytene, the second around the maturation division; continuing development with death of the cells as they reach the critical periods leads to a progressive reduction in the numbers of pachytene spermatocytes and young spermatids. With longer periods of heating a third critical period develops between the cap-phase and the acrosome-phase spermatids. Recovery begins after about 3 weeks when developing cells begin to pass safely through the 'critical periods' and the epithelium is repopulated. However, as the kinetics of spermatogenesis are not affected, the lesion proceeds through the various cell types by a process of 'maturation depletion'. This means that at different times after exposure to heat, there may be very few of a certain cell type, which were not directly damaged.

The identification of the cell types damaged from histological preparations is confirmed by the timing of the fall in the concentration of spermatozoa in rete testis fluid in both rams and rats (Figure 11.3; Setchell *et al.*, 1971; Setchell and Waites, 1972). Recovery from even moderate heating is usually not complete in rats for at least 60 days, as judged by testis weight which remains less than before heating. Repeated exposure to heating leads to a progressive decrease in testis weight. This suggests

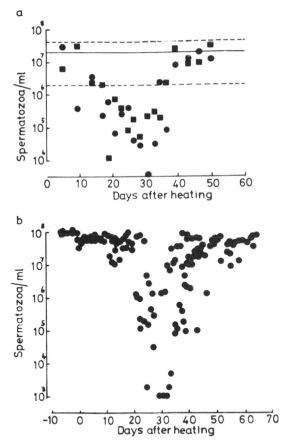

Figure 11.3 The concentration of spermatozoa in rete testis fluid of (a) rats and (b) rams at various times after locally heating the testis, the rams to 40°C for 3h, the rats to 41°C for 1 h ( ● ) or 1½ h ( ■ ). (Data of Setchell *et al.*, 1971; and Setchell and Waites, 1972)

that the spermatogonia which were undergoing mitosis at the time of heating were killed but those which were not dividing were more resistant (Bowler, 1967, 1972).

**11.1.2.2    Effect of heat on androgen production**    The interstitial tissue appears hypertrophied after the testis has been heated or made cryptorchid but this is probably due mainly to shrinkage of the tubules (Clegg, 1961; Huseby et al., 1961). Indirect tests of androgen production suggest a moderate fall with sometimes a transient increase 10 to 15 days after heating (Moore, 1944; Glover, 1955b, 1956; Clegg, 1960; Huseby et al., 1961; Moule and Waites, 1963; Moorehead and Morgan, 1967). Direct measurements of circulating testosterone levels confirm that a moderate decrease in testosterone production occurs (see Section 6.1.4.7; Liptrap and Raeside, 1970), which is often not reflected in changes in size of the prostate. The Leydig cells during heating can still respond to exogenous gonadotrophins, although the response by cryptorchid testes may be decreased (Eik-Nes, 1965). Libido is unaffected by heating sufficient to cause disruptions of spermatogenesis (Casady et al., 1953).

**11.1.2.3    Mechanism of heat damage**    No complete explanation has yet been found for the damaging effects of body temperature on the testis but several possible contributing factors have been identified.

When the testis or any other tissue is heated, its rate of metabolism increases, as does any chemical reaction. However this increase in oxygen consumption is not accompanied by a corresponding increase in blood flow, which is either unchanged or only slightly increased. The testis normally removes about half the oxygen from arterial blood, so that when it is heated, the oxygen content of the spermatic venous blood falls until the testis reaches 37°C. At temperatures above 37°C, the oxygen content of the spermatic venous blood falls no further (Figure 11.4) and therefore the testis enters into a state of self-induced hypoxia (Waites and Setchell, 1964). As the capillaries do not penetrate the seminiferous tubules, the oxygen tension at their centres will be about 7mm Hg lower than that in venous blood inside the capillaries, because of the oxygen consumption by the tubular cells and the limited rate of diffusion of oxygen through tissue (see Section 9.3.1.1).

A second possibility is concerned with the decrease in fluid secretion seen when the testis is heated (Setchell et al., 1971). If the germinal cells are dependent for nutrients on fluid secreted around them by the Sertoli cells, any reduction in the production of this fluid could lead to damage of the germinal cells. However fluid secretion in cryptorchid testes is normal for at least 24h (Section 8.5.3), suggesting that this mechanism may only be important at temperatures higher than body temperature. An effect of temperature on the blood–testis barrier has been reported but the change does not appear to be sufficient to allow any albumin or cholesterol into the tubules. It is not yet possible to say whether a subtle alteration in function of the barrier could lead to the cell damage known to occur after heating (Section 8.7.7).

The supply of glucose to the tubules is probably close to being limiting under normal conditions, since no glucose can be detected in rete testis fluid. Glucose is

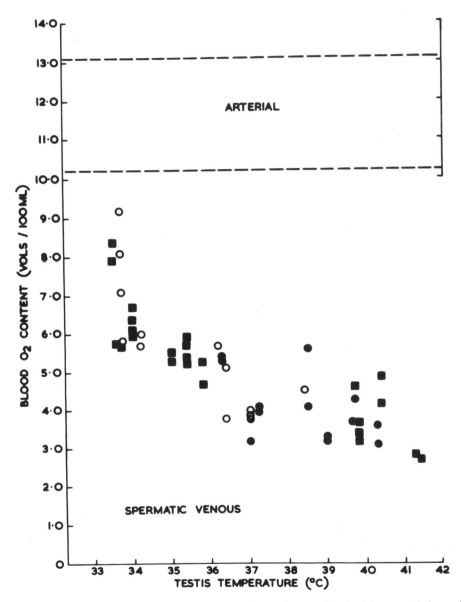

Figure 11.4 The oxygen content of blood from the internal spermatic vein of four rams during and after testicular heating, shown in relation to the range of arterial oxygen content in the same experiments. ■, steady temperature; ●, rising temperature; ○, falling temperature. (Reproduced from Waites and Setchell, 1964)

transported into the tubules by facilitated diffusion, presumably in amounts just sufficient to meet the metabolic needs of the tissue; nothing is known of the effects of temperature on this system (Section 9.4.1). It is interesting that the cells damaged by heat are those whose amino-acid incorporation into protein is most heavily dependent on glucose, and protein synthesis by testis slices is maximal at 32°C in the absence of glucose, but at 36°C in the presence of glucose (Section 9.6.2). However there is other evidence that DNA synthesis in teased out tubules from mouse testis is maximal at 32°C, whereas protein synthesis decreases and RNA synthesis increases as the temperature rises from 30 to 38°C (Nishimune and Komatsu, 1972). There is also evidence for release of lysosomal enzymes in heated testes (Turpeinen *et al.*, 1961; Blackshaw and Hamilton, 1970; Lee and Fritz, 1972; Lee, 1974).

The effects of cryptorchidism on testicular lipids is discussed in Section 9.1.3.

### 11.1.3    Cold
Cooling has much less effect on the testis than heating, unless temperatures close to freezing are reached (Hart, 1922; Chang, 1943, 1946; MacDonald and Harrison, 1954; Harris and Harrison, 1955; Goldzveig and Smith, 1956; Heroux and Campbell, 1959, 1960; Baillie, 1961; Dugal *et al.*, 1962; Dugal and Dunnigan, 1962). The damage to the testis caused by frost-bite of the scrotum in bulls (Faulkner *et al.*, 1967) is probably a result not of the cold but of oedema of the scrotal skin and the consequent increase in testicular temperature.

**11.1.4    Ultrasound**    Exposure of the testes of rats to high frequency sound waves ($1 \cdot 1 \times 10^6$cps, 1watt/cm$^2$ for 10min) caused testicular degeneration and sterility. Although the temperature of the testis rises to as high as 39°C during treatment, the amount of damage is much greater than one would expect from this degree of heating, and the sound waves are probably exerting some specific effect (Fahim *et al.*, 1975; Dumontier *et al.*, 1977).

### References to Chapter 11.1

Abbott, C. R. (1959) The effect of X-irradiation on the secretory capacity of the testes. *J. Endocrinol.* **19**, 33–43.

Albers-Schönberg, H. E. (1903) Uber eine bisher unbekannte Wirkung der Röntgenstrahlen auf den Organismus der Tiere. *Muench. Med. Wochschr.* **50**, 1859–60.

Asdell, S. A. and Salisbury, G. W. (1941) The rate at which spermatogenesis occurs in the rabbit. *Anat. Rec.* **80**, 145–54.

Baillie, A. H. (1961) Histochemical studies of the mouse testis following cold exposure. *Scot. Med. J.* **6**, 6–11.

Bascom, K. F. (1925) Quantitative studies of the testis. I. Some observations on the cryptorchid testes of sheep and swine. *Anat. Rec.* **30**, 225–41.

Bateman, A. J. (1958a) Mutagenic sensitivity of maturing germ cells in the male mouse. *Heredity* **12**, 213–32.

Bateman, A. J. (1958b) Sensitivity of the various germ cell stages of the male mouse to X-rays. *Radiation Res.* **9**, 90.

Beaumont, H. M. (1960) Changes in the radiosensitivity of the testis during fetal development. *Intern. J. Radiation Biol.* **2**, 247–56.

Bergonie, J. and Tribondeau, L. (1904) Action des rays X sur le testicule du rat blanc. *Compt. Rend. Soc. biol.* **56**, 400–2 and 592–5.

Bergonie, J. and Tribondeau, L. (1905) L'aspermatogénése experimentale complète obtenue par les rayons X, est-elle définitive? *Compt. Rend. Soc. Biol.* **57**, 678–80.

Bergonie, J. and Tribondeau, L. (1906) Action des rayons X sur le testicule. *Arch. electr. med.* **14**, 779–91, 823–46, 874–83 and 911–27.

Berliner, D. L. and Ellis, L. C. (1965) The effects of ionizing radiations on endocrine cells. IV. Increased production of 17α, 20α-dihydroxyprogesterone in rat testes after irradiation. *Radiation Res.* **24**, 368–73.

Berliner, D. L., Ellis, L. C. and Taylor, G. N. (1964) The effects of ionizing radiations on endocrine cells. II. Restoration of androgen production with a reduced nicotinamide adenine dinucleotide phosphate-generating system after irradiation of rat testes. *Radiation Res.* **22**, 345–56.

Binhammer, R. T. (1967) Effect of increased endogenous gonadotrophin on testis of irradiated immature and mature rats. *Radiation Res.* **30**, 676–86.

Blackshaw, A. W. and Hamilton, D. (1970) The effect of heat on hydrolytic enzymes and spermatogenesis in the rat testis. *J. Reprod. Fertil.* **22**, 569–71.

Botschi, A. (1929) Untersuchungen uber Kryptorchismus beim Pferd, Schwein, Hund und bei der Katze, unter besonderer Berucksichtigung der mikroskopischen Anatomie. *Z. Anat. Entw.-Gesch.* **89**, 727–53.

Bowler, K. (1967) The effects of repeated temperature applications to the testis on fertility in male rats. *J. Reprod. Fertil.* **14**, 171–3.

Bowler, K. (1972) The effect of repeated applications of heat on spermatogenesis in the rat: a histological study. *J. Reprod. Fertil.* **28**, 325–34.

Carter, T. C., Lyon, M. F. and Phillips, R. J. S. (1954) Induction of sterility in male mice by chronic gamma radiation. *Brit. J. Radiol.* **27**, 418–22.

Casady, R. B., Myers, R. M. and Legates, J. E. (1953) The effect of exposure to high ambient temperature on spermatogenesis in the dairy bull. *J. Dairy Sci.* **36**, 14–23.

Casady, R. B., Legates, J. E. and Myers, R. M. (1955) The effect of intermittent exposure to high ambient temperature on the physiological responses of dairy bulls. *J. Anim. Sci.* **14**, 1244.

Casady, R. B., Legates, J. E. and Myers, R. M. (1956) Correlations between ambient temperature varying from 60°–95°F and certain physiological responses in young dairy bulls. *J. Anim. Sci.* **15**, 141–52.

Casarett, G. W. and Hursh, J. B. (1956) Effects of daily doses of X-rays on spermatogenesis in dogs. *Radiation Res.* **5**, 473.

Chang, M. C. (1943) Disintegration of epididymal spermatozoa by application of ice to the scrotal testis. *J. Exptl. Biol.* **20**, 16–22.

Chang, M. C. (1946) Fertilizing capacity of spermatozoa following cold treatment of the scrotal testis of rabbits. *J. Exptl. Biol.* **22**, 95–100.

Chang, M. C., Hunt, D. M. and Romanoff, E. B. (1957) Effects of radiocobalt irradiation of rabbit spermatozoa *in vitro* on fertilization and early development. *Anat. Rec.* **129**, 211–29.

Chowdhury, A. K. and Steinberger, E. (1963) Selective damage induced by heat to the testicular germinal epithelium of rats. *Anat. Rec.* **145**, 217.

Chowdhury, A. K. and Steinberger, E. (1964) A quantitative study of the effect of heat on germinal epithelium of rat testes. *Am. J. Anat.* **115**, 509–24.

Chowdhury, A. K. and Steinberger, E. (1970) Early changes in the germinal epithelium of rat testes following exposure to heat. *J. Reprod. Fertil.* **22**, 205–12.

Chowdhury, A. K. and Steinberger, E. (1972) The influence of cryptorchid milieu on the initiation of spermatogenesis in the rat. *J. Reprod. Fertil.* **29**, 173–8.

Clegg, E. J. (1960) Some effects of artificial cryptorchidism on the accessory reproductive organs of the rat. *J. Endocrinol.* **20**, 210–19.

Clegg, E. J. (1961) Further studies on artificial cryptorchidism: Quantitative changes in the interstitial cells of the rat testis. *J. Endocrinol.* **21**, 433–41.

Clegg, E. J. (1963a) Studies on artificial cryptorchidism: morphological and quantitative changes in the Sertoli cells of the rat testis. *J. Endocrinol.* **26**, 567–75.

Clegg, E. J. (1963b) Studies on artificial cryptorchidism: Degenerative and regenerative changes in the germinal epithelium of the rat testis. *J. Endocrinol.* **27**, 241–51.

Clegg, E. J. (1965) Studies on artificial cryptorchidism: compensatory changes in the scrotal testis of unilaterally cryptorchid rats. *J. Endocrinol.* **33**, 259–68.

Collins, P. and Lacy, D. (1969) Studies on the structure and function of the mammalian testis. II. Cytological and histochemical observations on the testis of the rat after a single exposure to heat applied for different lengths of time. *Proc. Roy. Soc.* (London) B. **172**, 17–38.

Collins, P. and Lacy, D. (1974) Studies on the structure and function of the mammalian testis. IV. Steroid metabolism *in vitro* by isolated interstitium and seminiferous tubules of rat testis after heat sterilization. *Proc. Roy. Soc.* (London) B. **186**, 37–51.

Courot, M. (1963) Some results obtained in the irradiation with X-rays of testes of lambs. In *Effects of Ionizing Radiation on the Reproductive System*, eds. W. Carlson and F. Gassner, 270–86, New York: Pergamon Press.

Cox, D. F. and Willham, R. L. (1961) Influences on the patterns of sperm production in swine following X-irradiation. *J. Exptl. Zool.* **146**, 5–10.

Crew, F. A. E. (1922) A suggestion as to the cause of the aspermatic condition of the imperfectly descended testis. *J. Anat.* **56**, 98–106.

Cunningham, B. and Osborn, T. I. (1929) Infra-red sterility; preliminary report. *Endocrinology* **13**, 93–6.

Cunningham, J. T. (1927) Experiments on artificial cryptorchidism and ligature of the vas deferens in mammals. *J. Exptl. Biol.* **4**, 333–41.

Davis, J. R. and Firlit, C. F. (1966) The germinal epithelium of cryptorchid testes experimentally induced in prepubertal and adult rats. *Fert. Steril.* **17**, 187–200.

DeAlba, J. and Riera, S. (1966) Sexual maturity and spermatogenesis under heat stress in the bovine. *Anim. Prod.* **8**, 137–44.

Djanuar, R. (1965) Effect of high temperatures on spermiogenesis of rams. *Commun. Vet.* (Bogor, Indonesia) **9**, 13–18.

Dugal, L. P. and Dunnigan, J. (1962) Le poids de l'électro-éjaculat chez le cobaye soumis à une exposition cronique au froid. *Can. J. Biochem. Physiol.* **40**, 407–12.

Dugal, L. P., Saucier, G. and Desmaris, A. (1962) Impairment of testicular function in the white rat chronically exposed to a low ambient temperature. *Can. J. Biochem. Physiol.* **40**, 325–36.

Dumontier, A., Burdick, A., Ewigman, B. and Fahim, M. S. (1977) Effects of sonication on mature rat testes. *Fertil. Steril.* **28**, 195–204.

Dutt, R. H. and Hamm, P. T. (1957) Effect of exposure to high environmental temperature and shearing on semen production of rams in winter. *J. Anim. Sci.* **16**, 328–34.

Dutt, R. H. and Simpson, E. C. (1957) Environmental temperature and fertility of Southdown rams early in the breeding season. *J. Anim. Sci.* **16**, 136–43.

Dym, M. and Clermont, Y. (1970) Role of spermatogonia in the repair of the seminiferous epithelium following X-irradiation of the rat testis. *Am. J. Anat.* **128**, 265–82.

Edwards, R. G. and Sirlin, J. L. (1958) The effect of 200r of X-rays on the rate of spermatogenesis and spermiogenesis in the mouse. *Exptl. Cell Res.* **15**, 522–8.

Eik-Nes, K. B. (1965) Factors influencing the secretion of testosterone in the anesthetized dog. *Folia Endocrinol. Japan.* **40**, 1483–94.

Elfving, G. (1950) Effects of the local application of heat on the physiology of testis. An experimental study on rats. Dissertation, T. A. Sahalan Kirjapaino Oy, Helsinki.

Ellis, L. C. (1970) Radiation effects. In *The Testis*, eds. A. D. Johnson, W. R. Gomes and N. L. VanDemark, **III**, 333–76, New York: Academic Press.

Ellis, L. C. and Berliner, D. L. (1963) The effects of ionizing radiations on endocrine cells. I. Steroid biotransformations and androgen production by testes from irradiated mice. *Radiation Res.* **20**, 549–63.

Ellis, L. C. and Berliner, D. L. (1964) The effects of ionizing radiations on endocrine cells. III. Restoration of testosterone production with a reduced nicotinamide adenine dinucleotide phosphate-generating system after localized irradiation to heads of rats. *Radiation Res.* **23**, 156–64.

Ellis, L. C. and Beliner, D. L. (1967) The effect of ionizing radiations on endocrine cells. VI. Alterations in androgen biosynthesis by canine testicular tissue after the internal deposition of some radionuclides. *Radiation Res.* **32**, 520–37.

Ellis, L. C. and Kampen, K. R. van (1970) The effects of ionizing radiation on endocrine cells. VII. Androgen synthesis and metabolism by rat testicular minced and teased-tubular preparations after 450r of whole-body X-irradiation. *Radiation Res.* **48**, 146–63.

El Sheikh, A. S. and Casida, L. E. (1955) Motility and fertility of spermatozoa as affected by increased ambient temperature. *J. Anim. Sci.* **16**, 1146–50.

Erickson, B. H. (1963) Effects of $\gamma$-irradiation on the primitive germ cells of the bovine testis. *Int. J. Radiat. Biol.* **7**, 361–7.

Erickson, B. H. (1964) Effects of neonatal gamma irradiation on hormone production and spermatogenesis in the testis of the adult pig. *J. Reprod. Fertil.* **8**, 91–100.

Erickson, B. H. (1976) Effect of $^{60}$Co $\gamma$-radiation on the stem and differentiating spermatogonia of the postpubertal rat. *Radiat. Res.* **68**, 433–48.

Erickson, B. H. and Blend, M. J. (1976) Response of the Sertoli cell and stem germ cell to $^{60}$Co $\gamma$-radiation (dose and dose rate) in testes of immature rats. *Biol. Reprod.* **14**, 641–50.

Erickson, B. H. and Martin, P. G. (1973) Influence of age on the response of rat stem spermatogonia to γ-radiation. *Biol. Reprod.* **8**, 607–12.

Erickson, B. H., Reynolds, R. A. and Brooks, F. T. (1972) Differentiation and radioresponse (dose and dose rate) of the primitive germ cell of the bovine testis. *Radiat. Res.* **50**, 388–400.

Ershoff, B. H. (1959) Effects of prenatal X-irradiation on testicular function and morphology in the rat. *Am. J. Physiol.* **196**, 896–8.

Eschenbrenner, A. B., Miller, E. and Lorenze, E. (1948) Quantitative histological analysis of the effect of chronic whole-body irradiation with gamma rays on the spermatogenic elements and the interstitial tissue of testes of mice. *J. Nat. Cancer Inst.* **9**, 133–47.

Ewing, L. L. and VanDemark, N. L. (1963) II. Effect of temperature elevation *in vivo* on subsequent metabolic activity of rabbit testicular tissue *in vitro*. *J. Reprod. Fertil.* **6**, 9–16.

Fahim, M. S., Fahim, Z., Der, R., Hall, D. G. and Harman, J. (1975) Heat in male contraception (hot water 60°C, infrared, microwave and ultrasound) *Contraception* **11**, 549–62.

Faulkner, L. C., Hopwood, M. L., Masken, J. F., Kingman, H. E. and Stoddart, H. L. (1967) Scrotal frost-bite in bulls. *J. Am. vet. Med. Assn.* **151**, 602–5.

Felizet, G. and Branca, A. (1898) Histologie du testicule ectopique. *J. Anat. Physiol.* (Paris) **34**, 589–641.

Felizet, G. and Branca, A. (1902) Recherches sur le testicule en ectopie. *J. Anat. Physiol.* (Paris) **38**, 329–42.

Ferroux, C. R., Regaud, C. and Samsonow, N. (1938) Effets des rayons de roentgen administrés sans fractionnement de la dose, sur les testicules du rat, au point de vue de la stérilisation de l'épithélium séminal. *Compt. Rend. Soc. Biol.* **128**, 170–3.

Fogg, L. C. and Cowing, R. F. (1951) The changes in cell morphology and histochemistry of the testes following irradiation and their relation to other induced testicular changes. I. Quantitative random sampling of germinal cells at intervals following direct irradiation. *Cancer Res.* **11**, 23–8.

Fowler, D. G. (1968) Skin folds and Merino breeding. 7. The relations of heat applied to the testis and scrotal thermoregulation to fertility in the Merino ram. *Aust. J. exp. Agric. Anim. Husb.* **8**, 142–8.

Fowler, D. G. and Dun, R. B. (1966) Skin fold and Merino breeding. 4. The susceptibility of rams selected for a high degree of skin wrinkle to heat-induced infertility. *Aust. J. Exptl. Agric. Anim. Husb.* **6**, 121–7.

Free, M. J. and VanDemark, N. L. (1968) Gas tensions in spermatic and peripheral blood of rams with normal and heat treated testes. *Am. J. Physiol.* **214**, 863–5.

Free, M. J., Vera Cruz, N. C., Johnson, A. D. and Gomes, W. R. (1968) Metabolism of glucose-1-$^{14}$C and glucose-6-$^{14}$C by testis tissue from cryptorchid and testosterone propionate treated rabbits. *Endocrinology* **82**, 183–7.

Free, M. J., Massie, E. D. and VanDemark, N. L. (1969) Glucose metabolism in the cryptorchid rat testis. *Biol. Reprod.* **1**, 354–66.

French, D. J., Leeb, C. H., Fahrion, S. L., Law, O. T. and Jecht, E. W. (1973) Self-induced scrotal hyperthermia in man followed by a decrease in sperm output. *Andrologie* **5**, 311–16.

Freund, M. and Borrelli, F. J. (1964) The effects of X-irradiation on male fertility in the guinea pig. Effect of 75, 150 and 300 roentgens of whole-body X-irradiation on semen production. *Radiation Res.* **22**, 404–13.

Freund, M. and Murphree, R. L. (1960) Effect of whole-body $\gamma$-irradiation on the characteristics and metabolism of bull semen during the early post-irradiation period. *J. Dairy Sci.* **43**, 1130–4.

Fukui, N. (1923) Action of body temperature on the testicle. *Japan. Med. World* **3**, 160–3.

Glover, T. D. (1955a) The effect of a short period of scrotal insulation on the semen of the ram. *J. Physiol.* (London) **128**, 22P.

Glover, T. D. (1955b) Some effects of scrotal insulation on the semen of rams. *Proc. Soc. Study Fertil.* **7**, 66–75.

Glover, T. D. (1956) The effect of scrotal insulation and the influence of the breeding season upon fructose concentration in the semen of the ram. *J. Endocrinol.* **13**, 235–42.

Godard, M. E. (1856) Etudes sur la monorchidie et la cryptorchidie. *C. rend. Soc. Biol.* **3**, 315–460.

Goldzveig, S. A. and Smith, A. U. (1956) The fertility of male rats after moderate and after severe hypothermia. *J. Endocrinol.* **14**, 40–53.

Goubaux, A. and Follin, E. (1855) De la cryptorchidie chez l'homme et les principaux animaux domestiques. *C. rend. Soc. Biol.* **2**, 293–330.

Griffiths, J. (1893) Structural changes in the testicle of the dog when it is replaced within the abdominal cavity. *J. Anat. Physiol.* (London) **27**, 483–500.

Guieyesse-Pellissier, A. (1937) Etude de la dégénérescence et de la régénérescence des testicules après chauffage. *Arch. Anat. Microscop.* **33**, 5–47.

Gunn, R. M. C. (1936) Fertility in sheep. *Bull. Council Sci. Ind. Res.* (Australia) **94**, 1–116.

Gunn, R. M. C., Sanders, R. N. and Granger, W. (1942) Studies in fertility in sheep. 2. Seminal changes affecting fertility in rams. *Bull. Council Sci. Ind. Res.* (Australia) **148**, 1–140.

Gunn, S. A., Gould, T. C. and Anderson, W. A. D. (1960) The effect of X-irradiation on the morphology and function of the rat testes. *Am. J. Pathol.* **37**, 203–13.

Günsel, E. (1949a) Uber die Warmeempfindlichkeit des Keimepithels im Rattenhoden. *Strahlentherapie* **80**, 299–304.

Günsel, E. (1949b) Zur Frage der Roentgenatrophie des Hodens. *Strahlentherapie* **80**, 467–74.

Hagen, E. O. (1971) The effect of subsequent fertility of temporary cryptorchidization in Swiss white mice. *Can. J. Zool.* **49**, 587–90.

Hagenäs, L. and Ritzen, E. M. (1976) Impaired Sertoli cell function in experimental cryptorchidism in the rat. *Mol. cell Endocrinol.* **4**, 25–34.

Hagenäs, L., Plöen, L., Ritzen, E. M. and Ekwall, H. (1977) Blood-testis barrier: maintained function of inter-Sertoli cell junctions in experimental cryptorchidism in the rat, as judged by a simple Lanthanum-immersion technique. *Andrologia* **9**, 250–4.

Hall, E. R. and Hupp, E. W. (1970) Localization of irradiation damage in rat seminiferous tubules. *Nature* **225**, 85–6.

Hardberger, F. M. (1950) Effects of constant high temperature (32.2°C) on the testes and spermatozoa of the albino mouse. *Proc. Louisiana Acad. Sci.* **13**, 35–41.

Harding, L. K. (1961) The survival of germ cells after irradiation of the neonatal male rat. *Intern. J. Radiation Biol.* **3**, 539–41.

Harris, R. and Harrison, R. G. (1955) The effect of low temperature on the guinea pig testis. *Proc. Soc. Study Fertil.* **7**, 24–34.

Hart, C. (1922) Beitrage zur biologischen Bedeutung der innersekretorischen Organe. II. Der Einfluss abnormer Aussentemperaturen auf Schilddruse und Hoden. *Arch. Ges. Physiol.* **196**, 151–76.

Heald, A. H., Beard, C. and Lyons, W. R. (1939) The effects of roentgen irradiation and prolan upon the testes of immature rats. *Am. J. Roentgenol. Radium Therapy* **41**, 448–52.

Heller, C. G., Heller, G. V., Warner, G. A. and Rowley, M. J. (1968) Effects of graded doses of ionizing radiation on testicular cytology and sperm count in man. *Radiation Res.* **35**, 493–4.

Hertwig, P. (1938) Die Regeneration des Samenepithels der Maus nach Röntgenbestrahlung unter besonderer Berucksichtigung des Spermatogonien. *Arch. Exptl. Zellforsch. Gewebezucht.* **22**, 68–73.

Hovatta, O. and Kormano, M. (1974) Development of the rat seminiferous tubules following prepubertal whole-body X-irradiation. *Andrologia* **6**, 277–85.

Hughes, G. C. (1962) Radiosensitivity of male germ-cells in neonatal rats. *Intern. J. Radiation Biol.* **4**, 511–9.

Hunter, J. (1756) Observations on certain parts of the animal oeconomy. London.

Huseby, R. A., Dominguez, O. V. and Samuels, L. T. (1961) Function of normal and abnormal testicular interstitial cells in the mouse. *Rec. Progr. Horm. Res.* **17**, 1–42.

Idänpään-Heikkilä, P. (1966) Effect of local heat *in vivo* on the fine structure of the basement membrane and the Sertoli cells of the rat testis. *Fertil. Steril.* **17**, 689–95.

Igboeli, G. and Foote, R. H. (1969) Changes in epididymal spermatozoa and in the testes of rabbits after experimental cryptorchidism. *J. Exptl. Zool.* **170**, 489–98.

Imig, C. J., Thomson, J. D. and Hines, H. M. (1948) Testicular degeneration as a result of microwave irradiation. *Proc. Soc. exp. Biol. N.Y.* **69**, 382–6.

Joel, C. A. (1942) Zur Wirkung des Testosterone-propionate auf die Regeneration des Rontgengeschadigten Hodens. *Endokrinologie* **24**, 310–7.

Jones, E. A. (1960) Number of spermatogonia after X-irradiation of the adult rat. *Int. J. radiat. Biol.* **2**, 157–70.

Kochar, N. K. and Bateman, A. J. (1969) Post-irradiation changes in Sertoli cells. *J. Reprod. Fertil.* **18**, 265–74.

Kohn, H. I. (1955) On the direct and indirect effects of X-rays on the testis of the rat. *Radiation Res.* **3**, 153–6.

Kohn, H. I. and Kallman, R. F. (1955) The effect of fractionated X-ray dosage upon the mouse testis. 1. Maximum weight loss following 80 to 240 r given 2 to 5 fractions during 1 to 4 days. *J. Nat. Cancer Inst.* **15**, 891–99.

Kormano, M. and Hovatta, O. (1972) *In vitro* contractility of rat seminiferous tubules following 400 R whole body X-irradiation. *Strahlentherapie* **144**, 713–18.

Kormano, M., Harkonen, M. and Kontinen, E. (1964) Effect of experimental cryptochordism on the histochemically demonstrable dehydrogenases of the rat testes. *Endocrinology* **74**, 44–51.

Lagerlof, N. (1934) Morphologische Untersuchungen uber Veranderungen im Spermabild und in den Hoden bei Bullen, mit verminderter oder aufgehobener Fertilitat. *Acta Path. Microbiol Scand.* Suppl. **XIX**.

Lasowsky, J. M. (1931) Beobachtungen uber die Wirkung hoher Temperatur der Außenwelt auf die Hoden von weissen Maussen. *Virchows Arch. Path. Anat.* **280**, 311–29.

Lee, L. P. K. (1974) Temperature effect on the permeability of plasma membranes of advanced germinal cells of the rat testes. *Can. J. Biochem,* **52**, 586–93.

Lee, L. P. K. and Fritz, I. B. (1972) Studies on spermatogenesis in rats. V. Increased thermolability of lysosomes from testicular germinal cells and its possible relationship to impairments in spermatogenesis in cryptorchidism. *J. biol. Chem.* **247**, 7956–61.

Leonard, A., Imbaud, F. and Maisin, J. R. (1964) Testicular injury in rats irradiated during infancy. *Brit. J. Radiol.* **37**, 764–8.

Lindsay, D. R. (1969) Sexual activity and semen production of rams at high temperatures. *J. Reprod. Fertil.* **18**, 1–8.

Liptrap, R. M. and Raeside, J. I. (1970) Urinary steroid excretion in cryptorchidism in the pig. *J. Reprod. Fertil.* **21**, 293–301.

MacDonald, J. and Harrison, R. G. (1954) Effect of low temperature on rat spermatogenesis. *Fertil. Steril.* **5**, 205–16.

MacLeod, J. and Hotchkiss, R. S. (1941) The effect of hyperpyrexia upon spermatozoa counts in men. *Endocrinology* **28**, 780–4.

MacLeod, J., Hotchkiss, R. S. and Sitterson, B. W. (1964) Recovery of male fertility after sterilization by nuclear radiation. *J. Am. Med. Assoc.* **187**, 637–41.

Mandl, A. M. (1964) The radiosensitivity of germ cells. *Biol. Rev.* **39**, 288–371.

Maqsood, M. and Ashikowa, J. K. (1961) Fertility studies of X-irradiated male mice. *Fertil. Steril.* **12**, 452–9.

Maqsood, M. and Reineke, E. P. (1950) Influence of environmental temperature and thyroid status on sexual development in the male mouse. *Am. J. Physiol.* **162**, 24–30.

Mazzarri, G., du Mesnil du Buisson, F. and Ortavant, R. (1968) Action of temperature on spermatogenesis, sperm production, and fertility of the boar. *Proc. 6th Intern. Congr. Animal Reprod.* (Paris) **1**, 305.

McKenzie, F. F. and Berliner, V. (1937) The reproductive capacity of the ram. *Missouri Univ. Agr. Exp. Sta. Bull.* **265**, 1–143.

McNitt, J. I. and First, N. L. (1970) Effects of 72-hour heat stress on semen quality in boars. *Int. J. Biometeorol.* **14**, 373–80.

Momigliano, E. and Essenberg, J. M. (1944) Regenerative processes induced by gonadotropic hormones in irradiated testes of the albino rat. *Radiology* **42**, 273–82.

Monod, C. and Arthaud, G. (1887) Contribution a l'étude des altérations du testicule ectopique et de leurs conséquences (infécondité). *Arch. gen. Med.* **20**, 641–52.

Monesi, V. (1962) Relation between X-ray sensitivity and stages of the cell cycle in spermatogonia of the mouse. *Radiation Res.* **17**, 809–38.

Moore, C. R. (1922) Cryptorchidism experimentally produced. *Anat. Rec.* (*Proc. Soc. Zool.*) **24**, 383.

Moore, C. R. (1924a) Properties of the gonads as controllers of somatic and psychical characteristics. VI. Testicular reactions to experimental cryptorchidism. *Am. J. Anat.* **34**, 269–316.

Moore, C. R. (1924b) Properties of the gonads as controllers of somatic and psychical characteristics. VIII. Heat application and testicular degeneration; the function of the scrotum. *Am. J. Anat.* **34**, 337–58.

Moore, C. R. (1924c) The behaviour of the testis in transplantation, experimental cryptorchidism, vasectomy, scrotal insulation and heat application. *Endocrinology* **8**, 493–508.

Moore, C. R. (1926) The biology of the mammalian testis and scrotum. *Q. Rev. Biol.* **1**, 4–50.

Moore, C. R. (1944) Hormone secretion by experimental cryptorchid testes. *Yale J. Biol. Med.* **17**, 203–16.

Moore, C. R. and Chase, H. D. (1923) Heat application and testicular degeneration. *Anat. Rec.* **26**, 344–5.

Moore, C. R. and Oslund, R. (1924) Experiments on the sheep testis—cryptorchidism, vasectomy and scrotal insulation. *Am. J. Physiol.* **67**, 595–607.

Moorehead, J. R. and Morgan, C. F. (1967) Cryptorchidism: Its pre- and postpubertal effects on the hypophysis of the rat. *Fertil. Steril.* **18**, 232–7.

Moule, G. R. and Waites, G. M. H. (1963) Seminal degeneration in the ram and its relation to the temperature of the scrotum. *J. Reprod. Fertil.* **5**, 433–46.

Nebel, B. R. and Murphy, C. J. (1960) Damage and recovery of mouse testis after 1000 r acute localized X-irradiation, with reference to restitution cells, Sertoli cell increase, and type A spermatogonial recovery. *Radiation res.* **12**, 626–41.

Nelson, W. O. (1951) Mammalian spermatogenesis: Effects of experimental cryptorchidism in the rat and non-descent of the testis in man. *Rec. Progr. Horm. Res.* **6**, 29–56.

Niemi, M. and Kormano, M. (1965) Response of the cycle of the seminiferous epithelium of the rat testis to artificial cryptorchidism. *Fertil. Steril.* **16**, 236–42.

Nishimune, Y. and Komatsu, T. (1972) Temperature-sensitivity of mouse testicular DNA synthesis *in vitro. Exptl. cell Res.* **75**, 514–17.

Oakberg, E. F. (1955a) Sensitivity and time of degeneration of spermatogenetic cells irradiated in various stages of maturation in the mouse. *Radiation Res.* **3**, 369–91.

Oakberg, E. F. (1955b) Degeneration of spermatogonia of the mouse following exposure to X-rays and stages in the mitotic cycle at which cell death occurs. *J. Morphol.* **97**, 39–54.

Oakberg, E. F. (1957a) X-ray damage and recovery in the mouse testis. *Anat. Rec.* **128**, 597.

Oakberg, E. F. (1957b) Gamma-ray sensitivity of spermatogonia of the mouse. *J. Exptl. Zool.* **134**, 343–56.

Oakberg, E. F. (1959) Initial depletion and subsequent recovery of spermatogonia of the mouse after 20 r of gamma rays and 100, 300 and 600 r of X-rays. *Radiation Res.* **11**, 700–19.

Oakberg, E. F. (1975) Effects of radiation on the testis. *Handbk Physiol.* Section 7, *Endocrinology* V, 233–43.

Oakberg, E. F. and Di Minno, R. L. (1955) Gamma-ray sensitivity of different developmental stages of spermatogenesis of the mouse. *Genetics* **40**, 588.

Oakberg, E. F. and Di Minno, R. L. (1960) X-ray sensitivity of primary spermatocytes of the mouse. *Intern. J. Radiation Biol.* **2**, 196–209.

Oakes, W. R. and Lushbaugh, C. C. (1952) Course of testicular injury following accidental exposure to nuclear radiation: Report of a case. *Radiology* **59**, 737–43.

Okauchi, K. (1963) Effects of high environmental temperatures on sexual organs of male mice. 3. Effects of high and subsequent cool ambient temperatures on the incidence of abnormal spermatozoa. *Bull. Facul. Agric. Univ. Miyazaki.* **9**, 89–98.

Oloufa, M. M., Bogart, R. and McKenzie, F. F. (1951) Effect of environmental temperature and the thyroid gland in the male rabbit. *Oregon State Coll. Agric. exp. Sta. Tech. Bull.* **20**, 1–39.

Pace, H. B., Murphree, R. L. and Hupp, E. W. (1959) Effects of total body irradiation on semen production of boars. *J. Anim. Sci.* **18**, 1554.

Pace, H. B., Hupp, E. W. and Murphree, R. L. (1962) Changes in semen and blood of boars following total body gamma irradiation. *J. Anim. Sci.* **21**, 615–20.

Payne, J. M. (1956) The degenerative changes in the adult mouse testis returned to the abdominal cavity. *J. Pathol. Bacteriol.* **71**, 117–23.

Phillips, R. W. and McKenzie, F. F. (1934) The thermoregulatory function and mechanism of the scrotum. *Missouri Univ. Agr. exp. Sta. Res. Bull.* **217**, 1–73.

Piana, G. P. and Savarese, G. (1891) Su alcuni studii anatomo-patologici. *La Clinica Veterinaria* **14**, 50–1.

Procopé, B. J. (1965) Effect of repeated increase of body temperature on human sperm cells. *Intern. J. Fertil.* **10**, 333–9.

Rapaport, F. T., Sampath, A., Kano, K., McCluskey, R. T. and Milgrom, F. (1969) Immunological effects of thermal injury. I. Inhibition of spermatogenesis in guinea pigs. *J. exp. Med.* **130**, 1411–22.

Rathore, A. K. and Yeates, N. T. M. (1967) Morphological changes in ram spermatozoa due to heat stress. *Vet. Rec.* **81**, 343–4.

Regaud, C. and Blanc, J. (1906) Actions des rayons X sur les diverses générations de la lignée spermatique. Extrême sensibilité des spermatogonies à ces rayons. *C. rend. Soc. Biol.* **61**, 163–5.

Regaud, C. and Dubreuil, G. (1907) Actions des rayons de roentgen sur le testicule du lapin. I. Conservation de la puissance virile et stérilisation. *Compt. Rend. Soc. Biol.* **63**, 647–9.

Regaud, C. and Nogier, T. (1911) Sterilisation roentgenienne totale et definitive sans radiodermite, des testicules du Belier adulte. Conditions de sa realisation. *Compt. Rend. Soc. Biol.* **70**, 202 3.

Regaud, C. and Policard, A. (1901) Etude comparative du testicule du porc normal, impubere et ectopique, au point de vue des cellules interstitielles. *C. rend. Soc. Biol.* **53**, 450–2.

Robinson, D. and Rock, J. (1967) Intrascrotal hyperthermia induced by scrotal insulation: Effect on spermatogenesis. *Obstet. Gynecol.* **29**, 217–23.

Robinson, D., Rock, J. and Menkin, M. F. (1968) Control of human spermatogenesis by induced changes of intrascrotal temperature. *J. Am. Med. Assoc.* **204**, 290–7.

Rock, J. and Robinson, D. (1965) Effect of induced intrascrotal hyperthermia on testicular function in man. *Am. J. Obstet. Gynecol.* **93**, 793–801.

Rugh, R. (1950) The immediate and delayed morphological effects of X-radiations on meiotic chromosomes. *J. Cell. Comp. Physiol.* **36**, 185–203.

Rugh, R. and Jackson, S. (1958) Effect of fetal X-irradiation upon the subsequent fertility of the offspring. *J. Exptl. Zool.* **138**, 209–21.

Samuels, L. D. (1966) Depletion of mouse spermatogonia following exposure to Polonium-210. *Nature* **210**, 434–5.

Sand, K. (1921) Etudes expérimentales sur les glands sexuelles chez les mammifères. Cryptorchidie expérimentale. *J. Physiol.* (Paris) **19**, 515–27.

Sandemann, T. F. (1966) The effects of X-irradiation on male human fertility. *Brit. J. Radiol.* **39**, 901–7.

Schoen, E. J. (1964) Effect of local irradiation on testicular androgen biosynthesis. *Endocrinology* **75**, 56–65.

Setchell, B. P. and Waites, G. M. H. (1972) The effects of local heating on the flow and composition of rete testis fluid in the rat, with some observations on the effects of age and unilateral castration. *J. Reprod. Fertil.* **30**, 225–33.

Setchell, B. P., Voglmayr, J. K. and Hinks, N. T. (1971) The effect of local heating on the flow and composition of rete testis fluid in the conscious ram. *J. Reprod. Fertil.* **24**, 81–9.

Shaver, S. L. (1953a) X-irradiation injury and repair in the germinal epithelium of male rats. I. Injury and repair in adult rats. *Am. J. Anat.* **92**, 391–431.

Shaver, S. L. (1953b) X-irradiation injury and repair in the germinal epithelium of male rats. II. Injury and repair in immature rats. *Am. J. Anat.* **92**, 433–49.

Simpson, C. G. and Ellis, L. C. (1967) The direct effects of X-irradiation on the biosynthesis of androgens by rat testicular tissue *in vitro*. *Radiation Res.* **31**, 139–48.

Skinner, J. D. and Louw, G. N. (1966) Heat stress and spermatogenesis in *Bos indicus* and *Bos taurus* cattle. *J. Appl. Physiol.* **21**, 1784–90.

Soliman, F. A. and Abd-El-Malek, A. S. (1959) Influence of temperature and light on reproduction in male rats. *Nature* (London) **183**, 266–7.

Starkie, C. M. (1961) The effect of cysteamine on the survival of foetal germ cells after irradiation. *Int. J. Rad. Biol.* **3**, 609–17.

Steinberger, E. and Dixon, W. J. (1959) Some observations on the effect of heat on the testicular germinal epithelium. *Fertil. Steril.* **10**, 578–95.

Stieve, F. (1923) Der Erfolg hoherer Außentemperatur auf die Keimdrusen der Hausmaus. *Anat. Anz.* **57** (Erg-Bd.), 28–53.

Suvanto, O. and Kormano, M. (1970) Effect of experimental cryptorchidism and cadmium injury on the spontaneous contractions of the seminiferous tubules of the rat testis. *Virchows Arch. Abt. B. Zellpath.* **4**, 217–24.

Takewaki, K. (1960) Recovery of spermatogenesis following orchidopexy in cryptorchid rats. *Anat. Zool. Jap.* **33**, 115–23.

Tokoyama, I. (1963) Quoted by Leblond *et al.* (1963) Spermatogenesis. In *Mechanisms Concerned with Conception*, ed. C. G. Hartmann, 1–72, Oxford: Pergamon Press.

Turpeinen, P., Turpeinen, O. and Talanti, S. (1961) Effect of local heat *in vivo* on hyaluronidase, succinic dehydrogenase and phosphatases of the rat testis. *Endocrinology* **70**, 731–7.

VanDemark, N. L., Zogg, C. A. and Hays, R. L. (1968) Effect of hyper- and hypoglycemia accompanying cryptorchidism on testis function. *Am. J. Physiol.* **215**, 977–84.

Vankatachalam, P. S. and Ramanathan, K. S. (1962) Effects of moderate heat on the testes of rats and monkeys. *J. Reprod. Fertil.* **4**, 51–6.

Voglmayr, J. K., Setchell, B. P. and White, I. G. (1970) Metabolism and ultrastructure of testicular spermatozoa after local heating of the ram testis. *J. Reprod. Fertil.* **24**, 71–81.

Waites, G. M. H. and Ortavant, R. (1968) Effets précoces d'une brève élévation de la température testiculaire sur la spermatogénèse du bélier. *Ann. Biol. anim. Biochim. Biophys.* **8**, 323–31.

Waites, G. M. H. and Setchell, B. P. (1964) Effect of local heating on blood flow and metabolism in the testis of the conscious ram. *J. Reprod. Fertil.* **8**, 339–49.

Waites, G. M. H. and Setchell, B. P. (1969a) Some physiological aspects of the function of the testis. In *The Gonads*, ed. K. W. McKerns, 649–714, New York: Appleton Century Crofts.

Waites, G. M. H. and Setchell, B. P. (1969b) Physiology of the testis, epididymis and scrotum. In *Recent Advances in Reproductive Physiology*, ed. Anne McLaren, 1–63, London: Logos Press.

Wall, P. G. (1961) Effects of X-irradiation of differentiating Leydig cells of the immature rat. *J. Endocrinol.* **23**, 291 301.

Wangensteen, D. H. (1927) Undescended testis; experimental and clinical study. *A.M.A. Arch. Surg.* **14**, 663–731.

Watanabe, A. (1959) The effect of heat on the human spermatogenesis. *Kyushu J. Med. Sci.* **10**, 101–17.

Wattenwyl, H. von and Joel, C. A. (1941a) Die Wirkung der Rontgenbestrahlen auf Rattenhoden. II. Technik der experimentellen Rontgenbestrahlung des Rattenhodens und Methodik zur Prufung der Strahlenwirkung Allgemeine Darstellung der Veranderungen am Samenepithel nach Rontgenbestrahlung. *Strahlentherapie* **70**, 499–521.

Wattenwyl, H. von and Joel, C. A. (1941b) Die Wirkung der Rontgenstrahlen auf den Rattenhoden. III. Verlauf der Degeneration bzw. Regeneration des Samenepithels nach Bestrahlung mit 60 bis 2400r bis zu 50 Tagen nach der Bestrahlung. *Strahlentherapie* **70**, 588–631.

Wattenwyl, H. von and Joel, C. A. (1942) Die Wirkung der Rontgenstrahlen auf den Rattenhoden. IV. Verlauf der Degeneration bzw. Regeneration des Samenepithels nach Bestrahlung mit 150 bis 2400r von 75 bis zu 300 Tagen nach der Bestrahlung. *Strahlentherapie* **72**, 62–92.

Wettemann, R. P., Wells, M. E., Omtvedt, I. T., Pope, C. E. and Turman, E. J. (1976) Influence of elevated ambient temperature on reproductive performance of boars. *J. Anim. Sci.* **42**, 664–9.

Whitehead, R. H. (1908) A peculiar case of cryptorchism and its bearing upon the problem of the function of the interstitial cells of the testis. *Anat. Rec.* **2**, 177–8.

Widmaier, R. (1964) Uber den Einfluβ von Rontgenstrahlen auf die postnatale Hodenentwicklung und Keimzellreifung bei der Maus. *Zeitschr. f. mikroskop.-anatomische Forschung* **71**, 229–55.

Willham, R. L. and Cox, D. F. (1961) Sperm production in swine after exposure to x-irradiation. *Radiation Res.* **14**, 223–9.

Williams, W. L. and Cunningham, B. (1940) Histologic changes in the rat testis following heat treatment. *Yale J. Biol. Med.* **12**, 309–16.

Young, W. C. (1927) The influence of high temperature of the guinea pig testis. Histologic changes and effects on reproduction. *J. Exptl. Zool.* **49**, 459–99.

Zogg, C. A., Hays, R. L., VanDemark, N. L. and Johnson, A. D. (1968) Relation of time and surgery in experimental cryptorchidism in testis changes. *Am. J. Physiol.* **215**, 985–90.

## 11.2      Nutrient supply

If the effects of heat are mediated by lack of oxygen and/or glucose, reduction in the availability of these substances should also damage the testis. This is indeed so.

### 11.2.1      Ischaemia

In general the germinal elements are more sensitive to ischaemia than the hormone-producing tissue but in some species, especially human, collateral circulation can be very significant (Miflet, 1879; Griffiths, 1896; Martins, 1933; Hellner, 1933; Maatz, 1934; Iwasita, 1939; Smith, 1955; Fels and Bur, 1958; Fels, 1959). The sensitivity of the rat testis to ischaemia has been the subject of some dispute; the minimum time to produce lesions has been estimated as 20 minutes by Steinberger and Tjioe (1969) and 90 minutes by Oettle and Harrison (1952) (see also Hilscher, 1964). However the histological changes during moderate damage have been well characterized. Spermatogonia in mitosis are killed, as are pachytene spermatocytes. These are similar to the changes which occur after heat; the loss of young spermatids which follows heating of the testis does not seem to occur after ischaemia (Tjioe and Steinberger, 1970). Ischaemia following injections of serotonin (Boccabella *et al.*, 1962; Kormano *et al.*, 1968) or adrenaline (Chatterjee and Paul, 1968; Setchell *et al.*, 1966) causes atrophy and degeneration of the testes; the effect of adrenaline can be blocked by giving the rats ascorbic acid (Chatterjee, 1968).

### 11.2.2      Hypoxia

Hypoxia causes damage to the testis and while these changes have not been examined in detail, there appear to be some similarities to the effects of heat (Monge, 1942a, b, 1943, 1948; Gordon *et al.*, 1943; Dalton *et al.*, 1945; Walton and Uruski, 1946; Shettles, 1947; Altland, 1949; Baird and Cook, 1962; Altland and Highman, 1968). However, it is strange that exposure to high oxygen tensions for several days also causes damage to the testis; exposure up to 24h is without effect (Ozorio de Almeida, 1934; Matteo and Nahas, 1963; Gerschman, 1964; but cf. van den Brenk and Jamieson, 1967).

### 11.2.3      Hypoglycaemia and diabetes

Insulin hypoglycaemia produces histological changes in the testis of rats (Mancini *et al.*, 1960). Administration of 5-thio-D glucose to rodents leads to degeneration of the testis (Zysk *et al.*, 1975; Neumann and Schenck, 1977; Homm *et al.*, 1977), presumably by interfering with glucose transport or metabolism. Diabetes produced with alloxan or streptozotocin or by pancreatectomy has been associated with degeneration of the testis (Hunt and Baily, 1961; Folgia *et al.*, 1963, 1969; Schöffling *et al.*, 1967; Oksanen, 1975) which can be prevented by the administration of ascorbic acid (Deb and Chatterjee, 1963). However, there was no change in testis weight 8 or 21 days after treatment with alloxan despite reduced gonadotrophin and testosterone concentrations in serum. Rats made diabetic with streptozotocin for three weeks did have slightly smaller testes (Howland and Zebrowski, 1972, 1974, 1976). Human diabetics are often impotent and may show histological abnormalities of the testis (Babbott *et al.*, 1958; Schöffling *et al.*, 1963; Federlin *et al.*, 1965; Kent, 1966; Rauch-Strooman *et al.*, 1970; Faerman *et al.*, 1972).

## 11.2.4    Nutritional deficiencies

### 11.2.4.1    Energy and protein lack
There is no doubt that restriction of food intake can result in impairment of both endocrine and spermatogenic function of the testis, but in adult animals, this only occurs when the dietary restriction is severe (Siperstein, 1920–1; Leathem, 1970, 1975). The androgen production is reduced more than the spermatogenic function (Figure 11.5; Moore and Samuels, 1931; Davies *et al.*, 1957; Setchell *et al.*, 1965), presumably because the seminiferous tubules have first call on the reduced amount of testosterone being secreted. Libido is usually but not always depressed (see Setchell *et al.*, 1965). It has not been established whether there are direct effects on the testis or whether everything can be explained by the reduction in gonadotrophin secretion which undoubtedly occurs during underfeeding (Mason and Wolfe, 1930; Mulinos and Pomerantz, 1941; Srebnik and Nelson, 1962; Root and Russ, 1972; Stewart *et al.*, 1975; Millar and Fairall, 1976). The latter seems more likely because the effects on the testis can be, at least partially, reversed by injections of gonadotrophins but there may also be a reduced response of androgen target organs including the seminiferous tubules. In immature animals, as one would expect, the effects are more pronounced than in the adult and puberty can be appreciably delayed (Widdowson *et al.*, 1964).

Energy consumption seems to be more important than the protein content of the diet, except when practically no protein is consumed (Tilton *et al.*, 1964; Moule *et al.*, 1966; Braden *et al.*, 1974) but properly controlled experiments to test this are difficult to design.

### 11.2.4.2    Specific amino-acid deficiencies
Deficiencies of certain amino-acids lead to degeneration of the testis. Arginine deficiency in men caused a sharp reduction in sperm numbers in semen within 9 days and in rats, an effect on spermatogenesis could be found after 3 weeks (Shettles, 1960). Deficiencies of lysine, tryptophane, phenylalanine and histidine have been reported to cause impairment of spermatogenesis, and contradictory reports are found for leucine (Maun *et al.*, 1945, 1946a, b; Adamstone and Spector, 1950; Schwartz *et al.*, 1951; Scott and Schwartz, 1953; Scott, 1954, 1956, 1964).

Ethionine, a methionine analogue, when added to the diet of rats, produces severe testicular damage (Kaufman *et al.*, 1956; Goldberg *et al.*, 1959a) which can be

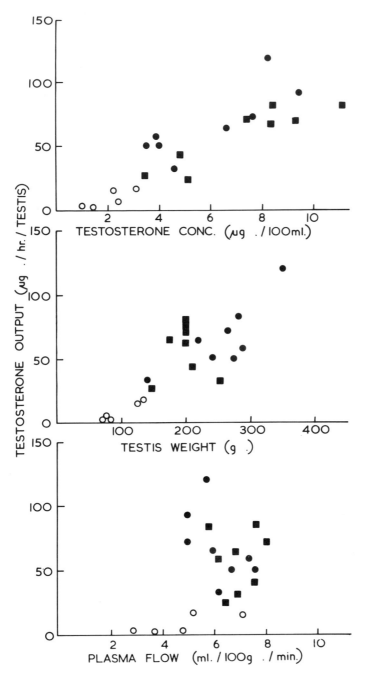

Figure 11.5 The output of testosterone, testosterone concentration in blood from the internal spermatic vein, testis weight and testicular plasma flow in well-fed conscious (■) and anaesthetized (●) rams, and in underfed anaesthetized rams (O) which had lost between 22 and 38 per cent of their body weight over 11 to 15 weeks. (Reproduced from Setchell *et al.*, 1965)

mitigated either by the simultaneous administration of methionine (Pfau and Eyal, 1963) or of testosterone (Goldberg *et al.*, 1959b, 1961). The inclusion of large amounts of tyrosine in the diet of rats inhibits spermatogenesis (Deb and Biswas, 1965; Hueper and Martin, 1943); the injection of ascorbic acid reduces its effect (Biswas and Deb, 1966).

It is also interesting to note that the inclusion of glutamine in the medium improves the function of pieces of testis in organ culture in chemically defined media (see Section 7.6.2).

### 11.2.4.3   Essential fatty acids

As long ago as 1929, Burr and Burr noticed that a fat-free diet decreased the 'mating potential' of rats and later it was shown that such rats showed testicular degeneration (Burr and Burr, 1930; Evans *et al.*, 1934). Rats and rabbits fed diets deficient in essential fatty acids had smaller testes which lacked spermatids and spermatozoa (Panos and Finerty, 1954; Ahluwalia *et al.*, 1967; Bieri *et al.*, 1969). Androgen production appeared to be normal (Ahluwalia *et al.*, 1968, 1971; van der Molen and Bijlveld, 1971) although injections of testosterone of hCG restored testicular weight to control levels and improved the histological appearance (Panos *et al.*, 1959). The effects of the deficiency could be intensified by including cholesterol in the diet (Holman and Peifer, 1960), or prevented by the inclusion of methyl arachidonate in the diet or daily injections of prostaglandin $E_2$ (Hafiez, 1974). When rats were fed a diet containing peanut oil as the only lipid, there was a marked impairment in spermatogenesis which could be prevented by adding ethyl linoleate to the diets (Aaes-Jorgensen *et al.*, 1967).

Changes in the pituitary resembling castration changes were seen in EFA-deficient rats but not rabbits (Panos *et al.*, 1959; Ahluwalia *et al.*, 1967). Erucic acid, which is thought to compete with or anatagonize essential fatty acids, produces testicular lesions in rats very like those in essential fatty acid deficiency (Carroll and Noble, 1957).

### 11.2.4.4   Zinc and other minerals

The testis is affected in many mineral deficiencies where there is general debility. One well-established specific effect on the testis has been found, namely zinc deficiency. This deficiency produces an atrophy of the testes of young rats and rams and a reduction in sperm production in rams which does not appear in pair-fed control animals. In rats, the lesion cannot be reversed by adding zinc to the diet after the atrophy has developed but in rams the effects of zinc deficiency on the testes are reversible (Millar *et al.*, 1958; Underwood and Somers, 1969). Effects of zinc deficiency on the testes have also been described in mice, cattle, goats, dogs and humans (Miller and Miller, 1962; Miller *et al.*, 1964; Pitts *et al.*, 1966; Prasad *et al.*, 1967; Sandstead *et al.*, 1967). Pituitary depression is involved in the lesions in young animals, but in adult animals, the tubular lesions develop in spite of treatment with gonadotrophins (Millar *et al.*, 1960) and indeed the pituitary of zinc deficient rats shows 'castration changes' suggesting enhanced secretion of gonadotrophins (Millar *et al.*, 1958; Lei *et al.*, 1976).

Zinc is an essential constituent of certain enzymes, notably carbonic anhydrase, lactate dehydrogenase and malate dehydrogenase and zinc inhibits the activity of ribonuclease (see Gunn and Gould, 1970). In zinc deficiency, ribonuclease activity is increased and this may lead to defective nucleic acid and protein metabolism (Macapinlac *et al.*, 1968; Somers and Underwood, 1969; Williams and Chesters, 1969). Zinc is present in high concentrations in ejaculated spermatozoa. However the concentrations in the secretions of the accessory glands and the glands themselves are remarkably high, much higher than in the testis and this may be the source of much of the zinc in ejaculated spermatozoa (see Mann, 1964). Nevertheless the testis does take up considerable amounts of $^{65}$Zn. This uptake was reduced after hypophysectomy although only periods longer than 9 days were studied when changes in the cellular composition of the testis would be evident. Treatment with pituitary hormones for 5 days restored zinc uptake (Gunn *et al.*, 1961). It was also reduced by adrenalectomy and restored by administration of cortisone (Rudzik and Reidel, 1960a). Surprisingly, ACTH caused a further decrease in uptake of zinc in hypophysectomized rats (Rudzik and Reidel, 1960b).

Zinc is incorporated into the spermatozoa in the testis during the later stages of spermiogenesis. However uptake is comparatively slow. Maximum values are not reached until 7 to 10 days after a single injection (Gunn and Gould, 1970), presumably because the zinc does not readily cross the blood–testis barrier and is transported into the tubules by some carrier mechanism.

### 11.2.4.5   Vitamin A deficiency

Diets deficient in vitamin A cause degeneration of the seminiferous tubules (Gross, 1924; Wolbach and Howe, 1925, 1928; Mason, 1930, 1933, 1939; Evans, 1932; Sapsford, 1951; Gunn *et al.*, 1942; Hodgson *et al.*, 1946; Madsen *et al.*, 1948; Lindley *et al.*, 1949; Gambal, 1966; Palludan, 1963, 1966; Scott and Scott, 1964; Thompson *et al.*, 1964) even when retinoic acid is included in the diet to prevent abnormalities elsewhere in the body except the retina (Coward *et al.*, 1966; Howell *et al.*, 1967; Ahluwalia and Bieri, 1970). Eventually only Sertoli cells and spermatogonia remain in the tubules, and the Leydig cells also appear to be affected. The effects of deficiency have been reported in rats, cats, bulls, boars and rams. The effects of vitamin deficiency can be prevented by intratesticular injections of retinol (Palludan, 1966; Ahluwalia and Bieri, 1971), so the effect appears to be directly on the testis. An induced vitamin A deficiency in the testis may occur in alcoholics because ethanol blocks the conversion by alcohol dehydrogenase of retinol to its active form retinal in the testis (van Thiel *et al.*, 1974). It is curious that retinol and retinoic acid bind specifically to the rat testis to approximately the same extent, whereas in several other tissues, the amount of retinoic acid bound was much less than that of retinol. The binding to the testis per mg of protein is higher for the testes than for any of the other tissues studied (Ong and Chytil, 1975).

Vitamin A deficient diets with added retinoic acid fed to pregnant rats and to their male offspring, produced animals with only Sertoli cells in their testes. The plasma testosterone concentration was slightly decreased in these animals, but rather surprisingly, plasma FSH was normal and LH slightly decreased, although these hormones increased normally after castration (Krueger *et al.*, 1974 but see Rich and de Kretser, 1977).

Excess vitamin A can also inhibit spermatogenesis (Biswas and Deb, 1965).

### 11.2.4.6    Vitamin B deficiencies

Deficiencies in several B-vitamins have noticeable effects on the testis. Thiamine deficiency leads to decreased androgen production by the Leydig cells of rats. However this appears to be largely due to depression of pituitary function since treatment with gonadotrophins restores the androgenic function of the testis to normal (Lutwak-Mann and Mann, 1950; Lutwak-Mann, 1958).

Mild pyridoxine deficiency in young rats causes a reduction in growth of tubules and Leydig cells; severe deficiency leads to tubular degeneration with only Sertoli cells remaining. Again the testis respond to gonadotrophins and testosterone so the primary lesion may be in the pituitary (Delost and Terroine, 1966). Pyridoxine deficiency reduced incorporation of linoleate into phospholipids and triglycerides (Goswami and Coniglio, 1966).

Biotin deficiency causes cryptorchidism, and retarded growth of the testis in young animals and degeneration of the testis in adult rats. Spermatocytes and spermatogonia appear to be mainly affected, and treatment with testosterone enhanced the severity of the effect (Communal, 1957). Deficiency of pantothenic acid also leads to testicular damage in rats (Fidanza *et al.*, 1956; Baboriak *et al.*, 1958).

### 11.2.4.7    Vitamin C deficiency

Ascorbic acid deficiency in guinea pigs is accompanied by degeneration of both Leydig cells and seminiferous tubules but many of the effects can be attributed to the associated debility (Lindsay and Medes, 1926; Mason, 1933, 1939; Mukerjee and Bannerjee, 1954; Cavazos *et al.*, 1961a, b; Terroine, 1965). Similarly, the lack of libido mentioned in clinical descriptions of scurvy in men is probably non-specific.

However, there are high concentrations of ascorbic acid in the testis of rats, where it is formed from glucose and there are high concentrations of ascorbic acid or some closely related compound in rete testis fluid (see Sections 8.2.1 and 9.1.2).

### 11.2.4.8    Vitamin E deficiency

In the rat, vitamin E deficiency produces specific and irreversible damage to the testis. The tubules often contain only Sertoli cells but the Leydig cells are not noticeably affected. The lesion in the testis develops only slowly but is completely irreversible when vitamin E is returned to the diet (Osborne and Mendel, 1919; Mattill *et al.*, 1924; Evans, 1925; Mason, 1925, 1926, 1933, 1939; Drummond *et al.*, 1939; Wu *et al.*, 1973). Similar lesions have been reported in the testis of guinea pigs and hamsters with little or no effect in rabbits, mice or in domestic ruminants (Leathem, 1970) although the last-mentioned do show other lesions of vitamin E deficiency. The fatty acid composition of the lipids of the testis is altered during vitamin E deficiency (see Section 9.1.3). Chronic administration of gonadotrophins restored the weights of the accessory sex organs of vitamin E deficient male rats (Greenberg and Ershoff, 1951).

However, castration-like changes can be seen in the pituitary during vitamin E deficiency (see Section 6.2) suggesting that the vitamin has a direct effect on the testis and that, as degeneration develops, the pituitary increases its production of gonadotrophins because of reduced negative feedback from the testes.

## References to Chapter 11.2

Aaes-Jorgensen, E. (1961) Essential fatty acids. *Physiol. Rev.* **41**, 1–51.

Aaes-Jorgensen, E. and Holman, R. T. (1958) Essential fatty acid deficiency. I. Content of polyenoic acids in testis and heart as an indicator of EFA status. *J. Nutr.* **65**, 633–41.

Aaes-Jorgensen, E., Funch, J. P., Engel, P. F. and Dam, H. (1956a) The role of fat in the diet of rats. 9. Influence of growth and histological findings of diets with hydrogenated arachis oil or no fat, supplemented with linoleic acid or raw skim milk, and of crude casein compared with vitamin test casein. *Brit. J. Nutr.* **10**, 292–304.

Aaes-Jorgensen, E., Funch, J. P. and Dam, H. (1956b) The role of fat in the diet of rats. 10. Influence on reproduction of hydrogenated arachis oil as the sole dietary fat. *Brit. J. Nutr.* **10**, 317–24.

Aaes-Jorgensen, E., Funch, J. P. and Dam, H. (1957) The role of fat in the diet of rats. 11. Influence of a small amount of ethyl linoleate on degeneration of spermatogenic tissue caused by hydrogenated arachis oil as sole dietary fat. *Brit. J. Nutr.* **11**, 298–304.

Adamstone, F. B. and Spector, H. (1950) Tryptophan deficiency in the rat. *Arch. Path.* **49**, 173–84.

Ahluwalia, B. and Bieri, J. G. (1970) Effects of exogenous hormones on the male reproductive organs of vitamin A-deficient rats. *J. Nutr.* **100**, 715–24.

Ahluwalia, B. and Bieri, J. G. (1971) Local stimulatory effect of vitamin A on spermatogenesis in the rat. *J. Nutr.* **101**, 141–52.

Ahluwalia, B., Pincus, G. and Holman, R. T. (1967) Essential fatty acid deficiency and its effects upon reproductive organs of male rabbits. *J. Nutr.* **92**, 205–14.

Ahluwalia, B., Shima, S. and Pincus, G. (1968) *In vitro* synthesis of androgens by testicular tissue of rat deficient in essential fatty acids. *J. Reprod. Fertil.* **17**, 263–73.

Alfin-Slater, R. B. and Bernick, S. (1958) Changes in tissue lipids and tissue histology resulting from essential fatty acid deficiency. *Am. J. Clin. Nutr.* **6**, 613–24.

Allen, E. (1919) Degeneration in the albino rat testis due to a diet deficient in the water soluble vitamine, with a comparison of similar degeneration in rats differently treated, and consideration of the Sertoli tissue. *Anat. Rec.* **16**, 93–117.

Altland, P. D. (1949) Effect of discontinuous exposure to 25,000 ft simulated altitude on growth and reproduction of the albino rat. *J. Exptl. Zool.* **110**, 1–17.

Altland, P. D. and Highman, B. (1968) Sex organ changes and breeding performance of male rats exposed to altitude: Effect of exercise and physical training. *J. Reprod. Fertil.* **15**, 215–22.

Baird, B. and Cook, S. F. (1962) Hypoxia and reproduction in Swiss mice. *Am. J. Physiol.* **202**, 611–5.

Babbott, D., Rubin, A. and Ginsburg, S. J. (1958) The reproductive characteristics of diabetic men. *Diabetes* **7**, 33–5.

Barboriak, J., Cowgill, G. and Whedon, A. (1958) Testicular changes in pantothenic acid deficient rats. *J. Nutr.* **66**, 457–63.

Bieri, J. G. and Prival, E. L. (1966) Effect of deficiencies of $\alpha$-tocopherol, retinol and zinc on the lipid composition of rat testes. *J. Nutr.* **89**, 55–61.

Bieri, J. G., Mason, K. E. and Prival, E. L. (1969) Essential fatty acid deficiency and the testis: lipid composition and the effect of preweaning diet. *J. Nutr.* **97**, 163–72.

Biswas, N. M. and Deb, C. (1965) Testicular degeneration in rats during hypervitaminosis A. *Endokrinologie* **49**, 64–9.

Biswas, N. M. and Deb, C. (1966) Role of ascorbic acid in testicular degeneration and adrenal hypertrophy in tyrosine-fed rats. *Endocrinology* **79**, 1157–9.

Boccabella, A. V., Salgado, E. and Alger, E. A. (1962) Testicular function and histology following serotonin administration *Endocrinology* **71**, 827–37.

Braden, A. W. H., Turnbull, K. E., Mattner, P. E. and Moule, G. R. (1974) Effect of protein and energy content of the diet on the rate of sperm production in rams. *Aust. J. Biol. Sci.* **27**, 67–73.

Brenk, H. A. S. van den and Jamieson, D. (1967) Hyperbaric oxygen and testicular damage and fertility. *Experientia* **23**, 302–3.

Burr, G. O. and Burr, M. M. (1929) A new deficiency disease produced by the rigid exclusion of fat from the diet. *J. Biol. Chem.* **82**, 345–67.

Burr, G. O. and Burr, M. M. (1930) On the nature and role of the fatty acids essential in nutrition. *J. Biol. Chem.* **86**, 587–621.

Butler, W. R., Johnson, A. D. and Gomes, W. R. (1968) Effect of short-term vitamin A deficiency on testicular lipids in the rat. *J. Reprod. Fertil.* **15**, 157–9.

Carroll, K. K. and Noble, R. L. (1957) Influence of a dietary supplement of erucic acid and other fatty acids on fertility in the rat. *Can. J. Biochem. Physiol.* **35**, 1093–105.

Carstensen, H., Marklund, S., Damber, J-E., Näsman, B. and Lindgren, S. (1976) No effect of oxygen *in vivo* on plasma or testis testosterone in rats and no induction of testicular superoxide dismutase. *J. Steroid Biochem.* **7**, 465–7.

Cavazos, L. F., Jeffrey, J. E., Manning, J. P. and Feagans, W. M. (1961a) Effects of avitaminosis C and inanition on guinea pig testes and seminal vesicles. *Anat. Rec.* **139**, 296–7.

Cavazos, L. R., Jeffrey, J. E., Manning, J. P. and Feagans, W. M. (1961b) Histochemical changes in testes and seminal vesicles of scorbutic guinea pigs. *Anat. Rec.* **140**, 71–6.

Chatterjee, A. (1968) Blockade of epinephrine induced gonadal inhibition in rats by ascorbic acid. *Endokrinologie* **53**, 242–8.

Chatterjee, A. and Paul, B. S. (1968) Testicular atrophy in rats following epinephrine administration. *Endokrinologie* **52**, 406–7.

Christian, J. J. (1959) Adrenocortical, splenic and reproductive responses of mice to inanition and grouping. *Endocrinology* **65**, 189–97.

Communal, R. (1957) Carence et subcarence en biotine; leurs rapports avec l'histopathologie testiculaire chez le rat albinos. *Compt. Rend. Acad. Sci. Paris* **245**, 372–3.

Coward, W. A., Howell, J. McC., Pitt, G. A. J. and Thompson, J. N. (1966) Effects of hormones on reproduction in rats fed a diet deficient in retinol (vitamin A alcohol) but containing methyl retinoate (vitamin A acid methyl ester). *J. Reprod. Fertil.* **12**, 309–18.

Dalton, A. J., Jones, B. F., Peters, V. B. and Mitchell, E. R. (1945) Organ changes in rats exposed repeatedly to lowered oxygen tension with reduced barometric pressure. *J. Nat. Cancer Inst.* **6**, 161–85.

Davies, D. V., Mann, T. and Rowson, L. E. A. (1957) Effect of nutrition on the onset of male sex hormone activity and sperm formation in monozygous bull-calves. *Proc. Roy. Soc.* (London) B. **147**, 332–51.

Deb, C. and Biswas, N. M. (1965) Testicular degeneration after 1-tyrosine feeding in rats, role of ascorbic acid. *Experientia* **21**, 73–4.

Deb, C. and Chatterjee, A. (1963) Role of ascorbic acid on testicular degeneration in alloxan diabetic rats. *Experientia* **19**, 595–6.

Delost, P. and Terroine, T. (1955) Etude des modifications pathologique du développement du testicule provoquées par la carence en biotine chez le rat. *C. rend. Soc. Biol.* **149**, 907–10.

Delost, P. and Terroine, T. (1966) Sur l'origine des troubles sexuels mâles de la carence en pyridoxine. Traitements hormonaux et vitaminiques. *Arch. Sci. Physiol.* **20**, 65–82.

Dickerson, J. W. T., Gresham, G. A. and McCance, R. A. (1964) The effect of undernutrition and rehabilitation on the development of the reproductive organs: Pigs. *J. Endocrinol.* **29**, 111–8.

Drummond, J. C., Noble, R. L. and Wright, M. D. (1939) Studies on the relationship of vitamin E (tocopherols) to the endocrine system. *J. Endocrinol.* **1**, 275–86.

Dutt, R. H. and Barnhart, C. E. (1959) Effect of plane of nutrition upon reproductive performance of boars. *J. Anim. Sci.* **18**, 3–13.

Elcoate, P. V., Fischer, M. I., Mawson, C. A. and Illar, M. T. (1955) The effect of zinc deficiency on the male genital system. *J. Physiol.* (London) **129**, 53P–54P.

Evans, H. M. (1925) Invariable occurrence of male sterility with dietaries lacking fat-soluble vitamin E. *Proc. Natl. Acad. Sci. USA* **11**, 373–77.

Evans, H. M. (1932) Testicular degeneration due to inadequate vitamin A in cases where E is adequate. *Am. J. Physiol.* **99**, 477–86.

Evans, H. M. and Burr, G. O. (1925) The antisterility vitamin fat-soluble E. *Proc. Natl. Acad. Sci. USA* **11**, 334–41.

Evans, H. M., Lepkovsky, S. and Murphy, E. A. (1934) Vital need of the body for certain unsaturated fatty acids. VI. Male sterility on fat free diet. *J. Biol. Chem.* **106**, 445–50.

Faerman, I., Vilar, O., Rivarola, M. A., Rosner, J. M., Jadzinsky, M. N., Fox, D., Perez Lloret, A., Bernstein-Hahn, L. and Saraceni, D. (1972) Impotence and diabetes. Studies of androgenic function in diabetic impotent males. *Diabetes* **21**, 23–30.

Federlin, K., Schoffling, K., Neubronner, P. and Pfeiffer, E. F. (1965) Histometric studies of the testes of diabetics with gonadal function disturbance. *Diabetologia* **1**, 85–90.

Fels, E. (1959) Greffe ovarienne intrasplénique chez des rats à pédicule testiculaire lié. *Compt. Rend. Soc. Biol.* **153**, 1277–8.

Fels, E. and Bur, G. E. (1958) Modification du testicule du rat par ligature du pédicule vasculaire. *Compt. Rend. Soc. Biol.* **152**, 1395–6.

Fidanza, A., Esposito, L. and Bonomolo, R. (1956) Acido pantotenico e gonadi. *Boll. Soc. Ital. Biol. Sper.* **32**, 1188–90.

Folgia, V. G., Borghelli, R. F., Chieri, R. A., Fernandez-Collazo, E. L., Spindler, I. and Wesely, O. (1963) Sexual disturbances in the diabetic rat. *Diabetes* **12**, 231–7.

Funch, J. P., Aaes-Jorgensen, E. and Dam, H. (1957) The role of fat in the diet of rats. 12. Effect on rats of type and quantity of dietary fat with and without linoleate supplementation. *Brit. J. Nutr.* **11**, 426–33.

Gambal, D. (1966) Effect of hormones on the testicular lipids of vitamin A deficient rats. *J. Nutr.* **89**, 203–9.

Gerschman, R. (1964) Biological effects of oxygen. In *Oxygen in the Animal Organism*, eds. F. Dickens and E. Neil, 475, New York: Macmillan.

Goldberg, G. M., Pfau, A. and Ungar, H. (1959a) Effect of testosterone on testicular lesions produced by DL-ethionine in rats. *Am. J. Pathol.* **35**, 649–57.

Goldberg, G. M., Pfau, A. and Ungar, H. (1959b) Testicular lesions following ingestion of DL-ethionine studied by a quantitative cytologic method. *Am. J. Pathol.* **35**, 383–91.

Goldberg, G. M., Goldberg, S. and Gold, J. J. (1961) An endocrine mechanism involved in testicular changes produced by a DL-ethionine supplemented diet. A study of morphologic changes and endocrine dysfunction. *Endocrinology* **69**, 430—7.

Gordon, A. S., Tornetta, F. J., D'Angelo, S. A. and Charipper, H. A. (1943) Effects of low atmospheric pressure on the activity of the thyroid, reproductive system and anterior lobe of the pituitary in the rat. *Endocrinology* **33**, 366–83.

Goswami, A. and Coniglio, J. G. (1966) Effect of pyridoxine deficiency on the metabolism of linoleic acid in the rat. *J. Nutr.* **89**, 210–6.

Greenberg, S. M. and Ershoff, B. H. (1951) Effects of chorionic gonadotropin on sex organs of male rats deficient in essential fatty acids. *Proc. Soc. Exptl. Biol. N.Y.* **78**, 552–4.

Griffiths, J. (1896) The effects upon the testes of ligature of the spermatic artery, spermatic veins, and of both artery and veins. *J. Anat. Physiol.* **30**, 81–105.

Gross, L. (1924) The effects of vitamin deficient diets on rats, with special reference to the motor functions of the intestinal tract *in vivo* and *in vitro*. *J. Path.* **27**, 27–50.

Gunn, R. M. C., Sanders, R. N. and Granger, W. (1942) Studies in fertility in sheep. 2. Seminal changes affecting fertility in rams. *Bull. C.S.I.R.O.* No. 148.

Gunn, S. A., Gould, T. C. and Anderson, W. A. D. (1961) Hormonal control of zinc in mature rat testis. *J. Endocrinol.* **23**, 37–45.

Hafiez, A. A. (1974) Prostaglandin E$_2$ prevents impairment of fertility in rats fed a diet deficient in essential fatty acids. *J. Reprod. Fertil.* **38**, 273–86.

Hellner, H. (1933) Die oertlichen Kreislaufstoerungen des Hodens. *Beitr. Klin. Chir.* **158**, 225–69.

Herrick, E. H., Mead, E. R., Egerton, B. W. and Hughes, J. S. (1952) Some effects of cortisone on vitamin-C deficient guinea pigs. *Endocrinology* **50**, 259–63.

Hodgson, R. E., Hall, S. R., Sweetman, W. J., Wiseman, H. G. and Converse, H. T. (1946) The effect of vitamin A deficiency on reproduction in dairy bulls. *J. dairy Sci.* **29**, 669–87.

Holman, R. T. (1968) Essential fatty acid deficiency, a long scaly tale. *Progr. Chem. Fats* **9**, 279–348.

Holman, R. T. and Aaes-Jorgensen, R. (1956) Effects of trans fatty acid isomers upon essential fatty acid deficiency in rats. *Proc. Soc. Exptl. Biol. N.Y.* **93**, 175–9.

Holman, R. T. and Peifer, J. J. (1960) Acceleration of essential fatty acid deficiency by dietary cholesterol. *J. Nutr.* **70**, 411–17.

Homm, R. E., Rusticus, C. and Hahn, D. W. (1977) The antispermatogenic effects of 5-thio-D-glucose in male rats. *Biol. Reprod.* **17**, 697–700.

Howell, J. McC., Thompson, J. N. and Pitt, G. A. J. (1967) Changes in the tissues of guinea pigs fed on a diet free from vitamin A but containing methyl retinoate. *Brit. J. Nutr.* **21**, 37–44.

Howland, B. E. and Zebrowski, E. J. (1972) Hyposecretion of gonadotrophins in alloxan-treated male rats. *J. Reprod. Fertil.* **31**, 115–18.

Howland, B. E. and Zebrowski, E. J. (1974) Serum and pituitary gonadotropin levels in alloxan-diabetic rats. *Horm. Metab. Res.* **6**, 121–4.

Howland, B. E. and Zebrowski, E. J. (1976) Some effects of experimentally-induced diabetes on pituitary-testicular relationships in rats. *Horm. Metab. Res.* **8**, 465–9.

Hunt, E. L. and Baily, D. W. (1961) The effect of alloxan diabetes on the reproductive system of young male rats. *Acta Endocrinol.* **38**, 432–40.

Hueper, W. C. and Martin, G. J. (1943) Tyrosine poisoning in rats. *Arch. Path. lab. Med.* **35**, 685–94.

Ierano, A. and Mascitelli, C. (1967) Effetto dei supplementi vitaminici A ed E ed acido pantotenico sul coenzima A testicolare. *Boll. Soc. Ital. Biol. Sper.* **43**, 1025–8.

Iwasita, K. (1939) Ortliche Blutzirkulationsstorung des Hodens: klinischexperimentelle Beitrage zur Kenntnis des Eiflusses der Samenstranggebäss-Absperrung auf den Hoden. *Japan J. Dermatol. Urol.* **45**, 126–35.

Kaufman, N., Klavins, J. V. and Kinney, T. D. (1956) Testicular damage following ethionine administration. *Am. J. Pathol.* **32**, 105–15.

Kent, J. R. (1966) Gonadal function in impotent diabetic males. *Diabetes* **15**, 537.

Krueger, P. M., Hodgen, G. D. and Sherins, R. J. (1974) New evidence for the role of the Sertoli cell and spermatogonia in feedback control of FSH secretion in male rats. *Endocrinology* **95**, 955–62.

Kormano, M., Karhunen, P. and Kahanpää, K. (1968) Effect of long-term 5-hydroxytryptamine treatment on the rat testis. *Ann. Med. Exptl. Fenn.* **46**, 474–8.

Leathem, J. H. (1970) Nutrition. In *The Testis*, eds. A. D. Johnson, W. R. Gomes and N. L. VanDemark, III, 169–205, New York: Academic Press.

Leathem, J. H. (1975) Nutritional influences on testicular composition and function in mammals. *Handbk. Physiol.* Sect. 7, Vol. V, 225–32.

Lei, K. Y., Abbasi, A. and Prasad, A. S. (1976) Function of pituitary-gonadal axis in zinc-deficient rats. *Am. J. Physiol.* **230**, 1730–2.

Lindley, C. E., Brugman, H. H., Cunha, T. J. and Warwick, E. J. (1949) The effect of vitamin A deficiency on semen quality and the effect of testosterone and pregnant mare serum on vitamin A deficient rams. *J. anim. Sci.* **8**, 590–602.

Lindsey, B. and Medes, G. (1926) Histological changes in the testis of the guinea pig during scurvy and inanition. *Am. J. Anat.* **37**, 213–35.

Lutwak-Mann, C. (1958) The dependence of gonadal function upon vitamins and other nutritional factors. *Vit. Horm.* **16**, 35–75.

Lutwak-Mann, C. and Mann, T. (1950) Restoration of secretory function in male accessory glands of vitamin B-deficient rats by means of chorionic gonadotrophin. *Nature* **165**, 556–7.

Maatz, R. (1934) Uber die anatomischen Folgen vorubergehender Gefassabklenmung am Hoden. *Zentr. Allgem. Pathol. Pathol. Anat.* **59**, 243–9.

Macapinlac, M. P., Pearson, W. N., Barney, G. H. and Darby, W. J. (1968) Protein and nucleic acid metabolism in the testis of zinc-deficient rats. *J. Nutr.* **95**, 569–77.

Madsen, L. L., Eaton, O. N., Heemstra, L., Davis, R. E., Cabell, C. A. and Knapp, B. (1948) Effectiveness of carotene and failure of ascorbic acid to increase sexual activity and semen quality of vitamin A deficient beef bulls. *J. anim. Sci.* **7**, 60–9.

Mancini, R. E., Penhos, J. C., Izquierdo, I. A. and Heinrich, J. J. (1960) Effects of acute hypoglycaemia on the rat testis. *Proc. Soc. exptl. Biol. N.Y.* **104**, 699–702.

Mann, T. (1960) Effect of nutrition on androgenic activity and spermatogenesis in mammals. *Proc. Nutr. Soc.* **19**, 15–18.

Mann, T. (1964) *The Biochemistry of Semen and of the Male Reproductive Tract*, London: Methuen.

Martins, T. (1933) Effets de l'ischémie du testicule après ligature de l'artère spermatique interne chez le rat. Destruction de l'épithélium séminal, maintien de la genitalia accessoria, altérations du lobe antérieur de l'hypophyse. *Compt. Rend. Soc. Biol.* **114**, 141–3.

Mason, K. E. (1925) Sterility in the albino rat due to a dietary deficiency. *Proc. Natl. Acad. Sci. USA* **11**, 377–82.

Mason, K. E. (1926) Testicular degeneration in albino rats fed a purified food ration. *J. exptl. Zool.* **45**, 159–229.

Mason, K. E. (1930) The specificity of vitamin E for the testis. I. Relation between vitamins A and E. *J. exptl. Zool.* **55**, 101–22.

Mason, K. E. (1933) Differences in testis injury and repair after vitamin A-deficiency, vitamin E-deficiency, and inanition. *Am. J. Anat.* **52**, 153–239.

Mason, K. E. (1939) Relation of the vitamins to the sex glands. In *Sex and Internal Secretion*, ed. E. Allen, 1149–212. London: Balliere, Tindall and Cox.

Mason, K. E. and Wolfe, J. M. (1930) The physiological activity of the hypophyses of rats under various experimental conditions. *Anat. Rec.* **45**, 232.

Matteo, R. S. and Nahas, G. G. (1963) Sodium bicarbonate: Increase in survival rate of rats inhaling oxygen. *Science* **141**, 719–20.

Mattill, H. A., Carman, J. S. and Clayton, M. M. (1924) The nutritive properties of milk. III. The effectiveness of the X substance in preventing sterility in rats on milk rations high in fat. *J. Biol. Chem.* **61**, 729–40.

Maun, M. E., Cahill, W. M. and Davis, R. M. (1945) Morphological studies of rats deprived of essential amino acids. I. Phenylalanine. *Arch. Path. Lab. Med.* **39**, 294–300.

Maun, M. E., Cahill, W. M. and Davis, R. M. (1946a) Morphological studies of rats deprived of essential amino acids. II. Leucine. *Arch. Path. Lab. Med.* **40**, 173–8.

Maun, M. E., Cahill, W. M. and Davis, R. M. (1946b) Morphological studies of rats deprived of essential amino acids. III. Histidine. *Arch. Path. Lab. Med.* **41**, 25–31.

Miflet, J. (1879) Uber die pathologischen Veranderungen des Hodens, welche durch Storungen der lokalen Blutzirknlation verursacht werden. *Arch. Klin. Chir.* **24**, 399–428.

Millar, M. J., Fischer, M. I., Elcoate, P. V. and Mawson, C. A. (1958) The effects of dietary zinc deficiency on the reproductive system of male rats. *Can. J. Biochem. Physiol.* **36**, 557–69.

Millar, M. J., Elcoate, P. V., Fischer, M. I. and Mawson, C. A. (1960) Effect of testosterone and gonadotrophin injections on the sex organ development of zinc-deficient male rats. *Can. J. Biochem. Physiol.* **38**, 1457–66.

Millar, R. and Fairall, N. (1976) Hypothalamic, pituitary and gonadal hormone production in relation to nutrition in the male hyrax (*Procavia capensis*). *J. Reprod. Fertil.* **47**, 339–41.

Miller, J. K. and Miller, W. J. (1964) Experimental zinc deficiency and recovery of calves. *J. Nutr.* **76**, 467–74.

Miller, W. J., Pitts, W. J., Clifton, C. M. and Schmittle, S. C. (1964) Experimentally produced zinc deficiency in the goat. *J. Dairy Sci.* **47**, 556–9.

Molen, H. J. van der and Bijlveld, M. J. (1971) Testosterone production by testis of essential fatty acid deficient rats. *Acta. Endocrinol.* Suppl. 155, 71.

Monge, C. (1942a) Life in the Andes and chronic mountain sickness. *Science* **95**, 79–84.

Monge, C. (1942b) Fisiologia de la reproducion en la altur: Aplicationes a la industria animal. *Anales Fac. Med. Univ. nacl. mayor San Marcos, Lima* **25**, 19–33.

Monge, C. (1943) Chronic mountain sickness. *Physiol. Rev.* **32**, 166–84.

Monge, C. (1948) *Acclimation in the Andes*, Baltimore: John Hopkins Press.

Moore, C. R. and Samuels, L. T. (1931) The action of testis hormone in correcting changes induced in the rat prostate and seminal vesicles by vitamin B deficiency or partial inanition. *Am. J. Physiol.* **96**, 278–88.

Moule, G. R., Braden, A. W. H. and Mattner, P. E. (1966) Effects of season, nutrition and hormone treatment on the fructose content of ram semen. *Aust. J. Agric. Res.* **17**, 923–31.

Mukherjee, A. K. and Banerjee, S. (1954) Studies on histological changes in experimental scurvy. *Anat. Rec.* **120**, 907–15.

Mulinos, M. G. and Pomerantz, L. (1941) The reproductive organs in malnutrition: Effects of chorionic gonadotropin upon atrophic genitalia of underfed male rats. *Endocrinology* **29**, 267–75.

Neumann, F. and Schenck, B. (1977) Formal genesis of giant cells in the germinal epithelium in the rat thioglucose model. *Andrologia* **9**, 323–8.

Oettle, A. G. and Harrison, R. G. (1952) The histological changes produced in the rat testis by temporary and permanent occlusion of the testicular artery. *J. Pathol. Bacteriol.* **64**, 273–97.

Oksanen, A. (1975) Testicular lesions of streptozotocin diabetes rats. *Hormone res.* **6**, 138–44.

Ong, D. E. and Chytil, F. (1975) Retinoic acid-binding protein in rat tissue. Partial purification and comparison to rat tissue retinol-binding protein. *J. biol. Chem.* **250**, 6113–7.

Osborne, T. B. and Mendel, L. B. (1919) The nutritive value of yeast protein. *J. biol. Chem.* **38**, 223–7.

Ozorio de Almeida, A. (1934) Recherches sur l'action toxiques das hautes pressions d'oxygène. *Compt. Rend. Soc. Biol.* **116**, 1225–7.

Palludan, B. (1963) Vitamin A deficiency and its effect on the sexual organs of the boar. *Acta Vet. Scand.* **4**, 136–55.

Palludan, B. (1966) Direct effect of vitamin A on boar testis. *Nature* **211**, 639–40.

Panos, T. C. and Finerty, J. C. (1954) Effects of a fat free diet on growing male rats with special reference to the endocrine system. *J. Nutr.* **54**, 315–31.

Panos, T. C., Klein, G. F. and Finerty, J. C. (1959) Effect of fat deficiency of pituitary-gonad relationships. *J. Nutr.* **68**, 509–40.

Pfau, A. and Eyal, Z. (1963) Effect of DL-methionine on testicular lesions produced by DL-ethionine in rats. *Lab. Invest.* **12**, 911–20.

Pitts, W. J., Miller, W. J., Fosgate, O. T., Morton, J. D. and Clifton, C. M. (1966) Effect of zinc deficiency and restricted feeding from two to five months of age on reproduction in Holstein bulls. *J. Dairy Sci.* **49**, 995–1000.

Popoff, J. C. and Okultilschew, G. Z. (1936) Der Einfluss der Ernahrung auf die Geschlects-function. *Z. Zuechtung,* Ser. B. **34**, 221–39.

Prasad, A. S., Oberleas, D., Wolf, P. and Horwitz, J. P. (1967) Studies on zinc deficiency: changes in trace elements and enzyme activities in tissues of zinc deficient rats. *J. Clin. Invest.* **46**, 549–57.

Rauch-Stroomann, J. G., Petry, R., Mauss, J., Heinz, H. A., Jakubowsli, H. D., Senge, T., Muller, K. M., Eckardt, B., Berthold, K. and Sauer, H. (1970) Studies on sexual function in diabetes. *Excerpta med. Found. Symp.* **209**, 112.

Rich, K. A. and Kretser, D. M. de (1977) Effect of differing degrees of destruction of the rat seminiferous epithelium on levels of serum follicle stimulating hormone and androgen binding protein. *Endocrinology* **101**, 959–68.

Root, A. W. and Russ, R. D. (1972) Short-term effects of castration and starvation upon pituitary and serum levels of luteinizing and follicle stimulating hormone in male rats. *Acta Endocrinol.* **70**, 665–75.

Rudzik, A. D. and Riedel, B. E. (1960a) The effects of adrenalectomy and cortisone on zinc metabolism in the sex glands and adrenal of the male rat. *Can. J. Biochem. Physiol.* **38**, 845–51.

Rudzik, A. D. and Riedel, B. E. (1960b) The effects of hypophysectomy and ACTH on zinc metabolism in the sex glands and adrenals of the male rat. *Can. J. Biochem. Physiol.* **38**, 1003–8.

Sandstead, H. H., Prasad, A. S., Schulert, A. R., Farid, Z., Miale, A., Bassilly, S. and Darby, W. J. (1967) Human zinc deficiency. Endocrine manifestations and response to treatment. *Am. J. Clin. Nutr.* **20**, 422–42.

Sapsford, C. S. (1951) Seasonal changes in spermatogenesis in rams: their relation to plane of nutrition and to vitamin A status. *Aust. J. agric. Res.* **2**, 331–61.

Schoffling, K., Federlin, K., Ditschuneit, H. and Pfeiffer, E. F. (1963) Disorder of sexual function in male diabetics. *Diabetes* **12**, 519–27.

Schoffling, K., Federlin, K., Schmitt, W. and Pfeiffer, E. F. (1967) Histometric investigation of the testicular tissue of rats with alloxan diabetes and Chinese hamsters with spontaneous diabetes. *Acta Endocrinol.* **54**, 335–46.

Schwartz, C., Scott, E. B. and Ferguson, R. L. (1951) Histopathology of amino acid deficiencies. I. Phenylalanine. *Anat. Rec.* **110**, 313–27.

Scott, E. B. (1954) Histopathology of amino acid deficiencies. III. Histidine. *Arch. Path.* **58**, 129–41.

Scott, E. B. (1956) Histopathology of amino acid deficiencies. V. Isoleucine. *Proc. Soc. Exptl. Biol. N.Y.* **92**, 134–40.

Scott, E. B. (1964) Histopathology of amino acid deficiencies. VII. Valine. *Exptl. Mol. Pathol.* **3**, 610–21.

Scott, E. B. and Schwartz, C. (1953) Histopathology of amino-acid deficiencies. II. Threonine. *Proc. Soc. exp. Biol. N.Y.* **84**, 271–6.

Scott, P. P. and Scott, M. G. (1964) Vitamin A and reproduction in the cat. *J. Reprod. Fertil.* **8**, 270–1.

Setchell, B. P., Waites, G. M. H. and Lindner, H. R. (1965) Effect of undernutrition on testicular blood flow and metabolism and the output of testosterone in the ram. *J. Reprod. Fertil.* **9**, 149–62.

Setchell, B. P., Waites, G. M. H. and Thorburn, G. D. (1966) Blood flow in the testis of the conscious ram measured with krypton-85; effects of heat, catecholamines and acetyl choline. *Circulation Res.* **18**, 755–65.

Sewell, R. F. and Miller, I. L. (1966) Fatty acid composition of testicular tissue from EFA-deficient swine. *J. Nutr.* **88**, 171–5.

Shettles, L. B. (1947) Effect of low oxygen tension on fertility in adult male guinea pigs. *Federation Proc.* **6**, 200.

Shettles, L. B. (1960) The relation of dietary deficiencies to male fertility. *Fertil. Steril.* **11**, 88–99.

Siperstein, D. M. (1920–1) The effects of acute and chronic inanition upon the development and structure of the testis in the albino rat. *Anat. Rec.* **20**, 355–81.

Smith, G. I. (1955) Cellular changes from graded testicular ischemia. *J. Urol.* **73**, 355–62.

Somers, M. and Underwood, E. J. (1969) Ribonuclease activity and nucleic acid and protein metabolism in the testes of zinc-deficient rats. *Aust. J. Biol. Sci.* **22**, 1277–82.

Srebnik, H. H. and Nelson, M. M. (1962) Anterior pituitary function in male rats deprived of dietary protein. *Endocrinology* **70**, 723–30.

Steinberger, E. and Tjioe, D. Y. (1969) Spermatogenesis in rat testes after experimental ischemia. *Fertil. Steril.* **20**, 639–49.

Stewart, S. F., Kopia, S. and Gawlak, D. L. (1975) Effect of underfeeding, hemigonadectomy, sex and cyproterone acetate on serum FSH levels in immature rats. *J. Reprod. Fertil.* **45**, 173–6.

Terroine, T. (1965) Le contrôle vitaminique de la richesse en acides nucleiques. II. Acide ascorbique. *Arch. Sci. Physiol.* **19**, 81–98.

Terroine, T. and Delost, P. (1961) Influence des carences vitaminiques sur le développement des organes sexuels. *Ann. Nutr. Aliment.* **15**, B291–B329.

Thiel, D. H. van, Gavaler, J. S. and Lester, R. (1974) Ethanol inhibition of vitamin A metabolism in the testes: possible mechanism for sterility in alcoholics. *Science,* New York. **186**, 941–2.

Thompson, J. N., Howell, J. McC. and Pitt, G. A. J. (1964) Vitamin A and reproduction in rats. *Proc. Roy. Soc.* B. **159**, 510–35.

Tjioe, D. Y. and Steinberger, E. (1970) A quantitative study of the effect of ischaemia on the germinal epithelium of rat testes. *J. Reprod. Fertil.* **21**, 489–94.

Tilton, W. A., Warnick, A. C., Cunha, J. J., Loggins, P. E. and Shirley, R. L. (1964) Effect of low energy and protein intake on growth and reproductive performance of young rams. *J. Anim. Sci.* **23**, 645–50.

Underwood, E. J. and Somers, M. (1969) Studies of zinc nutrition in sheep. I. The relation of zinc to growth, testicular development and spermatogenesis in young rams. *Aust. J. Agric. Res.* **20**, 889–97.

Walker, B. L. (1968) Recovery of rat tissue lipids from essential fatty acid deficiency: Brain, heart and testes. *J. Nutr.* **94**, 469–74.

Walton, A. and Uruski, W. (1946) The effects of low atmospheric pressure on the fertility of male rabbits. *J. Exptl. Biol.* **23**, 71–6.

Widdowson, E., Mavor, W. O. and McCance, R. A. (1964) The effect of undernutrition and rehabilitation on the development of the reproductive organs: Rats. *J. Endocrinol.* **29**, 119–26.

Williams, R. B. and Chesters, J. K. (1969) Effects of zinc deficiency on nucleic acid synthesis in the rat. *Proceedings of WAAP/IBP International Symposium,* 164–7, Aberdeen, Scotland.

Wolbach, S. B. and Howe, P. R. (1925) Tissue changes following deprivation of fat-soluble A vitamin. *J. exp. Med.* **42**, 753–77.

Wolbach, S. B. and Howe, P. R. (1928) Vitamin A deficiency in the guinea pig. *Arch. Path.* **5**, 239–53.

Wolf, R. C. and Leathem, J. H. (1955) Hormonal and nutritional influences on the biochemical composition of the rat testis. *Endocrinology* **57**, 286–90.

Wu, S. H., Oldfield, J. E., Whanger, P. D. and Weswig, P. H. (1973) Effect of selenium, vitamin E and antioxidants on testicular function in rats. *Biol. Reprod.* **8**, 625–9.

Zysk, J. R., Bushway, A. A., Whistler, R. L. and Carlton, W. W. (1975) Temporary sterility produced in male mice by 5-thio-D-glucose. *J. Reprod. Fertil.* **45**, 69–72.

## 11.3    Chemical damage

Only substances which affect the testis directly are discussed here, not those drugs which affect the function of the testis by depressing the function of the pituitary; anti-androgens are dealt with in Section 6.1.8.

### 11.3.1    Metal compounds

**11.3.1.1    Cadmium salts**    Cadmium is a reasonably toxic metal but Parizek and Zahor noticed in 1956 (see also Parizek, 1960) that small doses, about one-fiftieth of the toxic dose, produced a most striking effect in the testis. Within 24 hours, the whole organ becomes discoloured and eventually the testis shrinks as the cells become necrotic. A few interstitial cells lying under the tunica survive; these eventually recolonize the tissue and androgen production returns but most animals are permanently infertile (see also Kar and Das, 1960; Kar *et al.*, 1961; Allanson and Deanesly, 1962; Mason *et al.*, 1964; Mason and Young, 1967; Bouissou and Fabre, 1965, 1966a, b; Caujolle *et al.*, 1965). Interstitial cell tumours are common in cadmium treated animals (Gunn *et al.*, 1965a). The only other tissue in the body affected by these doses is the Gasserian ganglion (Gabbiani, 1966) although the ovary and placenta are more sensitive than most organs. The damage to the testis seems to originate in the testicular capillaries which have certain anatomical features in common with the vessels of the Gasserian ganglion (see Section 3.3.3). Electron microscopic change can be seen within 6 hours after a dose of cadmium and measurements of permeability with iodinated albumin, fluorescent gamma globulin or saccharated iron oxide show that vascular permeability is increased (Chiquoine, 1964; Waites and Setchell, 1966; Clegg and Carr, 1966, 1967; Maekawa *et al.*, 1965, 1966; Gupta *et al.*, 1967). The permeability of the blood–testis and blood–rete barriers is also increased at about the same time (Johnson, 1969; Setchell and Waites, 1970; Koskimies, 1973a, b) and intratesticular pressure is increased (Kormano and Suvanto, 1968). Blood flow is initially slightly increased but as the vascular permeability increases, the blood flow falls. In the rat, by 12 hours after a single subcutaneous injection of cadmium, testicular blood flow is about 5 per cent of normal (Niemi and Kormano, 1965; Waites and Setchell, 1966; Figure 11.6). Cadmium salts, injected directly into the testes have the same effect (Kar and Das, 1962; Cameron,

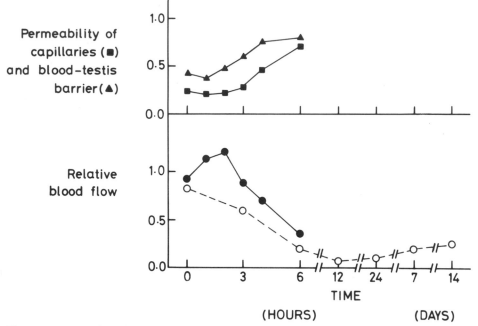

Figure 11.6 The effect of a subcutaneous injection of cadmium chloride (3 μmoles or 7mg CdCl₂. 2½H₂O/100g body weight in about 1·5ml physiological saline) on testicular blood flow (expressed as 'relative blood flow').○,experiment 1;●, experiment 2), capillary permeability measured with radioactive albumin (■) and the permeability of the blood–testis barrier, measured with radioactive rubidium (▲).
Note that the permeability of blood–testis barrier rises first, then permeability of the testicular capillaries. Blood flow at first rises, then falls to very low levels.
(Data from Waites and Setchell, 1966; and Setchell and Waites, 1970)

1965) and administration in this way removes some of the variability between species due to different rates of absorption (see Gunn and Gould, 1970a, 1975). The testes of most mammals tested (monkey, rat, mouse, gerbil, rabbit, goat, sheep, American possum) are sensitive to cadmium (Kar, 1961, 1962; Cameron and Foster, 1963; Girod, 1964; Girod and Chauvineau, 1964; Chiquoine, 1964; Gunn et al., 1965b; Ramaswami and Kaul, 1966) but the armadillo and house shrew appear not to be (Chiquoine and Suntzeff, 1965; Dryden and McAllister, 1970). The testes of very young animals are not affected (Clegg et al., 1969b; Gunn and Gould, 1970) and a second dose of cadmium has little effect on the blood vessels which grow into a testis regenerating after a previous dose (Gunn et al., 1966; Clegg et al., 1969a). Simultaneous treatment with zinc or selenium prevents the damage to the testis (Parizek, 1957, 1960; Gunn et al., 1963, 1968b; Bouissou and Fabre, 1966a; Ramaswami and Kaul, 1966; Gasiewicz and Smith, 1976) and it is interesting that the testicular capillaries as well as the formed spermatozoa in the tubules are rich in zinc (Timm and Schulz, 1966). Very little of the injected cadmium finds its way to the testis, most goes to the liver, kidney, pancreas, gut or muscle, and the cadmium in the testis is virtually confined to the interstitial tissue (Berlin and Ullberg,

1963; Kar et al., 1965; Gunn et al., 1968a; Johnson and Miller, 1970; Johnson and Sigman, 1971). Cadmium does damage isolated seminiferous tubules in vitro but only in much higher concentrations (Kar et al., 1966). Testes made cryptorchid are more sensitive to cadmium (Chatterjee and Ray, 1972) but heating the testis to about 42°C seems to lessen the effect of cadmium given just before (Bouissou and Fabre, 1966b). Preliminary exposure to a non-damaging dose reduces the effect of a damaging dose given 2 or 5 days later, but not up to 24 hours or more than 3 weeks later (Ito and Sawauchi, 1966; Gunn and Gould, 1970a, b).

**11.3.1.2    Mercury compounds**    Organic mercurial compounds reduce the uptake of thymidine by spermatogonia and of uridine and leucine by spermatids, and reduce the fertility of treated mice (Lee and Dixon, 1975).

**11.3.1.3    Lead salts**    Rats treated with lead acetate ($100\mu g$ per day for 60 days) showed atrophy and abnormal histological appearance of the testis (Hilderbrand et al., 1973).

**11.3.2    Sulphur-containing compounds**
The most important compounds in this group are esters of methane sulphonic acid; some are diesters and some are monoesters (Jackson, 1959, 1965, 1966, 1969, 1970, 1971; Fox and Fox, 1967).

**11.3.2.1    Diesters of methane sulphonic acid**    The most widely studied compound is butane dimethanesulphonate (Busulphan or Myleran), which is used for the treatment of chronic myelogenous leukemia, and has an effect on the germinal epithelium like that of radiation. In rats, it kills spermatogonia and the rest of the germ cells are then lost by a 'maturation depletion' but the drug does not affect androgen production. Larger doses kill a wider variety of spermatogonia and probably some spermatocytes, so a longer period of sterility is produced (Bollag, 1953; Jackson et al., 1959, 1962; Monoyer, 1962; Fox et al., 1963; Partington et al., 1964; Kar et al., 1968a; De Rooij and Kramer, 1968, 1970).

When given to rats between the 13th and 15th days of pregnancy, it kills the gonocytes in the fetal testes, and the young are born with no germ cells in the testis (Bollag, 1953; Hemsworth and Jackson, 1963; Forsberg and Olivecrona, 1966; Vanhems and Bousquet, 1972; Gillet and Laporte, 1973) rather like X-irradiated fetuses (see Section 11.1.1).

Busulphan kills spermatogonia in adult gerbils (Singh and Mathur, 1968) and monkeys (Kar et al., 1968b) and is a mutagenic agent in mice (Ehling and Malling, 1968).

Methylene dimethanesulphonate has an effect in adult rats at very low doses on epididymal spermatozoa. Small daily doses also produce a spermatogonial action like Busulphan (Fox and Jackson, 1965). Propylene 1,3-dimethane-sulphonate also has a Busulphan-like action (Jackson, 1969). By contrast, ethylene dimethanesulphonate produces sterility in rats beginning 1 week after injection and lasting for 8 to 10 weeks. It would therefore appear to destroy spermatids (Jackson, 1966; Jackson and Craig, 1969; Cooper and Jackson, 1970) but it may act by depressing androgen production by the testis, as the prostate and seminal vesicles in treated animals are smaller than normal, the pituitary shows castration changes and the animals can be protected by treatment with testosterone propionate (Jackson *et al.*, 1973). Several of these esters have been shown to penetrate readily through the blood–testis barrier (see Section 8.7.5).

**11.3.2.2 Monoesters of methanesulphonic acid** The simplest monoester, methyl methanesulphonate affects late spermatids and spermatozoa and even at high doses has no effect on spermatogonia (Jackson *et al.*, 1961). The activity of the esters diminishes as the carbon chain is lengthened, but branched chain esters, such as isopropyl methanesulphonate act on the spermatogonia and spermatocytes rather than the spermatids (Jackson, 1964, 1970).

Methyl methanesulphonate is a mutagenic substance (Partington and Jackson, 1963; Partington and Bateman, 1964) and produces appreciable methylation of the nucleic acids in the testis (Swann and Magee, 1968). It penetrates readily through the blood–testis barrier (see Section 8.7.5). Some of the other compounds are also mutagenic (Cattanach *et al.*, 1968; Generoso and Russell, 1969; Generoso, 1973).

Isopropyl esters of *p*-nitro-benzene sulphonic acid, *m*-mitro-benzene sulphonic acid and *p*-acetoamido-benzene sulphonic acid have an antifertility action in mice and rats, although surprisingly the methyl esters were ineffective even at near-toxic doses. The ortho-isomers of the isopropyl esters, and the ethylene, propylene and butylene diesters were also ineffective (Rooney and Jackson, 1974).

**11.3.3    Nitrogen-containing compounds**

**11.3.3.1    Nitrofurans** Nitrofurans were introduced originally as bacteriostatic compounds, but it was soon noticed that they have an effect on the testis. They prevent cells from developing beyond primary spermatocyte stage with the first changes appearing in spermatocytes about 72 hours after the beginning of treatment. Androgen production was apparently unaffected, but castration cells appeared in the pituitary suggesting that gonadotrophin secretion was increased (Section 6.2). The compounds so far examined include furacin, furadoxyl, furadantin and furazolidine (Prior and Ferguson, 1950; Nelson and Steinberger, 1953; Paul *et al.*, 1953; Featherstone *et al.*, 1955; Nelson and Bunge, 1957; Nelson and Patanelli, 1965; Hollinger and Davis, 1966, 1969; Heinke and Jaeschke, 1969).

Niridazole (Ambilhar, Ciba) is a related compound widely used as a schistosomicide. It disrupts spermatogenesis in rats, mice and guinea pigs, although it does not appear to be mutagenic, as judged by the dominant lethal test (Lambert *et al.*, 1965; Jackson *et al.*, 1974; Etribi *et al.*, 1976; Jones *et al.*, 1976).

**11.3.3.2    Dinitropyrroles** Of this series of compounds, the most effective in rats is ORF 1616 (1-(N-diethylcarbamylmethyl)-2,4-dinitropyrrole). A single dose (500mg/kg) produces infertility within 21 days, lasting for about 4 weeks. After this time fertility is completely restored. The site of damage is in the spermatocytes and spermatids, with no apparent changes in the spermatogonia (King *et al.*, 1963; Patanelli and Nelson, 1964; Debeljuk, 1975).

**11.3.3.3    Nitroimidazoles**  A large number of compounds of this type have been shown to have an effect on the testis. The most potent appears to be 1-methyl-5-nitroimidazole and this compound has been studied in some detail. In doses of 50mg/kg/day per os in propylene glycol, it appears to kill the pachytene spermatocytes and the more developed cells are removed by maturation depletion (Patanelli, 1975).

**11.3.3.4    Quinazolinone and benzimidazole derivatives**  2,3-dihydro-2-(1-naphthyl)-4(1H)-quinazolinone (U 29,409), urea 1,1-dimethyl-3-(5-methyl-2-benzimidazolyl)-hydrochloride hydrate (U 32 422E) and 2-benzimidazolecarbamic acid, methyl ester (U 32 104) cause the exfoliation of germ cells from the seminiferous tubules, due to premature release of spermatids (Ericsson, 1971; Parvinen and Kormano, 1974). The latter two drugs in smaller doses cause disturbances of cell division around meiosis.

**11.3.3.5    Aziridines**  Triethylene melamine destroys spermatogonia, but the rapidity of onset of sterility suggests that it also affects spermatids and spermatozoa. At lower doses (0·2mg/kg) the effect of spermatids remains, while that on spermatogonia and spermatozoa is lost; normal numbers of spermatozoa are found suggesting that the low dose effect is a subtle one, and no corresponding histological changes can be found. Androgen production appears to be normal (Jackson and Bock, 1955; Bock and Jackson, 1957; Cattanach, 1957; Cattanach and Edwards, 1958; Steinberger *et al.*, 1959; Steinberger, 1962; Kat *et al.*, 1968b; Paufler and Foote, 1969). Triethylene melamine is also mutagenic (Bateman, 1960; Ehling, 1971; Generoso, 1973; Matter and Generoso, 1974).

In the rat, a number of other compounds of this type also cause sterility (e.g. NN ethyleneurea, ethylenediamine, monoethyleneurea, carbethoxy ethyleneimine, methyl ethyleneurea diethyleneiminosulphoxide, tris (ethyleneimino) benzo quinone (Trenimone), chloro-bis (ethyleneimino) pyrimidine). In general, increasing the number of alkylating groups on the molecule increases the effect on the spermatogonia without decreasing the effect on the post-meiotic cells (Jackson *et al.*, 1959; Fox and Fox, 1967; Jackson, 1959, 1966, 1969). Some members of a series of bis-aziridine derivatives of certain diamines caused infertility in male mice (Skinner and Tong, 1968).

**11.3.3.6    Hydroxyurea**  Hydroxyurea (3mg/ml) in drinking water of sexually mature male rats caused testicular atrophy. Histological changes began to appear after about 14 days of treatment, the first abnormalities seen being eosinophilic cells with large dense nuclei at the level of the primary spermatocytes, and multinucleated eosinophilic cells nearer the centres of the tubules. Eventually the tubules contained only Sertoli cells and A type spermatogonia. Hydroxyurea blocks cell division by inhibiting the reduction of ribonucleoside diphosphate to the corresponding deoxyribonucleoside and thereby stopping the synthesis of DNA. It therefore presumably stops mitosis of the spermatogonia and the synthesis of DNA by the pre-leptotene spermatocytes. The other cells would then be lost by maturation depletion, but the occurrence of giant cells suggests that there is also a direct action on the spermatids.

After treatment was terminated, the epithelium was repopulated from the stem spermatogonia which were unaffected by the treatment (Mecklenburg et al., 1975).

**11.3.3.7   Pipecolinomethylhydroxyindane (PMHI)**   Treatment, by mouth, of rats with DL-6-(N-α-pipecolino methyl)-5-hydroxyindane maleate at a dose of 1·5mg/kg/day for 21 days caused a decrease in weight of the testes and interfered with spermatogenesis and fertility. With larger doses all the tubules contained only Sertoli cells, but with smaller doses, only some tubules were affected. There was some reduction in the size of the accessory organs but hypophysectomy or simultaneous treatment with testosterone or gonadotrophins did not prevent the effect of PMHI. The drug was also effective orally in mice, hamsters, guinea pigs and by subcutaneous injection in rabbits, dogs and monkeys (Boris et al., 1974a, b).

**11.3.3.8   Dimethylnitrosamine**   This carcinogen causes moderate to severe necrosis of the seminiferous epithelium (Hard and Butler, 1970). It is also a mutagen but only when hydroxylated, probably in the liver (Malling, 1971; Malling and Frantz, 1973). Unlike methyl methanesuphonate, it causes little methylation of DNA in the testis (Swann and Magee, 1968), but both compounds appear to enter the seminiferous tubules readily (see Section 8.7.5).

**11.3.3.9   Nitrogen mustards**   Testicular lesions involving spermatogonia and spermatocytes have been described in mice following parenteral administration of certain of these compounds (Landing et al., 1949), but Jackson (1966) came to the conclusion that, in rats, lethal or near lethal doses of these compounds must be given to affect spermatogenesis.

**11.3.4      Halogenated compounds**

**11.3.4.1   Bis (dichloroacetyl) diamines**   These drugs were introduced as amoebicides, but they also have antispermatogenic activity which is not correlated with their amoebicidal action. Three related compounds causing testicular damage were: N,N'-bis (dichloroacetyl)-N,N'-diethyl-1,4-xylylenediamine (WIN 13,099); N,N'-bis (dichloroacetyl)-N,N'-diethyl-1,6-hexanediamine (WIN 17,416); and N,N'-bis (dichloracetyl)-1,8-octanediamine (WIN 18,446). The last named is the most potent. They are active orally in rats, dogs, monkeys and man in daily doses of 100mg/kg. Some authors state that WIN 18,446 damages spermatids both immature and maturing, others claim that it damages spermatocytes and the other cells are lost by maturation depletion (Coulston et al., 1960; Beyler et al., 1961; Drobeck and Coulston, 1962; Kar et al., 1966; Reddy and Svoboda, 1967; Gomes et al., 1973; Debeljuk, 1975). In men, azoospermia occurred between 8 and 11 weeks after beginning treatment with WIN 18,446, and the resultant infertility was reversible (Heller et al., 1961, 1963). This drug seemed to be a promising contraceptive for men until it was shown that it also produced violent 'antabuse-like' reactions to alcohol!

**11.3.4.2    Thiophenes**    These compounds have not been extensively investigated but 5-chloro-2-acetyl thiophene produces lesions in the germinal epithelium very like those produced by nitrofuran (Steinberger *et al.*, 1956).

**11.3.4.3    Fluoroacetamide**    In small doses (250$\mu$g orally) this drug acts on the secondary spermatocytes and spermatids but in larger doses it causes a complete disappearance of the germinal cells. Regeneration occurs after the treatment is stopped (Mazzanti *et al.*, 1964, 1968; Novi and Mazzanti, 1967; Sud and Steinberger, 1969; Steinberger and Sud, 1970). Fluoroacetate has a similar effect (Mazzanti *et al.*, 1965).

**11.3.4.4    Ethylene dibromide**    This substance is used to fumigate grain and it has been shown that when cattle were fed it at a dose of 2mg/kg/day, it produced changes in the semen and in the histological appearance of the testes (Amir and Volcani, 1965, 1967; Amir, 1973, 1975).

**11.3.4.5    Hexachlorophene**    The oral administration of single or repeated doses (75mg/kg) of hexachlorophene (2,2-methylene bis [3,5,6-trichlorophenol]) to male rats resulted in lesions which appeared within 2 days in the seminiferous tubules. The lesions which developed after a single dose were reversible (Thorpe, 1967).

**11.3.4.6    Sodium 2-methyl-4-chlorophenoxyacetate**    This widely used herbicide, when given to rats in their drinking water, produced at concentrations above 1000ppm minor alterations in the seminiferous epithelium without affecting testis weight (Elo and Parvinen, 1976).

**11.3.5    Phosphorus-containing compounds**

**11.3.5.1    Trimethyl phosphate**    Five oral doses of 100mg/kg to rats make them sterile for 2 weeks, beginning after the second week. This suggests that the spermatids were affected. Some side-effects were noted, and there are indications of dominant lethal mutations. The testes of mice are much less susceptible, and the effect in this species appears to be on epididymal spermatozoa. Fertility returned to normal some weeks after treatment was withdrawn (Jackson and Jones, 1968; Jackson, 1970).

**11.3.5.2    Hexamethyl phosphoramide**    This drug in rats at a dose rate of 100mg/kg for 21 days leads to prolonged infertility but this amount is close to the toxic dose. Accessory organs are unaffected, and the drug appears to act by suppressing meiosis. It is also effective in mice and rabbits (Jackson and Craig, 1966, 1969; Jackson *et al.*, 1969).

**11.3.6    Miscellaneous substances**
The carcinogen 7,12-dimethylbenz [$\alpha$] anthracene (DMBA) causes severe but selective damage in the seminiferous tubules, affecting spermatogonia and spermatocytes. A similar effect was not found with chemically related carcinogens such as 3-methylcholanthrene, benzo [$\alpha$] pyrene or 2-aceto amino phenanthrene (Ford and

Huggins, 1963), or metabolites of DMBA such as 7- and 12-hydroxydimethylbenz [α] anthracene (Jackson, 1970).

The antibiotic, actinomycin D, disrupts spermatogenesis within 24 hours when injected intratesticularly into Chinese hamsters, by inhibiting RNA synthesis by leptotene and pachytene spermatocytes (Barcellona and Brinkley, 1973). This is curious in view of the virtual exclusion of actinomycin from the tubules (see Section 8.7.5).

There is one report that $m$-xylohydroquinone affects spermatogenesis in rats (Sanyal and Rama, 1959), but Kar et al. (1963) were unable to confirm these findings.

Exposure of rats to 10 per cent carbon dioxide in air for 4 hours leads to testicular changes and premature release of tubular spermatozoa, and it has been suggested that $CO_2$ may act as a local regulator in the spermiation process (VanDemark et al., 1972; Schanbacher et al., 1974).

Substitution of 20 per cent nitrous oxide for the equivalent volume of nitrogen in air also leads to changes in the testes of rats. After 14 days, the tubules show widespread abnormalities and by 35 days contain very few germ cells more advanced than spermatogonia. Recovery is rapid when the rats are returned to a normal atmosphere (Kripke et al., 1976).

Male guinea pigs exposed to air containing natural gas (methane) for two periods daily, each of about a minute, showed degeneration of the seminiferous epithelium beginning within a week. By five weeks, the tubules contained only Sertoli cells and spermatogonia (Cameron, 1951).

Men with alcoholic cirrhosis often have small testes and low serum testosterone concentrations (see Moore, 1926; Loyd and Williams, 1948; Galvao-Teles et al., 1973; Kent et al., 1973; Southren et al., 1973). Feeding alcohol to normal rats causes their testes to degenerate (Bouin and Garnier, 1900; Arlitt and Wells, 1917; Allen, 1919; Kostlich, 1921; van Thiel et al., 1974; but cf. Huttunen et al., 1976), and both acute and chronic intake of alcohol reduce plasma testosterone levels in humans, rats and mice (Bard and Bartke, 1974; Mendleson and Mello, 1974; Gordon et al., 1976; Mendleson et al., 1977; van Thiel et al., 1975; but cf. Dotson et al., 1976; Ylikakri et al., 1974). Alcohol may have a direct action on the testis by inducing a local vitamin A deficiency (see Section 11.2.4.5).

Substitution of 30 per cent of the water intake with deuterium oxide produces sterility in male mice, interfering with meiosis at a late stage of prophase (Hughes et al., 1960; Hughes and Glass, 1965; Oakberg and Hughes, 1968.

## References to Chapter 11.3

Allanson, M. and Deanesley, R. (1962) Observations on cadmium damage and repair in rat testes and the effects on the pituitary gonadotrophs. J. Endocrinol. 24, 453–62.

Allen, E. (1919) Degeneration in the albino rat due to a diet deficient in the water-soluble vitamine, with a comparison of similar degeneration in rats differently treated and a consideration of the Sertoli tissue. Anat. Rec. 16, 93–112.

Amir, D. (1973) The sites of the spermicidal action of ethylene dibromide in bulls. J. Reprod. Fertil. 35, 519–28.

Amir, D. (1975) Individual and age differences in the spermidicidal effect of ethylene dibromide in bulls. *J. Reprod. Fertil.* **44**, 561–5.

Amir, D. and Volcani, R. (1965) Effect of dietary ethylene dibromide on bull semen. *Nature* **206**, 99–100.

Amir, D. and Volcani, R. (1967) The effect of dietary ethylene dibromide (EDB) on the testes of bulls. *Fertil. Steril.* **18**, 144–8.

Arlitt, A. H. and Wells, H. G. (1917) The effect of alcohol on the reproductive tissues. *J. exp. Med.* **26**, 769–78.

Barcellona, W. J. and Brinkley, B. R. (1973) Effects of actinomycin D on spermatogenesis in the Chinese hamster. *Biol. Reprod.* **8**, 335–49.

Bard, F. and Bartke, A. (1974) Effect of ethyl alcohol on plasma testosterone level in mice. *Steroids* **23**, 921–8.

Bateman, A. J. (1960) The induction of dominant lethal mutations in rats and mice with triethylenemelamine (TEM). *Genet. Res. Camb.* **1**, 381–92.

Berlin, M. and Ullberg, S. (1963) The fate of Cd$^{109}$ in the mouse. *Arch. Environ. Health* **7**, 686–93.

Beyler, A. L., Potts, G. O., Coulston, F. and Surrey, A. R. (1961) The selective testicular effects of certain bis(dichloroacetyl) diamines. *Endocrinol.* **69**, 819–33.

Bock, M. and Jackson, H. (1957) The action of triethylene melamine on the fertility of male rats. *Brit. J. Pharmacol.* **12**, 1–7.

Bollag, W. (1953) Der Einfluss von Myleran auf die Keimdrusen von Ratten. *Experientia* **9**, 268.

Boris, A., DeMartino, L. and Cox, D. C. (1974a) Further studies on the antitesticular activity of a pipecolinomethylhydroxyindane. *J. Reprod. Fertil.* **38**, 395–400.

Boris, A., Ng, C. and Hurley, J. F. (1974b) Antitesticular and antifertility activity of a pipecolinomethylhydroxyindane in rats. *J. Reprod. Fertil.* **38**, 287–94.

Bouin, P. and Garner, C. (1900) Altérations du tube séminifère au cours de l'alcoolisme expérimental chez le rat blanc. *C. rend. Soc. Biol.* **52**, 23–5.

Bouissou, H. and Fabre, M-T. (1965) Lésions provoquées par le sulfate de cadmium sur le testicule du rat. *Arch. Malad. Professionnelles.* **26**, 127–38.

Bouissou, H. and Fabre, M-T. (1966a) Sulfate de cadmium et acétate de zinc: leur action sur la spermatogénèse. *Laval Medical* **37**, 576–8.

Bouissou, H. and Fabre, M-T. (1966b) Action du cadmium sur le testicule du rat. Essais d'explication pathogénique. *Arch. Anat. Path.* **14**, 158–65.

Bu'Lock, D. E. and Jackson, C. M. (1971–72) Suppression of testicular androgen synthesis in the rat by ethylene dimethanesulphonate. *Gynecol. Invest.* **2**, 305–8.

Cameron, E. (1965) The effects of intratesticular injections of cadmium chloride in the rabbit. *J. Anat.* **99**, 907–12.

Cameron, E. and Foster, C. L. (1963) Observations on the histological effects of sublethal doses of cadmium chloride in the rabbit. 1. The effect on the testis. *J. Anat.* **97**, 269–80.

Cameron, J. A. (1951) Effect of inhaled methane on the testes of guinea pigs. *Fertil. Steril.* **2**, 538–41.

Cattanach, B. M. (1957) Induction of translocations in mice by triethylenemelamine. *Nature* **180**, 1364–5.

Cattanach, B. M. and Edwards, R. G. (1958) The effect of triethylenemelamine on the fertility of male mice. *Proc. Roy. Soc. Edinburgh* B. **67**, 54–64.

Cattenach, B. M., Pollard, C. E. and Isaacson, J. H. (1968) Ethyl methanesulfonate-induced chromosome breakage in the mouse. *Mutation Res.* **6**, 297–307.

Caujolle, F., Bouissou, H., Fabre, M-T., Pham-Huu-Chanh, M. and Silve, G. (1965) Etude experimentale de l'action toxique du cadmium sur le testicule du rat. *Bull. Acad. Nat. Med.* (Paris) **149**, 146–50.

Chatterjee, A. and Ray, P. (1972) An early differential effect of cadmium on the scrotal and contralateral cryptorchid testes in the rat. *J. Reprod. Fertil.* **30**, 297–300.

Chiquoine, A. D. (1964) Observations on the early events of cadmium necrosis of the testis. *Anat. Rec.* **149**, 23–35.

Chiquoine, A. D. and Suntzeff, V. (1965) Sensitivity of mammals to cadmium necrosis of the testis. *J. Reprod. Fertil.* **10**, 455–7.

Clegg, E. J. and Carr, I. (1966) Increased vascular permeability in the reproductive organs of cadmium chloride-treated male rats. *J. Anat.* **100**, 696–7.

Clegg, E. J. and Carr, I. (1967) Changes in the blood vessels of the rat testis and epididymis produced by cadmium chloride. *J. Pathol. Bact.* **94**, 317–22.

Clegg, E. J., Carr, I. and Niemi, M. (1969a) The effect of a second dose of cadmium salts on vascular permeability in the rat testis. *J. Endocrinol.* **45**, 265–8.

Clegg, E. J., Niemi, M. and Carr, I. (1969b) The age at which the blood vessels of the rat testis become sensitive to cadmium salts. *J. Endocrinol.* **43**, 445–9.

Cooper, E. R. A. and Jackson, H. (1970) Comparative effects of methylene, ethylene and propylene dimethanesulphonates on the male rat reproductive system. *J. Reprod. Fertil.* **23**, 103–8.

Coulston, F., Beyler, A. L. and Drobeck, H. P. (1960) The biologic actions of a new series of bis(dichloroacetyl) diamines. *Toxicol. Appl. Pharmacol.* **2**, 715–31.

Debeljuk, L. (1975) Serum follicle-stimulating hormone and luteinizing hormone levels in male rats with experimentally induced damage of the germinal epithelium. *J. Endocrinol.* **66**, 53–60.

Debeljuk, L., Arimura, A. and Schally, A. V. (1973) Pituitary and serum FSH and LH levels after massive and selective depletion of the germinal epithelium in the rat testis. *Endocrinology* **92**, 48–54.

Dotson, L. E., Robertson, L. S. and Tuchfeld, B. (1975) Plasma alcohol, smoking, hormone concentrations and self-reported aggression: a study of social drinking situation. *J. Stud. Alcohol.* **36**, 578–86.

Drobeck, H. P. and Coulston, F. (1962) Inhibition and recovery of spermatogenesis in rats, monkeys and dogs medicated with bis(dichloracetyl) diamines. *Exptl. Molec. Pathol.* **1**, 251–74.

Dryden, G. L. and McAllister, H. Y. (1970) Sustained fertility after $CdCl_2$ injection by a non-scrotal mammal, *Suncus murinus* (Insectivora, Soricidae). *Biol. Reprod.* **3**, 23–30.

Ehling, U. H. (1971) Comparison of radiation- and chemically-induced dominant lethal mutations in male mice. *Mutation Res.* **11**, 35–44.

Ehling, U. H. and Malling, H. V. (1968) 1,4-di(methane-sulfonoxy)butane (myleran) as a mutagenic agent in mice. *Genetics* **60**, 174–5.

Ehling, U. H., Cumming, R. B. and Malling, H. V. (1968) Induction of dominant lethal mutations by alkylating agents in male mice. *Mutat. Res.* **5**, 417–28.

Elo, H. and Parvinen, M. (1976) Effect of sodium 2-methyl-4-chlorophenoxyacetate on sper-matogenesis in the rat. *J. Reprod. Fertil.* **48**, 243–4.

Ericsson, R. J. (1971) Antispermatogenic properties of 2,3-dihydro-2-(1-naphthyl)-4 (IH)-quinazolinone (U-29,409). *Proc. Soc. exp. Biol. N.Y.* **137**, 532–5.

Etribi, A., Ibrahim, A., El-Haggar, S., Awad, H. and Metawi, B. (1976) Effect of ambilhar (niridazole) on spermatogenesis in guinea pigs. *J. Reprod. Fertil.* **48**, 439–40.

Featherstone, R. M., Nelson, W. O., Welden, F., Marberger, E., Boccabella, A. and Boccabella, R. (1955) Pyruvate oxidation in testicular tissues during furadoxyl-induced spermatogenic arrest. *Endocrinology* **56**, 727–36.

Ford, E. and Huggins, C. (1963) Selective destruction in testis induced by 7,12-di-methylbenz(α) anthracene. *J. Exptl. Med.* **118**, 27–40.

Forsberg, J. G. and Olivecrona, H. (1966) The effect of prenatally administered busulphan on rat gonads. *Biol. Neonat.* **10**, 180–92.

Fox, B. E. and Fox, M. (1967) Biochemical aspects of the actions of drugs on spermatogenesis. *Pharmacol. Rev.* **19**, 21–57.

Fox, B. W. and Jackson, H. (1965) *In vivo* effects of methylene dimethanesulphonate on proliferating cell systems. *Brit. J. Pharmacol.* **24**, 24–8.

Fox, B. W., Jackson, H., Craig, A. W. and Glover, T. D. (1963) Effects of alkylating agents on spermatogenesis in the rabbit. *J. Reprod. Fertil.* **5**, 13–22.

Gabbiani, G. (1966) Action of cadmium chloride on sensory ganglia. *Experientia* **22**, 261–2.

Galvao-Teles, A., Anderson, D. C., Burke, C. W., Marshall, J. C., Corker, C. S., Brown, R. L. and Clark, M. L. (1973) Biologically active androgens and oestradiol in men with chronic liver disease. *Lancet* **I**, 173–7.

Gasiewicz, T. A. and Smith, J. C. (1976) Interactions of cadmium and selenium in rat plasma *in vivo* and *in vitro*. *Biochim. Biophys. Acta* **428**, 113–22.

Generoso, W. M. (1973) Evaluation of chromosome aberration effects of chemicals on mouse germ cells. *Environ. Health Perspec.* **6**, 13–22.

Generoso, W. M. and Russell, W. L. (1969) Strain and sex variations in the sensitivity of mice to dominant-lethal induction with ethyl methane-sulfonate. *Mutation Res.* **8**, 589–98.

Gillet, J. and Laporte, P. (1973) Action du busulphan et de FSH sur le tube séminifère du rat: effets morpologiques et cellulaires. *Arch. Anat. microsc. Morph. exp.* **62**, 385–98.

Girod, C. (1964) A propos de l'influence du chlorure de cadmium sur le testicule: recherches chez le singe *Macacus irus* F. Cuv. *Comp. Rend. Sean. Soc. Biol.* **158**, 297–301.

Girod, C. and Chauvineau, A. (1964) Nouvelles observations concernant l'influence du chlorure de cadmium sur le testicule du singe *Macacus irus* F. Cuv. *Compt. Rend. Soc. Biol.* **158**, 2113–5.

Gomes, W. R., Criblez, T. L., VanDemark, N. L. and Johnson, A. D. (1966) Effect of busulfan on testicular tissue metabolism. *J. Anim. Sci.* **25**, 1264.

Gomes, W. R., Hall, R. W., Jain, S. K. and Boots, D. R. (1973) Serum gonadotropin and testosterone levels during loss and recovery of spermatogenesis in rats. *Endocrinology* **93**, 800–9.

Gordon, G. G., Altman, K., Southren, A. L., Rubin, E. and Lieber, C. S. (1976) Effect of alcohol (ethanol) administration on sex-hormone metabolism in normal men. *N. Engl. J. Med.* **295**, 793–7.

Gunn, S. A. and Gould, T. C. (1970a) Cadmium and other mineral elements. In *The Testis*, eds.

A. D. Johnson, W. R. Gomes and N. L. VanDemark, III, 377–481. New York: Academic Press.

Gunn, S. A. and Gould, T. C. (1970b) Specificity of the vascular system of the male reproductive tract. *J. Reprod. Fertil.* Suppl. **10**, 75–95.

Gunn, S. A. and Gould, T. C. (1975) Vasculature of the testes and adnexa. *Handbk. Physiol. Sect.* 7, Vol. V, 117–42.

Gunn, S. A., Gould, T. C. and Anderson, W. A. D. (1963) The selective injurious response of testicular and epididymal blood vessels to cadmium and its prevention by zinc. *Am. J. Pathol.* **42**, 685–702.

Gunn, S. A., Gould, T. C. and Anderson, W. A. D. (1965a) Comparative study of interstitial cell tumors of rat testis induced by cadmium injection and vascular ligation. *J. Nat. Cancer Inst.* **35**, 329–35.

Gunn, S. A., Gould, T. C. and Anderson, W. A. D. (1965b) Strain differences in susceptibility of mice and rats to cadmium-induced testicular damage. *J. Reprod. Fertil.* **10**, 273–5.

Gunn, S. A., Gould, T. C. and Anderson, W. A. D. (1966) Loss of selective injurious vascular response to cadmium to regenerated blood vessels of testis. *Am. J. Pathol.* **48**, 959–69.

Gunn, S. A., Gould, T. C. and Anderson, W. A. D. (1968a) Failure of [109]Cd to traverse spermatogenic pathway. *J. Reprod. Fertil.* **16**, 125–8.

Gunn, S. A., Gould, T. C. and Anderson, W. A. D. (1968b) Selectivity of organ response to cadmium injury and various protective measures. *J. Pathol. Bacteriol.* **96**, 89–96.

Gupta, R. K., Barnes, G. W. and Skelton, F. R. (1967) Light-microscopic and immunopathologic observations of cadmium chloride-induced injury in mature rat testis. *Am. J. Pathol.* **51**, 191–205.

Hagenas, L., Ploen, L. and Ritzen, E. M. (1978) The effect of nitrofurazone on the endocrine secretory and spermatogenic functions of the rat testis. *Andrologia*, **10**, 107–26.

Hard, G. C. and Butler, W. H. (1970) Toxicity of dimethylnitrosamine for the rat testis. *J. Pathol.* **102**, 201–7.

Harkonen, M. and Kormano, M. (1968) Acute cadmium-induced changes in the energy metabolism of the rat testis. *J. Reprod. Fertil.* **21**, 221–6.

Heinke, E. and Jaeschke, H. (1969) Die Wirkung von Hydroxy-methylnitrofurantoin auf die Spermiogenese des Mannes. *Therapiewoche* **19**, 1664–7.

Heller, C. G., Moore, D. J. and Paulsen, C. A. (1961) Suppression of spermatogenesis and chronic toxicity in men by a new series of bis(dichloracetyl) diamines. *Toxicol. Appl. Pharmacol.* **3**, 1–11.

Heller, C. G., Flageole, B. Y. and Matson, L. J. (1963) Histopathology of the human testes as affected by bis(dichloroacetyl) diamines. *Exptl. Mol. Pathol.* Suppl. **2**, 107–14.

Hemsworth, B. M. and Jackson, H. (1963) Effect of busulphan on the developing gonad of the male rat. *J. Reprod. Fertil.* **5**, 187–94.

Hilderbrand, D. C., Der, R., Griffin, W. T. and Fahim, M. S. (1973) Effect of lead acetate on reproduction. *Am. J. Obstet. Gynecol.* **115**, 1058–65.

Hollinger, M. A. and Davis, J. R. (1966) Effect of nitrofurazone on the incorporation of L-lysine-U-C[14] into protein of rat testis. *Biochem. Pharmacol.* **15**, 1235–7.

Hollinger, M. A. and Davis, J. R. (1969) Effect of nitrofurazone on the aerobic metabolism of uniformly labelled [14C] glucose in tissue slices of rat testes. *J. Reprod. Fertil.* **19**, 585–9.

Hughes, A. M. and Glass, L. E. (1965) Histological investigations of the mechanism of sterility induced by deuterium oxide in mice. *Nature* **208**, 1119–20.

Hughes, A. M., Bennett, E. L. and Calvin, M. (1960) Further studies on sterility produced in male mice by deuterium oxide. *Ann. N. Y. Acad. Sci.* **84**, 763–9.

Huttunen, M. O., Härkönen, M., Niskanen, P., Leino, T. and Ylikahri, R. (1976) Plasma testosterone concentrations in alcoholics. *J. Stud. Alcohol* **37**, 1165–77.

Ito, T. and Sawauchi, K. (1966) Inhibitory effects on cadmium-induced testicular damage by pretreatment with smaller cadmium dose. *Okajimas Folia Anat. Japan.* **42**, 107–17.

Jackson, H. (1959) Antifertility substances. *Pharmacol. Rev.* **11**, 135–72.

Jackson, H. (1965) Problems in the chemical control of male fertility. In *Agents Affecting Fertility,* eds. C. R. Austin and J. S. Perry, 62–76. London: Churchill.

Jackson, H. (1966) *Antifertility Compounds in the Male and Female,* Springfield, Illinois: Thomas.

Jackson, H. (1969) Chemical interference with spermatogenesis and fertility. In *Advances in Reproductive Physiology,* ed. A. McLaren, **4**, 65–98, London: Logos Press.

Jackson, H. (1970) Antispermatogenic agents. *Brit. Med. Bull.* **26**, 79–86.

Jackson, H. (1971) Non-steroidal anti-fertility agents in the male. In Nobel Symposium, *Control of Human Fertility,* **15**, 119–36, Stockholm: Alunquist & Wiksell.

Jackson, H. and Bock, M. (1955) The effect of triethylenemelamine on the fertility of rats. *Nature* **175**, 1037–8.

Jackson, H. and Craig, A. W. (1966) Antifertility action and metabolism of hexamethylphosphoramide. *Nature* **212**, 86–7.

Jackson, H. and Craig, A. W. (1969) Effects of alkylating chemicals on reproductive cells. *Ann. N.Y. Acad. Sci.* **160**, 215–27.

Jackson, H. and Jones, A. R. (1968) Antifertility action and metabolism of trimethylphosphate in rodents. *Nature* **220**, 591–2.

Jackson, H. and Schnieden, H. (1968) Pharmacology of reproduction and fertility. *Ann. Rev. Pharmacol.* **8**, 467–90.

Jackson, H., Fox, B. W. and Craig, A. W. (1959) The effect of alkylating agents on male rat fertility. *Brit. J. Pharmacol.* **14**, 149–57.

Jackson, H., Fox, B. W. and Craig, A. W. (1961) Antifertility substances and their assessment in the male rodent. *J. Reprod. Fertil.* **2**, 447–65.

Jackson, H., Partington, M. and Fox, B. W. (1962) Effect of 'busulphan' ('myleran') on the spermatogenic cell population of the rat testis. *Nature* **194**, 1184–5.

Jackson, H., Jones, A. R. and Cooper, E. R. A. (1969) Effects of hexamethylphosphoramide on rat spermatogenesis and fertility. *J. Reprod. Fertil.* **20**, 263–70.

Jackson, H., Jackson, C. M. and Jones, P. (1973) Hormonal antagonism to the antispermatogenic effect of ethylenedimethanesulphonate in rats. *J. Reprod. Fertil.* **34**, 133–6.

Jackson, H., Jones, P. and Whiting, M. H. (1974) Assessment of dominant lethal mutations in Niridazole-treated male mice. *IRCS Med. Sci.* **2**, 1331.

Johnson, A. D. and Miller, W. J. (1970) Early actions of cadmium in the rat and domestic fowl testis. II. Distribution of injected [109]cadmium. *J. Reprod. Fertil.* **22**, 395–405.

Johnson, A. D. and Sigman, M. B. (1971) Early actions of cadmium in the rat and domestic fowl testis. IV. Autoradiographic location of [115m] cadmium. *J. Reprod. Fertil.* **24**, 115–18.

Johnson, A. D. and Walker, G. P. (1970) Early actions of cadmium in the rat and domestic fowl testis. V. Inhibition of carbonic anhydrase. *J. Reprod. Fertil.* **23**, 463–8.

Johnson, A. D., Gomes, W. R. and VanDemark, N. L. (1970) Early actions of cadmium in the rat and domestic fowl testis. I. Testis and body temperature changes caused by cadmium and zinc. *J. Reprod. Fertil.* **22**, 383–93.

Johnson, M. H. (1969) The effect of cadmium chloride on the blood–testis barrier of the guinea-pig. *J. Reprod. Fertil.* **19**, 551–3.

Jones, P., Jackson, H. and Whiting, M. H. S. (1976) Comparative effects of niridazole on spermatogenesis and reproductive capacity in the mouse, rat and Japanese quail. *J. Reprod. Fertil.* **46**, 217–24.

Kar, A. B. (1961) Chemical sterilization of male rhesus monkeys. *Endocrinology* **69**, 1116–9.

Kar, A. B. (1962) Chemical sterilization of male goats. *Indian J. Vet. Sci.* **32**, 70–3.

Kar, A. B. and Das, R. P. (1960) Testicular changes in rats after treatment with cadmium chloride. *Acta Biol. Med. Ger.* **5**, 153–73.

Kar, A. B. and Das, R. P. (1962) Sterilization of males by intratesticular administration of cadmium chloride. *Acta Endocrinol.* **40**, 321–31.

Kar, A. B. and Kamboj, V. P. (1965) Cadmium damage of the rat testis and its prevention. *Ind. J. exp. Biol.* **3**, 45–9.

Kar, A. B., Dasgupta, P. R. and Das, R. P. (1961) Effect of low dose of cadmium chloride on the genital organs and fertility of male rats. *J. Sci. Ind. Res.* (India) **20C**, 322–6.

Kar, A. B., Bose, A. R. and Das, R. P. (1963) Effect of *m*-xylohydroquinone on the genital organs and fertility of male rats. *J. Reprod. Fertil.* **5**, 77–81.

Kar, A. B., Chowdhury, A. R., Goswami, A. and Kamboj, V. P. (1965) Intertesticular distribution of cadmium in rats. *Ind. J. Exptl. Biol.* **3**, 139–41.

Kar, A. B., Dasgupta, P. R. and Jehan, Q. (1966a) *In vitro* action of cadmium chloride on isolated seminiferous tubules of the rat testis. *Acta Biol. Med. Ger.* **16**, 665–70.

Kar, A. B., Jehan, Q., Kamboj, V. P. and Chowdhury, A. R. (1966b) Effect of N,'-bis(dichloroacetyl)-1,8-octamethylenediamine on the chemical composition of the rat seminiferous tubules. *Intern. J. Fertil.* **11**, 291–6.

Kar, A. B., Jehan, Q., Kamboj, V. P., Chowdhury, S. R. and Chowdhury, A. R. (1968a) Effect of busulphan on biochemical composition of rat seminiferous tubules. *Ind. J. exp. Biol.* **6**, 9–72.

Kar, A. B., Kamboj, V. P. and Chandra, H. (1968b) Effect of some chemicals on spermatogenesis in rhesus monkeys. *J. Reprod. Fertil.* **16**, 165–70.

Kent, J. R., Scaramuzzi, R., Lauwers, W., Parlow, A., Hill, M., Penardi, R. and Hilliard, J. (1973) Plasma testosterone, oestradiol and gonadotrophins in hepatic insufficiency. *Gastroenterology* **64**, 111–15.

King, T. O., Berliner, V. R. and Blye, R. P. (1963) Pharmacology of 2,4-dinitropyrroles—a new class of anti-spermatogenic compounds. *Biochem. Pharmacol.* **12**, Suppl. 69 (abstr.).

Kormano, M. and Suvanto, O. (1968) Cadmium-induced changes in the intratesticular pressure in the rat. *Acta Pathol. Microbiol. Scand.* **72**, 444–5.

Koskimies, A. I. (1973a) Effect of cadmium on protein composition of fluids in rat rete testis and seminiferous tubules. *Ann. Med. Exptl. Biol. Fenniae* (Helsinki) **51**, 74–81.

Koskimies, A. I. (1973b) Rapid appearance of excess serum protein in the rete testis in response to Cd-induced increase of vascular permeability. In *Immunology of Reproduction*, 126, Sofia: Bulgarian Academy of Sciences Press.

Kostitch, A. (1921) Sur l'involution du processus spermatogénétique provoquée par l'alcoolisme expérimentale. *C. rend. Soc. Biol.* **84**, 674–6.

Kramer, M. F. and de Rooij, D. G. (1970) The effect of three alkylating agents on the seminiferous epithelium of rodents. II. Cytotoxic effects. *Virch. Arch. Abt. B. Zellpath.* **4**, 276–82.

Kripke, B. J., Kelman, A. D., Shah, N. K., Balogh, M. D. and Handler, A. H. (1976) Testicular reaction to prolonged exposure to nitrous oxide. *Anesthesiology* **44**, 104–13.

Lambert, C. R., Siniari, V. S. P. and Tripod, J. (1965) Effect of Ciba 32644 Ba on spermatogenesis in laboratory animals. *Acta trop.* **22**, 155–61.

Landing, B. H., Goldin, A. and Noe, H. A. (1949) The testicular lesion in mice following parenteral administration of nitrogen mustards. *Cancer* **2**, 1075–82.

Lee, D. J. W. and Barnes, M. McC. (1971) The effect of a single dose of cadmium chloride on the phospholipid fatty acids of the rat testis. *J. Reprod. Fertil.* **27**, 25–30.

Lee, I. P. and Dixon, R. L. (1975) Effects of mercury on spermatogenesis studied by velocity sedimentation cell separation and serial mating. *J. Pharmacol. Exptl. Therapeut.* **194**, 171–81.

Lloyd, C. W. and Williams, R. H. (1948) Endocrine changes associated with Laennec's cirrhosis of the liver. *Am. J. Med.* **4**, 315–30.

Light, A. E. (1967) Additional observations on the effects of busulfan on cataract formation, duration of anesthesia and reproduction in rats. *Toxicol. Appl. Pharmacol.* **10**, 459–66.

Maekawa, K. and Tsunenari, Y. (1967) Mechanism involved in selective injurious effect of cadmium on the testis. *Gunma Symp. Endocrinol.* **4**, 161–7.

Maekawa, K., Tsunenari, Y., Nokubi, K. and Waki, M. (1965) The earliest effect of cadmium on the testis; and increase of alkaline phosphatase activity in the capillary wall of the testicular interstitium. [In Japanese with English summary.] *Acta Anat. Nippon.* **40**, 200–8.

Maekawa, K., Tsunenari, Y. and Kurematsu, Y. (1966) Role of increased vascular permeability in cadmium injury of the testis. [In Japanese with English summary.] *Acta Anat. Nippon.* **41**, 327–33.

Malling, H. V. (19171) Dimethylnitrosamines: formation of mutagenic compounds by interaction with mouse liver microsomes. *Mutation Res.* **13**, 425–9.

Malling, H. V. and Frantz, C. N. (1973) *In vitro* versus *in vivo* metabolic activation of mutagens. *Environ. Health Perspec.* **6**, 71–82.

Mason, K. E. and Young, J. O. (1967) Effects of cadmium upon the excurrent duct system of the rat testis. *Anat. Rec.* **159**, 311–24.

Mason, K. E., Brown, J. A., Young, J. O. and Nesbit, R. R. (1964) Cadmium-induced injury of the rat testis. *Anat. Rec.* **149**, 135–48.

Matter, B. E. and Generoso, W. M. (1974) Effects of dose on the induction of dominant lethal mutations with triethylenemelamine in male mice. *Genetics* **77**, 753–63.

Mazzanti, L., Lopez, M. and Grazia Berti, M. (1964) Selective destruction in testes induced by fluoracetamide. *Experientia* **20**, 492–3.

Mazzanti, L., Lopez, M. and Grazia Berti, M. (1965) Atrofia del testicolo prodotta dal monofluoroacetato sodico nel ratto albino. *Experientia* **21**, 446–7.

Mazzanti, L., Lopez, M. and Del Tacca, M. (1968) La rigenerazione del testicolo atrofico da fluoroacetamide. *Experientia* **24**, 258–9.

Mecklenburg, R. S., Hetzel, W. D., Gulyas, B. J. and Lipsett, M. B. (1975) Regulation of FSH secretion: Use of hydroxyurea to deplete germinal epithelium. *Endocrinology* **96**, 564–70.

Mendelson, J. H. and Mello, N. K. (1974) Alcohol, aggression and androgens. *Res. Publ. Ass. Res. Nerv. Ment. Dis.* **52**, 225–47.

Mendelson, J. H., Mello, N. K. and Ellingboe, J. (1977) Effects of acute alcohol intake on pituitary-gonadal hormones in normal human males. *J. Pharmacol. Exp. Ther.* **202**, 676–82.

Monoyer, J. (1962) Altérations de la spermatogénèse chez des rats traits par le Myleran. *Comp. Rend. Soc. Biol.* **146**, 968–70.

Moore, C. R. (1926) The biology of the mammalian testis and scrotum. *Q. Rev. Biol.* **1**, 4–50.

Nelson, W. O. and Bunge, R. G. (1957) The effect of therapeutic dosages of nitrofurantoin (furadantin) upon spermatogenesis in man. *J. Urol.* **77**, 275–81.

Nelson, W. O. and Patanelli, D. J. (1965) Chemical control of spermatogenesis. In *Agents Affecting Fertility,* eds. C. R. Austin and J. S. Perry, 78–92. London: Churchill.

Nelson, W. O. and Steinberger, E. (1953) Effects of nitrofuran compounds on the testis of the rat. *Fed. Proc.* **12**, 103.

Niemi, K. and Kormano, M. (1965) An angiographic study of cadmium-induced vascular lesions in the testis and epididymis of the rat. *Acta. path. microbiol. Scand.* **63**, 513–21.

Novi, A. M. and Mazzanti, L. (1967) Azione bloccante della Fluoroacetamide sulla spermatogenesi nel ratto. *Boll. Soc. Ital. Biol. Sper.* **43** Suppl. 20, abst. 162.

Oakberg, E. F. and Hughes, A. M. (1968) Deuterium oxide effect on spermatogenesis in the mouse. *Exptl. Cell. Res.* **50**, 306–14.

Parizek, J. (1957) The destructive effect of cadmium ion on testicular tissue and its prevention by zinc. *J. Endocrinol.* **15**, 56–63.

Parizek, J. (1960) Sterilization of the male by cadmium salts. *J. Reprod. Fertil.* **1**, 294–309.

Parizek, J. and Zahor, Z. (1956) Effect of cadmium salts on testicular tissue. *Nature* **177**, 1036–7.

Partington, M. and Bateman, A. J. (1964) Dominant lethal mutations induced in male mice by methyl methanesulfonate. *Heredity* **19**, 191–220.

Partington, M. and Jackson, H. (1963) The induction of dominant lethal mutations in rats by alkane sulphonic esters. *Genet. Res. Camb.* **4**, 333–45.

Partington, M., Fox, B. W. and Jackson, H. (1964) Comparative action of some methane sulphonic esters on the cell population of the rat testis. *Exptl. Cell. Res.* **33**, 78–88.

Parvinen, M. and Kormano, M. (1974) Early effects of antispermatogenic benzimidazole derivatives U 32.422 E and U 32.104 on the seminiferous epithelium of the rat. *Andrologia* **6**, 245–53.

Patanelli, D. J. (1975) Supression of fertility in the male. Handbk. Physiol. Section 7, *Endocrinology* V, 245–58.

Patanelli, D. J. and Nelson, W. O. (1964) A quantitative study of inhibition and recovery of spermatogenesis. In *Recent Progress in Hormone Research*, ed. G. Pincus, **20**, 491–543, New York: Academic Press.

Paufler, S. K. and Foote, R. H. (1969) Effect of triethylenemelamine and cadmium chloride on spermatogenesis in rabbits. *J. Reprod. Fertil.* **19**, 309–19.

Paul, H. E., Paul, M. F., Kopko, F., Bender, R. C. and Everett, G. (1953) Carbohydrate metabolism studies on the testis of rats fed certain nitrofurans. *Endocrinology* **53**, 585–92.

Peyre, A., Joffre, M. and Debordes, E. (1968) L'activité succinodeshydrogenasique du plexus pampiniforme du testicule de rat après administration de cadmium. Etude histoenzymologique. *Compt. Rend. Soc. Biol.* **162**, 2220–2.

Prior, J. T. and Ferguson, J. H. (1950) Cytotoxic effects on a nitrofuran on the rat testis. *Cancer* **3**, 1062–72.

Ramaswami, L. S. and Kaul, D. K. (1966) The effect of cadmium and selenium on the testis of the desert gerbil *Meriones hurrianae* Jerdon. *J. Roy. Microscop. Sco.* **85**, 297–304.

Reddy, K. J. and Svoboda, D. J. (1967) Alterations in rat testes due to an antispermatogenic agent. Light and electron microscopic study. *Arch. Path.* **84**, 376–92.

Ribelin, W. E. (1963) Atrophy of rat testis as index of chemical toxicity. *Arch. Path.* **75**, 229–35.

Rooij, D. G. de and Kramer, M. F. (1968) Spermatogonial stem cell renewal in the rat, mouse and golden hamster. *Z. Zellforsch.* **92**, 400–5.

Rooij, D. G. de and Kramer, M. F. (1970) The effect of three alkylating agents on the seminiferous epithelium of rodents. I. Depletory effects. *Virch. Arch. Abt. B. Zellpath.* **4**, 267–75.

Rooney, F. R. and Jackson, H. (1974) Antifertility activity of alkyl esters of nitro- and acetamido-benzenesulphonic acids. *Andrologia* **6**, 263–8.

Sanyal, S. N. and Rana, M. (1959) Oral contraceptives—*m*-xylohydroquinone: Biological studies on males. *J. Med. Intern. Med. Abstr. Rev.* **22**, 19.

Schanbacher, B. D., VanDemark, N. L. and Gomes, W. R. (1974) Spermiation and blood gases in carbon dioxide-exposed rats. *Am. J. Physiol.* **226**, 588–91.

Setchell, B. P. and Waites, G. M. H. (1970) Changes in the permeability of the testicular capillaries and of the 'blood–testis barrier' after injection of cadmium chloride in the rat. *J. Endocrinol.* **47**, 81–6.

Singh, K. and Mathur, R. S. (1968) Enzyme changes in the gerbil testis produced by the administration of an alkylating agent busulphan (Myleran, 1,4,dimethanesulphonoxybutane). *Acta Anat.* **71**, 472–80.

Singh, K., Nath, R. and Chakravarti, R. N. (1974) Isolation and characterization of cadmium-binding protein from rat testes. *J. Reprod. Fertil.* **36**, 257–65.

Skinner, W. A. and Tong, H. C. (1968) Effect of alkylating agents derived from diamines on fertility of the male mouse. *Experientia* **24**, 924–5.

Southren, A. L., Gordon, G. C., Olwo, J., Rafii, F. and Rosenthal, W. S. (1973) Androgen metabolism in cirrhosis of the liver. *Metabolism* **22**, 695–702.

Steinberger, E. A. (1962) A quantitative study of the effect of an alkylating agent (Triethylenemelamine) on the seminiferous epithelium of rats. *J. Reprod. Fertil.* **3**, 250–9.

Steinberger, E. and Nelson, W. O. (1957) The effect of furadroxyl treatment and X-irradition on the hyaluronidase concentration of rat testes. *Endocrinology* **60**, 105–17.

Steinberger, E. and Sud, B. N. (1970) Specific effect of fluoroacetamide on spermiogenesis. *Biol. Reprod.* **2**, 369–75.

Steinberger, E., Nelson, W. O. and Boccabella, A. (1956) Cytotoxic effects of 5-chlor-2-acetyl thiophen (BA-11044) on the testis of the rat. *Anat. Rec.* **125**, 312–3.

Steinberger, E., Nelson, W. O., Boccabella, A. and Dixon, W. J. (1959) A radiomimetic effect of triethylenemelamine on reproduction in the male rat. *Endocrinology* **65**, 40–50.

Sud, B. N. and Steinberger, E. (1969) Specificity of the effect of fluoracetamide (FFA) on spermatogenesis. *Anat. Rec.* **163**, 271–2.

Swann, P. F. and Magee, P. N. (1968) Nitrosamine-induced carconogenesis. The alkylation of nucleic acids of the rat by N-methyl-N-nitrosourea, dimethylnitrosamine, dimethyl sulphate and methyl methanesulphonate. *Biochem. J.* **110**, 39–47.

Thiel, D. H. van, Lester, R. and Sherins, R. J. (1974) Hypogonadism in alcoholic liver disease: evidence for a double defect. *Gastroenterology* **67**, 1188–99.

Thiel, D. H. van, Gavaler, J., Lester, R. and Goodman, M. D. (1975) Alcohol-induced testicular atrophy. *Gastroenterology* **69**, 326–32.

Thorpe, E. (1967) Some pathological effects of hexachlorophene in the rat. *J. Comp. Path.* **77**, 137–43.

Timm, F. and Sculz, G. (1966) Hoden und Schwermetalle. *Histochemie* **7**, 15–21.

Uematsu, K. (1966) Testicular changes of rats induced by nitrofurazone. A light and electron microscopic study. *Med. J. Osaka Univ.* **16**, 287–320.

VanDemark, N. L., Schanbacher, B. D. and Gomes, W. R. (1972) Alterations in testes of rats exposed to elevated atmospheric carbon dioxide. *J. Reprod. Fertil.* **28**, 457–60.

Vanhems, E. and Bousquet, J. (1972) Influence du misulban sur le développement du testicule du rat. *Ann. Endocrinol.* (Paris) **33**, 119–28.

Waites, G. M. H. and Setchell, B. P. (1966) Changes in blood flow and vascular permeability of the testis, epididymis and accessory reproductive organs of the rat after the administration of cadmium chloride. *J. Endocrinol.* **32**, 329–42.

Webb, M. (1972) Biochemical effects of $Cd^{2+}$-injury in the rat and mouse testis. *J. Reprod. Fertil.* **30**, 83–98.

Ylikahri, R., Huttunen, M., Harkonen, M., Seuderling, U., Onikki, S., Karonen, S.-L. and Adlercreutz, H. (1974) Low plasma testosterone values in men during hangover. *J. steroid Biochem.* **5**, 655–8.

Zedeck, M. S., Sternberg, S. S., Poynter, R. W. and McGowan, J. H. (1969) Early biochemical and pathologic effects of methylazoxymethanol acetate (MAM), a potent carcinogen. *Proc. Am. Ass. Cancer Res.* **10**, 103.

## 11.4   Immunological damage

Several of the early immunologists, Landsteiner (1899), Metalnikoff (1900) and Metchnikoff (1900) noticed independently that antibodies could be raised against an animal's own spermatozoa, but immunized animals, as a rule, did not show any testicular abnormalities. Mild focal lesions can be produced by repeated injections of homologous testicular homogenates over several months (Bishop, 1961) but when animals are immunized with testis homogenates mixed with Freund's complete adjuvant, a widespread immunological reaction is produced in the testis; the germinal epithelium is completely destroyed in 6 or 8 weeks and only Sertoli cells remain (Voisin *et al.*, 1951; Delauney and Voisin, 1952; Freund *et al.*, 1953; 1954). Only mature animals are affected, in fact only animals with spermatozoa in the rete testis and excurrent ducts; animals with spermatids in the testis before the liberation of spermatozoa are not (Bishop *et al.*, 1961; Johnson, 1970a). In these immature animals, an autoallergic damage could only be produced if the peritubular barrier was weakened with cadmium, physical damage to the testis or non-specific inflammation of the testis. These treatments produced focal lesions in the tubules, around which in autoimmunized animals, large numbers of leucocytes congregated (Johnson, 1970a).

In the mature animals the immune response appears to begin at the rete testis and spread along the tubules (Waksman, 1959, 1960; Brown and Glynn, 1969; Johnson, 1970b, 1972, 1973; Tung *et al.*, 1970). The late spermatids and mature spermatozoa were affected before the early spermatids (Johnson, 1970c) and it would therefore appear that an antigen–antibody reaction occurs in the rete causing an increase in the permeability of the tubules, or that the antibodies enter at the rete, reach the cells from the luminal surface and the reaction increases the permeability of the tubules. The normal rete appears to be more permeable to dyes and protein than the seminiferous tubules (see Chapter 8) and the action of adjuvant may be to increase the permeability of the rete. An effect of adjuvants on the tubules is suggested by the observations that if adjuvant alone is injected into guinea pigs, collagen fibres disappear from the peritubular spaces, peroxidase penetrates more readily past the myoid cells and appeared in some Sertoli cells and spermatogonia although it was still stopped by the Sertoli cell–Sertoli cell junctions (Willson *et al.*, 1973). Other authors have suggested that an inflammatory reaction in the interstitial tissue is important (Waksman, 1959; Tung *et al.*, 1970).

The situation is not straightforward because immunological damage cannot be produced in the tubules by the passive transfer of antibodies, although this treatment does produce a reaction in the rete (Tung *et al.*, 1971b, but cf. Nagano and Okumura, 1973a). On the other hand, when lymph node cells from an actively immunized animal are injected into the testis of a non-immunized recipient, there is a cellular reaction inside the tubules as well as in the interstitial tissue, but little or no reaction in the rete (Tung *et al.*, 1971a). The significance of this last observation is obscured by the further observation that if macrophages from the peritoneal cavity of the actively immunized animal are used instead of lymph node cells, the reaction does begin in the rete and occurs more rapidly in the tubules (Kantor and Dixon, 1972). Perhaps there is the need for a dual attack on the tubules from both the rete and the tubular wall, and both a cellular and a humoral immunologic reaction may be involved.

Immunological damage in one testis can be produced by injecting hot water into the other testis, or by injuring it by compression (Boughton and Spector, 1963; Rapaport *et al.*, 1969; Fernandez Collazo *et al.*, 1972; Fainboim *et al.*, 1976). Immunological aspermatogenesis was also produced if the heated testis was removed and injected into a second animal without adjuvant but if injected into the same animal it was necessary to give multiple injections at the same site. A similar reaction was produced with a normal testis which had been heated after removal from the animal (Mancini *et al.*, 1972b; Mazzolli *et al.*, 1975). The lesion was associated with the development of delayed hypersensitivity (see Mancini, 1974, 1976) but its progression through the testis has not been recorded. Immunological aspermatogenesis apparently results when the immune system is exposed to testis antigen *and* a granulomatous reaction; the latter may result from the adjuvant or other causes.

It is interesting that brain (Katsh and Bishop, 1958; Katsh and Katsh, 1965) and parotid glands (Nagano and Okumura, 1973b) contain antigens which when injected with adjuvant produce autoimmune aspermatogenesis as it had been shown by Lewis (1934) that brain and testis contained cross-reacting antigens.

Some progress has been made in the isolation from the testis and characterization of a protein and two glycoproteins which are potent antigens for the induction of allergic aspermatogenesis (Katsh *et al.*, 1972; Jackson *et al.*, 1975, 1976; Hagopian *et al.*, 1975, 1976).

## References to Chapter 11.4

Albin, R. J., Soanes, W. A and Bronson, P. M. (1971) Autoantibodies to rabbit testes as a consequence of *in situ* freezing. *Folia Biol.* (Praha) **17**, 429–31.

Andrada, J. A., Andrada, E. C. and Witebsky, E. (1969) Experimental autoallergic orchitis in rhesus monkeys. *Proc. Soc. Exptl. Biol. Med.* **130**, 1106–13.

Attaran, S. E., Hodges, C. V., Crary, L. S., Vangalder, G. C., Lawson, R. K. and Ellis, L. R. (1966) Homotransplants of the testis. *J. Urol.* **95**, 387–9.

Bevilacqua, R. (1968) Rilievi istologici sulle gonadi di ratte tratti con immunsieri antiparotide ed antisottomascellare. *Boll. Soc. Ital. Biol. Sper.* **44**, 833–5.

Bishop, D. W. (1961) Aspermatogenesis induced by testicular antigen uncombined with adjuvant. *Proc. Soc. Exptl. Biol. Med.* **107**, 116–20.

Bishop, D. W. (1969) Sorbitol dehydrogenase: Enzymic antigen and assay for induced aspermatogenesis. In *International Convocation on Immunology*, eds. N. R. Rose and F. Milgrom, 334–8, Basel: Karger.

Bishop, D. W. and Carlson, G. L. (1965) Immunologically induced aspermatogenesis in guinea pigs. *Ann. N.Y. Acad. Sci.* **124**, 247–66.

Bishop, D. W., Narbaitz, R. and Lessof, M. (1961) Induced aspermatogenesis in adult guinea pigs injected with testicular antigen and adjuvant in neonatal stages. *Develop. Biol.* **3**, 444–85.

Boughton, B. and Spector, W. G. (1963) 'Auto-immune' testicular lesions induced by injury to the contralateral testis and intradermal injection of adjuvant. *J. Pathol. Bact.* **86**, 69–74.

Brown, P. C. and Glynn, L. E. (1969) The early lesion of experimental allergic orchitis in guinea-pigs: An immunological correlation. *J. Pathol.* **98**, 277–82.

Brown, P. C., Glynn, L. E. and Holborow, E. J. (1963) The pathogenesis of experimental allergic orchitis in guinea pigs. *J. Pathol. Bacteriol.* **86**, 505–20.

Brown, P. C., Holborow, E. J. and Glynn, L. E. (1965) The aspermatogenic antigen in experimental allergic orchitis in guinea pigs. *Immunol.* **9**, 255–60.

Brown, P. C., Glynn, L. E. and Holborow, E. J. (1967) The dual necessity for delayed hypersensitivity and circulating antibody in the pathogenesis of experimental allergic orchitis in guinea pigs. *Immunol.* **13**, 307–14.

Brown, P. C., Dorling, J. and Glynn, L. E. (1972) Ultrastructural changes in experimental allergic orchitis in guinea pigs. *J. Pathol.* **106**, 229–33.

Delaunay, A. and Voisin, G. A. (1952) Sur des lésions testiculaires provoquées chez le cobaye et chez le rat par l'endotoxine typhoïque. *C. rend. Acad. Sci.* **234**, 158–60.

Eyquem, A. and Krieg, H. (1965) Experimental autosensitization of the testis. *Ann. N.Y. Acad. Sci.* **124**, 270–8.

Fainboim, L., Barrera, C. N. and Mancini, R. E. (1976) Immunologic and testicular response in guinea pigs after unilateral traumatic orchitis. *Andrologia* **8**, 243–8.

Fernandez Collazo, E., Thierer, E. and Mancini, R. E. (1972) Immunologic and testicular response in guinea pigs after unilateral thermal orchitis. *J. Allergy Clin. Immunol.* **49**, 167–73.

Freund, J., Lipton, M. M. and Thompson, G. E. (1953) Aspermatogenesis in the guinea pig induced by testicular tissue and adjuvants. *J. Exptl. Med.* **97**, 711–26.

Freund, J., Lipton, M. M. and Thompson, G. E. (1954) Impairment of spermatogenesis in the rat after cutaneous injection of testicular suspension with complete adjuvants. *Proc. Soc. Exptl. Biol. Med.* **87**, 408–11.

Freund, J., Thompson, G. E. and Lipton, M. M. (1955) Aspermatogenesis, anaphylaxis, and cutaneous sensitization induced in the guinea pig by homologous testicular extract. *J. Exptl. Med.* **101**, 591–604.

Gunaya, K. P., Sheth, A. R. and Rao, S. S. (1970) Immunological studies with rat testis: antigenic characterization. *J. Reprod. Fertil.* **23**, 263–9.

Hagopian, A., Jackson, J. J., Carlo, D. J., Limjuco, G. A. and Eylar, E. H. (1975) Experimental allergic aspermatogenic orchitis. III. Isolation of spermatozoal glycoproteins and their role in allergic aspermatogenic orchitis. *J. Immunol.* **115**, 1731–43.

Hagopian, A., Limjuco, G. A., Jackson, J. J., Carlo, D. J. and Eylar, E. H. (1976) Experimental allergic aspermatogenic orchitis. IV. Chemical properties of sperm glycoproteins isolated from guinea pig testes. *Biochim. Biophys. Acta* **434**, 354–64.

Isidori, A. (1970) Antitesticular immunity. Role of the basement membrane. *Experientia* **26**, 1375–7.

Jackson, J. J., Hagopian, A., Carlo, D. J., Limjuco, G. A. and Eylar, E. H. (1975) Experimental allergic aspermatogenic orchitis. I. Isolation of a spermatozoal protein (AP 1) which induces allergic aspermatogenic orchitis. *J. Biol. Chem.* **250**, 6141–50.

Jackson, J. J., Hagopian, A., Limjuco, G. A., Carlo, D. J. and Eylar, E. H. (1976) Experimental allergic aspermatogenic orchitis. II. Some chemical properties of the AP1 protein of the sperm acrosome. *Biochim. Biophys. Acta.* **427**, 251–61.

Johnson, M. H. (1970a) An immunological barrier in the guinea pig testis. *J. Path.* **101**, 129–39.

Johnson, M. H. (1970b) Changes in the blood–testis barrier of the guinea pig in relation to histological damage following isoimmunization with testis. *J. Reprod. Fertil.* **22**, 119–27.

Johnson, M. H. (1970c) Selective damage to spermatogenic cells of high antigenicity during auto-allergic aspermatogenesis. *J. Pathol.* **102**, 131–8.

Johnson, M. H. (1972) The distribution of immunologloglobulin and spermatozoal autoantigen in the genital tract of the male guinea pig: its relationship to auto-allergic orchitis. *Fertil. Steril.* **23**, 383–92.

Johnson, M. H. (1973) Physiological mechanisms for the immunological isolation of spermatozoa. *Advan. Reprod. Physiol.* **6**, 279–324.

Kantor, G. L. and Dixon, F. J. (1972) Transfer of experimental allergic orchitis with peritoneal exudate cells. *J. Immunol.* **108**, 329–38.

Katsh, S. (1960a) Localization and identification of aspermatogenic factor in guinea pig testicles. *Intern. Arch. Allergy Appl. Immunol.* **16**, 241–75.

Katsh, S. (1960b) The anaphylactogenicity of testicular hyaluronidase and a species difference in testicular hyaluronidase demonstrated by isolated organ anaphylaxis. *Intern. Arch. Allergy Appl. Immunol.* **17**, 70–9.

Katsh, S. and Bishop, D. W. (1958) The effects of homologous testicular and brain and heterologous testicular homogenates combined with adjuvant upon the testis of guinea pigs. *J. Embryol. Exptl. Morphol.* **6**, 94–104.

Katsh, S. and Katsh, G. F. (1965) Aspermatogenic antigen from brain. *Experientia* **21**, 442–3.

Katsh, S., Aguirre, A. E., Leaver, F. W. and Katsh, G. F. (1972) Purification and partial characterization of aspermatogenic antigen. *Fertil. Steril.* **23**, 644–56.

Landsteiner, K. (1899) Zur Kenntnis der spezifisch auf Blutkörperchen wirkenden Sera. *Centr. Bakteriol.* **25**, 546–9.

Lee, S., Tung, K. S. and Orloff, M. J. (1971) Testicular transplantation in the rat. *Trans. Proc.* **3**, 586–90.

Levine, S. and Sowinski, R. (1970) Allergic inflammation, infarction and induced localization in the testis. *Am. J. Pathol.* **59**, 437–51.

Lewis, J. H. (1934) The antigenic relationship of alcohol-soluble fractions of brain and testicle. *J. Immunol.* **27**, 473–8.

Mancini, R. E. (1972) Immunological and testicular response of guinea pigs sensitized with homogenate from homologous thermal injured testis. *Proc. Soc. Exptl. Biol. Med.* **139**, 991–6.

Mancini, R. E. (1974) Immunologic and testicular response to a damage induced in the contralateral gland. In *Male Fertility and Sterility*, eds. R. E. Mancini and L. Martini, New York: Academic Press, 271–300.

Mancini, R. E. (1976) *Immunologic Aspects of Testicular Function*, Monographs on Endocrinology **9**, 1–114. Berlin, Heidelberg, New York: Springer-Verlag.

Mancini, R. E., Gallo Morando, G., Torres Aguero, M. and Pahul, G. (1972a) Homotransplantation of testis in dogs. I. Histologic study of the rejection phenomena. *Medicina* **32**, 215–22.

Mancini, R. E., Mazzolli, A. and Thierer, E. (1972b) Immunological and testicular response of guinea pigs sensitized with homogenate from homologous thermal injured testis. *Proc. Soc. Exptl. Biol. Med.* **139**, 991–6.

Mazzolli, A. B., Bustuoabad, O., Barrera, C. N. and Mancini, R. E. (1975) A new model for antisperm autoimmunity in guinea pigs. *Int. J. Fertil.* **21**, 49–54.

Menge, A. C. (1971) Effects of auto- and iso-immunization of bulls with semen and testis. *Int. J. Fertil.* **16**, 130–8.

Metalnikoff, S. (1900) Etudes sur la spermotoxine. *Ann. Inst. Pasteur* **14**, 577–89.

Metchnikoff, E. L. (1900) L'influence de l'organisme des toxines sur la spermotoxine et l'antispermotoxine. *Ann. Inst. Pasteur* **14**, 1–12.

Murthy, G. G., Peress, N. S. and Khan, S. A. (1976) Demonstration of antibodies to testicular basement membrane by immunofluorescence in a patient with multiple primary endocrine deficiencies. *J. Clin. Endocrinol. Metab.* **42**, 637–41.

Nagano, T. and Okumura, K. (1973a) Fine structural changes of allergic aspermatogenesis in the guinea pig. I. The similarity in the initial changes induced by passive transfer of anti-testis serum and by immunization with testicular tissue. *Virchows Arch. Abt. B. Zellpath.* **14**, 223–35.

Nagano, T. and Okumura, K. (1973b) Fine structural changes of allergic aspermatogenesis in the guinea pig. II. Induced by the homologous parotid gland as antigen. *Virchows Arch. Abt. B. Zellpath.* **14**, 237–45.

Pokorna, Z. and Vojtiskova, M. (1964a) Autoimmune damage of the testes induced with chemically modified organ specific antigen. *Folia Biol.* (Prague) **10**, 261–7.

Pokorna, Z. and Vojtiskova, M. (1964b) Ontogenic manifestation of the testicular antigen and the inducibility of autoimmune lesion by means of immature guinea pig testes. *Folia Biol.* (Prague) **10**, 392–7.

Pokorna, Z. and Vojtiskova, M. (1964c) Reduced inducibility of autoimmune condition by passive transfer of immune and normal sera. *Folia Biol.* (Prague) **10**, 405–8.

Pokorna, Z., Vojtiskova, M., Rychlikova, M. and Chutna, J. (1963) An isologous model of experimental autoimmune aspermatogenesis in mice. *Folia Biol.* (Prague) **9**, 203–9.

Raitsina, S. S., Davidova, A. I. and Gladkova, N. S. (1973) Blood testis barrier in the development of post traumatic and auto allergic aspermatogenesis. In *Immunol. Reprod.*, 83–93. Sofia: Bulgarian Academy Sciences Press.

Rapaport, F. T., Sampath, A., Kano, K., McCluskey, R. T. and Milgrom, F. (1969) Immunological effects of thermal injury. I. Inhibition of spermatogenesis in guinea pigs. *J. exp. Med.* **130**, 1411–22.

Rümke, P. (1965) Autospermagglutinins: A cause of infertility in men. *Ann. N.Y. Acad. Sci.* **124**, 696–701.

Rümke, P. and Titus, M. (1970) Spermagglutinin formation in male rats by subcutaneously injected syngeneic epididymal spermatozoa and by vasoligation or vasectomy. *J. Reprod. Fertil.* **21**, 69–79.

Spooner, R. L. (1964) Cytolytic activity of the serum of normal male guinea pigs against their own testicular cells. *Nature* (London) **202**, 915–6.

Toullet, F. and Voisin, G. A. (1969) Réactions d'hypersensibilité et anticorps seriques envers les auto-antigènes des spermatozoïdes. Relation avec le mécanisme de l'orchite aspermatogénétique autoimmune. *Ann. Inst. Pasteur* (Paris) **116**, 579–601.

Tung, K. S. K., Unanue, E. R. and Dixon, F. J. (1970) The immunopathology of experimental allergic orchitis. *Am. J. Pathol.* **60**, 313–29.

Tung, K. S. K., Unanue, E. R. and Dixon, F. J. (1971a) Pathogenesis of experimental allergic orchitis. 1. Transfer with immune lymph node cells. *J. Immunol.* **106**, 1453–62.

Tung, K. S. K., Unanue, E. R. and Dixon, F. J. (1971b) Pathogenesis of experimental allergic orchitis. II. The role of antibody. *J. Immunol.* **106**, 1463–72.

Voisin, G. A. and Toullet, F. (1968) Etude sur l'orchite aspermatogénétique autoimmune et les autoantigènes de spermatozoïdes chez le cobaye. *Ann. Inst. Pasteur* **114**, 727–55.

Voisin, G. A., Delaunay, A. and Barber, M. (1951) Sur des lésions testiculaires provoquées chez le cobaye par iso- et auto-sensibilisation. *Ann. Inst. Pastuer* (Paris) **81**, 48–63.

Vojtiskova, M. and Pokorna, Z. (1964) Prevention of experimental allergic aspermatogenesis by thymectomy in adult mice. *Lancet* **I**, 644–5.

Waksman, B. H. (1959) A histologic study of the auto-allergic testis lesion in the guinea pig. *J. Exptl. Med.* **109**, 311–24.

Waksman, B. H. (1960) The distribution of experimental auto-allergic lesions. *Am. J. Pathol.* **37**, 673–85.

Willson, J. T., Jones, N. A., Katsh, S. and Smith, S. W. (1973) Penetration of the testicular-tubular barrier by horse radish peroxidase induced by adjuvant. *Anat. Rec.* **176**, 85–100.

Willson, J. T., Jones, N. and Katsh, S. (1972) Induction of aspermatogenesis by passive transfer of immune sera or cells. *Int. Arch. Allergy* **43**, 172–81.

## 11.5   Naturally occurring abnormalities

The function of the testis is secondarily affected in a variety of illnesses, but the testis is also affected by several specific conditions which either involve only the testis or the testis and a few other parts of the body. The majority of these conditions have been described in man, but several syndromes have been reported in rats and domestic animals.

### 11.5.1   Klinefelter's syndrome

This condition which affects about $0 \cdot 2$ per cent of the population was first described by Klinefelter *et al.* in 1942. Patients with this syndrome have small firm non-functional testes, with poor development of secondary characteristics, azoospermia and gynecomastia and often show mental abnormalities. Histological examination of the testis reveals hyalinization and fibrosis of the seminiferous tubules, with a lack of elastic fibres in the boundary tissue of the tubules. The Leydig cells show decreased function but are hyperplastic and often grouped in adenomatous 'clumps' (see also Heller and Nelson, 1945; Becker *et al.*, 1966; Paulsen, 1974). The changes are often not apparent until puberty and the testis may be of normal size until then when hyalinization and fibrosis occur. The concentration of testosterone is usually low; sometimes total testosterone is normal but the free testosterone is low. Response to hCG is depressed. Serum and urinary FSH and LH are always elevated.

With the advent of chromosome analysis, these men were shown usually to have an extra X chromosome (Bradbury *et al.*, 1956; Ferguson-Smith *et al.*, 1960; Day *et al.*, 1963). Klinefelter's syndrome may also be found in other genetic abnormalities with XXXY or XXXXY chromosomes, or mosaicism involving XY and XXY lines or other combinations (Ferguson-Smith *et al.*, 1957; Day *et al.*, 1963; Kjessler, 1974; Paulsen, 1974; Cameron and Pugh, 1976). The infertility is irreversible and the gynecomastia does not respond to medical therapy. Treatment to remedy the deficiency in androgens is customary (see Paulsen, 1974). Klinefelter's syndrome has been described in cattle (Rieck *et al.*, 1969), sheep (Bruere *et al.*, 1969a), pigs

(Gluhouschi *et al.*, 1968), mice (Russell and Chu, 1961) and cats (Biggers and McFeely, 1966).

### 11.5.2    Tubular hyalinization without chromosomal abnormalities

It is now usual to consider those cases in which the seminiferous tubules are hyalinized, but with a normal chromosomal complement, as a separate entity from Klinefelter's syndrome. The histological picture in the testis is very similar and prognosis is equally poor (de Kretser *et al.*, 1972, 1974; Millet *et al.*, 1973; Baker *et al.*, 1976).

### 11.5.3    Sertoli-cell only syndrome

This relatively uncommon disorder usually involves just the seminiferous tubules with normal Leydig cell function. The seminiferous tubules contain no germinal cells. There may be some thickening of the boundary tissue but peritubular sclerosis and hyalinization are not present (Del Castillo *et al.*, 1947; Howard *et al.*, 1950; Heller *et al.*, 1952). This syndrome may be due to congenital absence of germ cells, and an XYY genetic abnormality is present in a small proportion of cases (Santen *et al.*, 1970; de Kretser *et al.*, 1972). Treatment is ineffective and androgen therapy unnecessary. Serum FSH is usually elevated and LH normal (see Paulsen, 1974). A similar condition can be produced experimentally in rats by X-rays or by drugs in pregnant females (see Sections 11.1.1 and 11.3.2.1).

### 11.5.4    Germinal cell arrest and hypoplasia

These conditions are common causes of infertility in men and usually are associated with oligospermia or azoospermia. They are of unknown aetiology and are really descriptions of the histological appearance of testicular biopsy specimens.

In germinal cell arrest spermatogenesis is normal up to a certain stage in development but no more mature cells are present. In hypoplasia, all stages of development are present but in reduced numbers. Leydig cells appear normal. Serum FSH is usually elevated, and LH and testosterone are normal (see Nelson and Heller, 1948; Johnsen, 1964, 1970a, b, 1972; Paulsen, 1974; Skakkebaek *et al.*, 1973a). Rational therapy is difficult in view of the unknown aetiology. Some success has been achieved in selected cases by 'rebound' after a course of therapy with testosterone (Heller *et al.*, 1950; Heckel *et al.*, 1951) and by gonadotrophin therapy (see Troen *et al.*, 1970; Schirren and Toyosi, 1970; Lunenfeld and Shalkovsky-Weissenberg, 1970; but cf. Sherins, 1974).

Two forms of congenital hypoplasia of the testis have been reported in Swedish cattle. In one there is a complete absence of the germinal epithelium, or the tubules contain only Sertoli cells and spermatogonia. The condition is not apparent until puberty, but thereafter becomes progressively more apparent. It can be bilateral or unilateral (Lagerlof, 1934, 1951, 1957, 1966). Genetic analysis suggests that it is due to a recessive autosomal gene (Eriksson, 1943; cf. Johansson, 1960). In the other form of heritable hypoplasia in bulls, there is arrest of spermatogenesis at the spermatocyte stage, and this has been attributed to either the development of multipolar spindles (Knudsen, 1961a) or 'sticky' chromosomes (Knudsen, 1961b). Spermatocytes with multipolar spindles have also been described in pituitary dwarf mice (Casaco and Martinazzi, 1963).

### 11.5.5   Hypogonadotrophic eunuchoidism and fertile eunuch syndrome

In patients with a deficiency in the production of gonadotrophins, testis function will obviously be abnormal and the patient will seem like a prepubertal boy. Cryptorchidism is not uncommon. This syndrome is often associated with anosmia or hyposmia (Kallman et al., 1944; Heller and Nelson, 1948; Howard et al., 1950; Bartter et al., 1952; Albert et al., 1954).

Low serum or urine FSH and LH are diagnostic and as a deficiency in LHRH is the probable cause, treatment with this or with gonadotrophins usually achieves normal testis function after prolonged treatment (see Paulsen, 1974).

A deficiency in gonadotrophin secretion can involve only LH, in the interesting group of patients known as fertile eunuchs (Pasqualini and Bur, 1950, 1955; McCullagh et al., 1953; Santen and Paulsen, 1973a, b). Serum LH and testosterone are low; serum FSH is normal. Secondary sexual characteristics are reduced and gynecomastia is common but presumably sufficient testosterone is being produced by the Leydig cells in close apposition to the tubules, that with normal concentrations of FSH in serum, spermatogenesis is possible (Makler et al., 1977).

### 11.5.6   Mumps orchitis

About 20 per cent of men with mumps also develop acute orchitis. If this occurs before puberty, there is usually no problem but in adults permanent damage to the tubules often results (Gall, 1947; Charny and Meranze, 1948; Werner, 1950; Ballew and Masters, 1954; Roseman, 1958; Scott, 1960a; Bartak et al., 1968; Morgan, 1976). During the acute phase, interstitial oedema and sloughing or degeneration of tubular cells are found. Long-term changes involve progressive tubular sclerosis and hyalinization and the full extent of the damage may not be apparent until 10 to 20 years after the acute infection. Leydig cell function is usually normal. Serum FSH is raised, and LH and testosterone are normal. Acute orchitis of a similar nature is often associated with gonorrheal infections and leprosy (see Morgan, 1976; Paulsen, 1974).

The testis and the parotid gland have some immunological similarities and immunization with parotid tissue in Freunds adjuvant can produce testicular lesions (see Section 11.6). Some of the testicular involvement in mumps may therefore be immunological in nature (see Andrada et al., 1977).

### 11.5.7   Testicular feminization, genetic abnormalities and intersexes

In animals with testicular feminization, there are small inguinal testes, lacking germ cells, but no other internal reproductive organs. In these animals, the tissues appear to be unable to respond to testosterone. It was originally suggested that this was because they could not convert testosterone to $5\alpha$-dihydrotestosterone but a lack of the appropriate androgen receptor in the cells now seems more likely (see Section 6.1.7). As a result, the Wolffian duct does not develop and the germ cells do not divide. The fetal testis presumably secretes the Müllerian duct inhibitor normally, and thus there are no Fallopian tubes or uterus.

This condition occurs in humans (Morris, 1953; Gordon et al., 1964; Simmer et al., 1965; Southren, 1965; French et al., 1965, 1966; Symington and Cameron, 1970; Faiman and Winter, 1974), cattle (Nes, 1966; Rieck, 1971), sheep (Bruere et al., 1969b), pigs (Wensing et al., 1975), mice (Lyon and Hawkes, 1970; Ohno and Lyon, 1970; Goldstein and Wilson, 1972; Blackburn et al., 1973) and rats—Stanley–Gumbreck male pseudohermaphrodite rats (Stanley and Gumbreck, 1964; Bardin et al., 1970, 1973; Stanley et al., 1973). In all these species, affected females produce normal females, normal males and pseudohermaphrodite males in the ratio 2:1:1. Sires of pseudohermaphrodites do not transmit the condition. The condition would therefore appear to be due to a sex-linked autosomal dominant or an X-linked recessive gene mutation.

The testes are small and contain only Sertoli cells and spermatogonia in the tubules. If a testis is transplanted to the anterior chamber of the eye more advanced stages of spermatogenesis are found, but no spermatozoa, suggesting that there is also a temperature effect (Chan et al., 1969). The interstitial cells appear normal morphologically (Vanha-Perttula et al., 1970) but testosterone concentrations in plasma are about a quarter of normal levels (Bardin et al., 1969). Serum FSH is normal and LH elevated (Sherins et al., 1971). The animals do not respond to physiological doses of exogenous androgens, but will respond to very large doses (Sherins and Bardin, 1971; Sherins et al., 1971; Bardin et al., 1973; Goldman and Klingele, 1975).

In another genetically abnormal rat (vestigal testis rat or vet-rat) there are tiny abdominal testes lying just inferior to the kidney. Adult Leydig cells cannot be seen and testosterone is undetectable in blood, but the secondary sex organs respond normally to injected hormone. As in the pseudo-hermaphrodite rats, the Müllerian duct has regressed, but near the testis the 'vet-rat' retains epididymis-like ductal structures, which respond to testosterone. This condition is apparently due to a failure of development of the Leydig cells, or an inherited lack of their receptors for LH (Stanley et al., 1973; Bardin et al., 1973).

'Restricted colour ($H^{re}$)' rats, become fertile at the normal time but remain fertile for only a few weeks. No ultrastructural abnormalities were noticed before the appearance of the first spermatids and spermatozoa, but after this time there was progressive vacuolation of the Sertoli cells and abnormalities of the spermatids. Abnormalities of the limiting membrane could also be seen but the Leydig cells appeared normal. 'Insensitivity' of the tubules to FSH has been suggested, on rather tenuous grounds, as the cause of these abnormalities (Russell and Gardner, 1973, 1974).

Abnormal spermiogenesis has also been reported in several mutant mice: 'T' (Yanagisawa, 1965; Olds, 1971), 'hop sterile' (Johnsen and Hunt, 1971), 'p-sterile' (Hunt and Johnson, 1971) and 'quaking' (Bennett et al., 1971; Coniglio et al., 1975) mutants; and in 'dag-defect' (Blom, 1966) and 'pseudodroplet' (Blom, 1968) bulls. The final shape and size of the spermatozoa is also under genetic control (Blom and Birch-Andersen, 1962, 1975; Buttle and Hancock, 1965). In 'steel' mutant mice, there appear to be no germ cells in the testis, but normal amounts of testosterone in the testis and plasma, with increased serum FSH and LH (Younglai and Chui, 1973). Hereditary obese-hyperglycaemic mice (Ao mice) have small testes and hypoplastic seminal vesicles. The tubules appear to be normal but the Leydig cells are atrophic (Hellman et al., 1963). Guinea pigs which are silver-white because of two recessive genes at two different loci, have small testes devoid of spermatogenic cells (Wright, 1959).

Small sterile 'testes' are also found in freemartin cattle, which are genetic females which have had a male co-twin. This condition is not due to the effects of steroid hormones from the male fetus (Tandler and Keller, 1911; Keller and Tandler, 1916; Lillie, 1916, 1917) or of the transfer of germ cells from the male twin (Fechheimer et al., 1963; Goodfellow et al., 1965; Stewart, 1965) as has been suggested, but is probably due to the failure of medullary inhibition of the fetal gonad which normally occurs in the female (see Short, 1970; Jost et al., 1972, 1973, 1975; Viguier et al., 1973, 1976). The 'testes' of the freemartin are capable of producing large amounts of testosterone, which could account for the masculinization. The bull co-twin is usually thought to be reproductively normal, but in one such animal testosterone production by the testis 11 months after birth was very low (Short et al., 1969).

Genetically female intersex goats, have histological 'testes' which sometimes descend to an inguinal or scrotal position. These 'testes' contain normal amounts of testosterone which they secrete and the duct system in adult animals varies from slight to almost complete masculinization. In these animals, the gene for polledness apparently acts like a Y-chromosome to cause the gonad to develop into a testis, but the XX germ cells cannot develop in the testicular environment (Eaton, 1943, 1945; Short et al., 1969; Hamerton et al., 1969).

### 11.5.8    Varicocoele
Varicocoele is the term used to describe a varicosity of the internal spermatic vein or cremasteric vein; the varicosity ranges from a small enlargement of which the patient is unaware to a large swelling above the testis. It is almost invariably found on the left side, but is occasionally bilateral. Many of these patients have abnormal semen and testis biopsies and many are infertile (Perier, 1864; Spotoft, 1942; Tulloch, 1955; Young, 1956; Russell, 1957; Scott, 1958; Meyhofer and Wolf, 1960; Charny, 1962; MacLeod, 1965, 1969; Davis et al., 1965; Harrison, 1966; Schirren and Klosterhalfen, 1966; Brown et al., 1967; Etriby et al., 1967, Haensch and Hornstein, 1968; Charny and Baum, 1968; Schellen and Canton, 1974; Fernando et al., 1976; Verstoppen and Steeno, 1977a). However it is very strange that a unilateral vascular abnormality should produce a bilateral testicular lesion.

It was suggested that the countercurrent exchange of heat in the spermatic cord is affected (Hanley and Harrison, 1962), but on the whole measurements of scrotal and testicular temperature have not substantiated this idea (Tessler and Krahn, 1966; Stephenson and O'Shaughnesey, 1968; Agger, 1971; Zorgniotti and MacLeod, 1973). However most of these measurements appear to have been made on nude men in a climate chamber and should be repeated under more normal circumstances.

It has also been suggested that there is retrograde flow of blood in the left internal spermatic vein, and because this vein usually opens into the left renal vein just near the adrenal vein (Anson *et al.*, 1948; El Sadr and Mina, 1950; Brown *et al.*, 1967; Volter *et al.*, 1975; Comhaire and Vermeulen, 1976) that the retrogradely flowing blood has an abnormally high concentration of corticosteroids or catecholamines (MacLeod, 1965). In fact, abnormally high concentrations of corticosteroids have not been found (Charny and Baum, 1968; Koumans *et al.*, 1969; Aggar, 1971; Lindholmer *et al.*, 1973) but high concentrations of catecholamines have been shown in one study (Comhaire and Vermeulen, 1974). None of the authors make it clear just where they imagine this retrogradely flowing blood goes to and how a high concentration of hormones in the venous blood reaches the testis. It is to be hoped that they are not advocating a return to pre-Harveian cardiovascular ideas or suggesting that blood can flow from vein to artery against the pressure gradient.

However, it is generally agreed that the main valve in the left spermatic vein in man is the one near the renal vein and that in varicocoele patients, this and the other valves are incompetent or congenitally absent. Then if an anastomosis between the spermatic vein and the cremasteric vein possibly via the varicocoele is present, renal blood could flow down the spermatic vein, through the cremasteric vein and thence into the general circulation. This would not take venous blood through the testis which would continue to receive arterial blood and contribute its venous drainage into the varicocoele and the cremasteric vein. Under these circumstances the testis may be affected in several ways. Firstly, the vascular pressure relationships within the testis could be affected leading to abnormalities in filtration and reabsorption of fluid by the capillaries (see Section 3.3.3). Secondly vasoactive substances in the venous component of the spermatic cord can affect the arterial supply to the testis (see Section 3.4). Lastly, there is the possibility that although abnormalities of the testis are often found in patients with varicocoele, these may be incidental and the main cause of infertility could be the failure of the sperm to mature in the epididymis. The head of the epididymis receives arterial blood from branches of the internal spermatic artery. These branches, like the main artery, are probably involved in counter-current exchange of substances in the cord (see Section 3.2.2) and because of this, probably contain higher concentrations of testosterone than ordinary arterial blood. If the direction of venous flow in the cord is reversed then counter-current exchange of testosterone from vein to artery would not occur but exchange of corticoids or catecholamines might. This may lead to dysfunction of the epididymis and infertility.

Usually it is considered that the circulation of the two testes is independent, but there is some evidence in humans of connections between the two sides (El Sadr and Mina, 1950; Brown *et al.*, 1967). It is not clear whether this involves the internal spermatic vessels or the cremasteric vessels; the latter are more likely. If there are usually anastomoses in humans between the internal spermatic and cremasteric circulations, and connections between the cremasteric circulations of the left and right sides, this could account for the bilateral effect of the unilateral vascular condition.

Empirically it has been found that ligation of the internal spermatic vein either in the abdominal cavity or near the varicocoele, improves semen quality, particularly sperm motility, and many patients treated in this way seem to have become fertile (Palomo, 1949; Robb, 1954; Scott, 1960, 1961; Scott and Young, 1962; Hanley and

Harrison, 1962; Hegemann and Schmitz, 1968; Steeno *et al.*, 1976; Verstoppen and Steeno, 1977b). If this is so, and no controlled trials have been reported, then it would seem that either changed venous pressure or the action on the artery of vasoactive compounds in the veins are involved.

### 11.5.9    Neoplasms of the testis

Neoplasms involving testicular cells are uncommon, affecting only 0·002 per cent of human males. They can involve germinal or non-germinal cells. Four types affecting germinal cells are recognized in humans: (1) seminoma; (2) teratoma, or differentiated teratoma; (3) embryonal carcinoma, or malignant undifferentiated teratoma; and (4) choriocarcinoma, or malignant trophoblastic teratoma. Combined tumours are common.

Testicular swelling is the commonest presenting symptom and biopsy is not recommended for diagnosis. The gross appearance of the testis at operation is often characteristic and the detection of hCG secreted by the choriocarcinomas is a useful aid in diagnosis. Treatment is by orchidectomy and X-irradiation and its success depends on the type of cancer, with a five-year mortality ranging from 10 to 100 per cent.

Leydig cell tumours are the most important of the non-germinal neoplasms. They can be either virilizing or feminizing and are usually more benign than the germinal cell tumours. Malignant lymphomas may also occur in the testis (see Brown, 1976a, b; Gowing, 1976; Pugh, 1976a, b, c; Pugh and Cameron, 1976; Thackray and Crane, 1976).

Sertoli cell tumours are not uncommon in dogs and usually secrete considerable amounts of oestrogens. Teratomata are occasionally found in horses. Other types of neoplasm have been reported from other animals but are quite rare (Cotchin, 1976).

### References to Chapter 11.5

Agger, P. (1971) Plasma cortisol in the left spermatic vein in patients with varicocele. *Fertil. Steril.* **22**, 270–4.

Albert, A., Underdahl, L. O., Green, L. F. and Lorenz, N. (1954) Male hypogonadism. IV. The testis in prepubertal or pubertal gonadotrophic failure. *Proc. Staff Meet. Mayo Clin.* **29**, 131–6.

Allison, J. E., Stanley, A. J., Gumbreck, L. G. and James, D. (1968) Germ cell migration in rats carrying mutant genes affecting fertility. *Proc. Soc. exp. Biol. N.Y.* **129**, 96–8.

Allison, J. E., Chan, F., Stanley, A. J. and Gumbreck, L. H. (1971) Androgen insensitivity in male pseudohermaphrodite rats. *Endocrinology* **89**, 615–17.

Andrada, J. A., von der Walde, F., Hoschoian, J. C., Comini, E. and Mancini, E. (1977) Immunological studies in patients with mumps orchitis. *Andrologia* **9**, 207–15.

Anson, B. J., Cauldwell, E. W., Pick, J. W. and Beaton, L. E. (1948) The anatomy of the pararenal system of veins, with comments on the renal arteries. *J. Urol.* **60**, 714–37.

Baker, H. W. G., Bremner, W. J., Burger, H. G., de Kretser, D. M., Dulmanis, A., Eddie, L.

W., Hudson, B., Keogh, E. J., Lee, V. W. K. and Rennie, G. C. (1976) Testicular control of FSH secretion. *Rec. Progr. Horm. Res.* **32**, 429–69.

Ballew, J. W. and Masters, W. H. (1954) Mumps: A cause of infertility. I. Present considerations. *Fertil. Steril.* **5**, 536–43.

Bardin, C. W., Allison, J. E., Stanley, A. J. and Gumbreck, L. G. (1969) Secretion of testosterone by the pseudohermaphrodite rat. *Endocrinology* **84**, 435–6.

Bardin, C. W., Bullock, L., Schneider, G., Allison, J. E. and Stanley, A. J. (1970) Pseudohermaphrodite rat: end organ insensitivity to testosterone. *Science* **167**, 1136–37.

Bardin, C. W., Bullock, L. P., Sherins, R. J., Mowszowicz, I. and Blackburn, W. R. (1973) Part II. Androgen metabolism and mechanism of action in male pseudohermaphroditism: a study of testicular feminization. *Rec. Progr. Horm. Res.* **29**, 65–105.

Bartak, V., Skalova, E. and Nevarilova, A. (1968) Spermiogram changes in adults and youngsters after parotitic orchitis. *Int. J. Fertil.* **13**, 226–232.

Bartter, C. F., Sniffen, R. C., Simmons, M. D., Albright, F. and Howard, R. P. (1952) Effect of chorionic gonadotrophin (APL) in male eunuchoidism with low 'follicle stimulatory hormones': aqueous solution versus oil and beeswax suspension. *J. Clin. Endocrinol.* **12**, 1532–50.

Becker, K. L., Hoffman, D. L., Albert, A., Underdahl, L. O. and Mason, H. L. (1966) Klinefelter's syndrome clinical and laboratory findings in 50 patients. *Arch. intern. Med.* **118**, 314–21.

Bennett, W. I., Gall, A. M., Southard, J. L. and Sidman, R. L. (1971) Abnormal spermiogenesis in quaking, a myelin-deficient mutant mouse. *Biol. Reprod.* **5**, 30–58.

Biggers, J. D. and McFealy, R. A. (1966) Intersexuality in domestic animals. *Adv. Reprod. Physiol.* **1**, 29–59.

Blackburn, W. R., Chung, K. W., Bullock, L. and Bardin, C. W. (1973) Testicular feminization in the mouse; studies of Leydig cell structure and function. *Biol. Reprod.* **9**, 9–23.

Blom, E. (1966) A new sterilizing and hereditary defect (the 'Dag-defect') located in the bull sperm tail. *Nature* **209**, 739–40.

Blom, E. (1968) A new sperm defect 'pseudo-droplets'—in the middle piece of the bull sperm. *Nord. Veterinar Med.* **20**, 279–83.

Blom, E. and Birch-Andersen, A. (1962) Ultrastructure of the sterilizing 'knobbed-sperm' defect in the bull. *Nature* **194**, 989–90.

Blom, E. and Birch-Andersen, A. (1975) The ultrastructure of a characteristic spermhead-defect in the boar: The SME-defect. *Andrologia* **7**, 199–209.

Bradbury, J. T., Bunge, R. G. and Boccabella, R. A. (1956) Chromatin test in Klinefelter's syndrome. *J. Clin. Endocrin.* **16**, 689.

Brown, J. S., Dubin, L. and Hotchkiss, R. S. (1967) The varicocele as related to fertility. *Fertil. Steril.* **18**, 46–56.

Brown, N. J. (1976a) Miscellaneous tumours of epithelial type. In *Pathology of the Testis*, ed. R. C. B. Pugh, 304–15. Oxford: Blackwell.

Brown, N. J. (1976b) Yolk-sac tumour (orchioblastoma) and other testicular tumours of childhood. In *Pathology of the Testis*, ed. R. C. B. Pugh, 356–70. Oxford: Blackwell.

Bruere, A. N. (1970) A request for sheep with testicular feminization. *N.Z. Vet. J.* **18**, 173–4.

Bruere, A. N., Marshall, R. B. and Ward, D. P. J. (1969a) Testicular hypoplasia and XXY sex chromosome complement in two rams: the ovine counterpart of Klinefelter's syndrome in man. *J. Reprod. Fertil.* **19**, 103–8.

Bruere, A. N., McDonald, M. F. and Marshall, R. B. (1969b) Cytogenetical analysis of an ovine male pseudohermaphrodite and the possible role of the Y chromosome in cryptochordism of sheep. *Cytogenetics* **8**, 148–57.

Bullock, L. P. and Bardin, C. W. (1972) Androgen receptors in testicular feminization. *J. Clin. Endocrinol.* **35**, 935–7.

Buttle, H. R. L. and Hancock, J. L. (1965) Sterile boars with 'knobbed' spermatozoa. *J. Agric. Sci.* **65**, 255–60.

Cameron, K. M. and Pugh, R. C. B. (1976) Miscellaneous lesions. In *Pathology of the Testis*, ed. R. C. B. Pugh, 448–68. Oxford: Blackwell.

Casaco, E. and Martinazzie, M. (1963) Aspetti istologici del testicolo e di alcune ghiandole sessuali accessorie nel topo con namismo preipofisario. *Arch. Ital. Anat. Embriol.* **68**, 121–34.

Del Castillo, E. B., Trabucco, A. and de la Balze, F. A. (1947). Syndrome produced by absence of the germinal epithelium without impairment of the Sertoli or Leydig cells. *J. Clin. Endocrinol.* **7**, 493–502.

Chan, F., Allison, J. E., Stanley, A. J. and Gumbreck, L. G. (1969) Reciprocal transplantation of testes between normal and pseudohermaphroditic male rats. *Fertil. Steril.* **20**, 482–94.

Charny, C. W. (1962) Effect of varicocele on fertility. Results of varicocelectomy. *Fertil. Steril.* **13**, 47–56.

Charny, C. W. and Baum, S. (1968) Varicocele and infertility. *J. Am. Med. Assn.* **204**, 1165–8.

Charny, C. W. and Meranze, D. R. (1948) Pathology of mumps and orchitis. *J. Urol.* (Baltimore) **60**, 140–6.

Chung, K. W. and Hamilton, J. B. (1975) Testicular lipids in mice with testicular feminization. *Cell tiss. Res.* **160**, 69–80.

Chung, K. W. and Hamilton, J. B. (1976) Further observations on the fine structure of Leydig cells in the testes of male pseudohermaphrodite rats. *J. ultrastruct. Res.* **54**, 68–75.

Clegg, E. (1970) The terminations of the left testicular and adrenal veins in man. *Fertil. Steril.* **21**, 36–8.

Comhaire, F. and Kunnen, M. (1976) Selective retrograde venography of the internal spermatic vein: A conclusive approach to the diagnosis of varicocele. *Andrologia* **8**, 11–24.

Comhaire, F. and Vermeulen, A. (1974) Varicocele sterility: cortisol and catecholamines. *Fertil. Steril.* **25**, 88–95.

Coniglio, J. G., Grogan, W. M., Harris, D. G. and Fitzhugh, M. L. (1975) Lipid and fatty acid composition of testes of quaking mice. *Lipids* **10**, 109–12.

Cotchin, E. (1976) Spontaneous and experimentally induced testicular tumours in animals. In *Pathology of the Testis*, ed. R. C. B. Pugh, 371–408. Oxford: Blackwell.

Davis, J. E., Clyman, M. E., Decker, A., Ober, W. B. and Roland, M. (1965) Varicocele as a contributing factor to male fertility. *Int. J. Fertil.* **10**, 359–72.

Day, R. W., Levinson, J., Larson, W. and Wright, S. W. (1963) An XXXXY male. Case report and review. *J. pediat.* **63**, 589–98.

Eaton, O. N. (1943) An anatomical study of hermaphroditism in goats. *Am. J. vet. Res.* **4**, 333–43.

Eaton, O. N. (1945) The relation between polled and hermaphroditic characters in dairy goats. *Genetics. Princeton* **30**, 51–61.

El-Sadr, A. R. and Mina, E. (1950) Anatomical and surgical aspects in the operative management of varicocele. *Urol. Cutan. Rev.* **54**, 257–62.

Eriksson, K. (1950) Heritability of reproduction disturbances in bulls of Swedish red and white cattle (SRB). *Nord. Veterinar. Med.* **2**, 943–66.

Etriby, A., Girgis, S. M., Hefnawy, H. and Ibrahim, A. A. (1967) Testicular changes in subfertile males with varicocele. *Fertil. Steril.* **18**, 666–71.

Faiman, C. and Winter, J. S. D. (1974) The control of gonadotropin secretion in complete testicular feminization. *J. Clin. Endocrinol. Metab.* **39**, 631–8.

Fechheimer, N. S., Herschler, M. S. and Gilmore, L. O. (1963) Sex chromosome mosaicism in unlike-sexed cattle twins. In *Genetics Today*, ed. S. J. Geerts, **I**, 265. New York: Macmillan.

Ferguson-Smith, M. A., Lennox, B., Mack, W. S. and Stewart, J. S. S. (1957) Klinefelter's syndrome. Frequency and testicular morphology in relation to nuclear sex. *Lancet* **II**, 167–9.

Ferguson-Smith, M. A., Lennox, B., Stewart, J. S. S. and Mack, W. S. (1960) Klinefelter's syndrome. *Mem. Soc. Endocrinol.* **7**, 173–81.

Fernando, N., Leonard, J. M. and Paulsen, C. A. (1976) The role of varicocele in male fertility. *Andrologia* **8**, 1–9.

French, F. S., Baggett, B., Van Wyk, J. J., Talbert, L. M., Hubbard, W. R., Johnston, F. R., Weaver, R. P., Forchielli, E., Rao, G. S. and Sarda, I. R. (1965) Testicular feminization: clinical morphological and biochemical studies. *J. Clin. Endocrinol.* **25**, 661–77.

French, F. S., van Wyk, J. J., Baggett, B., Easterling, W. E., Talbert, L. M., Johnson, F. R., Forchielli, E. and Dey, A. C. (1966) Further evidence of a target organ defect in the syndrome of testicular feminization. *J. Clin. Endocrinol. Metab.* **26**, 493–503.

Gall, E. A. (1947) The histopathology of acute mumps orchitis. *Am. J. Path.* **23**, 637–51.

Gluhovschi, N., Bistriceanu, M., Rosu, M. and Bratu, M. (1968) Contributions a l'étude des aspects chromosomiques de certaines perturbations de la reproduction chez les animaux domestiques. *Proc. VIe Congr. Reprod. Anim. Insem. Artif.*, 881–3.

Goldman, A. S. and Klingele, D. (1974) Developmental defects of testicular morphology and steroidogenesis in the male rat pseudohermaphrodite and response to testosterone and dihydrotestosterone. *Endocrinology* **94**, 1–16.

Goldstein, J. L. and Wilson, J. D. (1972) Studies on the pathogenesis of the pseudohermaphrodism in the mouse with testicular feminization. *J. Clin. Invest.* **51**, 1647–58.

Goodfellow, S. A., Strong, S. J. and Stewart, J. S. S. (1965) Bovine freemartins and true hermaphroditism. *Lancet* **I**, 1040–1.

Gordon, G. B., Miller, L. R. and Bensch, K. G. (1964) Electron microscopic observations of the gonad in the testicular feminization syndrome. *Lab. Invest.* **13**, 152–60.

Gosfay, S. (1959) Untersuchungen der Vena spermatica interna durch retrograde Phlebographie bei Kranken mit Varikozele. *Z. Urol.* **52**, 105–15.

Gowing, N. F. C. (1976) Malignant lymphoma of the testis. In *Pathology of the Testis*, ed. R. C. B. Pugh, 334–55. Oxford: Blackwell.

Gumbreck, L. G. (1964) New genetic factors that affect fertility in the male rat. *Proc. 5th Int. Congr. Reprod. Animale, Trento,* 1964 **2**, 319–25.

Gumbreck, L. G., Stanley, A. J., Allison, J. E. and Easley, R. B. (1972) Restriction of colour in the rat with associated sterility in the male and heterochromia in both sexes. *J. exp. Zool.* **180**, 333–50.

Haensch, R. and Hornstein, O. (1968) Varicocele und Fertilitats-storung. *Dermatologica* (Basel) **136**, 335–49.

Hamerton, J. L., Dickson, J. M., Pollard, C. E., Grieves, S. A. and Short, R. V. (1969) Genetic intersexuality in goats. *J. Reprod. Fertil.* Suppl. **7**, 25–51.

Hancock, J. L. and Trevan, D. J. (1967) The acrosome and post-nuclear cap of bull spermatozoa. *J. Roy. Microscop. Soc.* **76**, 77–83.

Hanley, H. G. and Harrison, R. G. (1962) The nature and surgical treatment of varicocele. *Brit. J. Surg.* **50**, 64–7.

Harrison, R. G. (1966) The anatomy of varicocele. *Proc. roy. soc. Med.* (London) **59**, 763–6.

Heckel, N. J., Rosso, W. A. and Kestel, L. (1951) Spermatogenic rebound phenomenon after administration of testosterone propionate. *J. Clin. Endocrinol.* **11**, 235–45.

Hegemann, G. and Schmitz, W. (1968) Die suprainguinale Gefaß-resektion bei Varikozelen aus chirurgischanatomischer Sicht. *Z. Urol.* **61**, 3–9.

Heller, C. G. and Nelson, W. O. (1945) Hyalinization of seminiferous tubules associated with normal or failing Leydig cell function. *J. Clin. Endocrinol.* **5**, 1–12.

Heller, C. G. and Nelson, W. O. (1948) The testis–pituitary relationship in man. *Rec. Progr. Horm. Res.* **3**, 229–55.

Heller, C. G., Nelson, W. O., Hill, I. B., Henderson, E., Maddock, W. O., Jungck, E. C., Paulsen, C. A. and Mortimore, G. E. (1950) Improvement in spermatogenesis following depression of the human testis with testosterone. *Fertil. Steril.* **1**, 415–20.

Heller, C. G., Paulsen, C. A., Mortimore, G. E., Junck, E. C. and Nelson, W. O. (1952) Urinary gonadotrophins, spermatogenic activity and classification of testicular morphology—their bearing on the utilization hypothesis. *Ann. N.Y. Acad. Sci.* **55**, 685–702.

Hellman, B., Jacobsson, L. and Täljedal, I.-B. (1963) Endocrine activity of the testis in obese-hyperglycaemic mice. *Acta. Endocrinol.* **44**, 20–6.

Henninger, H. (1934) Zur operativen Behandlung der idiopathischen Varikozele. *Zbl. Chir.* **61**, 695–7.

Howard, R. P., Sniffen, R. C., Simmons, F. A. and Albright, F. (1950) Testicular deficiency: a clinical and pathologic study. *J. Clin. Endocrinol.* **10**, 121–86.

Hunt, D. M. and Johnson, D. R. (1971) Abnormal spermiogenesis in two pink-eyed sterile mutants in the mouse. *J. Embryol. expl. Morphol.* **26**, 111–21.

Ishihara, T. (1956) Note on a hydrotestis found in the Wistar-King-A rat strain. *Rept. Natl. Inst. Genet. (Japan)* **6**, 15–18.

Ivanissevich, O. (1960) Left varicocele due to reflux; experience with 4470 operative cases in forty-two years. *J. int. Coll. Surg.* **34**, 742–55.

Johansson, I. (1960) Genetic causes of faulty germ cells and low fertility. *J. Dairy Sci.* **43**, Suppl. 1–27.

Johnsen, S. G. (1964) Studies on the testicular-hypophyseal feed-back mechanism in man. *Acta Endocrinol.* Suppl. **90**, 99–124.

Johnsen, S. G. (1970a) The stage of spermatogenesis involved in the testicular-hypophyseal feed-back mechanism in man. *Acta Endocrinol.* **64**, 193–210.

Johnsen, S. G. (1970b) Testicular biopsy score count—a method for registration of spermatogenesis in human testes: normal values and results in 335 hypogonal males. *Hormones* **1**, 1–24.

Johnsen, S. G. (1972) Studies on the pituitary-testicular axis in male hypogonadism, particularly in infertile men with 'cryptogenetic' hypospermatogenesis. In *Gonadotropins*, eds. B. B. Saxena, C. G. Beling and H. M. Gandy, 593–608. New York: Wiley.

Johnson, D. R. and Hunt, D. M. (1971) Hop-sterile, a mutant gene affecting sperm tail development in the mouse. *J. Embryol. exp. Morphol.* **25**, 223–36.

Jost, A., Vigier, B. and Prepin, J. (1972) Freemartins in cattle: the first steps of sexual organogenesis. *J. Reprod. Fertil.* **29**, 349–80.

Jost, A., Vigier, B., Prepin, J. and Perchellet, J. P. (1973) Le developpement de la gonade des freemartins. *Ann. Biol. Animal. Bioch. Biophys.* Suppl. **13**, 103–14.

Jost, A., Perchellet, J. P., Prepin, J. and Vigier, B. (1975) The prenatal development of bovine freemartins. In *Intersexuality in the Animal Kingdom*, ed. R. Reinboth, 392–406. Berlin: Springer.

Judd, H. L., Hamilton, C. R., Barlow, J. J., Yen, S. S. C. and Kliman, B. (1972) Androgen and gonadotropin dynamics in testicular feminization syndrome. *J. Clin. Endocr.* **34**, 229–34.

Kallman, F., Schonfield, W. A. and Barrera, S. E. (1944) Genetic aspects of primary eunuchoidism. *Am. J. ment. Defic.* **48**, 203–36.

Keller, K. and Tandler, J. (1916) Uber das Verhalten der Eihaute bei der Zwillingstachtigkeit des Rindes. *Wien. tierarztl. Monatschr.* **3**, 513–26.

Kjessler, B. (1974) Chromosomal constitution and male reproductive failure. In *Male Fertility and Sterility*, eds. R. E. Mancini and L. Martini, 231–47. New York: Academic Press.

Klinefelter, H. F., Reifenstein, E. C. and Albright, F. (1942) Syndrome characterized by gynecomastia, aspermatogenesis without a-leydigism, and increased excretion of follicle-stimulating hormone. *J. Clin. Invest.* **2**, 615–27.

Knudsen, O. (1954) Cytomorphological investigations into spermiocytogenesis of bulls with normal fertility and bulls with acquired disturbances in spermiogenesis. *Acta Pathol. Microbiol. Scand.* Suppl. 101.

Knudsen, O. (1958) Studies on spermiocytogenesis in the bull. *Int. J. Fertil.* **3**, 389–403.

Knudsen, O. (1961a) Testicular hypoplasia with multipolar spindle formation in the spermiocytes of the bull. *Acta Vet. Scand.* **2**, 199–209.

Knudsen, O. (1961b) Sticky chromosomes as a cause of testicular hypoplasia in bulls. *Acta Vet. Scand.* **2**, 1–14.

Koumans, J., Steeno, O. and Heyns, W. (1969) Dehydroepiandrosterone sulfate, androsterone sulfate and corticoids in spermatic vein blood of patients with left varicocele. *Andrologia* **1**, 87–9.

Kretser, D. M. de, Burger, H. G., Fortune, D., Hudson, B., Long, A. R., Paulsen, C. A. and Taft, H. P. (1972) Hormonal, histological and chromosomal studies in adult males with testicular disorders. *J. Clin. Endocrinol. Metab.* **35**, 392–401.

Kretser, D. M. de, Burger, H. G. and Hudson, B. (1974) Diagnostic aspects of male infertility. In *Male Fertility and Sterility*, eds. R. E. Mancini and L. Martini, 337–53. New York: Academic Press.

Lagerlöf, N. (1934) Morphologische Untersuchungen uber Veranderung im Spermabild und in den Hoden bei Bullen mit verminderter oder aufgehobener Fertilitat. *Acta Path. Microbiol. Scand.* Suppl. 19.

Lagerlöf, N. (1951) Hereditary forms of sterility in Swedish cattle breeds. *Fertil. Steril.* **2**, 230–9.

Lagerlöf, N. (1957) Biological aspects of infertility in male domestic animals. *Int. J. Fertil.* **2**, 99–129.

Lagerlöf, N. (1966) A history of cytological and histological examination of sperm and testis. *Mededel. Veeartsenijschool. Rijksuniv. Gent.* **10**, 5–19.

Lane, J. W. (1955) Radiographic studies in varicocoele. *U.S. Armed Forces Med. J.* **6**, 1589–96.

Lillie, F. R. (1916) The theory of the free-martin. *Science* **43**, 611–13.

Lillie, F. R. (1917) The freemartin: a study of the action of sex hormones in the foetal life of cattle. *J. exp. Zool.* **23**, 371–452.

Lindholmer, C., Thulin, L. and Eliasson, R. (1973) Concentrations of cortisol and renin in the internal spermatic vein of men with varicocele. *Andrologie* **5**, 21–2.

Lunenfeld, B. and Shalkovsky-Weissenberg, R. (1970) Assessment of gonadotrophin therapy in male infertility. *Adv. exp. Med. Biol.* **10**, 613–28.

Lyon, M. F. and Hawkes, S. G. (1970) X-linked gene for testicular feminization in the mouse. *Nature* (London) **227**, 1217–19.

McCullagh, E. P. and Schaffenburg, C. A. (1952) The role of the seminiferous tubules in the production of hormones. *Ann. N.Y. Acad. Sci.* **55**, 674–84.

McCullagh, E. P., Sirridge, W. T. and McIntosh, H. W. (1950) Gametogenic failure with high urinary gonadotropin (FSH). *J. Clin. Endocrinol.* **10**, 1533–46.

McCullagh, E. P., Beck, J. C. and Schaffenburg, C. A. (1953) A syndrome of eunuchoidism with spermatogenesis, normal urinary FSH and low or normal ICSH: ('Fertile eunuchs'). *J. Clin. Endocrinol.* **13**, 489–509.

MacLeod, J. (1965) Seminal cytology in the presence of varicocele. *Fertil. Steril.* **16**, 735–57.

MacLeod, J. (1969) Further observations on the role of varicocele in human male infertility. *Fertil. Steril.* **20**, 545–63.

Maddock, W. O. and Nelson, W. O. (1952) The effect of chorionic gonadotropin in adult men; increased estrogen and 17-keto-steroid excretion gynecomastia, Leydig cell stimulation and seminiferous tubule damage. *J. Clin. Endocrinol. Metab.* **12**, 985–1014.

Makler, A., Glezerman, M. and Lunenfeld, B. (1977) The Fertile Eunuch Syndrome—an isolated Leydig-cell failure? *Andrologia*, **9**, 163–70.

Meyhofer, W. and Wolf, J. (1960) Varikozele und Fertilitat. *Derm. Wschr.* **142**, 1116–21.

Millet, D., Tranchimont, P., Netter, J. P., Thevenet, M., Thieblot, P., Vendrely, E. and Netter, A. (1973) Corrélations histo-hormonales (FSH, LH, steroids) dans les sterilités masculines sécrétoires. *Ann. Endocrinol.* **34**, 377–90.

Mills, R. G. (1919) The pathological changes in the testes in epidemic pneumonia. *J. Exp. Med.* **30**, 505–29.

Morgan, A. D. (1976) Inflammation and infestation of the testis and paratesticular structures. In *Pathology of the Testis*, ed. R. C. B. Pugh, 79–138. Oxford: Blackwell.

Morris, J. M. (1953) Syndrome of testicular feminization male pseudohermaphrodites. *Am. J. Obst. Gynec.* **65**, 1192–211.

Nelson, W. O. and Heller, C. G. (1948) The testis in human hypogonadism. *Rec. Progr. Horm. Res.* **3**, 197–221.

Nes, N. (1966) Testicular feminisering hos storfe. *Nord. vet. med.* **18**, 19–29.

Notkovich, H. (1955) Testicular artery arching over renal vein: Clinical and pathological considerations with special reference to varicocele. *Brit. J. Urol.* **27**, 267–71.

Ohno, S. and Lyon, M. F. (1970) X-linked testicular feminization in the mouse as a non-inducible regulatory mechanism of the Jacob-Monod type. *Clin. Genet.* **1**, 121–7.

Ohno, S., Dofuku, R. and Tettenborn, U. (1971) More about X-linked testicular feminization of the mouse as a non-inducible (i$^s$) mutation of a regulatory locus: 5$\alpha$-androstan-3$\alpha$, 17$\beta$-diol as the true inducer of kidney alcohol dehydrogenase and $\beta$-glucuronidase. *Clin. Genet.* **2**, 128–40.

Olds, P. J. (1971) Effect of the T locus on sperm ultrastructure in the house mouse. *J. Anat.* **109**, 31–37.

Palomo, A. (1949) Radical cure of varicocele by a new technique: preliminary report. *J. Urol (Baltimore)* **61**, 604–7.

Pasqualini, R. Q. and Bur, G. E. (1950) Sindrome hipoandrogenico con gametogenesis conservada. *Rev. Assoc. med. Argentina.* **64**, 6–10.

Pasqualini, R. Q. and Bur, G. E. (1955) Hypoandrogenic syndrome with spermatogenesis. *Fertil. Steril.* **6**, 144–57.

Paulsen, C. A. (1974) The Testes. In *Textbook of Endocrinology*, ed. R. H. Williams, 323–67. Philadelphia: Saunders.

Perier, C. (1864) Considérations sur l'anatomie et la physiologie et sur un mode de traitement du varicocèle. Inaug.-Diss., Paris.

Pugh, R. C. B. (1976a) Testicular tumours—introduction. In *Pathology of the Testis*, ed. R. C. B. Pugh, 139–59. Oxford: Blackwell.

Pugh, R. C. B. (1976b) Combined tumours. In *Pathology of the Testis*, ed. R. C. B. Pugh, 245–58. Oxford: Blackwell.

Pugh, R. C. B. (1976c) Relative malignancy of tumours. In *Pathology of the Testis*, ed. R. C. B. Pugh, 441–7. Oxford: Blackwell.

Pugh, R. C. B. and Cameron, K. M. (1976) Teratoma. In *Pathology of the Testis*, ed. R. C. B. Pugh, 199–244. Oxford: Blackwell.

Rieck, G. W. (1971) Die testikulare Feminisierung beim Rind als Sterilitatsursache von Farsen. *Zuchthyg.* **6**, 145–54.

Rieck, G. W., Hohn, H. and Herzog, A. (1969) Hypogonadismus, intermittierender Kryptorchidismus und segmentäre Aplasie der Ductus Wolffii bei einem männlichen Rind mit XXY- Gonosomen-Konstellation bzw. XXY-/XX-/XY- Gonosomen-Mosaik. *Deutsch. Tierarztl. Wochenschr.* **76**, 133–8.

Robb, W. A. T. (1954) Operative treatment of varicocele. *Brit. med. J.* **II**, 355–6.

Roseman, B. D. (1958) Mumps orchitis. *Clin. Proc. Children's Hosp. (Washington)* **14**, 239–54.

Russell, J. K. (1957) Varicocele, age, and fertility. *Lancet* **273**, 222.

Russell, L. B. and Chu, E. H. Y. (1961) An XXY male in the mouse. *Proc. Natl. Acad. Sci. U.S.* **47**, 571–5.

Russell, L. D. and Gardner, P. J. (1974) Testicular ultrastructure and fertility in the *Z. Zellforsh.* **140**, 473–9.

Russell, L. D. and Gardner, P. J. (1974) Testicular Ultrastructure and Fertility in the Restricted Color (H$^{re}$) Rat. *Biol. Reprod.* **11**, 631–43.

Santen, R. J. and Paulsen, C. A. (1973a) Hypogonadotropic eunuchoidism. I. Clinical study of the mode of inheritance. *J. Clin. Endocrinol.* **36**, 47–54.

Santen, R. J. and Paulsen, C. A. (1973b) Hypogonadotropic eunuchoidism. II. Gonadal responsiveness to exogenous gonadotropins. *J. Clin. Endocrinol.* **36**, 55–63.

Santen, R. J., Kretser, D. M. de, Paulsen, C. A. and Vorhees, J. (1970) Gonadotrophins and testosterone in the XYY syndrome. *Lancet* **II**, 371.

Schellen, T. M. C. M. and Canton, M. (1974) Varikozele und Subfertilität. *Andrologia* **6**, 333–40.

Schirren, C. and Klosterhalfen, H. (1966) Die Spermatogenese bei Varicocele. *Z. Haut- u. Geschl.-Kr.* **40**, 372–7.

Schirren, C. and Toyosi, J. O. (1970) Assessment of gonadotropin therapy in male infertility. *Adv. exp. Med. Biol.* **10**, 605–11.

Schneider, G. and Bardin, C. W. (1970) Defective testicular testosterone synthesis by the pseudohermaphrodite rat: an abnormality of 17$\beta$-hydroxysteroid dehydrogenase. *Endocrinology* **87**, 864–73.

Scott, L. S. (1958) The effects of varicocele on spermatogenesis. *Stud. Fertil.* **10**, 33–40.

Scott, L. S. (1960a) Mumps and male fertility. *Brit. J. Urol.* **32**, 183–7.

Scott, L. S. (1960b) Varicocele ligation with improved fertility. *J. Reprod. Fertil.* **1**, 45–51.

Scott, L. S. (1961) Varicocele. A treatable cause of subfertility. *Brit. med. J.* **I**, 788–90.

Scott, L. S. and Young, D. (1962) Varicocele: A study of its effects on human spermatogenesis, and of the results produced by spermatic vein ligation. *Fertil. Steril.* **13**, 325–34.

Sherins, R. J. (1974) Clinical aspects of treatment of male infertility with gonadotropins: testicular response of some men given hCG with and without pergonal. In *Male Fertility and Sterility*, eds. R. E. Mancini and L. Martini, 545–65. New York: Academic Press.

Sherins, R. J. and Bardin, C. W. (1971) Preputial gland growth and protein synthesis in the androgen-insensitive male pseudohermaphrodite rat. *Endocrinology* **89**, 835–41.

Sherins, R. J., Bullock, L., Gay, V. L., Vanha-Perttula, T. and Bardin, C. W. (1971) Plasma LH and FSH levels in the androgen insensitive pseudohermaphrodite rat: response to steroid administration. *Endocrinology* **88**, 763–70.

Short, R. V. (1970) The bovine freemartin: a new look at an old problem. *Phil. Trans. Roy. Soc.* (London) B. **259**, 141–7.

Short, R. V., Smith, J., Mann, T., Evans, E. P., Hallett, J., Fryer, A. and Hamerton, J. L. (1969) Cytogenetic and endocrine studies of a freemartin heifer and its bull co-twin.

*Cytogenetics* **8**, 369–88.

Simmer, H. H., Pion, R. J. and Dignam, W. J. (1965) *Testicular feminization. Springfield, Illinois: Charles C. Thomas.*

Skakkebaek, N. E., Hammen, R., Philip, J. and Rebbe, H. (1973a) Quantification of human seminiferous epithelium. III. Histological studies in 44 infertile men with normal chromosome complements. *Acta Path. Microbiol. Scand. A.* **81**, 97–111.

Skakkebaek, N. E., Hulten, M. and Philip, J. (1973b) Quantification of human seminiferous epithelium. IV. Histological studies in 17 men with numerical and structural autosomal aberrations. *Acat. Path. Microbiol. Scand. A.* **81**, 112–24.

Southren, A. L. (1965) The syndrome of testicular feminization. *Adv. Metab. Disord.* **2**, 227–55.

Spotoft, J. (1942) Varicocele. *Acta chir. Scand.* **86**, 1–36.

Stanley, A. J. and Gumbreck, L. G. (1964) Male pseudohermaphroditism with feminizing testis in the male rat—a sex-linked recessive character. *Prog. Endocr. Soc.* 40.

Stanley, A. J., Gumbreck, L. G., Allison, J. E. and Easley, R. B. (1973) Part I. Male pseudohermaphroditism in laboratory Norway rat. *Rec. Progr. Horm. Res.* **29**, 43–64.

Steeno, O., Koumans, J. and Moor, P. de (1976) Adrenal cortical hormones in the spermatic vein of 95 patients with left varicocele. *Andrologia* **8**, 101–4.

Steeno, O., Knops, J., Declerck, L., Adimoelja, A. and Voorde, H. van de (1976) Prevention of fertility disorders by detection and treatment of varicocele at school and college age. *Andrologia* **8**, 47–53.

Stephenson, J. D. and O'Shaughnessy, E. J. (1968) Hypospermia and its relationship to varicocele and intrascrotal temperature. *Fertil. Steril.* **19**, 110–17.

Stewart, J. S. S. (1965) The clinical investigation of intersexuality. *Ann. R. Coll. Surg. Engl.* **37**, 374–88.

Symington, T. and Cameron, K. M. (1976) Endocrine and genetic lesions. In *Pathology of the Testis,* ed. R. C. B. Pugh, 259–303. Oxford: Blackwell.

Tandler, J. and Keller, K. (1911) Ueber das Verhalten des Chorions bei verschiedengeschlechtlicher Zwillingsgraviditat des Rindes und uber die Morphologie des Genitales der weiblichen Tiere. *Deutsch. tierartzl. Wochschr.* **19**, 148–9.

Tessler, A. N. and Krohn, H. P. (1966) Varicocele and testicular temperature. *Fertil. Steril.* **17**, 201–3.

Thackray, A. C. and Crane, W. A. J. (1976) Seminoma. In *Pathology of the Testis,* ed. R. C. B. Pugh, 164–98. Oxford: Blackwell.

Troen, P., Yanaihara, T., Nankin, H., Tominaga, T. and Lever, H. (1970) Assessment of gonadotropin therapy in infertile males. *Adv. exp. Med. Biol.* **10**, 591–602.

Tulloch, W. S. (1955) Varicocele. *Proc. roy. Soc. Med.* **55**, 1046–7.

Vanha-Perttula, T., Bardin, C. W., Allison, J. F., Gumbreck, L. G. and Stanley, A. J. (1970) 'Testicular feminization' in the rat: morphology of the testis. *Endocrinology* **87**, 611–19.

Vassilev, J. (1962) Etude radiographique de la veine spermatique gauche an cours des varicocèles idiopathiques. *Presse med.* **70**, 704.

Verstoppen, G. R. and Steeno, O. P. (1977a) Varicocele and the pathogenesis of the associated subfertility: a review of the various theories. I. Varicocelogenesis. *Andrologia* **9**, 133–40.

Verstoppen, G. R. and Steeno, O. P. (1977b) Varicocele and the pathogenesis of the associated subfertility: a review of the various theories. II. Results of surgery. *Andrologia* **9**, 293–305.

Vigier, B., Prepin, J. and Jost, A. (1973) Absence de chimerisme XX/XY dans les tissus somatiques chez les foetus de veau freemartins et leurs jumeaux mâles. *Ann. Genet. Paris.* **17**, 149–55.

Vigier, B., Locatelli, A., Prepin, J. du Mesnil du Buisson, F. and Jost, A. (1976) Les premieres manifestations du 'freemartinisme' chez le foetus de veau ne dependent pas du chimerisme chromosomique. *C. rend. Acad. Sci. Paris.* **282**, 1355–8.

Volter, D., Wurster, J., Aeikens, B. and Schubert, G. E. (1975) Untersuchungen zur Struktur und Funktion der Vena spermatica interna—Ein Beitrag zur Aetiologie ser Varikozele. *Andrologia* **7**, 127–33.

Wensing, C. J. G., Colenbrander, B. and Bosma, A. A. (1975) Testicular feminisation syndrome and gubernacular development in a pig. *Proc. K. Ned. Akad. Wetens.* **78**, 402–5.

Werner, C. A. (1950a) Mumps orchitis and testicular atrophy. I. Occurrence. *Ann. int. Med.* **32**, 1066–74.

Werner, C. A. (1950b) Mumps orchitis and testicular atrophy. II. A factor in male sterility. *Ann. Intern. Med.* **32**, 1075–86.

Wright, S. (1959) Silvering (si) and diminution (dm) of coat color of the guinea pig and male sterility of the white or near white combination of these. *Genetics* **44**, 563–90.

Yanagisawa, K. (1965) Studies on the mechanism of abnormal transmission ratios at the T-locus in the house mouse. IV. Some morphological studies on the mature sperm in males heterozygous for t-alleles. *Jap. J. Genetics* **40**, 97–104.

Young, D. H. (1956) Influence of varicocele on human spermatogenesis. *Brit. J. Urol.* **28**, 426.

Younglai, E. V. and Chui, D. H. K. (1973) Testicular function in sterile steel mice. *Biol. Reprod.* **9**, 317–23.

Zarate, A., Garrido, J., Canales, E. S., Soria, J. and Schally, A. V. (1974) Disparity in the negative gonadal feedback control for LH and FSH in cases of germinal aplasia or Sertoli-cell-only syndrome. *J. Clin. Endocrinol. Metab.* **38**, 1125–7.

Zorgniotti, A. W. and Macleod, J. (1973) Studies in temperature, human semen quality, and varicocele. *Fertil. Steril.* **24**, 854–63.

# Animal Index

This index lists animals, except man, monkey, rat, mouse, hamster, guinea pig, rabbit, dog, cat, sheep, goat, cattle, pig and horse. In the reference list to Chapter 1, there are references to breeding seasons and cycles in a number of species, which are not included in this index. See also the heading Marsupials in the subject index.

# Subject Index